Trends in Mathematics is a series devoted to the publication of volumes arising from conferences and lecture series focusing on a particular topic from any area of mathematics. Its aim is to make current developments available to the community as rapidly as possible without compromise to quality and to archive these for reference.

Proposals for volumes can be sent to the Mathematics Editor at either

Birkhäuser Verlag
P.O. Box 133
CH-4010 Basel
Switzerland

or

Birkhäuser Boston Inc.
675 Massachusetts Avenue
Cambridge, MA 02139
USA

Material submitted for publication must be screened and prepared as follows:

All contributions should undergo a reviewing process similar to that carried out by journals and be checked for correct use of language which, as a rule, is English. Articles without proofs, or which do not contain any significantly new results, should be rejected. High quality survey papers, however, are welcome.

We expect the organizers to deliver manuscripts in a form that is essentially ready for direct reproduction. Any version of TeX is acceptable, but the entire collection of files must be in one particular dialect of TeX and unified according to simple instructions available from Birkhäuser.

Furthermore, in order to guarantee the timely appearance of the proceedings it is essential that the final version of the entire material be submitted no later than one year after the conference. The total number of pages should not exceed 350. The first-mentioned author of each article will receive 25 free offprints. To the participants of the congress the book will be offered at a special rate.

C*-algebras and Elliptic Theory

Bogdan Bojarski
Alexander Mishchenko
Evgenij Troitsky
Andrzej Weber
Editors

in cooperation with Dan Burghelea, Richard Melrose and Victor Nistor

Birkhäuser Verlag
Basel · Boston · Berlin

Editors:

Bogdan Bojarski
Institute of Mathematics
Polish Academy of Sciences,
ul. Sniadeckih 8
00-956 Warszawa
Poland
e-mail: bojarski@impan.gov.pl

Evgenij V. Troitsky
Department of Mechanics and Mathematics
Moscow State University
Leninskie Gory
119992 Moscow
Russia
e-mail: troitsky@mech.math.msu.su

Alexander S. Mishchenko
Department of Mechanics and Mathematics
Moscow State University
Leninskie Gory
119992 Moscow
Russia
e-mail: asmish@mech.math.msu.su

Andrzej Weber
Uniwersytet Warszawski
Instytut Matematyki
Banacha 2
02-097 Warszawa
Poland
e-mail: aweber@mimuw.edu.pl

2000 Mathematical Subject Classification 19Kxx, 58Jxx, 57Rxx, 58J32, 19D55

A CIP catalogue record for this book is available from the
Library of Congress, Washington D.C., USA

Bibliographic information published by Die Deutsche Bibliothek
Die Deutsche Bibliothek lists this publication in the Deutsche Nationalbibliografie; detailed bibliographic data is available in the Internet at http://dnb.ddb.de

ISBN 3-7643-7686-4 Birkhäuser Verlag, Basel – Boston – Berlin

Part of Springer Science+Business Media
Printed on acid-free paper produced from chlorine-free pulp. TCF ∞
Printed in Germany
ISBN-10: 3-7643-7686-4 e-ISBN: 3-7643-7687-2
ISBN-13: 976-3-7643-7686-4

9 8 7 6 5 4 3 2 1 www.birkhauser.ch

Contents

Editorial Introduction

B. Bojarski, A.S. Mishchenko, E.V. Troitsky and A. Weber

The present volume is basically formed by contributions of participants of the International Conference "C^*-algebras and elliptic theory" hold in Stefan Banach International Mathematical Center in Będlewo (Poland) in February 2004. The history of this Conference goes back to the idea of Prof. Bogdan Bojarski to strengthen collaboration between mathematicians from Poland and Russia, especially from Moscow, on the base of common scientific interests in the field of noncommutative geometry.

This idea leaded very quickly to the organization of the mentioned Conference which brought together about 60 mathematicians not only from Russia and Poland, but from other leading centers and awarded a support from the European program "Geometric Analysis Research Training Network". The conference started a series of annual conferences in Będlewo and Moscow alternately. Up to the present time three conferences of this series were organized and the forth one is planned to take place in Moscow in 2007 (`http://higeom.math.msu.su/oat2007`).

The contributions are mainly concentrated on applications of C^*-algebraic technic to geometrical and topological problems and appropriately present the main actual problems in this field of noncommutative geometry and topology and indicate principal directions of its development.

To present the volume into perspective let us remind that the notion "noncommutative geometry" was coined out by Alain Connes in 1980's to indicate a new trend in mathematics. A naive look on this trend goes back to the prominent theorem of Gelfand and Naimark, which identifies the category of commutative unital C^*-algebras and the category of compact Hausdorff topological spaces. The passage to noncommutative algebras gives rise to the notion of "noncommutative topological space", which turned out to be fruitful despite the fact that they are not spaces in usual sense. The method and problems of this domain brought together a number of important achievements and open questions from topology, geometry, algebra and functional analysis. A most fruitful interference here is an enriching of the index theory of elliptic operators by the theory of C^*-algebras.

The papers from the present collection reflect some important actual problems and achievements of noncommutative geometry.

Index of elliptic operators: The paper *"Index Theory for Generalized Dirac Operators on Open Manifolds"* by **J. Eichhorn** is devoted to the index theory on open manifolds. In the first part of the paper, a short review of index theory on open manifolds is given. In the second part, a general relative index theorem admitting compact topological perturbations and Sobolev perturbations of all other ingredients is established. **V. Nazaikinskii and B. Sternin** in the paper *"Lefschetz Theory on Manifolds with Singularities"* extend the Lefschetz formula to the case of elliptic operators on the manifolds with singularities using the semiclassical asymptotic method. In the paper *"Pseudodifferential Subspaces and Their Applications in Elliptic Theory"* by **A. Savin and B. Sternin** the method of so called pseudodifferential projectors in the theory of elliptic operators is studied. It is very useful for the study of boundary value problems, computation of the fractional part of the spectral AtiyahPatodiSinger eta invariant and analytic realization of topological K-groups with finite coefficients in terms of elliptic operators. In the paper *"Residues and Index for Bisingular Operators"* **F. Nicola and L. Rodino** consider an algebra of pseudo-differential operators on the product of two manifolds, which contains, in particular, tensor products of usual pseudo-differential operators. For this algebra the existence of trace functionals like Wodzickis residue is discussed and a homological index formula for the elliptic elements is proved. **B. Bojarski** and **A. Weber** in their paper *"Correspondences and Index"* define a certain class of correspondences of polarized representations of C^*-algebras. These correspondences are modeled on the spaces of boundary values of elliptic operators on bordisms between two manifolds. In this situation an index is defined. The additivity of this index is studied in the paper.

Noncommutative aspects of Morse theory: In the paper *"New L2-invariants of Chain Complexes and Applications"* by **V.V. Sharko** homotopy invariants of free cochain complexes and Hilbert complex are studied. These invariants are applied to calculation of exact values of Morse numbers of smooth manifolds. **A. Connes** and **T. Fack** in their paper *"Morse Inequalities for Foliations"* outline an analytical proof of Morse inequalities for measured foliations obtained by them previously and give some applications. The proof is based on the use of a twisted Laplacian.

Riemannian aspects: The paper *"A Riemannian Invariant, Euler Structures and Some Topological Applications"* by **D. Burghelea** and **S. Haller** discusses a numerical invariant associated with a Riemannian metric, a vector field with isolated zeros, and a closed one form which is defined by a geometrically regularized integral. This invariant extends the ChernSimons class from a pair of two Riemannian metrics to a pair of a Riemannian metric and a smooth triangulation. They discuss a generalization of Turaevs Euler structures to manifolds with non-vanishing Euler characteristics and introduce the Poincare dual concept of co-Euler structures. The duality is provided by a geometrically regularized integral and involves the invariant mentioned above. Euler structures have been introduced because they permit to remove the ambiguities in the definition of the Reidemeister torsion. Similarly, co-Euler structures can be used to eliminate the metric dependence of the RaySinger torsion. The BismutZhang theorem can then be reformulated as a statement comparing two genuine topological invariants. The paper *"Semiclassical Asymptotics and Spectral Gaps for Periodic Magnetic Schrödinger Operators on Covering Manifolds"* by **Yu.A. Kordyukov** is devoted to an exposition of a method to prove the existence of gaps in the spectrum of periodic second-order elliptic partial differential operators, which was suggested by Kordyukov, Mathai and Shubin, and describes the applications of this method to periodic magnetic Schrödinger operators on a Riemannian manifold, which is the universal covering of a compact manifold.

K-theory, C^*-algebras, and groups: In the paper *"The Group of Unital C^*-extensions"* by **V. Manuilov** and **K. Thomsen** it is shown that there is a natural six-terms exact sequence which relates the group which arises by considering all semi-split C^*-extensions of an algebra A by B to the group which arises from unital semi-split extensions of A by B. The paper *"The Thom Isomorphism in Gauge-equivariant K-theory"* by **V. Nistor** and **E. Troitsky** is devoted to the study of gauge-equivariant K-theory. In particular, they introduce and study products, which help to establish the Thom isomorphism in gauge-equivariant K-theory. They construct push-forward maps and define the topological index of a gauge-invariant family. The paper *"Bundles of C^*-algebras and the $KK(X;-,-)$-bifunctor"* by **E. Vasselli** is an overview of C^*-algebra bundles with a $\mathbb{Z}$-grading, with particular emphasis on classification questions. In particular, author discusses the role of the representable $KK(X;-,-)$-bifunctor introduced by Kasparov. As an application, Cuntz-Pimsner algebras associated with vector bundles are considered, and a classification in terms of K-theoretical invariants is given in the case of the base space being an n-sphere. **J. Brodzki** and **G.A. Niblo** in the paper *"Approximation Properties for Discrete Groups"* give a short survey of approximation properties of operator algebras associated with discrete groups. Then they demonstrate directly that groups that satisfy the rapid decay property with respect to a conditionally negative length function have the metric approximation property. The paper *"On the Hopf-type Cyclic Cohomology with Coefficients"* by **I.M. Nikonov** and **G.I. Sharygin** is devoted to the Hopf-type cyclic cohomology with coefficients. They calculate it in a couple of examples and propose a general construction of a coupling between algebraic and coalgebraic versions of such cohomology with values in the usual cyclic cohomology of an algebra.

C^*-algebras and Elliptic Theory
Trends in Mathematics, 1–21

Correspondences and Index

Bogdan Bojarski and Andrzej Weber

Abstract. We define a certain class of correspondences of polarized representations of C^*-algebras. Our correspondences are modeled on the spaces of boundary values of elliptic operators on bordisms joining two manifolds. In this setup we define the index. The main subject of the paper is the additivity of the index.

Mathematics Subject Classification (2000). Primary: 58J32, Secondary: 35J55, 35Q15, 47A53, 55N15, 58J20.

Keywords. Index of an elliptic operator, Riemann–Hilbert problem, bordism, K-theory.

1. Introduction

Let X be a closed manifold. Suppose it is decomposed into a sum of two manifolds X_+, X_- glued along the common boundary

$$\partial X_+ = \partial X_- = M .$$

Let

$$D : C^\infty(X;\xi) \to C^\infty(X;\eta)$$

be an elliptic operator of the first order. We assume that it possesses the unique extension property: if $Df = 0$ and $f_{|M} = 0$ then $f = 0$. In what follows we will consider only elliptic operators of the first order such that D and D^* have the unique extension property.

One defines the spaces $H_\epsilon(D) \subset L^2(M;\xi)$ for $\epsilon \in \{+,-\}$, which are the closures of the spaces of boundary values of solutions of $Df = 0$ on the manifolds X_ϵ with boundary $\partial X_\epsilon = M$. The space $H_\epsilon(D)$ is defined to be the closure of :

$$\{f \in C^\infty(M;\xi) \,:\, \exists \tilde f \in C^\infty(X_\epsilon;\xi),\ f = \tilde f_{|M},\ D(\tilde f) = 0\,\}$$

in $L^2(M;\xi)$. The pair of spaces $H_\pm(D)$ is a Fredholm pair, [4]. There are associated Calderón projectors $P_+(D)$ and $P_-(D)$, see [19].

A.W. is supported by KBN grant 1 P03A 005 26.

To organize somehow the set of possible Cauchy data we will introduce a certain algebraic object. We fix a C^*-algebra B, which is the algebra of functions on M in our case. Suppose it acts on a Hilbert space H. Now we consider Fredholm pairs in H. In our case $H = L^2(M;\xi)$ and one of the possible Fredholm pairs is $H_\pm(D)$. Note that this pair is not arbitrary. It has a property which we called *good.* A Fredholm pair is good if (roughly speaking) it remains to be Fredholm after conjugation with functions, see §4. These pairs act naturally on $K_1(M)$. Nevertheless the concept of a good Fredholm pair is not convenient to manipulate, thus we restrict our attention to the pairs of geometric origin, see §5. We call them *admissible.* They are the pairs of subspaces which are images of projectors which almost commute with the actions of the algebra B. This concept allows to extract the relevant analytico-functional information out of the Cauchy data. Further a Morse decomposition of a manifold is translated into this language.

Our paper is devoted to the study of the cut and paste technique on manifolds and its effect on indices. The spirit of these constructions comes from the earlier papers [4]–[6] or [8]. According to the topological and conformal field theory we investigate the behavior of the index of a differential operator on a manifold composed from bordisms

$$X = X_0 \cup_{M_1} X_1 \cup_{M_2} \cdots \cup_{M_{m-1}} X_{m-1} \cup_{M_m} X_m \,.$$

We think of M_i's as objects and we treat bordisms of manifolds as morphisms. Starting from this geometric background we introduce a category **PR**, whose objects are *polarized representations.* The algebra B may vary. We keep in mind that such objects arise when:

- B is an algebra of functions on a manifold M,
- there is given a vector bundle ξ over M, then $H = L^2(M;\xi)$ is a representation of B,
- there is given a pseudodifferential projector in H.

The morphisms in **PR** are certain correspondences, i.e., linear subspaces in the product of the source and the target. A particular case of principal value for our theory are the correspondences coming from bordisms of manifolds equipped with an elliptic operator. Precisely: suppose we are given a manifold W with a boundary $\partial W = M_1 \sqcup M_2$. Moreover, suppose that there is given an elliptic operator of the first order acting on the sections of a vector bundle ξ over W. Then the space of the boundary values of the Cauchy data of solutions is a linear subspace in $L^2(M_1;\xi_{|M_1}) \oplus L^2(M_2;\xi_{|M_2})$. In other words it is a correspondence from $L^2(M_1;\xi_{|M_1})$ to $L^2(M_2;\xi_{|M_2})$.

Basic example: The following example is instructive and serves as the model situation (see [7]): Let $W = \{z \in \mathbf{C} : r_1 \geq |z| \geq r_2\}$ be a ring domain and let D be the Cauchy-Riemann operator. The space $L^2(M_i)$ for $i = 1, 2$ is identified with the space of sequences $\{a_n\}_{n\in\mathbf{Z}}$, such that $\sum_{n\in\mathbf{Z}} |a_n|^2 r_i^{2n} < \infty$. The sequence $\{a_n\}$ defines the function on M_i given by the formula $f(z) = \sum_{n\in\mathbf{Z}} a_n z^n$. The subspace of the boundary values of holomorphic functions on W is identified with

$$\Big\{(\{a_n\},\{b_n\})\,:\,\Sigma_{n\in\mathbf{Z}}|a_n|^2 r_1^{2n}<\infty\,,\ \Sigma_{n\in\mathbf{Z}}|b_n|^2 r_2^{2n}<\infty \text{ and } a_n=b_n\Big\}\,.$$

It can be treated as the graph of an unbounded operator $\Phi : L^2(M_1) \to L^2(M_2)$. When we restrict Φ to the space $L^2(M_1)^\sharp$ consisting of the functions with coefficients $a_n = 0$ for $n < 0$ we obtain a compact operator. On the other hand the inverse operator $\Phi^{-1} : L^2(M_2) \to L^2(M_1)$ is compact when restricted to $L^2(M_2)^\flat$, the space consisting of the functions with coefficients $a_n = 0$ for $n \geq 0$.

The Riemann-Hilbert transmission problem of the Cauchy data across a hypersurface is a model for another class of morphisms. These are called *twists*. Our approach allows us to treat bordisms and twists in a uniform way. We calculate the global index of an elliptic operator in terms of local indices depending only on the pieces of the decomposed manifold (see Theorems 9.6 and 11.1). An interesting phenomenon occurs. The index is not additive with respect to the composition of bordisms. Instead each composition creates a contribution to the global index (Theorem 10.2):

$$L_1,\, L_2 \rightsquigarrow L_2 \circ L_1 + \delta(L_1, L_2)\,.$$

In the geometric situation this contribution might be nonzero for example when a closed manifold is created as an effect of composition of bordisms. One can show that if the bordisms in **PR** come from connected geometric bordisms supporting elliptic operators with the unique extension property then the index is additive. The contributions coming from twists are equivalent to the effects of pairings in the odd K-theory, Theorem 9.7.

It is a good moment now to expose a fundamental role of the splitting of the Hilbert space into a direct sum. The need of introducing a splitting was clear already in [4]:

- It was used to the study of Fredholm pairs with application to the Riemann-Hilbert transmission problem in [4].
- Splitting also came into light in the paper of Kasparov [13], who introduced a homological K-theory built from the Hilbert modules. The program of noncommutative geometry of A.Connes develops this idea, [10, 11] .
- Splitting plays an important role in the theory of loop groups in [16].
- There is also a number of papers in which surgery of the Dirac operator is studied. Splitting serves as a boundary condition, see, e.g., [12], [17]. These papers originate from [2].

In the present paper we omit the technicalities and problems arising for a general elliptic operator. We concentrate on the purely functional calculus of correspondences. This is mainly the linear algebra.

2. Fredholm pairs

Let us first summarize some facts about Fredholm pairs. We will follow [4]–[6]. Suppose that H_+ and H_- are two closed subspaces of a Hilbert space, such that $H_+ + H_-$ is also closed and

- $H_+ \cap H_-$ is of finite dimension,
- $H_+ + H_-$ is of finite codimension.

We assume that both spaces have infinite dimension. Then we say that the pair $(H_+, H_-) = H_\pm$ is Fredholm. We define its index

$$\mathrm{Ind}(H_\pm) = \dim(H_+ \cap H_-) - \mathrm{codim}(H_+ + H_-)\,.$$

The following statements follow from easy linear algebra.

Proposition 2.1. *A pair $H_\pm$ is Fredholm if, and only if the map*

$$\iota : H_+ \oplus H_- \to H$$

induced by the inclusions is a Fredholm operator. Moreover the indices are equal:

$$\mathrm{Ind}(H_\pm) = \mathrm{ind}(\iota)\,.$$

Here Ind denotes the index of a pair, whereas ind stands for the index of an operator. Suppose that H is decomposed into a direct sum

$$H = H^\flat \oplus H^\sharp\,.$$

We may assume that this decomposition is given by a symmetry S: a "sign" or "signature" operator. Let $P^\flat$ and $P^\sharp$ be the corresponding projectors. We can write $S = P^\sharp - P^\flat$. We easily have:

Proposition 2.2. *If $H_\pm$ is a pair with $H_+ = H^\sharp$, then it is Fredholm if and only if the restriction $P^\flat_{|H_-} : H_- \to H^\flat$ is a Fredholm operator. Moreover the indices are equal:*

$$\mathrm{Ind}(H_\pm) = \mathrm{ind}(P^\flat_{|H_-})\,.$$

Let $I \subset L(H)$ be an ideal which lies between the ideal of finite rank operators and the ideal of compact operators

$$F \subset I \subset K\,.$$

Define $GL(P^\flat, I) \subset GL(H)$ to be the set of the invertible automorphisms of H commuting with $P^\flat$ up to the ideal I. We will say that ϕ almost commutes with $P^\flat$ or we will write $\phi P^\flat \sim P^\flat \phi$. Obviously $GL(P^\flat, I) = GL(P^\sharp, I) = GL(S, I)$. We have the following description of Fredholm pairs stated in [4]. (The proof is again an easy linear algebra.)

Theorem 2.3. *Let $H_\pm$ be a Fredholm pair with $H_+ = H^\sharp$. Then there exists a complement $H^\flat$ (that is $H^\flat \oplus H^\sharp = H$) and there exists $\phi \in GL(P^\flat, I)$, such that $H_- = \phi(H^\flat)$. If $H_\pm$ is given by a pair of projectors $P_\pm$ satisfying $P_- + P_- - 1 \in I$, then we can take $H^\flat = ker\, P_+$. Moreover, the operator $\phi P^\flat + P^\sharp$ is Fredholm and*

$$\mathrm{ind}(\phi P^\flat + P^\sharp) = \mathrm{Ind}(H_\pm)\,.$$

The map

$$\widetilde{\text{ind}} : GL(P^\flat, I) \to \mathbf{Z}$$
$$\widetilde{\text{ind}}(\phi) = \text{ind}(\phi P^\flat + P^\sharp)$$

is a group homomorphism.

It follows that $\text{ind}(\phi P^\flat + P^\sharp) = \text{ind}(P^\flat \phi : H^\flat \to H^\flat) = \text{ind}(P^\sharp \phi^{-1} : H^\sharp \to H^\sharp)$.

3. Index formula for a decomposed manifold

The main example of a Fredholm pair is the following. Let D be an elliptic operator on $X = X_+ \cup_M X_-$. Then the pair of boundary value spaces $H_\pm(D)$ (as defined in the introduction) is a Fredholm pair.

Assumption 3.1 (Unique Extension Property). Let $\epsilon = +$ or $-$ and let $f \in C^\infty(X_\epsilon; \xi)$. If $Df = 0$ and $f_{|M} = 0$ then $f = 0$.

If D has the unique extension property, then

$$\ker(D) \simeq H_+(D) \cap H_-(D).$$

This formula is easy to explain: a global solution restricted to M lies in $H_+(D) \cap H_-(D)$. On the other hand if a section f of ξ over M can be extended to both X_+ and X_-, such that the extensions are solutions of $Df = 0$ then we can glue them to obtain a global solution. The unique extension property is necessary, because we need to know that a solution is determined by its restriction to M. Following the reasoning in [4], with Assumption 3.1 for D and D^* we have:

Corollary 3.2. $$\text{Ind}(H_\pm(D)) = \text{ind}(D).$$

For a rigorous proof see [9], §24 for Dirac type operators.

Remark 3.3. It may happen that D does not have the unique extension property. This is so for example when X is not connected. Then the Cauchy data $H_\pm(D)$ do not say anything about the index of the operator D on the components of X disjoined with M. There are also known elliptic operators without the unique extension property on connected manifolds, [15], [1]. It is difficult to characterize the class of all operators D with the unique extension property. Nevertheless the most relevant are Cauchy-Riemann and Dirac type operators. These operators do have the unique extension property on connected manifolds.

4. Good Fredholm pairs

Suppose there is given an algebra B and its representation ρ in a Hilbert space H. For a Fredholm pair $H_\pm$ in H and an invertible matrix $A \in GL_n(B)$ we define a new pair of subspaces $A \bowtie H_\pm$ in $H^{\oplus n}$. We set

$$(A \bowtie H_\pm)_- = \rho A(H_-^{\oplus n}) \qquad (A \bowtie H_\pm)_+ = H_+^{\oplus n}.$$

(As usually we treat ρA as an automorphism of $H^{\oplus n}$.)

Definition 4.1. Let B be a C^*-algebra which acts on a Hilbert space H. A *good Fredholm pair* is a pair of subspaces (H_+, H_-) in H, such that for any invertible matrix $A \in GL(n; B)$ the pair $A \bowtie H_\pm$ is a Fredholm pair.

We will see that the pair of boundary values $H_\pm(D) \subset H = L^2(M; \xi)$ for the operator D considered in the introduction is good.

Example 1 (Main example: Riemann-Hilbert problem). Consider the following problem: there is given a matrix-valued function $A : M \to GL_n(\mathbf{C})$. We look for the sequence $(s^1_\pm, \dots, s^n_\pm)$ of solutions of $Ds = 0$ on $X_\pm$ satisfying the transmission condition on M

$$A(s^1_-, \dots, s^n_-) = (s^1_+, \dots, s^n_+) .$$

A Fredholm operator is related to this problem and we study its index, see §11. On the other hand the matrix A treated as the gluing data defines an n-dimensional vector bundle Θ^A_X over X. Then

$$\mathrm{Ind}(A \bowtie H_\pm(D)) = \mathrm{ind}(D \otimes \Theta^A_X) .$$

This formula was obtained in [8], §1 under the assumption that D has a product form along M.

Corollary 4.2. *For the elliptic operator D the pair $H_\pm(D) \subset L^2(M; \xi)$ is a good Fredholm pair.*

Remark 4.3. Consider the differential in the Mayer-Vietoris exact sequence of $X = X_+ \cup_M X_-$

$$\delta : K_0(X) \to K_{-1}(M) .$$

The operator D defines a class $[D] \in K_0(X)$. The element $\delta[D]$ can be recovered from the good Fredholm pair $H_\pm(D) \subset L^2(M; \xi)$. Note that the pair $H_\pm(D)$ encodes more information. One can recover the index of the original operator. We describe the map δ via duality, therefore we neglect the torsion of K-theory. The construction is the following: for an element $a \in K^1(M)$ we define the value of the pairing

$$\langle \delta[D], a \rangle = \langle [D], \partial a \rangle .$$

The element a is represented by a matrix $A \in GL_n(C^\infty(M))$. Then

$$\langle [D], \partial a \rangle = \mathrm{ind}(D \otimes \Theta^A_X) - n \, \mathrm{ind}(D) ,$$

where Θ^A_X is the bundle defined in Example 1. Now

$$\langle [D], \partial a \rangle = \mathrm{Ind}(A \bowtie H_\pm(D)) - n \, \mathrm{Ind}(H_\pm(D)) .$$

5. Admissible Fredholm pairs

The following can be related to the paper of Birman and Solomyak [3] who introduced the name *admissible* for the subspaces which are the images of pseudo-differential projectors. Suppose that ξ is a vector bundle over a manifold M. We consider Fredholm pairs $H_\pm$ in $H = L^2(M; \xi)$ such that the subspaces $H_\pm$ are images of pseudodifferential projectors $P_\pm$ with symbols satisfying

$$\sigma(P_+) + \sigma(P_-) = 1\,.$$

We would like to free ourselves from the geometric context and state admissibility condition in an abstract way. We assume that H is an abstract Hilbert space with a representation of an algebra B, which is the algebra of functions on M in the geometric case. The condition that $P_\pm$ is pseudodifferential we substitute by the condition: $P_\pm$ commutes with the algebra action up to compact operators. We are ready now to give a definition:

Definition 5.1. We say that a pair of subspaces $H_\pm$ is an *admissible Fredholm pair* if there exist a pair of projectors P_ϵ for $\epsilon \in \{+,-\}$, such that $H_\epsilon = \operatorname{im} P_\epsilon$ and P_ϵ commutes with the action of B up to compact operators. Moreover, we assume that $P_+ + P_- - 1$ is a compact operator.

Proposition 5.2. *Each admissible Fredholm pair is a good Fredholm pair.*

Proof. Set $K = P_+ + P_- - 1$. If $v \in H_+ \cap H_-$, then $K(v) = v$. Since K is a compact operator, $\dim(H_+ \cap H_-) < \infty$. To prove that $H_+ + H_-$ is closed and of finite codimension, note that $\operatorname{im}(P_+ + P_-) \subset H_+ + H_-$. Since $P_+ + P_-$ is Fredholm its image is closed and of finite codimension. This way we have shown that $H_\pm$ is a Fredholm pair. Now, if we conjugate $P_+^{\oplus n}$ by ρA we obtain again an almost complementary pair of projectors. Thus $A \bowtie H_\pm$ is a Fredholm pair as well. □

We denote by $AFP(B)$ the set of good Fredholm pairs divided by the equivalence relation generated by homotopies and stabilization with respect to the direct sum. We also consider as trivial the pairs associated to projectors strictly satisfying $P_+ + P_- = 1$ and commuting with the action of B. In other words these are just direct sums of two representations of B. It is not hard to show that

Proposition 5.3. $$AFP(B) \simeq K^1(B) \oplus \mathbf{Z}\,.$$

Proof. We have the following natural transformation:

$$\begin{array}{rccc} \beta: & AFP(M) & \to & K_1(M) \\ & (H, P_\pm) & \mapsto & (H, S_+) \end{array}.$$

Here $S_+ = 2P_+ - 1$ is just the symmetry defined by P_+. We remind that the objects generating $K_1(M)$ are odd Fredholm modules, see [11], pp. 287–289. This procedure is simply forgetting about P_-. We can recover P_- (up to homotopy) by fixing the index of the pair, i.e., $\beta \oplus \operatorname{Ind}$ is the isomorphism we are looking for. Precisely, the pseudodifferential projector is determined up to homotopy by its symbol and the index, see [9]. □

6. Splittings and polarization

We adopt the concepts of splitting and polarization to our situation.

Definition 6.1. Let H be a representation of a $\mathbf{C}^*$-algebra B in a Hilbert space. A *splitting* of H is a decomposition

$$H = H^\flat \oplus H^\sharp ,$$

such that the projectors on the subspaces $P^\flat$, $P^\sharp$ commute with the action of B up to compact operators.

The basic example of a splitting is the one coming from a pseudodifferential projector. Another equivalent way of defining a splitting (as in [5]) is to distinguish a symmetry S, almost commuting with the action of B. Then $H^\flat$ is the eigenspace of -1 and $H^\sharp$ is the eigenspace of 1. Then we may think of H as a superspace, but we have to remember that the action of B does not preserve the grading.

Definition 6.2. In the set of splittings we introduce an equivalence relation: two splittings are equivalent if the corresponding projectors coincide up to compact operators. An equivalence class of the above relation is called a *polarization* of H.

Informally we can say, that polarization is a generalization of the symbol of a pseudodifferential projector.

Example 2. Let $\xi \to M$ be a complex vector bundle over a manifold. Let $\widetilde{\xi}$ be the pull back of ξ to $T^*M \setminus \{0\}$. Suppose $p : \widetilde{\xi} \to \widetilde{\xi}$ is a bundle map which is a projector (hence p is homogeneous of degree 0). Then p defines a polarization of $L^2(M;\xi)$. Just take a pseudodifferential projector $P = P^\sharp$ with $\sigma(P) = p$ and set

$$H^\flat = \ker P\,, \qquad H^\sharp = \operatorname{im} P\,.$$

Example 3. Suppose (H_+, H_-) is an admissible Fredholm pair given by projectors (P_+, P_-). Then the polarizations associated with P_+ and $1-P_-$ coincide. This way an admissible Fredholm pair defines a polarization. Furthermore each polarization defines an element of $K_1(B)$.

Intuitively polarizations can be treated as a kind of orientations dividing H into the upper half and lower half. Such a tool was used in [12] to split the index of a family of Dirac operators. (In [12] splittings were called generalized spectral sections.) Polarizations were discussed in the lectures of G. Segal (see [18], Lecture 2).

7. Correspondences, bordisms, twists

Definition 7.1. We consider the category **PR** having the following objects and morphisms

- Ob(**PR**) = Hilbert spaces (possibly of finite dimension) with a representation of some $\mathbf{C}^*$-algebra B and with a distinguished polarization,

- $\mathrm{Mor}_{\mathbf{PR}}(H_1, H_2)$ = closed linear subspaces $L \subset H_1 \oplus H_2$, such that the pair $(L, H_1^\flat \oplus H_2^\sharp)$ is Fredholm.

We write also $H_1 \xrightarrow{L} H_2$.

In particular

$$\mathrm{Mor}_{\mathbf{PR}}(H, 0) \subset \mathrm{Grass}(H) \supset \mathrm{Mor}_{\mathbf{PR}}(0, H)\,.$$

By Proposition 2.2 a subspace $L \subset H_1 \oplus H_2$ is a morphism if and only if

$$\Pi = P_1^\sharp \oplus P_2^\flat : L \to H_1^\sharp \oplus H_2^\flat$$

is a Fredholm operator. The composition in **PR** is the standard composition of correspondences:

$$L_1 \subset H_1 \oplus H_2\,, \qquad L_2 \subset H_2 \oplus H_3\,,$$

$$L_2 \circ L_1 = \{(x, z) \in H_1 \oplus H_3 \ : \ \exists y \in H_2\,,\ (x, y) \in L_1\,,\ (y, z) \in L_2\,\}\,.$$

In other words the morphisms are certain correspondences or relations, as they were called in [4]. Our approach also fits to the ideas of the topological field theory as presented in [18].

Proposition 7.2. *The composition of morphism is a morphism.*

Proof. Let $L_1 \in \mathrm{Mor}_{\mathbf{PR}}(H_1, H_2)$ and $L_2 \in \mathrm{Mor}_{\mathbf{PR}}(H_2, H_3)$. A simple linear algebra argument shows that

- the kernel of
$$\Pi_{13} : L_2 \circ L_1 \to H_1^\sharp \oplus H_3^\flat$$
is a quotient of $\ker(\Pi_{12}) \oplus \ker(\Pi_{23})$,
- the cokernel of Π_{13} is a subspace of $\mathrm{coker}(\Pi_{23}) \oplus \mathrm{coker}(\Pi_{12})$.

□

The role of polarizations in the definition of morphisms is clear and the algebra actions are involved implicitly. In fact, the object which plays the crucial role is the algebra of operators commuting with $P^\sharp$ up to compact operators, i.e., the odd universal algebra. The role of this algebra was emphasized in [5]. However, in the further presentation we prefer to expose the geometric origin of our construction and keep the name B.

We have two special classes of morphisms in **PR**:

Definition 7.3. A subspace $L \subset H \oplus H$ is a *twist* if it is the graph of a linear isomorphism $\phi \in GL(P^\sharp, K) \subset GL(H)$ commuting with the polarization projectors up to compact operators.

Proposition 7.4. *For a twist* $L = \mathrm{graph}(\phi) \subset H \oplus H$ *the pair* $(L, H^\flat \oplus H^\sharp)$ *is Fredholm, i.e.,* $L \in \mathrm{Mor}_{\mathbf{PR}}(H, H)$.

Proof. To show that $(L, H^\flat \oplus H^\sharp)$ is a Fredholm pair let us show that the projection

$$\Pi = P^\sharp \oplus P^\flat : L \to H^\sharp \oplus H^\flat \subset H \oplus H$$

is a Fredholm operator. Indeed, L is parameterized by

$$(1, \phi) : H \to L \subset H \oplus H .$$

The composition of these maps is equal to

$$F = P^\sharp \oplus P^\flat \phi .$$

Since ϕ almost commutes with $P^\flat$ the map F has a parametrix $\widetilde{F} = P^\sharp \oplus P^\flat \phi^{-1}$. □

Definition 7.5. A subspace $L \subset H_1 \oplus H_2$ is a bordism if L is the image of a projector P_L, such that

$$P_L \sim P_1^\sharp \oplus P_2^\flat .$$

By 5.2 for any $P_L \sim P_1^\sharp \oplus P_2^\flat$ the pair $(L, H_1^\flat \oplus H_2^\sharp)$ is Fredholm. The motivation for Definition 7.5 is the following:

Example 4. Let X be a bordism between closed manifolds M_1 and M_2, i.e.,

$$\partial X = M_1 \sqcup M_2 .$$

Suppose that $D : C^\infty(X; \xi) \to C^\infty(X; \eta)$ is an elliptic operator of the first order. Then the symbols of Calderón projectors define polarizations of $H_1 = L^2(M_1; \xi)$ and $H_2 = L^2(M_2; \xi)$, see Example 2. We reverse the polarization on M_2, i.e., we switch the roles of $H^\flat$ and $H^\sharp$. Let $L \subset L^2(M_1; \xi) \oplus L^2(M_2; \xi)$ be the closure of the space of boundary values of solutions of $Ds = 0$. Then $L \in \mathrm{Mor}_{\mathbf{PR}}(H_1, H_2)$ is a bordism in **PR**. This procedure indicates the following:

- the space $L \subset L^2(M_1 \sqcup M_2; \xi) = L^2(M_1; \xi) \oplus L^2(M_2; \xi)$ and the associated Calderón projector are *global* objects. One cannot recover them from the separated data in $L^2(M_1; \xi)$ and $L^2(M_2; \xi)$.
- but up to compact operators one can *localize* the projector P_L and obtain two projectors acting on $L^2(M_1; \xi)$ and $L^2(M_2; \xi)$.

We note that the following proposition holds:

Proposition 7.6.

1. *The composition of bordisms is a bordism.*
2. *The composition of a bordism and a twist is a bordism.*
3. *The composition of twists is a twist.*

Remark 7.7. Let $H_1 \xrightarrow{L_1} H_2 \xrightarrow{L_2} H_3$ be a pair of bordisms in **PR** coming from geometric bordisms

$$M_1 \sim_{X_1} M_2\,, \quad M_2 \sim_{X_2} M_3$$

and an elliptic operator on $X_1 \cup_{M_2} X_2$, as in Example 4. Then $L_2 \circ L_1$ coincides with the space of the Cauchy data along $\partial(X_1 \cup_{M_2} X_2) = M_1 \sqcup M_3$ of the solutions of $Ds = 0$ on $X_1 \cup_{M_2} X_2$.

8. Chains of morphisms

Now we introduce the notion of a chain. This is a special case of a Fredholm fan considered in [5] and in §12 below.

A chain of morphisms is a sequence correspondences

$$0 \xrightarrow{L_0} H_1 \xrightarrow{L_1} H_2 \xrightarrow{L_2} \cdots \xrightarrow{L_{m-1}} H_m \xrightarrow{L_m} 0\,.$$

Example 5. Let (H_+, H_-) be an admissible Fredholm pair in H. Then we have a sequence

$$0 \xrightarrow{H_-} H \xrightarrow{H_+} 0$$

which is a chain of bordisms with respect to the polarization defined by $P^\sharp = P_+$ (or $1 - P_-$), see Example 3.

Example 6. Each morphism in $L \in Mor_{\mathbf{PR}}(H_1, H_2)$ can be completed to a chain

$$0 \xrightarrow{L_1} H_1 \xrightarrow{L} H_2 \xrightarrow{L_2} 0\,.$$

Just take $L_1 = (0 \oplus H_1^\flat) \subset (0 \oplus H_1)$ and $L_2 = (H_2^\sharp \oplus 0) \subset (H_2 \oplus 0)$.

Example 7. It is proper to explain why we are interested in chains of morphisms. Suppose there is given a closed manifold which is composed of usual bordisms

$$X = X_0 \cup_{M_1} X_1 \cup_{M_2} \cdots \cup_{M_{m-1}} X_{m-1} \cup_{M_m} X_m\,.$$

We treat the manifolds M_i as objects and bordisms

$$M_{i-1} \sim_{X_i} M_i$$

as morphisms. In particular

$$\emptyset \sim_{X_1} M_1 \qquad \text{and} \qquad M_m \sim_{X_m} \emptyset\,.$$

Let $D : C^\infty(X;\xi) \to C^\infty(X;\eta)$ be an elliptic operator of the first order. This geometric situation gives rise to a chain of bordisms in the category **PR**:

- $H_i = L^2(M_i;\xi)$ with the action of $B_i = C(M_i)$ and the polarization defined by the symbol of Calderón projector, as in 4,
- $L_i \subset L^2(M_i;\xi) \oplus L^2(M_{i+1};\xi)$ is the space of boundary values of the solutions of $Ds = 0$ on X_i.

9. Indices in PR

Definition 9.1. Fix the splittings S of the objects of **PR**. The pair $(L, H_1^\flat \oplus H_2^\sharp)$ in $H_1 \oplus H_2$ is Fredholm by Definition 7.1. Define the index of a morphism $L \in Mor_{\mathbf{PR}}(H_1, H_2)$ by the formula:

$$\mathrm{Ind}_{S_1,S_2}(L) \stackrel{\mathrm{def}}{=} \mathrm{Ind}(L, H_1^\flat \oplus H_2^\sharp) = \mathrm{ind}(P_1^\sharp \oplus P_2^\flat : L \to H_1^\sharp \oplus H_2^\flat)\,.$$

Proposition 9.2. *We have the equality of indices for a twist*

1. $\mathrm{Ind}_{S,S}(\mathrm{graph}\ \phi)$,
2. *index of* $\begin{pmatrix} 1 & P^\flat \\ \phi & P^\sharp \end{pmatrix} : H \oplus H \to H \oplus H$,
3. $\widetilde{\mathrm{ind}}(\phi) = \mathrm{ind}(\phi P^\flat + P^\sharp) = \mathrm{Ind}(\phi(H^\flat), H^\sharp)$ (*compare Theorem* 2.3),

Proof. The graph of ϕ is parameterized by $(1, \phi)$ and $H^\flat \oplus H^\sharp$ is parameterized by $(P^\flat, P^\sharp)$. Thus by Theorem 2.1 the first equality follows. Now we multiply the matrix (2.) from the left by the symmetry $\begin{pmatrix} P^\sharp & P^\flat \\ P^\flat & P^\sharp \end{pmatrix}$ and we obtain $\begin{pmatrix} P^\flat\phi + P^\sharp & 0 \\ P^\sharp\phi + P^\flat & 1 \end{pmatrix} \sim \begin{pmatrix} \phi P^\flat + P^\sharp & 0 \\ \phi P^\sharp + P^\flat & 1 \end{pmatrix}$. The second equality follows. □

Remark 9.3. The index of a twist depends only on the polarization, not on the particular splitting. This is clear from 9.2.2. It is worthwhile to point out that if the twist $\phi = \widetilde{A} : H^{\oplus n} \to H^{\oplus n}$ is given by a matrix $A \in GL_n(B)$, then

$$\widetilde{\mathrm{ind}}(\widetilde{A}) = \langle [\widetilde{A}], [S_{H^\flat}] \rangle\,,$$

where $S_{H^\flat}$ is the symmetry with respect to $H^\flat$ and the bracket is the pairing in K-theory of $K^1(B)$ with $K_1(B)$.

On the other hand $\mathrm{Ind}_{S_1,S_2}(L)$ does depend on the splitting for general morphisms.

Remark 9.4. The index in Example 4 is equal to the index of the operator D with the boundary conditions given by the splittings, as in [2].

Remark 9.5. There are certain morphisms in **PR** which are interesting from the point of view of composition. We will say that L is a *special* correspondence if:

- L is the graph of an injective function ϕ defined on a subspace of H_1,
- the images of the projections of L onto H_1 and H_2 are dense.

(The second condition is equivalent to the first one for the adjoint correspondence defined as the orthogonal complement $L^\perp$.) If L is special, then

$$\mathrm{Ind}_{S_1,S_2}(L) = \mathrm{Ind}(L(H_1^\flat), H_2^\sharp)\,,$$

where

$$L(H_1^\flat) = \{y \in H_2 \,:\, \exists x \in H_1^\flat\ (x,y) \in L\}\,.$$

Indeed in this case we have

$$L \cap (H_1^\flat \oplus H_2^\sharp) \simeq L(H_1^\flat) \cap H_2^\sharp \quad \text{and} \quad L^\perp \cap (H_1^{\flat\perp} \oplus H_2^{\sharp\perp}) \simeq L^\perp(H_1^{\flat\perp}) \cap H_2^{\sharp\perp}\,.$$

Of course each twist is a special morphism. Another example of a special morphism is the one which comes from the Cauchy-Riemann operator. In general, we obtain a special morphism if the operator (and its adjoint) satisfies the following:

- if $s = 0$ on a hypersurface M and $Ds = 0$, then $s = 0$ on the whole component containing M.

In the set of morphisms we can introduce an equivalence relation: we say that $L \sim L'$ if L and L' are images of embeddings $i, i' : H \hookrightarrow H_1 \oplus H_2$ of a Hilbert space H, such that $i - i'$ is a compact operator. If $L \sim L'$, then $\mathrm{Ind}_{S_1,S_2}(L) = \mathrm{Ind}_{S_1,S_2}(L')$. If L is a bordism, then L is equivalent to a direct sum of subspaces in coordinates: $L \sim L_1 \oplus L_2$, $L_i \subset H_i$, such that L_1 is a finite-dimensional perturbation of $H_1^\sharp$ and L_2 is a finite-dimensional perturbation of $H_2^\flat$. Then $\mathrm{Ind}_{S_1,S_2}(L) = \mathrm{Ind}(H_1^\flat, L_1) + \mathrm{Ind}(L_2, H_2^\sharp)$.

Suppose, as in Example 7, we have an elliptic operator on a closed manifold X which is composed of geometric bordisms. Fix $n \in \mathbf{N}$ and a sequence of matrices

$$A_i \in GL_n(B_i)\,.$$

Define the bundle $\Theta_X^{\{A_i\}}$ obtained from trivial ones on X_i's and twisted along M_i's. Define the bordism $L_i(D) \in \mathrm{Mor}_{\mathbf{PR}}(H_i, H_{i+1})$ as in Example 4.

Theorem 9.6. *Suppose that* 3.1 *holds for D and D^* on each X_i for $i = 0, \ldots, n$. Then*

$$\mathrm{ind}(D \otimes \Theta_X^{\{A_i\}}) = n\left(\sum_{i=0}^{m} \mathrm{Ind}_{S_i,S_{i+1}}(L_i(D))\right) + \sum_{i=1}^{m} \widetilde{\mathrm{ind}}(\widetilde{A}_i)\,.$$

Here, as it was denoted before, $\widetilde{A} : H^{\oplus n} \to H^{\oplus n}$ is the operator associated to the matrix $A \in GL_n(B)$. This theorem is a special case of Theorem 11.1 proved below.

Taking into account Remark 9.3, the difference between the indices of the original and twisted operator can be expressed through the pairing in K-theory.

Theorem 9.7.

$$\mathrm{ind}(D \otimes \Theta_X^{\{A_i\}}) - n\,\mathrm{ind}(D) = \sum_{i=1}^{m} \widetilde{\mathrm{ind}}(\widetilde{A}_i) = \sum_{i=1}^{m} \langle [A_i], [S_{H_i^\flat}] \rangle\,.$$

The braked is the pairing between $[A_i] \in K^1(M_i)$ and $[S_{H_i^\flat}] \in K_1(M_i)$.

10. Indices of compositions

In 9.3 we have made some remarks about the dependence of indices on the particular splitting. Now let us see how indices behave under compositions of correspondences. From the considerations in §9 it is easy to deduce:

Proposition 10.1. *For the composition*

$$H_1 \xrightarrow{\phi} H_1 \xrightarrow{L} H_2\,,$$

where ϕ is a twist and L is a morphism we have

$$\mathrm{Ind}_{S_1,S_2}(L\circ\phi) = \mathrm{Ind}_{S_1,S_2}(L) + \widetilde{\mathrm{ind}}(\phi)\,.$$

The same holds for the opposite type composition

$$H_1 \xrightarrow{L} H_2 \xrightarrow{\phi} H_2\,,$$

$$\mathrm{Ind}_{S_1,S_2}(\phi\circ L) = \widetilde{\mathrm{ind}}(\phi) + \mathrm{Ind}_{S_1,S_2}(L)\,.$$

On the other hand $\mathrm{Ind}_{S_0,S_2}(L_2\circ L_1)$ differs from $\mathrm{Ind}_{S_0,S_1}(L_1)+Ind_{S_1,S_2}(L_2)$ in general. This is clear due to the basic example that comes from a decomposition $X = X_-\cup_M X_+$. The space $L_1 = H_-(D)$ is a correspondence $0\to L^2(M;\xi)$ and $L_2 = H_+(D)$ a correspondence $L^2(M;\xi)\to 0$. By 9.6 we have

$$\mathrm{Ind}_{Id,S_1}(L_1) + \mathrm{Ind}_{S_1,Id}(L_2) = \mathrm{ind}(D)\,,$$

while $L_2\circ L_1 : 0\to 0$ and $\mathrm{Ind}_{Id,Id}(L_2\circ L_1) = 0$.

Instead we have the following interesting property of indices:

Theorem 10.2. *The difference*

$$\delta(L_1,L_2) = \mathrm{Ind}_{S_0,S_1}(L_1) + \mathrm{Ind}_{S_1,S_2}(L_1) - \mathrm{Ind}_{S_0,S_2}(L_2\circ L_1)$$

does not depend on the particular splittings.

Proof. Since

$$\mathrm{Ind}_{S_{i-1},S_i}(L_i) = \mathrm{ind}(H^\flat_{i-1}\oplus L_i\oplus H^\sharp_i \to H_{i-1}\oplus H_i)$$

we have to compare indices of the operators

$$\alpha : H^\flat_0\oplus L_1\oplus H^\sharp_1\oplus H^\flat_1\oplus L_2\oplus H^\sharp_2 \to H_0\oplus H_1\oplus H_1\oplus H_2$$

and

$$\beta : H^\flat_0\oplus L_2\circ L_1\oplus H^\sharp_2 \to H_0\oplus H_2\,.$$

The kernel of α is isomorphic to the kernel of the operator which is induced by inclusions

$$H^\flat_0\oplus L_1\oplus L_2\oplus H^\sharp_2 \to H_0\oplus H_1\oplus H_2\,.$$

The former operator factors through

$$H^\flat_0\oplus (L_1+L_2)\oplus H^\sharp_2 \to H_0\oplus H_1\oplus H_2\,.$$

Here the direct sum is replaced by the algebraic sum inside $H_0\oplus H_1\oplus H_2$. The difference of the dimensions of the kernels is equal to the dimension of the intersection

$$(L_1\oplus 0)\cap(0\oplus L_2)\subset H_0\oplus H_1\oplus H_2$$

Now we observe that the kernel of the last operator is isomorphic to

$$H^\flat_0\oplus L_2\circ L_1\oplus H^\sharp_2 \to H_0\oplus H_2\,.$$

Therefore the difference of the dimensions of the kernels of α and β is equal to $\dim((L_1 \oplus 0) \cap (0 \oplus L_2))$, hence it does not depend on the splittings. We have the dual formula for cokernels and $L_i^\perp$, also not depending on the splittings. $\square$

We obtain a procedure of computing the sum of indices

$$\sum_{i=0}^{m} \mathrm{Ind}_{S_i,S_{i+1}}(L_i)$$

which would not involve splittings. We choose a pair of consecutive morphisms L_i, L_{i+1} and replace them by their compositions. The composition produces a number $\delta(L_i, L_{i+1})$ and the sequence of morphisms is shorter:

$$(L_0, L_1, \dots, L_m) \rightsquigarrow (L_0, L_1, \dots, L_i \circ L_{i+1}, \dots, L_m) + \delta(L_i, L_{i+1})\,.$$

We pick another composition and add its contribution to the previous one. We continue until we get $0 \to 0$. The sum of the contributions does not depend on the splittings. One can perform compositions in various ways. The sum of contributions stays the same.

Example 8. If D and D^* on X_i and X_{i+1} have the unique extension property 3.1, then $\delta(L_i, L_{i+1}) = 0$ as long the gluing process along M_{i+1} does not create a closed component of X. If it does then $\delta(L_i, L_{i+1})$ equals to the index of D restricted to this component.

11. Weird decompositions of manifolds

Let $\{M_e\}_{e\in E}$ be a configuration of disjoined hypersurfaces in a manifold X. We assume that orientations of the normal bundles are fixed. For simplicity assume that X and M_e's are connected. Let

$$X \setminus \bigsqcup_{e\in E} M_e = \bigsqcup_{v\in V} X_v$$

be the decomposition of X into connected components. Our situation is well described by an oriented graph

- the vertices (corresponding to open domains in X) are labelled by the set V
- the edges (corresponding to hypersurfaces) are labelled by E. The edge e starts at the vertex $v = s(e)$ corresponding to X_v which is on the negative side of M_e. It ends at $v' = t(e)$, such that $X_{v'}$ lies on the positive side of M_e. The functions $s, t : E \to V$ are the *source* and *target* functions.

For example the configuration of the cutting circles on the surface (Fig. 1) is described by the graph (Fig. 2).

A sequence of bordisms leads to the linear graph

$$\bullet_{X_0} \xrightarrow{M_1} \bullet_{X_1} \xrightarrow{M_2} \cdots \xrightarrow{M_{n-1}} \bullet_{X_{n-1}} \xrightarrow{M_n} \bullet_{X_n}\ .$$

Note that this is a dual description with respect to the one presented in Example 7. Suppose there is given an elliptic operator $D : C^\infty(X;\xi) \to C^\infty(X;\eta)$ and a

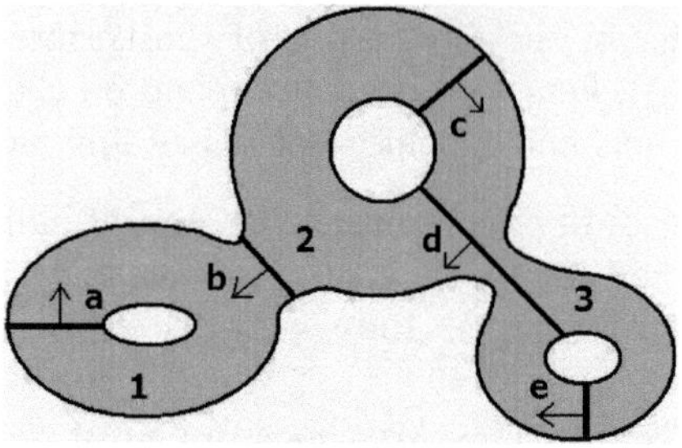

Fig. 1

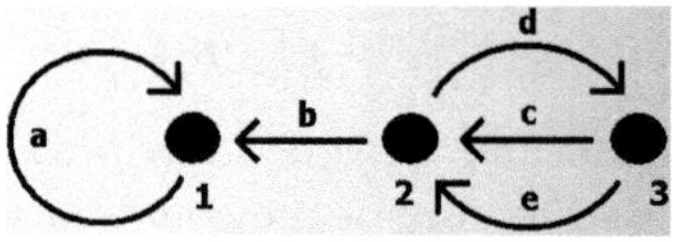

Fig. 2

set of transmission data $\{\phi_e\}_{e\in E}$, that is for each hypersurface M_e we are given a matrix-valued function $M_e \to GL_n(\mathbf{C})$. The Riemann-Hilbert problem gives rise to the operator

$$D^{[\phi]} : \bigoplus_{v\in V} C^\infty(X_v;\xi)^n \to \bigoplus_{v\in V} C^\infty(X_v;\eta)^n \oplus \bigoplus_{e\in E} C^\infty(M_e;\xi)^n$$

$$D^{[\phi]}(f_v) \stackrel{\text{def}}{=} \left(Df_v, \sum_{e:\, t(e)=v} f_{v|M_e} - \sum_{e:\, s(e)=v} \phi_e(f_{v|M_e}) \right), \quad \text{for} \quad f_v \in C^\infty(X_v;\xi)^n .$$

For $e \in E$ let us set $H(e) = L^2(M_e;\xi)$. The symbol of D together with the choice of orientations of the normal bundles define polarizations of $H(e)$. Let us fix particular splittings of the spaces $H(e)$ encoded in the symetries S_e. Set

$$\begin{aligned} H^{\text{bd}}(v) &= \bigoplus_{e:\, s(e)=v} H(e) \oplus \bigoplus_{e:\, t(e)=v} H(e), \\ H^{\text{in}}(v) &= \bigoplus_{e:\, s(e)=v} H^\sharp(e) \oplus \bigoplus_{e:\, t(e)=v} H^\flat(e), \\ H^{\text{out}}(v) &= \bigoplus_{e:\, s(e)=v} H^\flat(e) \oplus \bigoplus_{e:\, t(e)=v} H^\sharp(e). \end{aligned}$$

Let $L(v) \subset H^{\text{bd}}(v)$ be the space of boundary values of solutions of $Df_v = 0$ on X_v. It is a perturbation of $H^{\text{in}}(v)$. For each vertex v (i.e., for each open domain X_v) the pair of subspaces

$$L(v), H^{\text{out}}(v) \subset H^{\text{bd}}(v),$$

is Fredholm. Let Ind_v be its index with respect to the polarizations S_e. Moreover, let $\text{Ind}_e = \text{Ind}_{S_e,S_e}(\phi_e) = \widetilde{\text{ind}}(\phi_e)$ denote the index of ϕ_e, see Theorem 2.3.

Theorem 11.1. *Assume that D and D^* have unique extension property* (3.1) *on each X_v. Then*

$$\operatorname{ind}(D^{[\phi]}) = \sum_{v \in V} \operatorname{Ind}_v + \sum_{e \in E} \operatorname{Ind}_e \,.$$

In particular:

Corollary 11.2. *If there are no twists, i.e., each $\phi_e = 1 \in GL_1(C^\infty(M_e))$, then*

$$\operatorname{ind}(D) = \sum_{v \in V} \operatorname{Ind}_v \,.$$

Proof. (of 11.1.) The general result follows from the case when we have one vertex and one edge starting and ending in it. We just sum up all X_v's and all M_e's. Say that X is obtained from $\hat{X}$ with $\partial \hat{X} = M_s \sqcup M_t$ by identification M_s with M_t as presented on Fig. 3.

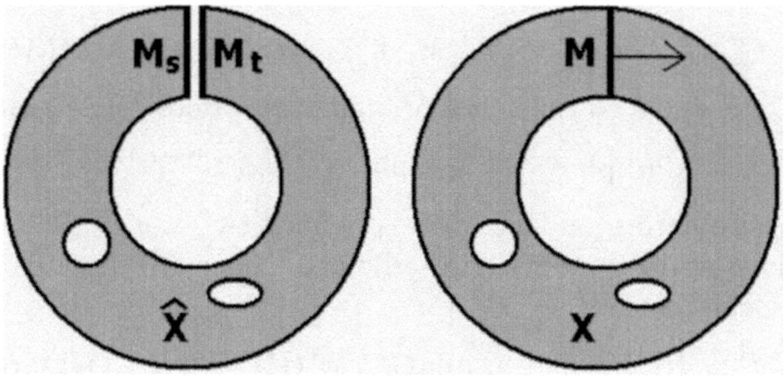

Fig. 3

Then our operator $D^{[\phi]}$ is of the form:

$$D^{[\phi]} : C^\infty(\hat{X}; \xi)^n \to C^\infty(\hat{X}; \eta)^n \oplus C^\infty(M; \xi)^n$$
$$D^{[\phi]}(u) = \left(Du, u_{|M_t} - \phi(u_{|M_s})\right) \,.$$

We replace $\xi^{\oplus n}$ by ξ and treat ϕ as an automorphism of ξ. The index of the operator is equal to the index of a Fredholm pair:

Theorem 11.3. *Let $L \subset L^2(M_s \sqcup M_t; \xi) = L^2(M; \xi) \oplus L^2(M; \xi)$ be the space of boundary values of the operator D on $\hat{X}$. Then*

$$\operatorname{ind}(D^{[\phi]}) = \operatorname{Ind}(L, \operatorname{graph}(\phi)) \,.$$

The proof of Theorem 11.1 relies on this formula. Our proof is based on the principle that the index can be computed by restricting the argument to the spaces of smooth sections. The precise argument demands introduction and consecutive use of the whole scale of Sobolev spaces with all usual technicalities involved. The reader may also take this formula as the definition of the index of the problem considered above. We calculate the kernel and cokernel of $D^{[\phi]}$:

- the kernel consist of solutions of $Du = 0$ on $\hat{X}$ satisfying $\phi(u_{|M_s}) = u_{|M_t}$. By our assumption u is determined by its boundary value. Thus

$$\ker D^{[\phi]} \simeq L \cap \operatorname{graph} \phi \,.$$

The cokernel consists of

$$\Big\{(v,w)\in C^\infty(\hat{X};\eta^*)\oplus C^\infty(M;\xi^*):\\ \forall u\in C^\infty(X_+;\xi)\quad \langle Du,v\rangle+\langle u_{|M_t}-\phi(u_{|M_s}),w\rangle=0\Big\}\,.$$

Let $G:\xi_{|M}\to\eta_{|M}$ be the isomorphism of the bundles defined by the symbol of D as in [14]. It follows that

- $D^*v=0$ (since we can take any u with support in $int\,\hat{X}$)
- by Green formula $\langle Du,v\rangle=\langle Gu_{|M_s},v_{|M_s}\rangle+\langle Gu_{|M_t},v_{|M_t}\rangle$
- since $u_{|M_s}$ and $u_{|M_t}$ may be arbitrary it follows that
  $G^*(v_{|M_s})=-\phi^*w,$
  $G^*(v_{|M_t})=w,$
- therefore $v_{|M_s}=-G^{*-1}\phi^*G^*(v_{|M_t})$.

Now we use the identification

$$G^*\times G^*:L^2(M_s;\eta^*)\times L^2(M_t;\eta^*)\to L^2(M_s;\xi^*)\times L^2(M_t;\xi^*)$$

under which $L^\perp$ is equal to the space of boundary values $H(D^*)$ and

$$(\text{graph }\phi)^\perp=(\text{graph}(-G^{*-1}\phi^*G^*))^{op}\,.$$

(Here the opposite correspondence R^{op} is defined by $(x,y)\in R^{op}\equiv(y,x)\in R$.) In other words ϕ and $G^{*-1}\phi^*G^*$ are adjoined. Since the boundary values of v determine v we can identify

$$\text{coker }D^{[\phi]}\simeq H(D^*)\cap(-\text{graph}(G^{*-1}\phi^*G^*))^{\text{op}}\simeq L^\perp\cap(\text{graph }\phi)^\perp\,.\qquad\square$$

Proof. (Continuation of 11.1.) After fixing a splitting of $L^2(M;\xi)=H_e$, we have in our notation $H_v^{\text{in}}=H^\flat\oplus H^\sharp$, $H_v^{\text{out}}=H^\sharp\oplus H^\flat$. By 2.3 there exists a linear isomorphism $\Psi:H\oplus H\to H\oplus H$ almost commuting with $P^\flat\oplus P^\sharp$, such that $L=\Psi(H^\flat\oplus H^\sharp)$. We parameterize the graph of ϕ by $H^\sharp\oplus H^\flat$ using the composition $\Phi=\begin{pmatrix}1&0\\ \phi&1\end{pmatrix}\circ\begin{pmatrix}P^\sharp&P^\flat\\ P^\flat&P^\sharp\end{pmatrix}$. Thus

$$\text{Ind}(\text{graph }\phi,L)=\text{ind}\left(\Phi\circ\begin{pmatrix}P^\sharp&0\\0&P^\flat\end{pmatrix}+\Psi\circ\begin{pmatrix}P^\flat&0\\0&P^\sharp\end{pmatrix}\right).$$

Since Ψ almost commutes with $P^\flat\oplus P^\sharp$, the considered operator is almost equal to the composition

$$\left(\Phi\circ\begin{pmatrix}P^\sharp&0\\0&P^\flat\end{pmatrix}+\begin{pmatrix}P^\flat&0\\0&P^\sharp\end{pmatrix}\right)\circ\left(\begin{pmatrix}P^\sharp&0\\0&P^\flat\end{pmatrix}+\Psi\circ\begin{pmatrix}P^\flat&0\\0&P^\sharp\end{pmatrix}\right).$$

Now we use additivity of indices. The index of the second term is equal to Ind_v. It remains to compute the first index, that is ind $\begin{pmatrix}1&P^\flat\\ \phi P^\sharp&\phi P^\flat+P^\sharp\end{pmatrix}$. If we conjugate the above matrix by the symmetry $\begin{pmatrix}P^\sharp&P^\flat\\ P^\flat&P^\sharp\end{pmatrix}$ we obtain $\begin{pmatrix}P^\sharp+P^\flat\phi&0\\ P^\flat+P^\sharp\phi&1\end{pmatrix}$. Its index is equal to $\text{ind}(P^\sharp+P^\flat\phi)=\text{Ind}_e$. $\square$

The additivity of the index is not a surprise due to the well-known integral formula for the analytic index. What is interesting in Theorem 11.2 is that the contribution coming from separate pieces of X is also an integer number. This partition into local indices depends only on the choice of splittings along hypersurfaces.

12. Index of a fan

We will give another formula for the index of $D^{[\phi]}$ which is expressed in terms of the twisted fan $\{L(i)\}$. The general reference for fans is [5]. Let us first say what we mean by a fan: it is a collection of spaces

$$L_1, L_2, \ldots, L_n \subset H$$

which is obtained from a direct sum decomposition

$$H_1 \oplus H_2 \oplus \cdots \oplus H_n = H$$

by a sequence of twists $\Psi_1, \Psi_2, \ldots, \Psi_n \in GL(H)$, i.e. $L_i = \Psi_i(H_i)$. We assume that each Ψ_i almost commutes with each projection P_j of the direct sum. We say that the fan $\{L(i)\}$ is a perturbation of the direct sum decomposition $H = \oplus H_i$.

Theorem 12.1 (Index of a Fredholm fan). *Let $L_1, L_2, \ldots, L_n \subset H$ be a fan. Then the following numbers are equal:*

1. *the index of the map $\iota : L_1 \oplus L_2 \oplus \cdots \oplus L_n \to H$, which is the sum of inclusions,*
2. *the index of the operator $\Psi_1 P_1 + \Psi_2 P_2 + \cdots + \Psi_n P_n : H \to H$,*
3. *the sum*
$$\sum_{i=1}^{n} \operatorname{ind}(P_i \Psi_i : H_i \to H_i) = \sum_{i=1}^{n} \operatorname{ind}(P_i : L_i \to H_i)\,,$$
4. *the difference*
$$\sum_{i=1}^{n-1} \dim(L_1 + \cdots + L_i) \cap L_{i+1} - \operatorname{codim}(L_1 + \cdots + L_n)\,.$$

Proof. The equality (1.=2.) follows from the fact that $\Psi_i : H_i \to P_i$ is a parameterization of L_i. The equality (2.=3.) follows since

$$\Psi_1 P_1 + \Psi_2 P_2 + \cdots + \Psi_n P_n \sim \prod_{i=1}^{n} (P_1 + \cdots + \Psi_i P_i + \cdots + P_n)\,.$$

To prove the equality (1.=4.) one checks that

$$\dim(\ker \iota) = \sum_{i=1}^{n-1} (L_1 + \cdots + L_i) \cap L_{i+1}\,.$$

This is done by induction with respect to n. □

Let us assume that the graph associated to our configuration does not contain edges starting and ending in the same vertex (e.g., the situation on Fig. 1 is not allowed). Then $H^{\mathrm{bd}}(v)$ is a summand in $H = \bigoplus_{e\in E} H(e)$ (there are no terms $H(e)$ appearing twice). Moreover, $\{L(v)\}_{v\in V}$ is a fan in H which is a perturbation of the direct sum decomposition

$$H = \bigoplus_{v\in V} H^{\mathrm{in}}(v)\,.$$

Consider a fan, which is twisted with respect to $\{L(v)\}_{v\in V}$. Set $(\phi\bowtie L)(v) = \widetilde{\phi}_v(L(v))$, where $\widetilde{\phi}_v$ is an automorphisms of H:

$$\widetilde{\phi}_v(f) \stackrel{\mathrm{def}}{=} \begin{cases} \phi_e(f) & \text{if } f\in H(e),\ s(e)=v\,,\\ f & \text{if } f\in H(e),\ s(e)\neq v\,. \end{cases}$$

Theorem 12.2. *Assume that D and D^* have unique extension property* (3.1) *on each X_v. The index of $D^{[\phi]}$ is equal to the index of the Fredholm fan $\phi\bowtie L$.*

Proof. Combining Theorem 11.1 with 12.1.3 it remains to prove that for each vertex v

$$\mathrm{ind}(P^{\mathrm{in}}_v : (\phi\bowtie L)(v) \to H^{\mathrm{in}}(v)) = \mathrm{Ind}_v + \sum_{e\,:\,s(e)=v} \mathrm{Ind}_e\,.$$

If there are no twists, then the equality follows from Proposition 2.2. In general the proof follows from additivity of $\widetilde{\mathrm{ind}}$, see Theorem 2.3. □

References

[1] S. Alinhac, *Non unicité du problème de Cauchy pour des opérateurs de type principal II.* Ann. of Math. **117** (1983), 77–108.

[2] M.F. Atiyah, V.K. Patodi, I.M. Singer, *Spectral asymmetry and Riemannian geometry. I.* Math. Proc. Camb. Phil. Soc. **77** (1975), 43–69.

[3] M.Sh. Birman, M.Z. Solomyak, *On subspaces admitting pseudodifferential projections.* (Russian) Vestnik Leningrad. Univ. Mat. Mekh. Astronom. **1** (1982), 18–25. (English) Vestn. Leningr. Univ., Math. **15** (1983), 17–27.

[4] B. Bojarski, *The abstract linear conjugation problem and Fredholm pairs of subspaces.* Volume in Memoriam I.N. Vekua: Differential and Integral Equations. Boundary Value Problems. Publications of I.N. Vekua Institute of Applied Mathematics, Tibilisi 1979, 45–60.

[5] B. Bojarski, *The geometry of the Riemann-Hilbert problem.* Booss-Bavnbek, Bernhelm (ed.) et al., Geometric aspects of partial differential equations. Proceedings of a minisymposium on spectral invariants, heat equation approach, Roskilde, Denmark, September 18-19, 1998. Providence, RI: American Mathematical Society. Contemp. Math. **242** (1999), 25–33.

[6] B. Bojarski, *The geometry of the Riemann-Hilbert problem II.* in Boundary value problems, integral equations and related problems (Beijing/Chengde, 1999), 41–48, World Sci. Publishing, River Edge, NJ, 2000.

[7] B. Bojarski, A. Weber, *Generalized Riemann-Hilbert Transmission and Boundary Value Problems, Fredholm Pairs and Bordisms.* Bull. Polish Acad. Sci. Math. **50**, No 4 (2002), 479–496.

[8] B. Booss-Bavnbek, K.P. Wojciechowski, *Desuspension of splitting elliptic symbols, I.* Ann. Glob. Anal. Geom. **3**, No. 3, 337–383, (1985); *II.* Ann. Glob. Anal. Geom. **4**, No. 3, 349–400, (1986).

[9] B. Booss-Bavnbek, K.P. Wojciechowski, *Elliptic boundary problems for Dirac operators.* Mathematics: Theory & Applications. Boston, MA: Birkhäuser. (1993).

[10] A. Connes, *Non-commutative differential geometry.* Publ. Math., Inst. Hautes Etud. Sci. **62** (1985), 257–360.

[11] A. Connes, *Noncommutative geometry.* Academic Press, Inc., San Diego, CA, 1994.

[12] X. Dai, W. Zhang, *Splitting of the family index.* Comm. Math. Phys. **182** (1996), no. 2, 303–317.

[13] G.G. Kasparov, *Topological invariants of elliptic operators. I. K-homology.* (Russian) Math. USSR-Izv. **9** (1975), no. 4, 751–792; (English) Izv. Akad. Nauk SSSR Ser. Mat. **39** (1975), no. 4, 796–838.

[14] R.S. Palais, R.T. Seeley, *Cobordism invariance of the analytical index.* in *Seminar on the Atiyah-Singer index theorem.* ed. R.S. Palais, Annals of Mathematics Studies **57**, Princeton University Press, Princeton.

[15] A. Pliś, *Non-uniqueness in Cauchy's problem for differential equations of elliptic type.* J. Math. Mech. **9** (1960), 557–562.

[16] A. Pressley, G. Segal, *Loop groups.* Oxford Mathematical Monographs. Oxford Science Publications. The Clarendon Press, Oxford University Press, New York, 1986.

[17] S.G. Scott, K.P. Wojciechowski, *The ζ-determinant and Quillen determinant for a Dirac operator on a manifold with boundary.* Geom. Func. Anal. Vol. **10** (2000), 1202–1236.

[18] G. Segal, *Topological Field Theory ('Stanford Notes').* Available at http://www.cgtp.duke.edu/ITP99/segal/

[19] R.T. Seeley, *Singular Integrals and Boundary Value Problems.* Amer. J. Math **88** (1966), 781–809.

Bogdan Bojarski
Institute of Mathematics PAN
ul. Śniadeckich 8
PL-00-950 Warszawa, Poland
e-mail: bojarski@impan.gov.pl

Andrzej Weber
Institute of Mathematics
Warsaw University
ul. Banacha 2
PL-02-097 Warszawa, Poland
e-mail: aweber@mimuw.edu.pl

C^*-algebras and Elliptic Theory
Trends in Mathematics, 23–35

Approximation Properties for Discrete Groups

Jacek Brodzki and Graham A. Niblo

Abstract. We provide an illustration of an interesting and nontrivial interaction between analytic and geometric properties of a group. We provide a short survey of approximation properties of operator algebras associated with discrete groups. We then demonstrate directly that groups that satisfy the property RD with respect to a conditionally negative length function have the metric approximation property.

Mathematics Subject Classification (2000). Primary: 58B34; Secondary: 20F69, 46L89.

Keywords. Rapid decay, metric approximation property.

1. Introduction

The reduced C^*-algebra $C_r^*(\Gamma)$ of a group Γ (which we shall assume to be discrete) arises from the study of the left regular representation λ of the group ring $\mathbb{C}\Gamma$ on the Hilbert space of square-summable functions on the group. Various important properties of the group can be expressed in terms of analytic properties of this algebra. We give a survey of the main points below but let us just mention the result of Lance [16] that the reduced C^*-algebra is nuclear if and only if the group is amenable.

From the point of view of noncommutative geometry, a C^*-algebra is always considered as an algebra of continuous functions on some space. In the case of the reduced C^*-algebra that space is a space of representations of the group. Unfortunately it is not easy to understand the structure of this algebra in general, though partial information is obtained by studying topological invariants of this algebra, for instance its K-theory. This, too, is complicated although the Baum-Connes conjecture postulates a possible way to compute it.

In some cases one can find an interesting smooth subalgebra of $C_r^*(\Gamma)$, that would play a role similar to the algebra of smooth functions on a manifold. This algebra of Schwartz-type functions is most useful when it has the same K-theory as $C_r^*(\Gamma)$, while being more accessible to homological methods. Algebras of this kind are normally defined by imposing a suitable growth condition on the space of functions. A very interesting example of when this can be done is provided by the class that satisfy the rapid decay property (RD), introduced by Jolissaint [12]. Here the control on growth is derived from a length function which turns the group Γ into a metric space with interesting geometry.

Definition 1.1. A *length function* on a discrete group Γ is a function $\ell : \Gamma \to \mathbb{R}$ taking values in the non-negative reals which satisfies the following conditions:

1. $\ell(1) = 0$, where 1 is the identity element of the group;
2. For every $g \in \Gamma$ $\ell(g) = \ell(g^{-1})$.
3. For every $g, h \in \Gamma$, $\ell(gh) \leq \ell(g) + \ell(h)$ for all $g, h \in \Gamma$,

A group equipped with a length function becomes a metric space with the left-invariant metric $d(g,h) = \ell(h^{-1}g)$.

For any length function ℓ and a positive real number s we define a Sobolev norm on the group ring $\mathbb{C}\Gamma$

$$\|f\|_{\ell,s} = \sqrt{\sum_{\gamma \in \Gamma} |f(\gamma)|^2 (1 + \ell(\gamma))^{2s}}$$

Following Jolissaint [12] (see also [3]) we say that Γ has the *rapid decay property* (*property RD*) with respect to the length function ℓ if and only if it satisfies the following property: There exist a $C > 0$ and $s > 0$ such that for all $f \in \mathbb{C}\Gamma$

$$\|\lambda(f)\| \leq C \|f\|_{\ell,s},$$

where the norm on the left-hand side is the operator norm in $\mathfrak{L}(\ell^2(\Gamma))$. This inequality indicates how the operator norm, which is in general difficult to compute, can be controlled by a more computable Sobolev norm. Examples of RD groups include hyperbolic groups [10], groups acting on CAT(0)-cube complexes [3] and co-compact lattices in $SL_3(\mathbb{R})$ or $SL_3(\mathbb{C})$ [15] as well as co-compact lattices in $SL_3(\mathbb{H})$ [2].

The purpose of this note is to provide an illustration of an interesting and nontrivial interaction between analytic and geometric properties of a group. We provide a short survey of approximation properties of operator algebras associated with discrete groups. We then demonstrate directly that groups that satisfy the property RD with respect to a conditionally negative length function have the metric approximation property, which is defined below. We obtain this result by combining two important ingredients. The RD property provides us with estimates for operator norms, while the properties of conditionally negative kernels allow us to define multipliers, i.e., operators M_ϕ induced by pointwise multiplication by a function ϕ that map $C^*_r(\Gamma)$ into itself. These two properties together combine to control norm inequalities in the reduced C^*-algebra. Although this result is implied by a result of Jolissaint and Valette [13] we feel that this direct approach illustrates the important role played by the RD property rather well. For another short introduction to the Rapid Decay property and its interaction with multipliers (which play a key role in this note) we would like to draw the reader's attention to a short article by Indira Chatterji which appears as an appendix to [18].

This note represents an extended version of the talk delivered by the first author at the meeting on 'C^*-algebras and elliptic theory', which took place in Będlewo in February 2004. J. Brodzki would like to thank the organizers for providing a very stimulating environment for exchanging ideas.

2. Algebras associated with groups

It is well known that all topological information about a compact Hausdorff space X can be recovered from the unital abelian C^*-algebra $C(X)$ of continuous functions on X. Moreover, it is known that any commutative C^*-algebra is isomorphic to an algebra of continuous functions on a locally compact space X. This point of view has been developed with great success within noncommutative geometry, which provides the geometric, analytic and homological tools for the study of 'quantum spaces'. In this approach, C^*-algebras and their topological invariants are studied using methods modeled on classical topology and geometry.

When the space X is equipped with some algebraic structure, for instance when X is a locally compact group, one would hope to have a way of encoding, in operator-algebraic terms, both the topology and algebra of X. We shall outline briefly how this might be done.

Let us assume that Γ is a discrete group. The group ring $\mathbb{C}\Gamma$ consists of all finitely supported complex-valued functions on Γ, that is of all finite combinations $f = \sum_{\gamma \in \Gamma} f_\gamma \delta_\gamma$ with complex coefficients f_γ where δ_γ is the characteristic function of the set $\{\gamma\}$.

If we equip the group ring with the pointwise product of functions then the resulting $*$-algebra contains information, such as there is, about the topology of Γ but completely ignores its group structure. To encode that information we need to use the convolution product defined for any $\gamma, \eta \in \Gamma$ by

$$\delta_\gamma * \delta_\eta = \delta_{\gamma\eta}$$

The left-regular representation λ of the group ring $\mathbb{C}\Gamma$ assigns to each element $f \in \mathbb{C}\Gamma$ a bounded operator $\lambda(f)$ which acts on any $\xi \in \ell^2(\Gamma)$ by convolution:

$$\lambda(f)(\xi) = f * \xi.$$

The image $\lambda(\mathbb{C}\Gamma)$ of the group ring under the left-regular representation is a $*$-subalgebra of the algebra $\mathfrak{L}(\ell^2(\Gamma))$ of bounded operators on $\ell^2(\Gamma)$.

Definition 2.1. The closure of $\lambda(\mathbb{C}\Gamma)$ in the C^*-norm topology of $\mathfrak{L}(\ell^2(\Gamma))$ is by definition the reduced C^*-algebra of Γ denoted $C_r^*(\Gamma)$.

The reduced C^*-algebra of the group Γ does not arise from the topological structure of the group Γ but rather contains information about the representation theory of Γ. The case of abelian groups illustrates this point rather well. For an abelian group Γ the Pontryagin dual $\widehat{\Gamma}$ is by definition the group of characters, that is group homomorphisms from Γ with values in the circle group $\mathbb{T}$. Then one has

$$C_r^*(\Gamma) = C_0(\widehat{\Gamma}).$$

For example, when $\Gamma = \mathbb{Z}^k$, the dual group $\widehat{\mathbb{Z}^k}$ is the k-dimensional torus $(S^1)^k$.

The reduced C^*-algebra of a group collects the information about the irreducible unitary representations that make up the left regular representation of the group ring. Dually, the data concerning the representation theory of a group is encoded in its Fourier algebra [5], [9] which we will now describe.

We recall first that a complex-valued function ϕ on Γ is a coefficient function of the left regular representation iff

$$\phi(\gamma) = \langle \lambda(\delta_\gamma)\xi, \eta \rangle$$

for all $\gamma \in \Gamma$ and some vectors $\xi, \eta \in \ell^2(\Gamma)$.

Definition 2.2. The Fourier algebra $A(\Gamma)$ is the completion of $\mathbb{C}\Gamma$ in the norm

$$\|u\|_{A(\Gamma)} = \inf\{\|\xi\|\|\eta\| \mid u(\gamma) = \langle \lambda(\delta_\gamma)\xi, \eta \rangle\}.$$

With this norm, $A(\Gamma)$ is a Banach algebra with the pointwise multiplication.

3. A brief survey of approximation properties

The study of approximation properties was initiated by Grothendieck in relation to the notion of nuclearity that he introduced in [7]. His fundamental ideas have been applied to the study of groups; in this case one discovers that important properties of groups, like amenability or exactness, can be expressed in terms of approximation properties of the associated operator algebras introduced in the previous section. We give here a brief overview of the main facts. Our main references are Wassermann's lecture notes [23] and Paulsen's text [21].

Let A and B be two C^*-algebras and $\phi : A \to B$ be a linear map. Then

$$\phi \otimes \mathrm{id}_{M_n} : M_n(A) \to M_n(B), \quad (a_{ij}) \mapsto (\phi(a_{ij}))$$

is a linear map, denoted by ϕ_n. If ϕ is a $*$-homomorphism then ϕ_n is also a $*$-homomorphism.

The map ϕ is said to be *completely bounded* iff ϕ_n is bounded for all $n \geq 1$. Let

$$\|\phi\|_{cb} = \sup\{\|\phi_n\| \mid n \in \mathbb{N}\}$$

When this expression is finite, it is called the *completely bounded norm* of ϕ.

Similarly, the map ϕ is *completely positive* iff ϕ_n is positive for all n; it is *completely isometric* iff ϕ_n is isometric for all $n \geq 1$ and ϕ is *completely contractive* iff the maps ϕ_n are contractions ($\|\phi_n\| \leq 1$) for all n.

Completely positive maps are completely bounded [23, 1.5]. When A is unital and ϕ is a complete contraction, then ϕ is completely positive if and only if $\|\phi(1)\| = \|\phi\|$.

With these notions at hand we can introduce one of the main approximation properties.

Definition 3.1. A C^*-algebra A is *nuclear* iff it has the following completely positive approximation property (CPAP): The identity map id : $A \to A$ can be approximated in the point-norm topology by finite rank completely positive contractions. This means that that there exist nets of operators $T_\alpha : A \to M_{n_\alpha}(\mathbb{C})$ and $S_\alpha : M_{n_\alpha}(\mathbb{C}) \to A$ such that for all $a \in A$

$$\lim_\alpha \|S_\alpha T_\alpha(a) - a\| = 0$$

Equivalently (and more traditionally) one can say that a C^*-algebra A is nuclear if and only if the minimal and maximal C^*-norms on the algebraic tensor product $A \odot B$ are the same for any C^*-algebra B.

Nuclear algebras satisfy the metric approximation property of Grothendieck (MAP) which is stated as follows.

Definition 3.2. A C^*-algebra A has the *metric approximation property* iff the identity map on A can be approximated in the point-norm topology by a net of finite rank contractions.

It is clear that CPAP implies MAP.

One of the most important examples of how approximation properties of algebras relate to properties of groups is provided by the following theorem of Lance [16].

Theorem 3.3. *A discrete group Γ is amenable if and only if its reduced C^*-algebra $C^*_r(\Gamma)$ is nuclear.*

So we see that our group Γ is amenable if and only if its reduced C^*-algebra has the CPAP. An alternative way to characterise amenability is via an approximation property for the Fourier algebra $A(\Gamma)$. Leptin proved in [17] that a locally compact group G is amenable if and only if the Fourier algebra $A(G)$ has an approximate identity which is bounded in the norm $\| - \|_{A(G)}$.

In the case of free groups, Haagerup showed that the Fourier algebra $A(\mathbb{F}_n)$ has an approximate unit that is unbounded in the norm of the Fourier algebra, but is bounded in the so-called multiplier norm:

Definition 3.4. A complex-valued function u on Γ is a multiplier for $A(\Gamma)$ if the linear map $m_u(v) = uv$ maps $A(\Gamma)$ into $A(\Gamma)$. The set of multipliers of $A(\Gamma)$ is denoted $MA(\Gamma)$. If $u \in MA(\Gamma)$ then u is a bounded continuous function and m_u is a bounded operator on the space $A(\Gamma)$.

We say that u is a completely bounded multiplier if and only if the operator m_u is completely bounded. The set $M_0(A(\Gamma))$ of completely bounded multipliers is equipped with the norm

$$\|u\|_{M_0A(\Gamma)} = \|m_u\|_{cb},$$

which we shall call the multiplier norm.

By analogy with Leptin's result we have the following definition of weak amenability.

Definition 3.5. We say that a group Γ is *weakly amenable* iff $A(\Gamma)$ has an approximate identity that is bounded in the multiplier norm.

Hence a group is weakly amenable if there is a net $\{u_\alpha\}$ in $A(\Gamma)$ and a constant C such that $\|u_\alpha v - v\| \to 0$ for all $v \in A(\Gamma)$ and such that $\|u_\alpha\|_{M_0 A(\Gamma)} \leq C$ for all α.

We have thus defined a weak form of amenability via the Fourier algebra. Closing the circle it turns out that this property can also be formulated in terms of the completely bounded approximation property for the reduced C^*-algebra.

Definition 3.6. A C^*-algebra A is said to have the *completely bounded approximation property* (CBAP) if there is a positive number C such that the identity map on A can be approximated in the point-norm topology by a net $\{T_\alpha\}$ of finite rank completely bounded maps whose completely bounded norms are bounded by C.

We have the following important result of Haagerup (see [9, p. 669]).

Theorem 3.7. *Let Γ be a discrete group. Then the following are equivalent:*

1. *Γ is weakly amenable.*
2. *$C_r^*(\Gamma)$ has the CBAP.*

We have seen that a discrete group Γ is amenable if and only if $C_r^*(\Gamma)$ is nuclear. It is natural to ask if there is a property of groups that corresponds to the CBAP, and the answer is provided by the notion of exactness, introduced by Kirchberg and Wassermann in [14].

Definition 3.8. We say that a discrete group Γ is *exact* iff $C_r^*(\Gamma)$ is exact as a C^*-algebra: this means that the operation of taking the minimal tensor product with this algebra is an exact functor in the category of C^*-algebras.

Exact groups are known to admit uniform embeddings in a Hilbert space and therefore to satisfy the Novikov conjecture by an important result of Yu [25]. Here we have a concrete application of non-commutative geometry to a classical problem in topology. It is important in our context because of the following theorem, due to Kirchberg and Wassermann.

Theorem 3.9. *If a C^*-algebra A satisfies the CBAP then A is exact.*

We provide a proof that was kindly communicated to us by Ozawa.

Proof. Let A be a C^*-algebra with the CBAP; this means that there exists a uniformly bounded family of completely bounded finite rank operators $T_n : A \to A$ such that for any $a \in A$, $\|T_n(a) - a\| \to 0$.

We need to show that for any exact sequence

$$0 \to I \xrightarrow{i} B \xrightarrow{q} Q \to 0$$

of C^*-algebras, the sequence

$$0 \to A \otimes I \to A \otimes B \xrightarrow{\mathrm{id}_A \otimes q} A \otimes Q \to 0$$

is also exact, where $\otimes$ stands for the minimal tensor product.

We note first that for any $x \in A \otimes B$ we have $\|(T_n \otimes \mathrm{id}_B)(x) - x\| \to 0$. (Since the maps $T_n \otimes \mathrm{id}_B$ are uniformly bounded in n it is enough to check this assertion on simple tensors $x = a \otimes b$.). Assume now that x is an element of the kernel $\ker(\mathrm{id}_A \otimes q)$ of the quotient map $\mathrm{id}_A \otimes q : A \otimes B \to A \otimes Q$. Then clearly

$$(T_n \otimes q)(x) = (T_n \otimes \mathrm{id}_Q)(\mathrm{id}_A \otimes q)(x) = 0.$$

Given that

$$(T_n \otimes \mathrm{id}_Q)(\mathrm{id}_A \otimes q)(x) = (\mathrm{id}_A \otimes q)(T_n \otimes \mathrm{id}_B)(x)$$

we have that $(\mathrm{id}_A \otimes q)(T_n \otimes \mathrm{id}_B)(x) = 0$. Since every operator T_n is of finite rank, $(T_n \otimes \mathrm{id}_B)(x)$ belongs to the algebraic tensor product of $A \odot B$ of the algebras A and B. The algebraic tensor product is an exact functor, so the vanishing condition

$$(\mathrm{id}_A \otimes q)(T_n \otimes \mathrm{id}_B)(x) = 0$$

implies that $(T_n \otimes \mathrm{id}_B)(x)$ is an element of $A \otimes I$. Therefore

$$x = \lim(T_n \otimes \mathrm{id}_B)(x)$$

is also in $A \otimes I$.

This proves that the kernel of the map $\mathrm{id}_A \otimes q$ is identical to $A \otimes I$ for any algebras B and I, which implies that A is exact. □

Thus if the reduced C^*-algebra of a discrete group has the CBAP then the group is exact and so it satisfies the Novikov conjecture. In particular, since the CPAP implies the CBAP, Lance's theorem implies that amenable groups satisfy the Novikov conjecture. On the other hand there exist exact groups that are not amenable, for example the free groups [14] and the word hyperbolic groups [25].

4. The Metric Approximation Property

In this section we will prove the following theorem.

Theorem 4.1. *Let Γ be a discrete group satisfying the rapid decay property with respect to a length function ℓ which is conditionally negative. Then the reduced C^*-algebra $C^*_r(\Gamma)$ has the metric approximation property.*

The central point of our proof is an observation that the proof of the same property for free groups due to Haagerup [8] transfers directly to this more general situation. We also note that under the same hypotheses, the Fourier algebra $A(\Gamma)$ has a bounded approximate identity, which implies that it too has the MAP.

Following Haagerup [8, Def. 1.6] we say that a function $\phi : \Gamma \longrightarrow \mathbb{C}$ is a multiplier of $C^*_r(\Gamma)$ if and only if there exists a unique bounded operator $M_\phi : C^*_r(\Gamma) \to C^*_r(\Gamma)$ such that

$$M_\phi \lambda(\gamma) = \phi(\gamma)\lambda(\gamma) \tag{1}$$

for all $\gamma \in \Gamma$. This condition can be written equivalently as:

$$M_\phi \lambda(f) = \lambda(\phi \cdot f). \tag{2}$$

An important situation in which such operators arise is given by the following lemma, which is a generalisation of [8, Lemma 1.7]; the proof is essentially identical to the original.

Lemma 4.2. *Let Γ be a discrete group equipped with a length function ℓ. Assume that (Γ, ℓ) satisfies the rapid decay inequality for given $C, s > 0$.*

Let ϕ be any function on Γ such that

$$K = \sup_{\gamma \in \Gamma} |\phi(\gamma)|(1 + \ell(\gamma))^s < \infty.$$

Then ϕ is a multiplier of $C_r^*(\Gamma)$ and $\|M_\phi\| \leq CK$.

In particular this holds for any element $f \in \mathbb{C}\Gamma$ and for any such element M_f has finite rank.

Proof. We start by showing that property RD allows us to construct a family of multipliers for $C_r^*(\Gamma)$ with controlled operator norms. For any discrete group Γ the characteristic function δ_e of the identity element e of Γ is the identity of the group ring $\mathbb{C}\Gamma$. Since δ_e is a unit vector in $\ell^2(\Gamma)$ we have that for any $f \in \mathbb{C}\Gamma$, $\|\lambda(f)\| \geq \|\lambda(f)(\delta_e)\|_2 = \|f * \delta_e\|_2 = \|f\|_2$.

Then for any $f \in \mathbb{C}\Gamma$, the pointwise product $\phi \cdot f$ is also an element of $\mathbb{C}\Gamma$, so we can apply the rapid decay inequality to get:

$$\begin{aligned}
\|\lambda(\phi \cdot f)\| &\leq C \sqrt{\sum_{\gamma \in \Gamma} |\phi(\gamma) f(\gamma)|^2 (1 + \ell(\gamma))^{2s}} \\
&\leq C \sup_{\gamma \in \Gamma} \{|\phi(\gamma)|(1 + \ell(\gamma))^s\} \sqrt{\sum_{\gamma \in \Gamma} |f(\gamma)|^2} = CK\|f\|_2.
\end{aligned}$$

Putting together the two inequalities we have that

$$\|\lambda(\phi \cdot f)\| \leq CK\|f\|_2 \leq CK\|\lambda(f)\|.$$

This shows that the map from $\mathbb{C}\Gamma$ to $C_r^*(\Gamma)$ which sends $\lambda(f)$ to $\lambda(\phi \cdot f)$ is continuous and so extends to a unique map $M_\phi : C_r^*(\Gamma) \to C_r^*(\Gamma)$ with the property that $M_\phi \lambda(f) = \lambda(\phi \cdot f)$.

It is also clear that $\|M_\phi\| \leq CK$. Finally it is clear that if ϕ has finite support then M_ϕ has finite rank. □

We recall the definition of a conditionally negative kernel:

Definition 4.3. A *conditionally negative kernel* on a set V is a function $f : V \times V \longrightarrow \mathbb{R}$ such that for any finite subset $\{v_1, \ldots, v_n\} \subset V$ and any real numbers $\{\lambda_1, \ldots, \lambda_n\}$ such that $\sum_i \lambda_i = 0$ the following inequality holds:

$$\sum_{i,j} \lambda_i \lambda_j f(v_i, v_j) \leq 0$$

A *conditionally negative kernel on a group G* is a conditionally negative kernel on the set of elements of G such that for any g, h, k in G, $f(gh, gk) = f(h, k)$.

We can now prove the following.

Theorem 4.4. *Let Γ be a discrete group with a conditionally negative length function ℓ, which satisfies the property (RD) for $C, s > 0$. Then there exists a net $\{\phi_\alpha\}$ of functions on Γ with finite support such that*

1. *For each α, $\|M_{\phi_\alpha}\| \leq 1$;*
2. *$\|M_{\phi_\alpha}(x) - x\| \to 0$ for all $x \in C_r^*(\Gamma)$.*

Proof. Since the length function ℓ is conditionally negative, it follows from Schoenberg's lemma that for any $r > 0$ the function $\phi_r(\gamma) = e^{-r\ell(\gamma)}$ is of positive type. Thus by [8, Lemma 1.1] (see also [11, Lemma 3.2 and 3.5]), for every r there exists a unique completely positive operator $M_{\phi_r} : C_r^*(\Gamma) \to C_r^*(\Gamma)$ such that $M_{\phi_r}(\lambda(\gamma)) = \phi_r(\gamma)\lambda(\gamma)$ for all $\gamma \in \Gamma$ and $\|M_{\phi_r}\| = \phi_r(e) = 1$.

Let us now define a family $\phi_{r,n}$ of finitely supported functions on Γ by truncating the functions ϕ_r to balls of radius n with respect to the length function ℓ. For every $\gamma \in \Gamma$ we put:

$$\phi_{r,n}(\gamma) = \begin{cases} e^{-r\ell(\gamma)}, & \text{if } \ell(\gamma) \leq n \\ 0, & \text{otherwise.} \end{cases}$$

Since $e^{-x}(1+x)^s \to 0$ for any positive s and $x \to \infty$, we have that

$$\sup_{\gamma \in \Gamma} |\phi_r(\gamma)|(1 + \ell(\gamma))^s$$

is finite. If we denote this finite number by K, then clearly $\sup_{\gamma \in \Gamma} |\phi_{r,n}(\gamma)|(1 + \ell(\gamma))^s \leq K$. Thus, for every r and n, these functions are multipliers of $C_r^*(\Gamma)$, and the corresponding operators M_{ϕ_r} and $M_{\phi_{r,n}}$ have norms bounded by CK. Since the functions $\phi_{r,n}$ have finite support, the corresponding operators $M_{\phi_{r,n}}$ are of finite rank.

On the other hand, since

$$(\phi_r - \phi_{r,n})(\gamma) = \begin{cases} 0, & \ell(\gamma) \leq n \\ e^{-r\ell(\gamma)}, & \ell(\gamma) > n \end{cases}$$

we have that

$$\begin{aligned} &\sup_{\gamma \in \Gamma} |(\phi_r - \phi_{r,n})(\gamma)|(1 + \ell(\gamma))^s \\ &\qquad = \sup_{\ell(\gamma) > n} |(\phi_r - \phi_{r,n})(\gamma)|(1 + \ell(\gamma))^s \\ &\qquad \leq K_n < \infty \end{aligned}$$

where $K_n \to 0$ as $n \to \infty$. Thus these functions are multipliers of $C_r^*(\Gamma)$ and the corresponding operators $M_{\phi_r - \phi_{r,n}}$ are such that $\|M_{\phi_r - \phi_{r,n}}\| \leq CK_n \to 0$, as $n \to \infty$.

Since

$$\|M_{\phi_r} - M_{\phi_{r,n}}\| = \|M_{\phi_r - \phi_{r,n}}\|$$

we have $\|M_{\phi_r} - M_{\phi_{r,n}}\| \to 0$ as $n \to \infty$. This implies that $\|M_{\phi_{r,n}}\| \to \|M_{\phi_r}\| = \phi_r(e) = 1$.

To get the correct bound on the norm of these operators we introduce scaled functions:

$$\rho_{r,n} = \frac{1}{\|M_{\phi_{r,n}}\|}\phi_{r,n}.$$

The algebraic identity satisfied by the multipliers, as stated in (2), guarantees that on $\lambda(\mathbb{C}\Gamma)$ we have the following identity

$$M_{\rho_{r,n}} = \frac{1}{\|M_{\phi_{r,n}}\|}M_{\phi_{r,n}}. \tag{3}$$

We now want to show that each operator $M_{\rho_{r,n}}$ is a finite rank contraction on $C_r^*(\Gamma)$ and that the strong operator closure of the family $\{M_{\rho_{r,n}}\}$ contains the identity map id : $C_r^*(\Gamma) \to C_r^*(\Gamma)$. This means that for every positive ϵ there exists an operator $M_{\rho_{r,n}}$ such that

$$\|M_{\rho_{r,n}}x - x\| < \epsilon$$

for all $x \in C_r^*(\Gamma)$.

First, a simple use of the triangle inequality leads to the following argument.

$$\begin{aligned} \|M_{\rho_{r,n}} - M_{\phi_r}\| &\le \|M_{\rho_{r,n}} - M_{\phi_{r,n}}\| + \|M_{\phi_{r,n}} - M_{\phi_r}\| \\ &= \|(1 - 1/\|M_{\phi_{r,n}}\|)M_{\phi_{r,n}}\| + \|M_{\phi_{r,n}} - M_{\phi_r}\| \\ &\to 0 \qquad \text{as} \quad n \to \infty. \end{aligned} \tag{4}$$

Let $x \in C_r^*(\Gamma)$. Then x is a limit of a sequence of elements $x_m \in \lambda(\mathbb{C}\Gamma)$ so that $\|M_{\rho_{r,n}}(x)\| = \lim_{m\to\infty} \|M_{\rho_{r,n}}(x_m)\|$, and equation (3) implies that $\|M_{\rho_{r,n}}(x_m)\| = \|\frac{1}{\|M_{\phi_{r,n}}\|}M_{\phi_{r,n}}(x_m)\|$.

This leads to the following estimate:

$$\begin{aligned} \|M_{\rho_{r,n}}(x)\| &= \lim_{m\to\infty} \|\frac{1}{\|M_{\phi_{r,n}}\|}M_{\phi_{r,n}}(x_m)\| \\ &\le \lim_{m\to\infty} \|\frac{1}{\|M_{\phi_{r,n}}\|}M_{\phi_{r,n}}\|\|x_m\| = \lim_{m\to\infty} \|x_m\| = \|x\|. \end{aligned} \tag{5}$$

It follows that $\|M_{\rho_{r,n}}\| \le 1$.

Finally, it is clear that for any $\gamma \in \Gamma$, $e^{-r\ell(\gamma)} \to 1$ as $r \to 0$. Thus for any $x = \sum_{\gamma\in\Gamma} \mu_\gamma \lambda(\gamma) \in \mathbb{C}\Gamma$ we have

$$M_{\phi_r}(x) = \sum \mu_\gamma \phi_r(\gamma)\lambda(\gamma)$$

so that

$$\begin{aligned} \lim_{r\to 0} M_{\phi_r}(x) &= \lim_{r\to 0} \sum \mu_\gamma \phi_r(\gamma)\lambda(\gamma) \\ &= \sum \mu_\gamma (\lim_{r\to 0} \phi_r(\gamma))\lambda(\gamma) = \sum \mu_\gamma \lambda(\gamma) = x \end{aligned} \tag{6}$$

Since any $x \in C_r^*(\Gamma)$ can be approximated by a sequence $x_m \in \lambda(\mathbb{C}\Gamma)$ we have

$$\begin{aligned}\|M_{\phi_r}(x) - x\| \leq &\|M_{\phi_r}(x) - M_{\phi_r}(x_m)\|\\ &+ \|M_{\phi_r}(x_m) - x_m\| + \|x_m - x\|\end{aligned}$$

Given that $\|M_{\phi_r}\| \leq 1$ for all $r > 0$, $\|M_{\phi_r}(x) - M_{\phi_r}(x_m)\| \leq \|x - x_m\| < \epsilon/3$ for all large enough n and independently of r. Thus the sum of the first and third term of this sum can be made smaller than $(2/3)\epsilon$, for all $r > 0$, and independently of m. Now equation (6) shows that, as $r \to 0$, $M_{\phi_r}(x_m)$ tends to x_m so the middle term will be smaller than $\epsilon/3$ for all sufficiently small r. Thus, for all sufficiently small $r > 0$, $\|M_{\phi_r}(x) - x\| < \epsilon$ and so

$$\|M_{\phi_r}(x) - x\| \to 0$$

as $r \to 0$ for all $x \in C_r^*(\Gamma)$.

Let $\epsilon > 0$. Then it follows from (4) that for every $r > 0$ and all sufficiently large n, $\|M_{\rho_{r,n}} - M_{\phi_r}\| < \epsilon/2$. Secondly, as we have just shown, for all sufficiently small r, $\|M_{\phi_r}(x) - x\| < \epsilon/2$. Given that

$$\|M_{\rho_{r,n}}x - x\| \leq \|M_{\rho_{r,n}}x - M_{\phi_r}x\| + \|M_{\phi_r}(x) - x\|$$

for every $x \in C_r^*(\Gamma)$, the norm on the left-hand side can be made smaller than ϵ by taking a sufficiently large n and a sufficiently small $r > 0$.

This means that the strong closure of the family $\mathfrak{M} = \{M_{\rho_{r,n}}\}$ of finite rank contractions contains the identity map on the algebra $C_r^*(\Gamma)$. This implies that there exists a net of finitely supported functions ϕ_α with corresponding finite rank contractions $M_{\phi_\alpha} \in \mathfrak{M}$ such that $\|M_{\phi_\alpha}x - x\| \to 0$. This concludes the proof. □

As a corollary we obtain the main result of this note.

Theorem 4.5. *Let Γ be a discrete group satisfying the rapid decay property with respect to a length function ℓ which is conditionally negative. Then the reduced C^*-algebra $C_r^*(\Gamma)$ has the metric approximation property.*

The class of CAT(0) cube complexes plays in important role in geometry and geometric group theory. A CAT(0) cube complex is a cell complex in which each cell is isometric to a unit Euclidean cube, the glueing maps are isometries and such that the natural path metric obtained by integrating path length piecewise satisfies the CAT(0) inequality described in [1]. Intuitively this last condition ensures that the geodesic triangles in the path metric space are no fatter than they would be in Euclidean space. This condition ensures (among many other things) that the space is uniquely geodesic and contractible. Now according to Niblo and Reeves [19] given a group acting on a CAT(0) cube complex we obtain a conditionally negative kernel on the group which gives rise to a conditionally negative length function. By results of Chatterji and Ruane [3] the group will have the rapid decay property with respect to this this length function provided that the action is properly discontinuous, stabilisers are uniformly bounded and the cube complex has finite dimension.

Hence we obtain:

Corollary 4.6. *Groups acting properly discontinuously on a finite-dimensional CAT(0) cube complex with uniformly bounded stabilisers have the metric approximation property.*

This class of examples includes free groups, finitely generated Coxeter groups [20], and finitely generated right angled Artin groups for which the Salvetti complex is a CAT(0) cube complex. A rich class of interesting examples is furnished by Wise, [24], in which it is shown that many small cancellation groups act properly and co-compactly on CAT(0) cube complexes. The examples include every finitely presented group satisfying the B(4)-T(4) small cancellation condition and all those word-hyperbolic groups satisfying the B(6) condition.

Another class of examples where the main theorem applies is furnished by groups acting co-compactly and properly discontinuously on real or complex hyperbolic space. According to a result of Faraut and Harzallah [6] the natural metrics on these hyperbolic spaces are conditionally negative and they give rise to conditionally negative length functions on the groups. See [22] for a discussion and generalisation of this fact. The fact that these metrics satisfy rapid decay for the group was established by Jolissaint in [12].

Finally we remark that the net ϕ_α of Theorem 4.4 provides an approximate identity for the Fourier algebra $A(\Gamma)$ of the group Γ which is bounded in the multiplier norm. This implies, as in [8, Corollary 2.2], that if a group Γ satisfies the (RD) property with respect to a conditionally negative length function then its Fourier algebra $A(\Gamma)$ has the metric approximation property.

References

[1] M.R. Bridson, A. Haefliger, *Metric spaces of non-positive curvature.* Springer Verlag, Berlin 1999.

[2] I. Chatterji, *Property (RD) for cocompact lattices in a finite product of rank one Lie groups with some rank two Lie groups.* Geom. Dedicata **96** (2003), 161–177.

[3] I. Chatterji, K. Ruane, *Some geometric groups with rapid decay.* Geom. Funct. Anal. **15** (2005), no. 2, 311–339.

[4] J. de Canniere, U. Haagerup, *Multipliers of the Fourier algebra of some simple Lie groups and their discrete subgroups.* Amer. J. Math. **107** (1984), 455–500.

[5] P. Eymard, *L'algèbre de Fourier d'un groupe localement compact.* (French) Bull. Soc. Math. France **92** (1964), 181–236.

[6] J. Faraut and K. Harzallah, *Distances hilbertiennes invariantes sur un espace homogène.* Ann. Inst. Fourier (Grenoble) **24** (1974), no. 3, xiv, 171–217.

[7] A. Grothendieck, *Produits tensoriels topologiques et espaces nucléaires.* Mem. Amer. Math. Soc. (1955), no. **16**, 140 pp.

[8] U. Haagerup, *An example of a non-nuclear C^*-algebra which has the metric approximation property.* Inventiones Math. **50** (1979), 279–293.

[9] U. Haagerup, J. Kraus, *Approximation properties for group C^*-algebras and group von Neumann algebras.* Trans. Amer. Math. Soc. **344** (1994), no. 2, 667–699.

[10] P. de la Harpe, *Groupes hyperboliques, algèbres d'opérateurs et un théorème de Jolissaint.* C. R. Acad. Sci. Paris Ser. I **307** (1988), 771–774

[11] P. de la Harpe, A. Valette, *La propriété (T) de Kazhdan pour les groupes localement compacts.* Astérisque **175** (1989), Soc. Mathématique de France.

[12] P. Jolissaint, *Rapidly decreasing functions in reduced C^*-algebras of groups.* Trans. Amer. Math. Soc. **317** (1990), 167–196.

[13] P. Jolissaint, A. Valette, *Normes de Sobolev et convoluteurs bornés sur $L^2(G)$.* Ann. Inst. Fourier (Grenoble) **41** (1991), no. 4, 797–822.

[14] E. Kirchberg, S. Wassermann, *Exact groups and continuous bundles of C^*-algebras.* Math. Ann. **315** (1999), no. 2, 169–203.

[15] V. Lafforgue, *A proof of property (RD) for cocompact lattices of* $\mathrm{SL}(3, \mathbf{R})$ *and* $\mathrm{SL}(3, \mathbf{C})$. J. Lie Theory **10** (2000), no. 2, 255–267

[16] E.C. Lance, *On nuclear C^*-algebras.* J. Functional Analysis **12** (1973), 157–176

[17] H. Leptin, *Sur l'algèbre de Fourier d'un groupe localement compact.* (French) C.R. Acad. Sci. Paris Sér. A-**B 266** (1968) A1180–A1182.

[18] V. Mathai, *Heat kernels and the range of the trace on completions of twisted group algebras.* Contemporary Mathematics **398** (2006), 321–346.

[19] G.A .Niblo, L.D. Reeves, *The geometry of cube complexes and the complexity of their fundamental groups.* Topology, Vol. **37**, No 3, (1988) 621–633.

[20] G.A. Niblo and L.D. Reeves, *Coxeter groups act on $CAT(0)$ cube complexes.* Journal of Group Theory, **6**, (2003), 309–413.

[21] V. Paulsen, *Completely bounded maps and dilations.* Pitman Research Notes in Mathematics Series, **146** Longman, New York, 1986.

[22] G. Robertson, *Crofton formulae and geodesic distance in hyperbolic spaces.*J. Lie Theory **8** (1998), no. 1, 163–172.

[23] S. Wassermann, *Exact C^*-algebras and related topics.* Lecture Notes Series, **19**. Seoul National University, Research Institute of Mathematics, Global Analysis Research Center, Seoul, 1994.

[24] Daniel T. Wise, *Cubulating Small Cancellation Groups.* Geometric and Functional Analysis **14**, no. 1, 150–214.

[25] G. Yu, *The coarse Baum-Connes conjecture for spaces which admit a uniform embedding into Hilbert space.* Invent. Math. **139** (2000), no. 1, 201–240.

Jacek Brodzki and Graham A. Niblo
School of Mathematical Sciences
University of Southampton
Southampton SO17 1BJ, UK
e-mail: `j.brodzki@soton.ac.uk`
e-mail: `g.a.niblo@soton.ac.uk`.

C^*-algebras and Elliptic Theory
Trends in Mathematics, 37–60

A Riemannian Invariant, Euler Structures and Some Topological Applications

Dan Burghelea and Stefan Haller

Abstract. First we discuss a numerical invariant associated with a Riemannian metric, a vector field with isolated zeros, and a closed one form which is defined by a geometrically regularized integral. This invariant, extends the Chern–Simons class from a pair of two Riemannian metrics to a pair of a Riemannian metric and a smooth triangulation. Next we discuss a generalization of Turaev's Euler structures to manifolds with non-vanishing Euler characteristics and introduce the Poincaré dual concept of co-Euler structures. The duality is provided by a geometrically regularized integral and involves the invariant mentioned above. Euler structures have been introduced because they permit to remove the ambiguities in the definition of the Reidemeister torsion. Similarly, co-Euler structures can be used to eliminate the metric dependence of the Ray–Singer torsion. The Bismut–Zhang theorem can then be reformulated as a statement comparing two genuine topological invariants.

Mathematics Subject Classification (2000). 57R20, 58J52.

Keywords. Euler structure; co-Euler structure; combinatorial torsion; analytic torsion; theorem of Bismut–Zhang; Chern–Simons theory; geometric regularization .

1. Introduction

This paper follows entirely the lecture the first author gave at Bedlewo's workshop in February 2004 and is a survey of some of the results in [5] and [3]. We discuss in details two concepts, the invariant R which is a number associated with a Riemannian metric g, a vector field with isolated zeros X, and a closed one form ω, and the Euler resp. co-Euler structures which are affine versions of $H_1(M;\mathbb{Z})$ resp. $H^{n-1}(M;\mathcal{O}_M)$. They play an important role in our recent work about relating

---

The second author is supported by the *Fonds zur Förderung der wissenschaftlichen Forschung* (Austrian Science Fund), project number `P14195-MAT`.

the topology of non-simply connected manifolds to the complex geometry/analysis of the variety of complex representations of their fundamental group.

Both concepts existed in literature prior to our work, cf. [1] and [14]. We have extended, generalized and Poincaré dualized them because of our needs, cf. [5], but we also believe that they have independent interest.

Euler and co-Euler structures represent the additional topological data necessary to remove the geometric ambiguity from the Reidemeister torsion, resp. from the Ray–Singer torsion when extended to arbitrary representations, and provide genuine topological invariants. The invariant R, among other things, relates Euler and co-Euler structures.

We use the opportunity of having these two concepts presented in details to clarify the difference between the related concepts of (*combinatorial*) *torsion, Milnor metric* and (*modified*) *Ray–Singer metric* and to reformulate with their help the results of Bismut–Zhang, see [1].

The Bismut–Zhang theorem as formulated is about flat real vector bundles. The appendix completes the discussion with the case of flat complex vector bundles.

The invariant $R(X, g, \omega)$

Let M be a closed manifold and $\omega \in \Omega^1(M)$ a closed one form with real or complex coefficients.

(i) A pair of two Riemannian metrics g_1, g_2 determines the Chern–Simons class $\mathrm{cs}(g_1, g_2) \in \Omega^{n-1}(M; \mathcal{O}_M)/d\Omega^{n-2}(M; \mathcal{O}_M)$ and then the numerical invariant

$$R(g_1, g_2, \omega) := \int_M \omega \wedge \mathrm{cs}(g_1, g_2).$$

(ii) A pair of two vector fields without zeros X_1, X_2 determines a homology class $\mathrm{cs}(X_1, X_2) \in H_1(M; \mathbb{Z})$, see Section 3 below, and then a numerical invariant

$$R(X_1, X_2, \omega) = \langle [\omega], \mathrm{cs}(X_1, X_2) \rangle. \tag{1}$$

(iii) A pair consisting of a vector field without zeros and a Riemannian metric g determines a degree $n-1$ form $X^*\Psi(g) \in \Omega^{n-1}(M; \mathcal{O}_M)$ and therefore a numerical invariant

$$R(X, g, \omega) = \int_M \omega \wedge X^*\Psi(g). \tag{2}$$

Here $\Psi(g) \in \Omega^{n-1}(TM \setminus 0_M; \mathcal{O}_M)$ is the Mathai–Quillen form introduced in [9, Section 7] and discussed in details in [1], cf. Section 2 below.

One can extend the invariant (iii) to the case of vector fields with isolated zeros, not necessarily non-degenerate. Both smooth triangulations and Euler structures provide examples of such vector fields, cf. Sections 4 and 5. If X has zeros then the integrand in (2) is defined only on $M \setminus \mathcal{X}$, $\mathcal{X}$ the set of zeros of X, and the integral might be divergent. Fortunately it can be regularized by a procedure we will refer to as *geometric regularization* as described in Section 3 and this leads to the numerical invariant $R(X, g, \omega)$ from the title, cf. Theorem 1 below. This invariant for $X = -\operatorname{grad} f$, f a Morse function was considered in [1] in terms of currents. One can also extend the invariant (ii) to vector fields with isolated zeros, cf. Section 3.

A pleasant application of the invariant R and of the extension of (ii) is the extension of the Chern–Simons class from a pair of two Riemannian metrics g_1 and g_2 to a pair of two smooth triangulations τ_1 and τ_2 or to a pair of a Riemannian metric g and a smooth triangulation τ, cf. Section 5. These classes permit to treat on "equal foot" a Riemannian metric and a smooth triangulation when comparing subtle invariants like "torsion" defined using a Riemannian metric, and using a triangulation, in analogy with the comparison of such invariants for two metrics or two triangulations.

Euler structures

Euler structures were introduced by Turaev cf. [14] for manifolds M with vanishing Euler–Poincaré characteristic, $\chi(M) = 0$. We define the Euler structures for an arbitrary base pointed manifold (M, x_0) and show that the definition is independent of the base point provided $\chi(M) = 0$. The set of Euler structures $\mathfrak{Eul}_{x_0}(M;\mathbb{Z})$ is an affine version of $H_1(M;\mathbb{Z})$ in the sense that $H_1(M;\mathbb{Z})$ acts freely and transitively on $\mathfrak{Eul}_{x_0}(M;\mathbb{Z})$. Similarly there is the set of Euler structures with real coefficients $\mathfrak{Eul}_{x_0}(M;\mathbb{R})$ which is an affine space over $H_1(M;\mathbb{R})$, and there is a homomorphism $\mathfrak{Eul}_{x_0}(M;\mathbb{Z}) \to \mathfrak{Eul}_{x_0}(M;\mathbb{R})$ which is affine over the homomorphism $H_1(M;\mathbb{Z}) \to H_1(M;\mathbb{R})$.

We also introduce the set $\mathfrak{Eul}^*_{x_0}(M;\mathbb{R})$ of co-Euler structures on which the cohomology group $H^{n-1}(M;\mathcal{O}_M)$ acts freely and transitively.

$\mathfrak{Eul}^*_{x_0}(M;\mathbb{R})$ represents a smooth version (deRham version) of a dual aspect of $\mathfrak{Eul}_{x_0}(M;\mathbb{R})$. In the case of a closed manifold M we show the existence of an affine version of Poincaré duality map $P : \mathfrak{Eul}^*_{x_0}(M;\mathbb{R}) \to \mathfrak{Eul}_{x_0}(M;\mathbb{R})$. This can equivalently be described with the help of a coupling

$$\mathbb{T} : \mathfrak{Eul}_{x_0}(M;\mathbb{R}) \times \mathfrak{Eul}^*_{x_0}(M;\mathbb{R}) \to H_1(M;\mathbb{R})$$

based on a regularization very similar to the one for R, see Section 3.[1]

Primarily, the interest of Euler and co-Euler structures comes from the following. Suppose F is a flat real or complex vector bundle, and let F_{x_0} denote the fiber over the base point x_0. A co-Euler structure $\mathfrak{e}^* \in \mathfrak{Eul}^*_{x_0}(M;\mathbb{R})$ removes the metric ambiguity of the Ray–Singer torsion and provides a Hermitian scalar product, *the analytic scalar product*, in the complex line:

$$\det H^*(M;F) \otimes (\det F_{x_0})^{-\chi(M)} \tag{3}$$

[1]The concept of Euler and co-Euler structures can be extended from the tangent bundle to arbitrary rank k bundles. This is particularly easy if the Euler class of the bundle vanishes. The set of Euler structures of a vector bundle will be an affine version of $H_{n-k+1}(M;\mathbb{Z})$ resp. $H_{n-k+1}(M;\mathbb{R})$ and the set of co-Euler structures will be an affine version of $H^{k-1}(M;\mathcal{O}_E)$. There again is an affine version of Poincaré duality, based on a regularized integral. This permits to consider Euler and co-Euler structures as a functorial concept.

An Euler structure with real coefficients $\mathfrak{e} \in \mathfrak{Eul}_{x_0}(M;\mathbb{R})$ removes the triangulation ambiguity[2] and provides a Hermitian scalar product, *the combinatorial scalar product*, in the line (3), see also [7].

As an application of Euler and co-Euler structures we present a reformulation of a result of Bismut–Zhang, proved in [1], referred to as the Bismut–Zhang theorem, see Theorem 3 in Section 6. Precisely, the analytic scalar product associated to $\mathfrak{e}^*$ is the same as the combinatorial scalar product associated to $\mathfrak{e}$ multiplied by $e^{\langle(\log|\cdot|)_*\Theta_F,\mathbb{T}(\mathfrak{e},\mathfrak{e}^*)\rangle}$. Here $\Theta_F \in H^1(M;\mathbb{C}^*)$ is the cohomology class corresponding to $\det\circ\rho : H_1(M;\mathbb{Z}) \to \mathbb{C}^*$, and $(\log|\cdot|)_*\Theta_F \in H^1(M;\mathbb{R})$ denotes its image under the homomorphism $(\log|\cdot|)_* : H^1(M;\mathbb{C}^*) \to H^1(M;\mathbb{R})$ which is induced from the homomorphism of coefficients $\log|\cdot| : \mathbb{C}^* \to \mathbb{R}$. This cohomology class is known as Kamber–Tandeur class.

The results

Suppose M is a closed manifold of dimension n. Given a Riemannian metric g denote by $E(g) \in \Omega^n(M;\mathcal{O}_M)$ the Euler form and by $\Psi(g) \in \Omega^{n-1}(TM \setminus M;\mathcal{O}_M)$ the Mathai–Quillen form associated to g. If X_1 and X_2 are two vector fields with isolated zeros we get an element

$$\operatorname{cs}(X_1, X_2) \in C_1(M;\mathbb{Z})/\partial(C_2(M;\mathbb{Z}))$$

whose boundary equals the zeros of X_1 and X_2, weighted with their indices, see Section 2.

Theorem 1. *Let M be a closed connected manifold.*

(i) *Suppose $\omega \in \Omega^1(M)$ is a real- or complex-valued closed one form, g a Riemannian metric and X a vector field with isolated zeros. Let f be a smooth real or complex-valued function with $\omega = df$ in the neighborhood of the zero set $\mathcal{X}$ of X. Then the number*

$$R(X, g, \omega; f) := \int_{M\setminus\mathcal{X}} (\omega - df) \wedge X^*\Psi(g) - \int_M fE(g) + \sum_{x\in\mathcal{X}} \mathrm{IND}(x)f(x)$$

is independent of f and will therefore be denoted by $R(X, g, \omega)$.

(ii) *If g_1 and g_2 are two Riemannian metrics, then*

$$R(X, g_2, \omega) - R(X, g_1, \omega) = \int_M \omega \wedge \operatorname{cs}(g_1, g_2).$$

(iii) *If X_1 and X_2 are two vector fields with isolated zeros then*

$$R(X_2, g, \omega) - R(X_1, g, \omega) = \int_{\operatorname{cs}(X_1,X_2)} \omega.$$

(iv) *If ω_1 and ω_2 are two closed one forms so that $\omega_2 - \omega_1 = dh$ then*

$$R(X, g, \omega_2) - R(X, g, \omega_1) = -\int hE(g) + \sum_{x\in\mathcal{X}} \mathrm{IND}(x)h(x).$$

[2] and the additional ambiguity produced by the choice of a lift of each cell of the triangulation to the universal cover of the manifold

In Section 3 we will verify statements (i) through (iv). More precisely they are the contents of Lemma 1, Proposition 1 and Proposition 2.

An Euler structure on a base pointed manifold (M, x_0) is an equivalence class of pairs (X, c), where X is a vector field with isolated singularities and c is a singular one chain with integral coefficients whose boundary equals $\sum_{x\in\mathcal{X}} \mathrm{IND}(x)x - \chi(M)x_0$, where $\mathcal{X}$ denotes the zero set of X. Two such pairs (X_1, c_1) and (X_2, c_2) are equivalent if c_2 differs from $c_1 + \mathrm{cs}(X_1, X_2)$ by a boundary. We will write $\mathfrak{Eul}_{x_0}(M;\mathbb{Z})$ for the set of Euler structures based at x_0. This is an affine version of $H_1(M;\mathbb{Z})$ in the sense that $H_1(M;\mathbb{Z})$ acts freely and transitively on it. Considering chains c with real coefficients we get an affine version of $H_1(M;\mathbb{R})$ which we denote by $\mathfrak{Eul}_{x_0}(M;\mathbb{R})$.

The set $\mathfrak{Eul}^*_{x_0}(M;\mathbb{R})$ of co-Euler structures is defined as the set of equivalence classes of pairs (g, α) where $\alpha \in \Omega^{n-1}(M \setminus x_0; \mathcal{O}_M)$ satisfies $d\alpha = E(g)$. Two pairs (g_1, α_1) and (g_2, α_2) are equivalent iff $\alpha_2 - \alpha_1 = \mathrm{cs}(g_1, g_2)$. Here $\mathrm{cs}(g_1, g_2)$ denotes the Chern–Simons class, cf. Section 2 for definition. The cohomology $H^{n-1}(M;\mathcal{O}_M)$ acts on $\mathfrak{Eul}^*_{x_0}(M;\mathbb{R})$ freely and transitively by $[g,\alpha]+[\beta]:=[g,\alpha-\beta]$.

Theorem 2. *Let (M, x_0) be a closed connected base pointed manifold.*

(i) *Let $\pi_0(\mathfrak{X}(M, x_0))$ denote the set of connected components of vector fields which vanish only at x_0 equipped with the C^∞ topology, or any C^r topology, $r \geq 0$. If $\dim M > 2$ then the assignment $[X] \mapsto [X, 0]$ defines a bijection:*

$$\pi_0(\mathfrak{X}(M, x_0)) \to \mathfrak{Eul}_{x_0}(M;\mathbb{Z}).$$

(ii) *Let $\pi_0(\mathfrak{X}_0(M))$ denote the set of connected components of nowhere vanishing vector fields equipped with the C^∞ topology, or any C^r topology, $r \geq 0$. If $\chi(M) = 0$ and $\dim M > 2$ then the assignment $[X] \mapsto [X, 0]$ defines a surjection:*

$$\pi_0(\mathfrak{X}_0(M)) \to \mathfrak{Eul}_{x_0}(M;\mathbb{Z}).$$

(iii) *There exists an isomorphism*

$$P : \mathfrak{Eul}^*_{x_0}(M;\mathbb{R}) \to \mathfrak{Eul}_{x_0}(M;\mathbb{R}),$$

which is affine over the Poincaré duality $\mathrm{PD} : H^{n-1}(M;\mathcal{O}_M) \to H_1(M;\mathbb{R})$. That is $P(\mathfrak{e}^* + \beta) = P(\mathfrak{e}^*) + \mathrm{PD}(\beta)$, for all $\beta \in H^{n-1}(M;\mathcal{O}_M)$.

(iv) *The assignment $\mathbb{T}(\mathfrak{e}, \mathfrak{e}^*) := P(\mathfrak{e}^*) - \mathfrak{e}$*

$$\mathbb{T} : \mathfrak{Eul}_{x_0}(M;\mathbb{R}) \times \mathfrak{Eul}^*_{x_0}(M;\mathbb{R}) \to H_1(M;\mathbb{R})$$

is a corrected version of the invariant R. More precisely, if $\mathfrak{e} = [X, c]$, $\mathfrak{e}^* = [g, \alpha]$ and $[\omega] \in H^1(M;\mathbb{R})$ we have

$$\langle [\omega], \mathbb{T}(\mathfrak{e}, \mathfrak{e}^*)\rangle = \int_M \omega \wedge (X^*\Psi(g) - \alpha) - \int_c \omega$$

where $\omega \in \Omega^1(M)$ is any representative of $[\omega]$ which vanishes locally around x_0 and locally around the zeros of X.

Statements (i) and (ii) are essentially due to Turaev and are the contents of Propositions 3 and 4 in Section 4. The proof of (iii) and (iv) can be found at the end of Section 4, cf. Proposition 5.

The theorem of Bismut–Zhang in our reformulation is contained in Section 6 as Theorem 3.

2. Mathai–Quillen form

Let $\pi : E \to M$ be a rank k real vector bundle, and let $\tilde{\nabla} := (\nabla, \mu)$ be a pair consisting of a connection ∇ and a parallel Hermitian structure, i.e. , fiber wise scalar product, μ. Such pair will be called Euclidean connection. Let $\mathcal{O}_E$ denote the orientation bundle of E, a flat real line bundle over M.

For $\tilde{\nabla}$ an Euclidean connection denote by $E(\tilde{\nabla}) \in \Omega^k(M; \mathcal{O}_E)$ the Euler form of $\tilde{\nabla}$, i.e. the Pfaffian of the curvature of ∇, cf. [8]. Observe that a smooth path of Euclidean connections $\tilde{\nabla}_t$, $t \in I = [1, 2]$, can be interpreted as an Euclidean connection $\overline{\tilde{\nabla}}$ in the bundle $\overline{E} = E \times I \to \overline{M} = M \times I$. Indeed, a smooth vector field X on $\overline{M}$ can be regarded as a pair $X = (X'_t, f_t)$, $t \in I$, with X'_t a smooth family of vector fields on M, f_t a smooth family of functions on M and a section s in $\overline{E}$ as a smooth family of sections s_t in E. Define $\overline{\tilde{\nabla}}_X s := (\nabla_t)_{X_t} s_t + f_t d/dt s_t$.

Then for two Euclidean connections $\tilde{\nabla}_1$ and $\tilde{\nabla}_2$ and a smooth path of Euclidean connections $\tilde{\nabla}_t$, $t \in I$, consider the differential form $\mathrm{cs}(\tilde{\nabla}_1, \tilde{\nabla}_2, \tilde{\nabla}_t) \in \Omega^{k-1}(M; \mathcal{O}_E)$ obtained from $E(\overline{\tilde{\nabla}})$ by "integration along the fiber" in the trivial smooth bundle $M \times I \to M$. Change of the path ∇_t changes $\mathrm{cs}(\tilde{\nabla}_1, \tilde{\nabla}_2, \tilde{\nabla}_t)$ by an exact form so the class $\mathrm{cs}(\tilde{\nabla}_1, \tilde{\nabla}_2) \in \Omega^{k-1}(M; \mathcal{O}_E)/\Omega^{k-2}(M; \mathcal{O}_E)$ referred to as Chern–Simons class is independent on the path cf. [6].

In [9] (see also [1]) Mathai and Quillen have introduced the differential form

$$\Psi(\tilde{\nabla}) \in \Omega^{k-1}(E \setminus M; \pi^* \mathcal{O}_E)$$

called Mathai–Quillen form with the following properties.

(i) $\Psi(\tilde{\nabla})$ is the pullback of a form on $(E \setminus M)/\mathbb{R}_+$.

(ii) One has

$$d\Psi(\tilde{\nabla}) = \pi^* E(\tilde{\nabla}). \tag{4}$$

(iii) Modulo exact forms

$$\Psi(\tilde{\nabla}_2) - \Psi(\tilde{\nabla}_1) = \pi^* \mathrm{cs}(\tilde{\nabla}_1, \tilde{\nabla}_2). \tag{5}$$

(iv) Suppose $E = TM$ is equipped with a Riemannian metric g, $\tilde{\nabla}_g$ is the Levi–Civita pair and X is a vector field with isolated zero x. Let B_ϵ denote the ball of radius ϵ around x, with respect to some chart. Then

$$\lim_{\epsilon \to 0} \int_{\partial(M \setminus B_\epsilon)} X^* \Psi(\tilde{\nabla}_g) = \mathrm{IND}(x), \tag{6}$$

where $\mathrm{IND}(x)$ denotes the Hopf index of X at x.

(v) For $M = \mathbb{R}^n$, $E := TM$ equipped with $g_{ij} = \delta_{ij}$, $\tilde{\nabla}_g$ the Levi–Civita pair and in the coordinates $x_1, \ldots, x_n, \xi_1, \ldots, \xi_n$ one has:

$$\Psi(\tilde{\nabla}_g) = \frac{\Gamma(n/2)}{(2\pi)^{n/2}} \sum_{i=1}^{n} (-1)^i \frac{\xi_i}{(\sum \xi_i^2)^{n/2}} d\xi_1 \wedge \cdots \wedge \widehat{d\xi_i} \wedge \cdots \wedge d\xi_n.$$

We will consider the above definitions only for $E = TM$ and $\tilde{\nabla} = \tilde{\nabla}_g$, g a Riemannian metric, and use the notation $E(g)$ for $E(\tilde{\nabla}_g)$, $\mathrm{cs}(g_1, g_2)$ for $\mathrm{cs}(\tilde{\nabla}_{g_1}, \tilde{\nabla}_{g_2})$ and $\Psi(g)$ for $\Psi(\tilde{\nabla}_g)$.

Let $\xi : E \to M$ be a complex vector bundle equipped with a flat connection. Given two Hermitian structures μ_1 and μ_2 denote by $V(\mu_1, \mu_2)$ the positive real-valued function given at $y \in M$ by the volume with respect to the scalar product defined by $(\mu_2)_y$ of a parallelepiped provided by an orthonormal frame with respect to $(\mu_1)_y$.

Suppose that the bundle ξ is equipped with a flat connection ∇. To any Hermitian structure in $\xi : E \to M$ following Kamber–Tondeur one associates the real-valued closed (hence locally exact) differential form $\omega(\nabla, \mu) \in \Omega^1(M)$ defined as follows. For any $x \in M$ choose a contractible open neighborhood U, and denote by $\tilde{\mu}_x$ the Hermitian structure on $E|_U \to U$ obtained by parallel transport of μ_x. This Hermitian structure is well defined since U is one connected and the connection is flat.

Define $\omega(\nabla, \mu) := -\frac{1}{2} d \log V_x$ as being the logarithmic differential of the non-zero function $V_x : U \to \mathbb{R}$ defined by $V_x = V(\tilde{\mu}_x, \mu)$. The following property holds:

$$\omega(\nabla, \mu_2) - \omega(\nabla, \mu_1) = -\frac{1}{2} d \log(V(\mu_1, \mu_2))$$

3. The invariant $R(X, g, \omega)$. The geometric regularization

Suppose M is a closed manifold of dimension n, g a Riemannian metric and $X : M \to TM \setminus M$ a vector field without zeros. Suppose ω is a closed one form with real or complex coefficients. Define

$$R(X, g, \omega) := \int_M \omega \wedge X^* \Psi(g), \tag{7}$$

which will be a real or complex number. For every function h we have

$$R(X, g, \omega + dh) - R(X, g, \omega) = -\int_M hE(g),$$

and for any two Riemannian metrics g_1 and g_2 we have

$$R(X, g_2, \omega) - R(X, g_1, \omega) = \int_M \omega \wedge \mathrm{cs}(g_1, g_2).$$

These properties are straightforward consequences of (4), Stokes' theorem and (5).

Suppose X_1 and X_2 are two vector fields without zeros. Let $p : I \times M \to M$ denote the projection, where $I = [1, 2]$. Consider a section $\mathbb{X}$ of p^*TM which is transversal to the zero section and which restricts to X_i on $\{i\} \times M$, $i = 1, 2$. The zero set $\mathbb{X}^{-1}(0)$ is a closed one-dimensional canonically oriented submanifold of $I \times M$. Hence it defines a homology class in $I \times M$, which turns out to be independent of the chosen homotopy $\mathbb{X}$. We thus define $\mathrm{cs}(X_1, X_2) := p_*(\mathbb{X}^{-1}(0)) \in H_1(M; \mathbb{Z})$. One can show that

$$R(X_2, g, \omega) - R(X_1, g, \omega) = \int_{\mathrm{cs}(X_1, X_2)} \omega.$$

This property will be verified below in a slightly more general situation.

The above properties suggest the definition of the invariant R in the case X has isolated zeros even when the integral in (7) is divergent. This definition will be referred to as the *geometric regularization* of (7). We do not assume that the zeros of X are non-degenerate. Let $\mathcal{X}$ denote the zero set of X. Choose a function f so that $\omega' := \omega - df$ vanishes on a neighborhood of $\mathcal{X}$. Then

$$R(X, g, \omega; f) := \int_{M \setminus \mathcal{X}} \omega' \wedge X^* \Psi(g) - \int_M f E(g) + \sum_{x \in \mathcal{X}} \mathrm{IND}(x) f(x)$$

makes perfect sense. The next lemma establishes the proof of Theorem 1(i).

Lemma 1. *The quantity $R(X, g, \omega; f)$ does not depend on the choice of f.*

Proof. Suppose f_1 and f_2 are two functions such that $\omega_i' := \omega - df_i$, $i = 1, 2$ both vanish in a neighborhood U of $\mathcal{X}$, $i = 1, 2$. For every $x \in \mathcal{X}$ we choose a chart and let $B_\epsilon(x)$ denote the disk of radius ϵ around x. Put $B_\epsilon := \bigcup_{x \in \mathcal{X}} B_\epsilon(x)$.

For ϵ small enough $B_\epsilon \subset U$ and $f_2 - f_1$ is constant on each $B_\epsilon(x)$. Using (4), Stokes' theorem and (6) we get

$$\begin{aligned}
& R(X, g, \omega; f_2) - R(X, g, \omega; f_1) \\
&= -\int_{M \setminus \mathcal{X}} d((f_2 - f_1) \wedge X^*(\Psi(g)) + \sum_{x \in \mathcal{X}} \mathrm{IND}(x)(f_2 - f_1)(x) \\
&= -\lim_{\epsilon \to 0} \int_{\partial(M \setminus B_\epsilon)} (f_2 - f_1) \wedge X^* \Psi(g) + \sum_{x \in \mathcal{X}} \mathrm{IND}(x)(f_2 - f_1)(x) \\
&= -\sum_{x \in \mathcal{X}} (f_2 - f_1)(x) \lim_{\epsilon \to 0} \int_{\partial(M \setminus B_\epsilon(x))} X^* \Psi(g) + \sum_{x \in \mathcal{X}} \mathrm{IND}(x)(f_2 - f_1)(x) \\
&= 0
\end{aligned}$$

and thus $R(X, g, \omega; f_1) = R(X, g, \omega; f_2)$. □

Definition 1. In view of the previous lemma we define $R(X, g, \omega) := R(X, g, \omega; f)$, where f is any function so that $\omega - df$ vanishes locally around $\mathcal{X}$.

From the very definition we immediately verify Theorem 1(iv) which we restate as

Proposition 1. *For every function h we have:*

$$R(X,g,\omega+dh)-R(X,g,\omega)=-\int_M hE(g)+\sum_{x\in\mathcal{X}}\mathrm{IND}(x)h(x) \tag{8}$$

For any vector field with isolated zeros $\mathcal{X}$ we set

$$e_X:=\sum_{x\in\mathcal{X}}\mathrm{IND}(x)x,$$

a singular zero chain in M.

Suppose we have two vector fields X_1 and X_2 with non-degenerate zeros. Consider the vector bundle $p^*TM\to I\times M$, where $I:=[1,2]$ and $p:I\times M\to M$ denotes the natural projection. Choose a section $\mathbb{X}$ of p^*TM which is transversal to the zero section and which restricts to X_i on $\{i\}\times M$, $i=1,2$. The zero set of $\mathbb{X}$ is a canonically oriented one-dimensional submanifold with boundary. Hence it defines a singular one chain which, when pushed forward via p, is a one chain $c(\mathbb{X})$ in M, satisfying

$$\partial c(\mathbb{X})=e_{X_2}-e_{X_1}.$$

Suppose $\mathbb{X}_1$ and $\mathbb{X}_2$ are two non-degenerate homotopies from X_1 to X_2. Then certainly $\partial(c(\mathbb{X}_2)-c(\mathbb{X}_1))=0$, but we actually have

$$c(\mathbb{X}_2)-c(\mathbb{X}_1)=\partial\sigma, \tag{9}$$

for a two chain σ. Indeed, consider the vector bundle $q^*TM\to I\times I\times M$, where $q:I\times I\times M\to M$ denotes the natural projection. Choose a section of q^*TM which is transversal to the zero section, restricts to $\mathbb{X}_i$ on $\{i\}\times I\times M$, $i=1,2$ and which restricts to X_i on $\{s\}\times\{i\}\times M$ for all $s\in I$ and $i=1,2$. The zero set of such a section then gives rise to σ satisfying (9).

So for two vector fields with non-degenerate zeros this construction yields a one chain $\mathrm{cs}(X_1,X_2)$, well defined up to a boundary, satisfying $\partial\,\mathrm{cs}(X_1,X_2)=e_{X_2}-e_{X_1}$.

Let us extend this to vector fields with isolated singularities. Suppose X is a vector field with isolated singularities. For every zero $x\in\mathcal{X}$ we choose an embedded ball B_x centered at x, assuming all B_x are disjoint. Set $B:=\bigcup_{x\in\mathcal{X}}B_x$. Choose a vector field with non-degenerate zeros X' that coincides with X on $M\setminus B$. Let $\mathcal{X}'$ denote its zero set. For every $x\in\mathcal{X}$ we have

$$\mathrm{IND}_X(x)=\sum_{y\in\mathcal{X}'\cap B_x}\mathrm{IND}_{X'}(y).$$

So we can choose a one chain $\tilde{c}(X,X')$ supported in B which satisfies $\partial\tilde{c}(X,X')=e_{X'}-e_X$. Since $H_1(B;\mathbb{Z})$ vanishes the one chain $\tilde{c}(X,X')$ is well defined up to a boundary.

Given two vector fields X_1 and X_2 with isolated zeros we choose perturbed vector fields X_1' and X_2' as above and set

$$\mathrm{cs}(X_1,X_2):=\tilde{c}(X_1,X_1')+\mathrm{cs}(X_1',X_2')-\tilde{c}(X_2,X_2').$$

Then obviously $\partial\,\mathrm{cs}(X_1, X_2) = e_{X_2} - e_{X_1}$. Using $H_1(B;\mathbb{Z}) = 0$ again, one checks that different choices for X_1' and X_2' yield the same $\mathrm{cs}(X_1, X_2)$ up to a boundary.

Summarizing, for every pair of vector fields X_1 and X_2 with isolated zeros we have constructed a one chain

$$\mathrm{cs}(X_1, X_2) \in C_1(M;\mathbb{Z})/\partial(C_2(M;\mathbb{Z})),$$

which satisfies $\partial\,\mathrm{cs}(X_1, X_2) = e_{X_2} - e_{X_1}$.

Definition 2. For two Riemannian metrics g_1, g_2 and a closed one form ω set

$$R(g_1, g_2, \omega) := \int_M \omega \wedge \mathrm{cs}(g_1, g_2). \tag{10}$$

For two vector fields X_1, X_2 and a closed one form ω set

$$R(X_1, X_2, \omega) := \int_{\mathrm{cs}(X_1, X_2)} \omega. \tag{11}$$

Remark 1. Even though $\mathrm{cs}(g_1, g_2)$ is only defined up to an exact form this ambiguity does not affect the integral (10). Similarly, even though $\mathrm{cs}(X_1, X_2)$ is only defined up to a boundary this ambiguity does not affect the integral (11).

The next proposition is a reformulation of Theorem 1(ii) and Theorem 1(iii).

Proposition 2. *Let M be a closed manifold, ω a closed one form, g, g_1, g_2 Riemannian metrics and let X, X_1, X_2 be vector fields with isolated zeros. Then*

$$R(X, g_2, \omega) - R(X, g_1, \omega) = R(g_1, g_2, \omega) \tag{12}$$

and

$$R(X_2, g, \omega) - R(X_1, g, \omega) = R(X_1, X_2, \omega). \tag{13}$$

Proof. Let us prove (12). Choose f so that $\omega' := \omega - df$ vanishes on a neighborhood of $\mathcal{X}$, the zero set of X. Using $X^*(\Psi(g_2) - \Psi(g_1)) = \mathrm{cs}(g_1, g_2)$ modulo exact forms, Stokes' theorem and $d\,\mathrm{cs}(g_1, g_2) = E(g_2) - E(g_1)$ we conclude

$$\begin{aligned}
R(X, g_2, \omega) &- R(X, g_1, \omega) = \\
&= \int_{M\setminus\mathcal{X}} \omega' \wedge X^*\big(\Psi(g_2) - \Psi(g_1)\big) - \int_M f\big(E(g_2) - E(g_1)\big) \\
&= \int_M \omega \wedge \mathrm{cs}(g_1, g_2) - \int_M df \wedge \mathrm{cs}(g_1, g_2) - \int_M f\big(E(g_2) - E(g_1)\big) \\
&= \int_M \omega \wedge \mathrm{cs}(g_1, g_2) \\
&= R(g_1, g_2, \omega).
\end{aligned}$$

Now let us turn to (13). Let $\mathcal{X}_i$ denote the zero set of X_i, $i = 1, 2$. Assume first that the vector fields X_1 and X_2 are non-degenerate and that there exists a non-degenerate homotopy $\mathbb{X}$ from X_1 to X_2 whose zero set is contained in a simply connected $I \times V \subseteq I \times M$. Choose a function f such that $\omega' := \omega - df$ vanishes on V. Then

$$R(X_1, X_2, \omega) = \int_{\mathbb{X}^{-1}(0)} p^* df = \sum_{x \in \mathcal{X}_2} \mathrm{IND}_{X_2}(x) f(x) - \sum_{x \in \mathcal{X}_1} \mathrm{IND}_{X_1}(x) f(x),$$

where $p : I \times M \to M$ denotes the natural projection. Let $\tilde{p} : p^*TM \to TM$ be the natural vector bundle homomorphism over p. Using the last equation, Stokes' theorem and $d(\mathbb{X}^* \tilde{p}^* \Psi(g)) = p^* E(g)$ we get:

$$\begin{aligned} R(X_2, g, \omega) - R(X_1, g, \omega) &= \\ &= \int_{I \times (M \setminus V)} d\big(p^* \omega' \wedge \mathbb{X}^* \tilde{p}^* \Psi(g)\big) + R(X_1, X_2, \omega) \\ &= -\int_{I \times M} p^* (\omega' \wedge E(g)) + R(X_1, X_2, \omega) \\ &= R(X_1, X_2, \omega) \end{aligned}$$

For the last equality note that $\omega' \wedge E(g) = 0$ for dimensional reasons.

Still assuming that X_1 and X_2 have non-degenerate zeros we next treat the case of a general non-degenerate homotopy $\mathbb{X}$, whose zero set is not necessarily contained in a simply connected subset. Perturbing the homotopy slightly we may assume that no component of its zero set lies in a single $\{s\} \times M$. Then we certainly find $0 = t_0, \dots, t_k = 1$ so that Y_{t_i}, the restriction of $\mathbb{X}$ to $\{t_i\} \times M$, is transversal to the zero section, and so that $\mathbb{X}^{-1}(0) \cap ([t_{i-1}, t_i] \times M)$ is contained in a simply connected subset for every $1 \leq i \leq k$. The previous paragraph tells us

$$R(Y_{t_i}, g, \omega) - R(Y_{t_{i-1}}, g, \omega) = R(Y_{t_{i-1}}, Y_{t_i}, \omega)$$

for every $1 \leq i \leq k$. Therefore:

$$R(X_2, g, \omega) - R(X_1, g, \omega) = \sum_{i=1}^{k} R(Y_{t_{i-1}}, Y_{t_i}, \omega) = R(X_1, X_2, \omega)$$

It remains to deal with vector fields having degenerate but isolated singularities. Let X be such a vector field and let X' denote a perturbation as used before. Let $\mathcal{X}$ and $\mathcal{X}'$ denote their zero sets, respectively. Choose a function f such that $\omega' := \omega - df$ vanishes on the set B. Recall that B was the union of small balls covering $\mathcal{X}$. Since X and X' agree on $M \setminus B$ we have

$$\begin{aligned} R(X', g, \omega) - R(X, g, \omega) &= \sum_{x \in \mathcal{X}'} \mathrm{IND}_{X'}(x) f(x) - \sum_{x \in \mathcal{X}} \mathrm{IND}_X(x) f(x) \\ &= \int_{\tilde{\mathrm{cs}}(X, X')} df \\ &= R(X, X', \omega). \end{aligned}$$

This completes the proof of (13). □

Remark 2. A similar definition of $R(X, g, \omega)$ works for any vector field X with arbitrary singularity set $\mathcal{X} := \{x \in M \mid X(x) = 0\}$ provided ω is exact when restricted to a sufficiently small neighborhood of $\mathcal{X}$.

4. Euler and co-Euler structures

Let (M, x_0) be a base pointed closed connected manifold of dimension n. Let X be a vector field and let $\mathcal{X}$ denote its zero set. Suppose the zeros of X are isolated and define

$$e_X := \sum_{x \in \mathcal{X}} \mathrm{IND}(x)x \in C_0(M; \mathbb{Z}),$$

a singular zero chain. An Euler chain for X is a singular one chain $c \in C_1(M; \mathbb{Z})$ so that

$$\partial c = e_X - \chi(M)x_0.$$

Since $\sum_{x \in \mathcal{X}} \mathrm{IND}_X(x) = \chi(M)$ every vector field with isolated zeros admits Euler chains.

Consider pairs (X, c) where X is a vector field with isolated zeros and c is an Euler chain for X. We call two such pairs (X_1, c_1) and (X_2, c_2) equivalent if

$$c_2 = c_1 + \mathrm{cs}(X_1, X_2) \in C_1(M; \mathbb{Z})/\partial(C_2(M; \mathbb{Z})).$$

For the definition of $\mathrm{cs}(X_1, X_2)$ see Section 3. We will write $\mathfrak{Eul}_{x_0}(M; \mathbb{Z})$ for the set of equivalence classes as above and $[X, c] \in \mathfrak{Eul}_{x_0}(M; \mathbb{Z})$ for the element represented by the pair (X, c). Elements of $\mathfrak{Eul}_{x_0}(M; \mathbb{Z})$ are called (integral) Euler structures of M based at x_0. There is an obvious $H_1(M; \mathbb{Z})$ action on $\mathfrak{Eul}_{x_0}(M; \mathbb{Z})$ defined by

$$[X, c] + [\sigma] := [X, c + \sigma],$$

where $[\sigma] \in H_1(M; \mathbb{Z})$ and $[X, c] \in \mathfrak{Eul}_{x_0}(M; \mathbb{Z})$. Obviously this action is free and transitive. In this sense $\mathfrak{Eul}_{x_0}(M; \mathbb{Z})$ is an affine version of $H_1(M; \mathbb{Z})$.

Considering Euler chains with real coefficients one obtains in exactly the same way an affine version of $H_1(M; \mathbb{R})$ which we will denote by $\mathfrak{Eul}_{x_0}(M; \mathbb{R})$. There is an obvious map $\mathfrak{Eul}_{x_0}(M; \mathbb{Z}) \to \mathfrak{Eul}_{x_0}(M; \mathbb{R})$ which is affine over the homomorphism $H_1(M; \mathbb{Z}) \to H_1(M; \mathbb{R})$.

Remark 3. Another way to understand the $H_1(M; \mathbb{Z})$ action on $\mathfrak{Eul}_{x_0}(M; \mathbb{Z})$ is the following. Suppose $n > 2$ and represent $[\sigma] \in H_1(M; \mathbb{Z})$ by a simple closed curve σ. Choose a tubular neighborhood N of S^1 considered as vector bundle $N \to S^1$. Choose a fiber metric and a linear connection on N. Choose a representative of $[X, c] \in \mathfrak{Eul}(M, x_0)$ such that $X|_N = \frac{\partial}{\partial \theta}$, the horizontal lift of the canonic vector field on S^1. Choose a function $\lambda : [0, \infty) \to [-1, 1]$, which satisfies $\lambda(r) = -1$ for $r \leq \frac{1}{3}$ and $\lambda(r) = 1$ for $r \geq \frac{2}{3}$. Finally choose a function $\mu : [0, \infty) \to \mathbb{R}$ satisfying $\mu(r) = r$ for $r \leq \frac{1}{3}$, $\mu(r) = 0$ for $r \geq \frac{2}{3}$ and $\mu(r) > 0$ for all $r \in (\frac{1}{3}, \frac{2}{3})$. Now construct a new vector field $\tilde{X}$ on M by setting

$$\tilde{X} := \begin{cases} X & \text{on } M \setminus N \\ \lambda(r)\frac{\partial}{\partial\theta} + \mu(r)\frac{\partial}{\partial r} & \text{on } N, \end{cases}$$

where $r : N \to [0, \infty)$ denotes the radius function determined by the fiber metric on N and $-r\frac{\partial}{\partial r}$ is the Euler vector field of N. This construction is known as Reeb surgery, see, e.g. , [11]. If the zeros of X are all non-degenerate the homotopy $X_t := (1-t)X + t\tilde{X}$ is a non-degenerate homotopy from $X_0 = X$ to $X_1 = \tilde{X}$ from which one easily deduces that

$$[\tilde{X}, c] = [X, c] + [\sigma].$$

Particularly all the choices that entered the Reeb surgery do not effect the outcoming Euler structure $[\tilde{X}, c]$.

Let us consider a change of base point. Let $x_0, x_1 \in M$ and choose a path σ from x_0 to x_1. Define

$$\mathfrak{Eul}_{x_0}(M;\mathbb{Z}) \to \mathfrak{Eul}_{x_1}(M;\mathbb{Z}), \quad [X, c] \mapsto [X, c - \chi(M)\sigma]. \tag{14}$$

This is an $H_1(M;\mathbb{Z})$ equivariant bijection but depends on the homology class of σ.

Remark 4. So the identification $\mathfrak{Eul}_{x_0}(M;\mathbb{Z})$ with $\mathfrak{Eul}_{x_1}(M;\mathbb{Z})$ does depend on the choice of a homology class of paths from x_0 to x_1. However, different choices will give identifications which differ by the action of an element in $\chi(M)H_1(M;\mathbb{Z})$. So the quotient $\mathfrak{Eul}_{x_0}(M;\mathbb{Z})/\chi(M)H_1(M;\mathbb{Z})$ does not depend on the base point. Particularly, if $\chi(M) = 0$ then $\mathfrak{Eul}_{x_0}(M;\mathbb{Z})$ does not depend on the base point.

Let $\mathfrak{X}(M, x_0)$ denote the space of vector fields which vanish at x_0 and are non-zero elsewhere. We equip this space with the C^∞ topology, or any C^r topology, $r \geq 0$. Let $\pi_0(\mathfrak{X}(M, x_0))$ denote the space of homotopy classes of such vector fields. If $X \in \mathfrak{X}(M, x_0)$ we will write $[X]$ for the corresponding class in $\pi_0(\mathfrak{X}(M, x_0))$. The following proposition (due to Turaev in the case $\chi(M) = 0$) establishes the proof of Theorem 2(i).

Proposition 3. *Suppose $n > 2$. Then there exists a natural bijection*

$$\pi_0(\mathfrak{X}(M, x_0)) = \mathfrak{Eul}_{x_0}(M;\mathbb{Z}), \qquad [X] \mapsto [X, 0]. \tag{15}$$

Proof. Clearly (15) is well defined. Let us prove that it is onto. So let $[X, c]$ represent an Euler class. Choose an embedded disk $D \subseteq M$ centered at x_0 which contains all zeros of X and the Euler chain c. For this we may have to change c, but without changing the Euler structure $[X, c]$. Choose a vector field X' which equals X on $M \setminus D$ and vanishes just at x_0. Since $H_1(D;\mathbb{Z}) = 0$ we clearly have $[X', 0] = [X, c] \in \mathfrak{Eul}_{x_0}(M;\mathbb{Z})$ and thus (15) is onto.

Let us prove injectivity of (15). Let $X_1, X_2 \in \mathfrak{X}(M, x_0)$ and suppose that $\mathrm{cs}(X_1, X_2) = 0 \in H_1(M;\mathbb{Z})$. Let $D \subseteq M$ denote an embedded open disk centered at x_0. Consider the vector bundle $p^*TM \to I \times M$ and consider the two vector fields as a nowhere vanishing section of p^*TM defined over the set $\partial I \times \dot{M}$, where

$\dot{M} := M \setminus D$. We would like to extend it to a nowhere vanishing section over $I \times \dot{M}$. The first obstruction we meet is an element in

$$\begin{aligned} H^n(I \times \dot{M}, \partial I \times \dot{M}; \{\pi_{n-1}\}) &= H_1(I \times \dot{M}, I \times \partial D; \mathbb{Z}) \\ &= H_1(M, \bar{D}; \mathbb{Z}) \\ &= H_1(M; \mathbb{Z}) \end{aligned}$$

which corresponds to $\mathrm{cs}(X_1, X_2) = 0$. Here $\{\pi_{n-1}\}$ denotes the system of local coefficients determined by the sphere bundle of p^*TM with $\pi_{n-1} = \pi_{n-1}(S^{n-1})$. Since this obstruction vanishes by hypothesis the next obstruction is defined and is an element in:

$$\begin{aligned} H^{n+1}(I \times \dot{M}, \partial I \times \dot{M}; \{\pi_n\}) &= H_0(I \times \dot{M}, I \times \partial D; \pi_n(S^{n-1})) \\ &= H_0(M, \bar{D}; \pi_n(S^{n-1})) \\ &= 0 \end{aligned}$$

Since there is no other obstructions, obstruction theory, see, e.g. , [16], tells us that we find a nowhere vanishing section of p^*TM defined over $I \times \dot{M}$, which restricts to X_i on $\{i\} \times \dot{M}$, $i = 1, 2$. Such a section can easily be extended to a globally defined section of $p^*TM \to I \times M$, which restricts to X_i on $\{i\} \times M$, $i = 1, 2$ and whose zero set is precisely $I \times \{x_0\}$. Such a section can be considered as homotopy from X_1 to X_2 showing $[X_1] = [X_2]$. Hence (15) is injective. □

Remark 5. If $n > 2$ Reeb surgery defines an $H_1(M;\mathbb{Z})$ action on $\pi_0(\mathfrak{X}(M, x_0))$ which via (15) corresponds to the $H_1(M;\mathbb{Z})$ action on $\mathfrak{Eul}_{x_0}(M;\mathbb{Z})$, cf. Remark 3.

Let $\mathfrak{X}_0(M)$ denote the space of nowhere vanishing vector fields on M equipped with the C^∞ topology, or any C^r topology, $r \geq 0$. Let $\pi_0(\mathfrak{X}_0(M))$ denote the set of its connected components. The next proposition is a restatement of Theorem 2(ii).

Proposition 4. *If $n > 2$ then we have a surjection:*

$$\pi_0(\mathfrak{X}_0(M)) \to \mathfrak{Eul}_{x_0}(M;\mathbb{Z}), \qquad [X] \mapsto [X, 0]. \tag{16}$$

Proof. The assignment (16) is certainly well defined. Let us prove surjectivity. Let $[X, c]$ be an Euler structure. Choose an embedded disk $D \subseteq M$ which contains all zeros of X and its Euler chain c, cf. proof of Proposition 3. Since $\chi(M) = 0$ the degree of $X : \partial D \to TD \setminus 0_D$ vanishes. Modifying X only on D we get a nowhere vanishing X' which equals X on $M \setminus D$. Certainly X' has an Euler chain c' which is also contained in D and satisfies $[X, c] = [X', c']$. Since X' has no zeros we get $\partial c' = 0$ and since $H_1(D;\mathbb{Z}) = 0$ we arrive at $[X, c] = [X', c'] = [X', 0]$ which proves that (16) is onto. □

We will now describe another approach to Euler structures which is in some sense Poincaré dual to the other approach. We still consider a closed connected n-dimensional manifold with base point (M, x_0). Consider pairs (g, α) where g is a Riemannian metric on M and $\alpha \in \Omega^{n-1}(M \setminus x_0; \mathcal{O}_M)$ with $d\alpha = E(g)$. Here

$E(g) \in \Omega^n(M; \mathcal{O}_M)$ denotes the Euler class of g which is a form with values in the orientation bundle $\mathcal{O}_M$. We call two pairs (g_1, α_1) and (g_2, α_2) equivalent if

$$\mathrm{cs}(g_1, g_2) = \alpha_2 - \alpha_1 \in \Omega^{n-1}(M \setminus x_0; \mathcal{O}_M)/d\Omega^{n-2}(M \setminus x_0; \mathcal{O}_M).$$

We will write $\mathfrak{Eul}^*_{x_0}(M; \mathbb{R})$ for the set of equivalence classes and $[g, \alpha]$ for the equivalence class represented by the pair (g, α). Elements of $\mathfrak{Eul}^*_{x_0}(M; \mathbb{R})$ are called co-Euler structures based at x_0. There is a natural $H^{n-1}(M; \mathcal{O}_M)$ action on $\mathfrak{Eul}^*_{x_0}(M; \mathbb{R})$ given by

$$[g, \alpha] + [\beta] := [g, \alpha - \beta]$$

with $[\beta] \in H^{n-1}(M; \mathcal{O}_M)$. Since $H^{n-1}(M; \mathcal{O}_M) = H^{n-1}(M \setminus x_0; \mathcal{O}_M)$ this action is obviously free and transitive.

For a pair (g, α) as above and a closed one form ω we define a regularization of $\int_M \omega \wedge \alpha$ as follows. Choose a function f such that $\omega' := \omega - df$ vanishes locally around the base point x_0 and set:

$$S(g, \alpha, \omega; f) := \int_M \omega' \wedge \alpha - \int_M fE(g) + \chi(M) f(x_0)$$

Lemma 2. *The quantity $S(g, \alpha, \omega; f)$ does not depend on the choice of f and will thus be denoted by $S(g, \alpha, \omega)$. If $[g_1, \alpha_1] = [g_2, \alpha_2] \in \mathfrak{Eul}^*_{x_0}(M; \mathbb{R})$ then*

$$S(g_2, \alpha_2, \omega) - S(g_1, \alpha_1, \omega) = \int_M \omega \wedge \mathrm{cs}(g_1, g_2). \tag{17}$$

Moreover, for a function h we have

$$S(g, \alpha, \omega + dh) - S(g, \alpha, \omega) = -\int_M hE(g) + \chi(M) h(x_0). \tag{18}$$

Proof. Suppose we have two functions f_1 and f_2 so that both $\omega_1' := \omega - df_1$ and $\omega_2' := \omega - df_2$ vanish locally around x_0. Let B_ϵ denote a ball of radius ϵ around x_0. Then $f_2 - f_1$ will be constant on B_ϵ for ϵ sufficiently small. Using Stokes' theorem, $d\alpha = E(g)$ and $\int_M E(g) = \chi(M)$ we get:

$$\begin{aligned}
&S(g, \alpha, \omega; f_2) - S(g, \alpha, \omega; f_1) \\
&= -\int_{M \setminus \mathcal{X}} d((f_2 - f_1) \wedge \alpha) + \chi(M)(f_2 - f_1)(x_0) \\
&= -\lim_{\epsilon \to 0} \int_{\partial(M \setminus B_\epsilon)} (f_2 - f_1)\alpha + \chi(M)(f_2 - f_1)(x_0) \\
&= -(f_2 - f_1)(x_0) \lim_{\epsilon \to 0} \int_{\partial(M \setminus B_\epsilon)} \alpha + \chi(M)(f_2 - f_1)(x_0) \\
&= -(f_2 - f_1)(x_0) \lim_{\epsilon \to 0} \int_{M \setminus B_\epsilon} E(g) + \chi(M)(f_2 - f_1)(x_0) = 0
\end{aligned}$$

The second statement follows immediately from $\alpha_2 - \alpha_1 = \mathrm{cs}(g_1, g_2)$, Stokes' theorem and $d\,\mathrm{cs}(g_1, g_2) = E(g_2) - E(g_1)$. The last property is obvious. □

In view of (8), (12), (13), (17) and (18) the quantity

$$R(X,g,\omega) - S(g,\alpha,\omega) - \int_c \omega \tag{19}$$

does only depend on $[X,c] \in \mathfrak{Eul}_{x_0}(M;\mathbb{R})$, $[g,\alpha] \in \mathfrak{Eul}^*_{x_0}(M;\mathbb{R})$ and the cohomology class $[\omega] \in H^1(M;\mathbb{R})$. Thus (19) defines a coupling

$$\mathbb{T} : \mathfrak{Eul}_{x_0}(M;\mathbb{R}) \times \mathfrak{Eul}^*_{x_0}(M;\mathbb{R}) \to H_1(M;\mathbb{R}).$$

From the very definition we have

$$\langle [\omega], \mathbb{T}([X,c],[g,\alpha])\rangle = \int_M \omega \wedge (X^*\Psi(g) - \alpha) - \int_c \omega, \tag{20}$$

where ω is any representative of $[\omega]$ which vanishes locally around the zeros of X and vanishes locally around the base point x_0. Moreover, we have

$$\mathbb{T}(\mathfrak{e} + \sigma, \mathfrak{e}^* + \beta) = \mathbb{T}(\mathfrak{e}, \mathfrak{e}^*) - \sigma + \mathrm{PD}(\beta) \tag{21}$$

for all $\mathfrak{e} \in \mathfrak{Eul}_{x_0}(M;\mathbb{R})$, $\mathfrak{e}^* \in \mathfrak{Eul}^*_{x_0}(M;\mathbb{R})$, $\sigma \in H_1(M;\mathbb{R})$ and $\beta \in H^{n-1}(M;\mathcal{O}_M)$. Here PD is the Poincaré duality isomorphism $\mathrm{PD} : H^{n-1}(M;\mathcal{O}_M) \to H_1(M;\mathbb{R})$.

We have the following affine version of Poincaré duality, which establishes the proof of Theorem 2(iii) and (iv).

Proposition 5. *There is a natural isomorphism of affine spaces*

$$P : \mathfrak{Eul}^*_{x_0}(M;\mathbb{R}) \to \mathfrak{Eul}_{x_0}(M;\mathbb{R})$$

which is affine over the Poincaré duality $\mathrm{PD} : H^{n-1}(M;\mathcal{O}_M) \to H_1(M;\mathbb{R})$. *In other words, for every* $\beta \in H^{n-1}(M;\mathcal{O}_M)$ *and every* $\mathfrak{e}^* \in \mathfrak{Eul}^*_{x_0}(M;\mathbb{R})$ *we have*

$$P(\mathfrak{e}^* + \beta) = P(\mathfrak{e}^*) + \mathrm{PD}(\beta). \tag{22}$$

Moreover, $\mathbb{T}(\mathfrak{e}, \mathfrak{e}^*) = P(\mathfrak{e}^*) - \mathfrak{e}$.

Proof. Given $\mathfrak{e}^* = [g,\alpha] \in \mathfrak{Eul}^*_{x_0}(M;\mathbb{R})$ we choose a vector field X with isolated singularities $\mathcal{X}$. Then $X^*\Psi(g) - \alpha$ is closed and thus defines a cohomology class in $H^{n-1}(M \setminus (\mathcal{X} \cup \{x_0\});\mathcal{O}_M)$. We would like to define $P(\mathfrak{e}^*) := [X,c]$ where c be a representative of its Poincaré dual in $H_1(M,\mathcal{X} \cup \{x_0\};\mathbb{R})$. That is, we ask

$$\int_c \omega = \int_{M \setminus (\mathcal{X} \cup \{x_0\})} \omega \wedge (X^*\Psi(g) - \alpha)$$

to hold for every closed compactly supported one form ω on $M \setminus (\mathcal{X} \cup \{x_0\})$. In view of (20) this is equivalent to ask for $\mathbb{T}(P(\mathfrak{e}^*), \mathfrak{e}^*) = 0$. So we take the latter one as our definition of P. Because of (21) this has a unique solution. The equivariance property and the last equation follow at once. □

5. Smooth triangulations and extension of Chern–Simons theory

Smooth triangulations

Smooth triangulations provide a remarkable source of vector fields with isolated singularities.

To any smooth triangulation τ of the smooth manifold M one can associate a Lipschitz vector field X_τ called *Euler vector field*, with the following properties:

P1: The zeros of X_τ are all non-degenerate and are exactly the barycenters x_σ of the simplexes σ.

P2: For each zero x_σ the unstable set with respect to $-X_\tau$ coincides in a neighborhood of x_σ to the open simplex σ, consequently the zeros are hyperbolic. The Morse index of $-X_\tau$ at x_σ equals $\dim(\sigma)$ and the (Hopf) index of X_τ at x_σ equals $(-1)^{\dim(\sigma)}$.

P3: The piecewise differential function $f_\tau : M \to \mathbb{R}$, defined by $f_\tau(x_\sigma) = \dim(\sigma)$ and extended by linearity on each simplex of the baricentric subdivision of τ, is a Lyapunov function for $-X_\tau$, i.e. , strictly decreasing on non-constant trajectories of $-X_\tau$.

Such a vector field X_τ is unique up to an homotopy of vector fields which satisfy P1–P3. The convex combination provides the homotopy between any two such vector fields.

To construct such a vector field we begin with a standard simplex Δ_n of vectors $(t_0, \dots, t_n) \in \mathbb{R}^{n+1}$ satisfying $0 \leq t_i \leq 1$ and $\sum t_i = 1$.

(i) Let E_n denote the Euler vector field of the corresponding affine space ($\sum t_i = 1$) centered at the barycenter O (of coordinates $(1/(n+1), \dots, 1/(n+1))$ and restricted to Δ_n.

(ii) Let $e : \Delta_n \to [0, 1]$ denote the function which is 1 on the barycenter O and zero on all vertices.

(iii) Let $r : \Delta_n \setminus \{O\} \to \partial\Delta_n$ denote the radial retraction to the boundary.

Set $X'_n := e \cdot E_n$, which is a vector field on Δ_n.

By induction we will construct a canonical vector field X_n on Δ_n which at any point $x \in \Delta_n$ is tangent to the open face the point belongs to and vanishes only at the barycenter of each face. We proceed as follows:

Suppose we have constructed such canonical vector fields on all $\Delta(k)$, $k \leq n-1$. Using the canonical vector fields X_{n-1} we define the vector field X_n on the boundary $\partial\Delta_n$ and extend it to the vector field X''_n by taking at each point $x \in \Delta_n$ the vector parallel to $X_n(r(x))$ multiplied by the function $(1 - e)$ and at O the vector zero. Clearly such vector field vanishes on the radii $\overline{OP}$ (P the barycenter of any face). We finally put

$$X_n := X'_n + X''_n.$$

The vector field X_n is continuous and piecewise differential (actually Lipschitz) and has a well defined continuous flow.

Putting together the vector fields X_n on all simplexes (cells) we provide a piecewise differential (and Lipschitz) vector field X on any simplicial (cellular) complex or polyhedron and in particular on any smoothly triangulated manifold. The vector field X has a flow and f_τ is a Lyapunov function for $-X$. The vector field X is not necessary smooth but by a small (Lipschitz) perturbation we can approximate it by a smooth vector field X_τ which satisfies P1–P3. Any of the resulting vector fields is referred to as the Euler vector field of a smooth triangulation τ. It was pointed out to us that the vector field X_τ has first appeared in [12].

Extension of Chern–Simons theory

Let M be a closed manifold of dimension n. We equip $\Omega^k(M;\mathbb{R})$ with the C^∞ topology. The continuous linear functionals on $\Omega^k(M;\mathbb{R})$ are called k currents and denoted by $\mathcal{D}_k(M)$. Consider $\delta : \mathcal{D}_k(M) \to \mathcal{D}_{k-1}(M)$ given by $(\delta\varphi)(\alpha) := \varphi(d\alpha)$. Clearly $\delta^2 = 0$.[3]

We have a morphism of chain complexes

$$C_*(M;\mathbb{R}) \to \mathcal{D}_*(M), \quad \sigma \mapsto \hat{\sigma}, \quad \hat{\sigma}(\alpha) := \int_\sigma \alpha.$$

Here $C_*(M;\mathbb{R})$ denotes the space of singular chains with real coefficients. Moreover, we have a morphism of chain complexes

$$\Omega^{n-*}(M;\mathcal{O}_M) \to \mathcal{D}_*(M), \quad \beta \mapsto \hat{\beta}, \quad \hat{\beta}(\alpha) := (-1)^{\frac{1}{2}|\alpha|(|\alpha|+1)} \int_M \alpha \wedge \beta.$$

Here $|\alpha|$ denotes the degree of α. The sign is necessary so that this mappings actually intertwines the two differentials d and δ.

Every vector field with isolated singularities X gives rise to a zero chain e_X, cf. Section 4. Via the first morphism we get a zero current $\hat{E}(X)$. More explicitly $(\hat{E}(X))(h) = \sum_{x\in\mathcal{X}} \mathrm{IND}(x)h(x)$ for a function $h \in \Omega^0(M;\mathbb{R})$.

A Riemannian metric g has an Euler form $E(g) \in \Omega^n(M;\mathcal{O}_M)$. Via the second morphism we get a zero current $\hat{E}(g)$. More explicitly $(\hat{E}(g))(h) = \int_M hE(g)$ for a function $h \in \Omega^0(M;\mathbb{R})$.

Let $\mathcal{Z}^k(M;\mathbb{R}) \subseteq \Omega^k(M;\mathbb{R})$ denote the space of closed k forms on M equipped with the C^∞ topology. The continuous linear functionals on $\mathcal{Z}^k(M;\mathbb{R})$ are referred to as k *currents rel. boundary* and identify to $\mathcal{D}_k(M)/\delta(\mathcal{D}_{k+1}(M))$. The two chain morphisms provide mappings

$$C_k(M;\mathbb{R})/\partial(C_{k+1}(M;\mathbb{R})) \to \mathcal{D}_k(M)/\delta(\mathcal{D}_{k+1}(M)) \tag{23}$$

and

$$\Omega^{n-k}(M;\mathcal{O}_M)/d(\Omega^{n-k-1}(M;\mathcal{O}_M)) \to \mathcal{D}_k(M)/\delta(\mathcal{D}_{k+1}(M)). \tag{24}$$

For two vector fields with isolated zeros X_1 and X_2 we have constructed $\mathrm{cs}(X_1,X_2) \in C_1(M;\mathbb{Z})/\partial(C_2(M;\mathbb{Z}))$, cf. Section 3. This gives rise to $\mathrm{cs}(X_1,X_2) \in C_1(M;\mathbb{R})/\partial(C_2(M;\mathbb{R}))$, and via (23) we get a one current rel. boundary which we will denote by $\hat{\mathrm{cs}}(X_1,X_2)$. More precisely, $(\hat{\mathrm{cs}}(X_1,X_2))(\omega) = \int_{\mathrm{cs}(X_1,X_2)} \omega$ for a

[3]The chain complex $(\mathcal{D}_*(M),\delta)$ computes the homology of M with real coefficients.

closed one form $\omega \in \mathcal{Z}^1(M;\mathbb{R})$. Recall that we have $\operatorname{cs}(X_2, X_1) = -\operatorname{cs}(X_1, X_2)$, $\operatorname{cs}(X_1, X_3) = \operatorname{cs}(X_1, X_2) + \operatorname{cs}(X_2, X_3)$, $\partial \operatorname{cs}(X_1, X_2) = e_{X_2} - e_{X_1}$ and thus

$$\begin{aligned}
\hat{\operatorname{cs}}(X_2, X_1) &= -\hat{\operatorname{cs}}(X_1, X_2) \\
\hat{\operatorname{cs}}(X_1, X_3) &= \hat{\operatorname{cs}}(X_1, X_2) + \hat{\operatorname{cs}}(X_2, X_3) \\
\delta\hat{\operatorname{cs}}(X_1, X_2) &= \hat{E}(X_2) - \hat{E}(X_1).
\end{aligned}$$

For two Riemannian metrics g_1 and g_2 we have the Chern–Simons form $\operatorname{cs}(g_1, g_2) \in \Omega^{n-1}(M;\mathcal{O}_M)/d(\Omega^{n-2}(M;\mathcal{O}_M))$. Via (24) we get a one current rel. boundary which we will denote by $\hat{\operatorname{cs}}(g_1, g_2)$. More precisely $(\hat{\operatorname{cs}}(g_1, g_2))(\omega) = -\int_M \omega \wedge \operatorname{cs}(g_1, g_2)$ for a closed one form $\omega \in \mathcal{Z}^1(M;\mathbb{R})$. Recall that $\operatorname{cs}(g_2, g_1) = -\operatorname{cs}(g_1, g_2)$, $\operatorname{cs}(g_1, g_3) = \operatorname{cs}(g_1, g_2) + \operatorname{cs}(g_2, g_3)$, $d\operatorname{cs}(g_1, g_2) = E(g_2) - E(g_1)$ and thus

$$\begin{aligned}
\hat{\operatorname{cs}}(g_2, g_1) &= -\hat{\operatorname{cs}}(g_1, g_2) \\
\hat{\operatorname{cs}}(g_1, g_3) &= \hat{\operatorname{cs}}(g_1, g_2) + \hat{\operatorname{cs}}(g_2, g_3) \\
\delta\hat{\operatorname{cs}}(g_1, g_2) &= \hat{E}(g_2) - \hat{E}(g_1).
\end{aligned}$$

Suppose X is a vector field with isolated zeros and g is a Riemannian metric. We define one currents rel. boundary by $(\hat{\operatorname{cs}}(g, X))(\omega) := R(X, g, \omega)$ and $\hat{\operatorname{cs}}(X, g) := -\hat{\operatorname{cs}}(g, X)$. Proposition 1 and Proposition 2 tell that

$$\begin{aligned}
\delta\hat{\operatorname{cs}}(g, X) &= \hat{E}(X) - \hat{E}(g) \\
\hat{\operatorname{cs}}(g_1, X) &= \hat{\operatorname{cs}}(g_1, g_2) + \hat{\operatorname{cs}}(g_2, X) \\
\hat{\operatorname{cs}}(g, X_2) &= \hat{\operatorname{cs}}(g, X_1) + \hat{\operatorname{cs}}(X_1, X_2).
\end{aligned}$$

We summarize these observations in

Proposition 6. *Let any of the symbols x, y, z denote either a Riemannian metric g or a vector field with isolated zeros. Then one has:*

(i) $\hat{\operatorname{cs}}(y, x) = -\hat{\operatorname{cs}}(x, y)$
(ii) $\hat{\operatorname{cs}}(x, z) = \hat{\operatorname{cs}}(x, y) + \hat{\operatorname{cs}}(y, z)$
(iii) $\delta\hat{\operatorname{cs}}(x, y) = \hat{E}(y) - \hat{E}(x)$.

Suppose τ is a smooth triangulation. We define its Euler current by $\hat{E}(\tau) := \hat{E}(X_\tau)$, where X_τ is the Euler vector field. Similarly for two triangulations τ_1 and τ_2 we define a one current rel. boundary by $\hat{\operatorname{cs}}(\tau_1, \tau_2) := \hat{\operatorname{cs}}(X_{\tau_1}, X_{\tau_2})$.

Corollary 1. *Let any of the symbols x, y, z denote either a Riemannian metric g or a smooth triangulation. Then one has:*

(i) $\hat{\operatorname{cs}}(y, x) = -\hat{\operatorname{cs}}(x, y)$
(ii) $\hat{\operatorname{cs}}(x, z) = \hat{\operatorname{cs}}(x, y) + \hat{\operatorname{cs}}(y, z)$
(iii) $\delta\hat{\operatorname{cs}}(x, y) = \hat{E}(y) - \hat{E}(x)$.

6. Theorem of Bismut–Zhang

Let (M, x_0) be a closed connected manifold with base point. Let $\mathbb{K}$ be a field of characteristic zero, and suppose F is a flat $\mathbb{K}$ vector bundle over M, that is F is equipped with a flat connection ∇. Let F_{x_0} denote the fiber over the base point x_0. Holonomy at the base point provides a right $\pi_1(M, x_0)$ action on F_{x_0} and when composed with the inversion in $\mathrm{GL}(F_{x_0})$ a representation $\rho_F : \pi_1(M, x_0) \to \mathrm{GL}(F_{x_0})$. So we get a homomorphism $\det \circ \rho_F : \pi_1(M, x_0) \to \mathbb{K}^*$ which descends to a homomorphism $H_1(M; \mathbb{Z}) \to \mathbb{K}^*$ and thus determines a cohomology class $\Theta_F \in H^1(M; \mathbb{K}^*)$.

Suppose we have a smooth triangulation τ of M. It gives rise to a cellular complex $C^*_\tau(M; F)$ which computes the cohomology $H^*(M; F)$. Let $\mathcal{X}_\tau$ denote the set of barycenters of τ. For a cell σ of τ we let x_σ denote the barycenter of σ. Let X_τ denote the Euler vector field of τ, cf. Section 5. Then $\mathcal{X}_\tau$ is the zero set of X_τ. Moreover, for a cell σ we have $\mathrm{IND}_{X_\tau}(\sigma_x) = (-1)^{\dim \sigma}$. As a graded vector space we have $C^k_\tau(M; F) = \bigoplus_{\dim \sigma = k} F_{x_\sigma}$. So we get a canonical isomorphism of $\mathbb{K}$ vector spaces:

$$\det C^*_\tau(M; F) = \det H(C^*_\tau(M; F)) = \det H^*(M; F) \tag{25}$$

Recall that the determinant line of a vector space W is by definition $\det W := \Lambda^{\dim W} W$. For a $\mathbb{Z}$ graded vector space V^* one sets $V^{\text{even}} := \bigoplus_{k \text{ even}} V^k$, $V^{\text{odd}} := \bigoplus_{k \text{ odd}} V^k$ and defines its determinant line by $\det V^* := \det V^{\text{even}} \otimes (\det V^{\text{odd}})^*$.

Suppose we have given an Euler structure $\mathfrak{e} \in \mathfrak{Eul}_{x_0}(M; \mathbb{Z})$. For every $x \in \mathcal{X}_\tau$ choose a path π_x from x_0 to x, so that with $c := \sum_{x \in \mathcal{X}_\tau} \mathrm{IND}_{X_\tau}(x_\sigma)\pi_x$ we have $\mathfrak{e} = [X_\tau, c]$.[4] Let f_0 be a non-zero element in $\det F_{x_0}$. Note that a frame (basis) in F_{x_0} determines such an element in $\det F_{x_0}$. Using parallel transport along π_x we get a non-zero element in every $\det F_{x_\sigma}$. If the barycenters x_σ where ordered we would get a well defined non-zero element in $\det C^*_\tau(M; F)$.

Suppose $\mathfrak{o}$ is a cohomology orientation of M, i.e. , $\mathfrak{o}$ is an orientation of $\det H^*(M; \mathbb{R})$. We say an ordering of the zeros x_σ is compatible with $\mathfrak{o}$ if the non-zero element in $\det C^*_\tau(M; \mathbb{R})$ provided by this ordered base is compatible with the orientation $\mathfrak{o}$ via the canonic isomorphism

$$\det C^*_\tau(M; \mathbb{R}) = \det H(C^*_\tau(M; \mathbb{R})) = \det H^*(M; \mathbb{R}).$$

So an integral Euler structure $\mathfrak{e}$, a cohomology orientation $\mathfrak{o}$ and an element $f_0 \in \det F_{x_0}$ provide a non-zero element in $\det C^*_\tau(M; F)$ which corresponds to a non-zero element in $\det H^*(M; F)$ via (25). We thus get a mapping

$$\det F_{x_0} \setminus 0 \to \det H^*(M; F) \setminus 0. \tag{26}$$

This mapping is obviously homogeneous of degree $\chi(M)$. A straight forward calculation shows that it does not depend on the choice of π_x. As a matter of fact this mapping does not depend on τ either, only on the Euler structure $\mathfrak{e}$ and the cohomology orientation $\mathfrak{o}$. This is a non-trivial fact, and its proof is contained in

[4] Such a representative for the Euler structure is called spray or Turaev spider.

[10] and [13] for acyclic case and implicit in the existing literature cf. [7] and [3]. We define the *combinatorial torsion* to be the element

$$\tau^{\text{comb}}_{F,\mathfrak{e},\mathfrak{o}} \in \det H^*(M;F) \otimes (\det F_{x_0})^{-\chi(M)}$$

corresponding to the homogeneous mapping (26). Note that we also have

$$\tau^{\text{comb}}_{F,\mathfrak{e}+\sigma,\mathfrak{o}} = \tau^{\text{comb}}_{F,\mathfrak{e},\mathfrak{o}} \cdot \langle \Theta_F, \sigma \rangle^{-1},$$

for all $\sigma \in H_1(M;\mathbb{Z})$. Here $\langle \cdot, \cdot \rangle$ denotes the natural pairing of homology with integer coefficients and cohomology with coefficients in the Abelian group $\mathbb{K}^*$. Moreover

$$\tau^{\text{comb}}_{F,\mathfrak{e},-\mathfrak{o}} = (-1)^{\operatorname{rank} F} \tau^{\text{comb}}_{F,\mathfrak{e},\mathfrak{o}}.$$

Clearly, if $\chi(M) = 0$ then $\tau^{\text{comb}}_{F,\mathfrak{e},\mathfrak{o}} \in \det H^*(M;F)$.

Now consider the case when $\mathbb{K}$ is $\mathbb{R}$ or $\mathbb{C}$. Let μ be a Hermitian structure, i.e. , fiber wise Hermitian scalar product, on F. It induces a scalar product on $\det C^*_\tau(M;F)$ and via (25) a scalar product $||\cdot||^{\mathcal{M}}_{F,\tau,\mu}$ on the line $\det H^*(M;F)$. This is exactly what is called *Milnor metric* in [1]. The Hermitian structure μ also defines a scalar product on $(\det F_{x_0})^{-\chi(M)}$ which we will denote by $||\cdot||_{\mu_{x_0}}$. Moreover, μ gives rise to a closed one form $\omega(\nabla, \mu)$, where ∇ is the flat connection of F, see [1] and Section 2. For its cohomology class we have $[\omega(\nabla,\mu)] = (\log|\cdot|)_*\Theta_F$. Here $(\log|\cdot|)_* : H^1(M;\mathbb{C}^*) \to H^1(M;\mathbb{R})$ in the complex case, and $(\log|\cdot|)_* : H^1(M;\mathbb{R}^*) \to H^1(M;\mathbb{R})$ in the real case. Given an Euler structure with real coefficients $\mathfrak{e} \in \mathfrak{Eul}_{x_0}(M;\mathbb{R})$ we choose an Euler chain c so that $[X_\tau, c] = \mathfrak{e}$, and define a metric on $\det H^*(M;F) \otimes (\det F_{x_0})^{-\chi(M)}$ by:

$$||\cdot||^{\text{comb}}_{F,\mathfrak{e}} := ||\cdot||^{\mathcal{M}}_{F,\tau,\mu} \otimes ||\cdot||_{\mu_{x_0}} \cdot e^{\int_c \omega(\nabla,\mu)} \tag{27}$$

As the notation indicates this does not depend on the cohomology orientation, is independent of μ and does only depend on the Euler structure $\mathfrak{e} = [X, c]$. This follows from known anomaly formulas for the Milnor torsion, implicit in [1], or can be seen as a consequence of (28) and (29) below. Note that

$$||\cdot||^{\text{comb}}_{F,\mathfrak{e}+\sigma} = ||\cdot||^{\text{comb}}_{F,\mathfrak{e}} \cdot e^{\langle (\log|\cdot|)_*\Theta_F, \sigma \rangle} \tag{28}$$

for all $\sigma \in H_1(M;\mathbb{R})$. For an integral Euler structure $\mathfrak{e} \in \mathfrak{Eul}_{x_0}(M;\mathbb{Z})$ we have

$$||\tau^{\text{comb}}_{F,\mathfrak{e},\mathfrak{o}}||^{\text{comb}}_{F,\mathfrak{e}} = 1. \tag{29}$$

Here, abusing notation, $\mathfrak{e}$ at the same time denotes its image in $\mathfrak{Eul}_{x_0}(M;\mathbb{R})$.

Now let g be a Riemannian metric on M. Then we also have the Ray–Singer metric $||\cdot||^{\text{RS}}_{F,g,\mu}$ on $\det H^*(M;F)$, cf. [1]. Let $\mathfrak{e}^* \in \mathfrak{Eul}^*_{x_0}(M;\mathbb{R})$ and suppose $[g,\alpha] = \mathfrak{e}^*$, i.e. , $d\alpha = E(g)$. Define a metric on $\det H^*(M;F) \otimes (\det F_{x_0})^{-\chi(M)}$ by:

$$||\cdot||^{\text{an}}_{F,\mathfrak{e}^*} := ||\cdot||^{\text{RS}}_{F,g,\mu} \otimes ||\cdot||_{\mu_{x_0}} \cdot e^{-S(g,\alpha,\omega(\nabla,\mu))} \tag{30}$$

We call this metric the *modified Ray–Singer metric.*

The known anomaly formulas for the Ray–Singer torsion, see [1], imply that this is independent of μ and only depends on the co-Euler structure $\mathfrak{e}^*$. Note that

$$||\cdot||^{\text{an}}_{F,\mathfrak{e}^*+\beta} = ||\cdot||^{\text{an}}_{F,\mathfrak{e}^*} \cdot e^{\langle (\log|\cdot|)_*\Theta_F, \text{PD}(\beta) \rangle} \tag{31}$$

for all $\beta \in H^{n-1}(M; \mathcal{O}_M)$. The main theorem of Bismut–Zhang, see [1], can now be reformulated as follows:

Theorem 3 (Bismut–Zhang). *Suppose (M, x_0) is a closed connected manifold with base point and F a flat real or complex vector bundle over M. Let $\mathfrak{e} \in \mathfrak{Eul}_{x_0}(M;\mathbb{R})$ be an Euler structure with real coefficients, and let $\mathfrak{e}^* \in \mathfrak{Eul}^*_{x_0}(M;\mathbb{R})$ be a co-Euler structure, both based at x_0. Then one has:*

$$|| \cdot ||^{\mathrm{an}}_{F,\mathfrak{e}^*} = || \cdot ||^{\mathrm{comb}}_{F,\mathfrak{e}} \cdot e^{\langle (\log|\cdot|)_* \Theta_F, \mathbb{T}(\mathfrak{e},\mathfrak{e}^*)\rangle}$$

Particularly, if $\mathfrak{e} = P(\mathfrak{e}^*)$ then $|| \cdot ||^{\mathrm{an}}_{F,\mathfrak{e}^*} = || \cdot ||^{\mathrm{comb}}_{F,\mathfrak{e}}$.

For an alternative proof of the (original) Bismut–Zhang theorem see also [4].

Appendix A. Complex versus real torsion

Suppose V is a finite-dimensional complex vector space. Let $V_{\mathbb{R}}$ denote the vector space V considered as real vector space. We have a mapping

$$\begin{aligned} \theta_V : \det V &\to \det(V_{\mathbb{R}}) \\ v_1 \wedge v_2 \wedge \cdots \wedge v_n &\mapsto v_1 \wedge iv_1 \wedge v_2 \wedge iv_2 \wedge \cdots \wedge v_n \wedge iv_n. \end{aligned}$$

It has the property

$$\theta_V(z\alpha) = |z|^2 \theta_V(\alpha),$$

for all $z \in \mathbb{C}$ and $\alpha \in \det V$. If $f : V \to W$ is a complex linear mapping then the following diagram commutes:

$$\begin{array}{ccc} \det V & \xrightarrow{\theta_V} & \det(V_{\mathbb{R}}) \\ {\scriptstyle \det f}\downarrow & & \downarrow{\scriptstyle \det(f_{\mathbb{R}})} \\ \det W & \xrightarrow{\theta_W} & \det(W_{\mathbb{R}}) \end{array}$$

After identifying $\det \mathbb{C} = \mathbb{C}$ and $\det(\mathbb{C}_{\mathbb{R}}) = \Lambda^2\mathbb{R}^2 = \mathbb{R}$ we have

$$\theta_{\mathbb{C}} : \mathbb{C} \to \mathbb{R}, \qquad \theta_{\mathbb{C}}(z) = |z|^2.$$

Suppose L is a complex line, R a real line and $\theta : L \to R$ a mapping which satisfies

$$\theta(z\lambda) = |z|^2 \theta(\lambda), \tag{32}$$

for all $z \in \mathbb{C}$ and all $\lambda \in L$. If L' is another complex line, R' another real line and $\theta' : L' \to R'$ another mapping which satisfies (32) we can define

$$\theta \otimes \theta' : L \otimes L' \to R \otimes R', \qquad (\theta \otimes \theta')(\lambda \otimes \lambda') := \theta(\lambda) \otimes \theta'(\lambda')$$

which again satisfies (32) Note that

$$\begin{array}{ccc} L \otimes \mathbb{C} & = & L \\ {\scriptstyle \theta\otimes\theta_{\mathbb{C}}}\downarrow & & \downarrow{\scriptstyle \theta} \\ R \otimes \mathbb{R} & = & R \end{array}$$

commutes. If $0 \to V \to W \to U \to 0$ is a short exact sequence of complex vector spaces we have a commutative diagram:

$$\begin{array}{ccc} \det V \otimes \det U & = & \det W \\ \downarrow{\scriptstyle \theta_V \otimes \theta_U} & & \downarrow{\scriptstyle \theta_W} \\ \det(V_{\mathbb{R}}) \otimes \det(U_{\mathbb{R}}) & = & \det(W_{\mathbb{R}}) \end{array}$$

Note that for a complex vector space V we have a canonic isomorphism

$$(V^*)_{\mathbb{R}} = (V_{\mathbb{R}})^*, \qquad \varphi \mapsto \Re \circ \varphi.$$

Using this identification we get a commutative diagram:

$$\begin{array}{ccccc} \det V \otimes \det(V^*) & = & \det V \otimes (\det V)^* & = & \mathbb{C} \\ \downarrow{\scriptstyle \theta_V \otimes \theta_{V^*}} & & & & \downarrow{\scriptstyle \theta_{\mathbb{C}}} \\ \det(V_{\mathbb{R}}) \otimes \det((V^*)_{\mathbb{R}}) & = & \det(V_{\mathbb{R}}) \otimes (\det(V_{\mathbb{R}}))^* & = & \mathbb{R} \end{array}$$

Putting all this together we obtain

Proposition 7. *Let C^* be a finite-dimensional chain complex over $\mathbb{C}$. Let $C^*_{\mathbb{R}}$ denote the same chain complex viewed as chain complex over $\mathbb{R}$. Then $H(C^*_{\mathbb{R}}) = H(C^*)_{\mathbb{R}}$, and we have a commutative diagram:*

$$\begin{array}{ccc} \det C^* & = & \det H(C^*) \\ \downarrow{\scriptstyle \theta_{C^*}} & & \downarrow{\scriptstyle \theta_{H(C^*)}} \\ \det(C^*_{\mathbb{R}}) & = & \det H(C^*_{\mathbb{R}}) \end{array}$$

Now suppose F is a flat complex vector bundle over a closed manifold (M, x_0) with base point. Let $F_{\mathbb{R}}$ denote the vector bundle F considered as real bundle. Recall the mappings (26) from Section 6. Clearly $H^*(M;F)_{\mathbb{R}} = H^*(M;F_{\mathbb{R}})$. Let $A := \theta_{H^*(M;F)} \otimes (\theta_{F_{x_0}})^{-\chi(M)}$ denote the canonical mapping:

$$\det H^*(M;F) \otimes (\det F_{x_0})^{-\chi(M)} \xrightarrow{A} \det H^*(M;F_{\mathbb{R}}) \otimes (\det(F_{\mathbb{R}})_{x_0})^{-\chi(M)}$$

In this situation we obviously we have

Proposition 8. *a) For an integral Euler structure $\mathfrak{e} \in \mathfrak{Eul}_{x_0}(M;\mathbb{Z})$ and a cohomology orientation $\mathfrak{o}$ we have $A(\tau^{\mathrm{comb}}_{F,\mathfrak{e},\mathfrak{o}}) = \tau^{\mathrm{comb}}_{F_{\mathbb{R}},\mathfrak{e},\mathfrak{o}}$. b) For an Euler structure with real coefficients $\mathfrak{e} \in \mathfrak{Eul}_{x_0}(M;\mathbb{R})$ we have $||\cdot||^{\mathrm{comb}}_{F_{\mathbb{R}},\mathfrak{e}} \circ A = (||\cdot||^{\mathrm{comb}}_{F,\mathfrak{e}})^2$. c) For a co-Euler structure $\mathfrak{e}^* \in \mathfrak{Eul}^*_{x_0}(M;\mathbb{R})$ we have $||\cdot||^{\mathrm{an}}_{F_{\mathbb{R}},\mathfrak{e}^*} \circ A = (||\cdot||^{\mathrm{an}}_{F,\mathfrak{e}^*})^2$.*

Note that the previous proposition and the real version of Theorem 3 imply the complex version of Theorem 3.

References

[1] J.M. Bismut and W. Zhang, *An extension of a theorem by Cheeger and Müller*, Astérisque **205**, Société Mathématique de France, 1992.

[2] R. Bott and L.W. Tu, *Differential forms in algebraic topology*. Graduate texts in Mathematics **82**. Springer Verlag, New York–Berlin, 1982.

[3] D. Burghelea, *Removing Metric Anomalies from Ray–Singer Torsion*, Lett. Math. Phys. **47**(1999), 149–158.

[4] D. Burghelea, L. Friedlander and T. Kappeler, *Relative Torsion*, Commun. Contemp. Math. **3**(2001), 15–85.

[5] D. Burghelea and S. Haller, *Euler structures, variety of representations and the Milnor–Turaev torsion*, preprint Max-Planck-Institut für Mathematik in Bonn, **121**(2004).

[6] S.S. Chern and J. Simons, *Characteristic forms and geometric invariants,* Ann. Math. **99**(1974), 48–69.

[7] M. Farber and V. Turaev, *Poincaré–Reidemeister metric, Euler structures and torsion*, J. Reine Angew. Math. **520**(2000), 195–225.

[8] S. Kobayashi and K. Nomizu, *Foundations of Differential Geometry*. John Willey and Sons Inc., New York, 1969.

[9] V. Mathai and D. Quillen, *Superconnections, Thom Classes, and Equivariant Differential Forms*, Topology **25**(1986), 85–110.

[10] J. Milnor, *Whitehead torsion*, Bull. Amer. Math. Soc. **72**(1966), 358–426.

[11] L. Nicolaescu, *The Reidemeister Torsion of* 3*-manifolds*, de Gruyter Studies in Mathematics **30**, Walter de Gruyter & Co., Berlin, 2003.

[12] N. Steenrod, *The topology of fiber bundles*. Reprint of the 1957 edition. Princeton University Press, Princeton, NJ, 1999.

[13] V. Turaev, *Reidemeister torsion in Knot theory*, Uspekhi Mat. Nauk **41**(1986), 119–182.

[14] V. Turaev, *Euler structures, nonsingular vector fields, and Reidemeister-type torsions*, Math. USSR–Izv. **34**(1990), 627–662.

[15] V. Turaev, *Introduction to combinatorial torsions*. Notes taken by Felix Schlenk. Lectures in Mathematics ETH Zürich. Birkhäuser Verlag, Basel, 2001.

[16] G.W. Whitehead, *Elements of homotopy theory*, Graduate Texts in Mathematics **61**, Springer Verlag, New York–Berlin, 1978.

Dan Burghelea
Dept. of Mathematics, The Ohio State University,
231 West 18th Avenue
Columbus, OH 43210, USA.
e-mail: `burghele@mps.ohio-state.edu`

Stefan Haller
Department of Mathematics, University of Vienna,
Nordbergstraße 15
A-1090 Vienna, Austria.
e-mail: `stefan.haller@univie.ac.at`

C^*-algebras and Elliptic Theory
Trends in Mathematics, 61–72

Morse Inequalities for Foliations

Alain Connes and Thierry Fack

Abstract. We outline the analytical proof of the Morse inequalities for measured foliations obtained in [2] and give some applications. The proof is based on the use of a twisted Laplacian.

Mathematics Subject Classification (2000). Primary 57R30; Secondary 58E05, 46L51.

Keywords. Measured foliations, leafwise Morse inequalities.

Introduction

Let (M, F) be a p-dimensional smooth foliation on a compact manifold M and assume the existence of a holonomy invariant transverse measure Λ. For any Euclidean structure on F, the ith *Betti number* $\beta_i = \dim_\Lambda(\mathrm{Ker}(\Delta^i))$ $(i = 0, 1, \ldots, p)$ of (M, F, Λ) is by definition [1] the Murray-Von Neumann dimension $\dim_\Lambda$ of the square integrable field $H^i = \mathrm{Ker}(\Delta^i)$ of Hilbert spaces, where $\Delta^i = \left(\Delta_L^i\right)_L$ is the leafwise Laplace operator acting on forms of degree i along the leaves $L \in M/F$ of the foliation. These Betti numbers are finite and independent of the choice of the Euclidean structure on F (cf. [1]). Note that we have $\dim_\Lambda(\mathrm{Ker}(\Delta^i)) = \mathrm{Tr}_\Lambda(P^i)$, where the orthogonal projection $P^i = \left(P_L^i\right)$ on $\mathrm{Ker}(\Delta^i)$ belongs to the von Neumann algebra $N = W^*(M, F, \mathrm{End}(\Lambda^* F))$ associated with the foliation. This von Neumann algebra is equipped with a normal semifinite trace Tr_Λ associated with the transverse measure Λ, and which is formally defined for any $T \in N_+$ by $\mathrm{Tr}_\Lambda(T) = \int_{M/F} \mathrm{Trace}(T_L) d\Lambda(L)$. Classical pseudodifferential estimates insure [1] that the leafwise pseudodifferential operator $\left(I + \Delta^i\right)^{-m}$ of order $-2m$ is Hilbert-Schmidt with respect to the trace Tr_Λ for any integer $m > \frac{p}{4}$, so that $P^i = \left(I + \Delta^i\right)^{-m} P^i$ is trace class and hence $\dim_\Lambda(\mathrm{Ker}\,\Delta^i) = \mathrm{Tr}_\Lambda(P_i)$ is finite.

The foliated Morse inequalities obtained in [2] yield, for any suitably generic smooth function φ on M, a relationship between the Betti numbers β_i and the transverse measures $c_i = \Lambda\left(A_1^i(\varphi)\right)$ of the sets $A_1^i(\varphi)$ of Morse critical points of index i for the restriction of φ to the leaf manifold:

Theorem. (*Foliated Morse inequalities* [2]). *Let (M, F, Λ) be a measured p-dimensional smooth foliation on a compact manifold M and assume that the set of leaves with non-zero holonomy is Λ-negligible. Let $\varphi \in C^\infty(M, \mathbb{R})$ be a generalized foliated Morse function on (M, F). For any $i = 0, 1, \dots, p$, the critical Morse index $c_i = \Lambda\left(A_1^i(\varphi)\right)$ is well defined and we have:*

$$\beta_i - \beta_{i-1} + \cdots + (-1)^i \beta_o \leq c_i - c_{i-1} + \cdots + (-1)^i c_o$$

with equality for $i = p$. In particular, we have $\beta_i \leq c_i$ for any $i = 0, 1, \dots, p$.

This shows in particular that the *Euler characteristic* $\chi(F, \Lambda) = \sum_{i=0}^{p} (-1)^i \beta_i$ of (M, F, Λ), which is equal to $\sum_{i=0}^{p} (-1)^i c_i$, is computable from the leafwise singularities of φ. We present here an outline of the proof of the foliated Morse inequalities. The text of the present paper circulated as preprint [2]. The main difficulty is to define a right notion of *generalized foliated Morse function.* On one hand, we cannot expect in general that the restriction of φ to any leaf will be a Morse function, since the condition of being leafwise Morse is not generic. On the other hand, thanks to results by Igusa [7] and Eliashberg-Mishachev [3], we may assume that the restriction of φ to any leaf has only Morse or birth-death (i.e., cubic) type singularities.

To avoid unnecessary technicalities, we shall assume in the sequel that the holonomy groupoid of the foliation is Hausdorff, that Λ is absolutely continuous with respect to the Lebesgue class and that Λ-almost every leaf has no holonomy. For all technical details, we refer to [2].

1. Generalized foliated Morse functions

1.1. Singularities of smooth maps

Let V be a smooth p-dimensional manifold and $\varphi \in C^\infty(V, \mathbb{R})$. Recall that $m \in V$ is a *critical point* of φ if $d\varphi(m) = 0$. The set of critical points of φ is called the *critical manifold* and is denoted by $C(\varphi)$. A point $m \in C(\varphi)$ is called a *non-degenerate* (or *Morse type* or *A_1-type) singularity* if the Hessian form

$$H\varphi(m) = \sum_{1 \leq i,j \leq p} \frac{\partial^2 \varphi}{\partial x_i \partial x_j}(m) dx_i \otimes dx_j$$

of φ at m is non singular. Then, the number of negative signs in the signature of $H\varphi(m)$ is called the *index* of the Morse singularity. If $m \in C(\varphi)$ is a Morse type singularity of index i, one can find a neighborhood U of m and a local system of coordinates $(x_1, \dots, x_p) : U \to \mathbb{R}^p$ with $x_1(m) = \cdots = x_p(m) = 0$ such that we have:

$$\varphi = \varphi(m) - \sum_{1 \leq j \leq i} x_j^2 + \sum_{i+1 \leq j \leq p} x_j^2 \text{ on } U.$$

A critical point of φ which is not of Morse type is called a *degenerate singularity*. A singularity $m \in C(\varphi)$ is said to be of *multiplicity* k ($k = 1, 2, \dots$) if the quotient $\mathfrak{R}/\mathfrak{I}$ of the space $\mathfrak{R}$ of germs at m of smooth functions vanishing at m by the ideal $\mathfrak{I}$ generated by the germs at m of the partial derivatives $\frac{\partial\varphi}{\partial x_i}$ ($i = 1, 2, \dots, p$) has dimension k. Singularities of multiplicity k are isolated in $C(\varphi)$ and have the property that, after perturbation, they bifurcate into at most k Morse type critical points. If $m \in C(\varphi)$ is a singularity of finite multiplicity, there exists a neighborhood U of m and a local system of coordinates $(x_1, \dots, x_p) : U \to \mathbb{R}^p$ with $x_1(m) = \dots = x_p(m) = 0$ such that we have:

$$\varphi(u) = P(x_1(u), x_2(u), \dots, x_p(u)) \quad \text{on } U,$$

where P is the Taylor polynomial of φ of order the "Milnor number" of the singularity m. Morse type singularities are singularities of multiplicity 1. Recall that a point $m \in C(\varphi)$ is called a *birth-death type* (or A_2*-type*) *singularity* if there exists a neighborhood U of m and a local system of coordinates $(x_1, \dots, x_p) : U \to \mathbb{R}^p$ with $x_1(m) = \dots = x_p(m) = 0$ such that the map $x \to (D\varphi(x), det\left(D^2\varphi(x)\right)$ has rank p at m. For such a birth-death type singularity, one can find a local system of coordinates $(x_1, \dots, x_p)$ in a neighborhood U of m such that:

$$x_1(m) = \dots = x_p(m) = 0 \text{ and } \varphi = \varphi(m) + x_1^3 + \sum_{2 \le i \le p} \pm x_i^2 \quad \text{on } U.$$

Birth-death singularities are of multiplicity 2; they are the simplest kind of degenerate singularities. A function $\varphi \in C^\infty(V, \mathbb{R})$ with only Morse or birth-death type of singularities is called a *generalized Morse function* on V.

Finally, recall that a smooth map $\pi : N \to Q$ between two q-dimensional manifolds is called *folded* if there exists, at any point $m \in N$ where $\text{rank}(d\pi(m)) < q$, local coordinates centered around m and $\pi(m)$ respectively such that we have $\pi(x_1, x_2, \dots, x_q) = (x_1^2, x_2, \dots, x_q)$ in a neighborhood of m.

1.2. Foliated Morse functions

Let (M, F) be a p-dimensional smooth foliation on a compact n-dimensional manifold M. Any function $\varphi \in C^\infty(M, \mathbb{R})$ can be viewed in a foliation chart $U \cong \mathbb{R}^p \times \mathbb{R}^q$ ($q = n - p = \text{codim}(F)$) as a smooth q-parameter family $\varphi_t : u \in \mathbb{R}^p \to \varphi_t(u) = \varphi(u, t) \in \mathbb{R}$ of smooth functions, and hence as an element of the space $C^o(\mathbb{R}^q, C^k(\mathbb{R}^p, \mathbb{R}))$ for any $k = 0, 1, \dots$ This allows to define, by using finite coverings of M by foliation charts, the C^k topology on $C^\infty(M, \mathbb{R})$.

For any $\varphi \in C^\infty(M, \mathbb{R})$, denote by d_F the de Rham derivative in the leaf direction. A point $m \in M$ will be called a *leafwise singularity* (resp. a *leafwise Morse type singularity of index* i, a *leafwise birth-death type singularity*) if it is a critical point (resp. a Morse type singularity of index i, a birth-death type singularity) for the restriction of φ to the leaf through m. We shall denote by $C(\varphi)$ (resp. $A_1^i(\varphi)$, $A_2(\varphi)$) the set of all leafwise singularities (resp. leafwise Morse type singularities of index i, leafwise birth-death type singularities) of φ.

If we assume that $\varphi \in C^\infty(M,\mathbb{R})$ has only leafwise Morse type singularities, the map $m \in M \to (m, d_F\varphi(m)) \in T^*F$ will be transverse to the zero section of T^*F, and hence $C(\varphi)$ will be a closed q-dimensional ($q = \operatorname{codim}(F)$) submanifold of M transverse to the foliation. Since most interesting foliations do not admit closed transversals, we see that a good notion of foliated Morse function should allow degenerate leafwise singularities. Another reason to allow degenerate leafwise singularities is that the condition of being leafwise Morse is not generic. We can not, however, allow any kind of degenerate leafwise singularities otherwise the structure of $C(\varphi)$ will be very complicate.

It was noticed by Thom [9] that, for a generic φ, the critical locus $C(\varphi)$ is a smooth q-dimensional ($q = \operatorname{codim}(F)$) submanifold of M transverse to almost every leaf. In fact, one may assume generically not only that $C(\varphi)$ is a smooth q-dimensional submanifold of M but also that the restriction of φ to each leaf has only singularities of finite multiplicity. Indeed, the set of smooth functions satisfying these conditions is open and dense for the C^∞ topology by Tougeron's multijet transversality theorem [10]. Such a generic φ will be called a *generic foliated function.* Since the leafwise singularities of a generic foliated function φ are isolated in the leaf manifold, the set $A_1^i(\varphi)$ is a Borel transversal for any $i = 0, 1, \ldots, p$. Note however that the critical manifold $C(\varphi)$ of a generic foliated function φ may have a very complicate structure, since generic q-parameter families of smooth functions on $\mathbb{R}^p$ can have a zoo of complicated singularities for q large. Fortunately, it is possible to eliminate all leafwise singularities of multiplicity strictly larger than 2:

Theorem. [7], [3]. *Let (M,F) be a smooth p-dimensional foliation on a compact manifold M.*

(i) *Any function $\varphi_o \in C^\infty(M,\mathbb{R})$ can be C^o approximated by a smooth function $\varphi \in C^\infty(M,\mathbb{R})$ whose restriction to any leaf has only Morse or birth-death type singularities;*

(ii) *If $\operatorname{codim}(F) \leq \dim(F)$, any function $\varphi_o \in C^\infty(M,\mathbb{R})$ can be C^1 approximated by a smooth function $\varphi \in C^\infty(M,\mathbb{R})$ whose restriction to each leaf has only Morse or birth-death type singularities.*

Assertion (ii) was proved by Igusa [7]. Note that we can (cf. [6], Lemma 3.5, p. 313) choose an approximation φ of φ_o which has a normal form in the neighborhood of any point $m \in A_2(\varphi)$. The existence of a normal form (which will be described in the next subsection) uses the fact that generic q-parameter families $\varphi_t : u \in \overline{\mathbb{D}}^p \to \varphi_t(u) = \varphi(u,t) \in \mathbb{R}$ ($t \in \overline{\mathbb{D}}^q$) of smooth functions (see [6], p. 311 for the definition) form an open dense subset of the space of smooth families of generalized Morse functions on $\overline{\mathbb{D}}^p$, topologized as a subspace of $C^\infty(\overline{\mathbb{D}}^p \times \overline{\mathbb{D}}^q, \mathbb{R})$.

Assertion (i) is a consequence of a theorem of Eliashberg and Mishatchev [3]. They noticed that the higher leafwise singularities of φ inside any foliation chart $U \cong \mathbb{R}^p \times \mathbb{R}^q$ ($q = \operatorname{codim}(F)$) coincide with the singularities of the projection $(u,t) \in U \cap C(\varphi) \to t \in \mathbb{R}^q$. When the projection $(u,t) \in U \cap C(\varphi) \to t \in \mathbb{R}^q$ is folded, the functions $\varphi_t : u \to \varphi(u,t)$ have only Morse or birth-death type singularities. In the proof of (i), the approximation φ is constructed in such a way that, for any foliation chart $U \cong \mathbb{R}^p \times \mathbb{R}^q$, the projection on the transversal $(u,t) \in U \cap C(\varphi) \to t \in \mathbb{R}^q$ is folded. Finally, note that it was proved by Eliashberg that the C^o approximation φ of a generic foliated function φ_o can be chosen in such a way that $\Lambda(A_1^i(\varphi))$ is arbitrarily close to $\Lambda(A_1^i(\varphi_o))$ for $i = 0, 1, \ldots, p$. We shall note use this remark, which allows to get the foliated Morse inequalities for generic foliated functions from the case of generalized Morse functions as defined below:

Definition 1. Let (M, F) be a smooth p-dimensional foliation on a compact manifold M. A generic foliated function $\varphi \in C^\infty(M, \mathbb{R})$ is called a *generalized foliated Morse function* if its restriction to any leaf is a generalized Morse function.

1.3. Critical locus of a generalized foliated Morse function.

Let (M, F, Λ) be a measured p-dimensional smooth foliation on a compact manifold M and $\varphi \in C^\infty(M, \mathbb{R})$ a generalized foliated Morse function. From [10] and [6], we get the following properties:

(i) The set of leaves containing leafwise birth-death type singularities of φ is Λ-negligible;

(ii) The critical locus $C(\varphi)$ is a smooth compact q-dimensional manifold which decomposes into the disjoint union of the $A_1^i(\varphi)$ $(i = 0, 1, \ldots, p)$ and $A_2(\varphi)$;

(iii) Each $A_1^i(\varphi)$ is a (possibly empty) q-dimensional open submanifold of $C(\varphi)$ transverse to the foliation. Moreover, by the parametric Morse lemma (see for instance [5]), there exists for any compact subset K of $A_1^i(\varphi)$ a pair (E_+^{p-i}, E_-^i) of Euclidean vector bundles over K(corresponding to the eigenspace bundles of $D_F^2\varphi$), a foliated neighborhood U of K diffeomorphic to the zero section of K in $E_+^{p-i} \oplus E_-^i$ (foliated by the projection $p : U \to K$) such that we have for any $(t, u_+, u_-) \in U \subset E_+^{p-i} \oplus E_-^i$ lying over $t \in K$:

$$\varphi(t, u_+, u_-) = \varphi(t) + \|u_+\|^2 - \|u_-\|^2 ;$$

(iv) $A_2(\varphi)$ is a compact $(q-1)$-submanifold of $C(\varphi)$. Moreover, there exist for any $m \in A_2(\varphi)$ a compact neighborhood K of m in $A_2(\varphi)$, a neighborhood V of the zero section of K in the total space of the vector bundle $E_+^{p-i-1} \oplus E_-^i$ corresponding to the $(+)$ and $(-)$ eigenspaces of the leafwise Hessian of φ, and an open foliated neighborhood U of m isomorphic to a product $I_\epsilon \times V \times I_\delta$ $(\varepsilon, \delta > 0;\ I_\epsilon =]-\varepsilon, +\varepsilon[)$ such that:

(a) $\{0\} \times K \times I_\delta \subset I_\epsilon \times V \times I_\delta$ is a transversal for U;

(b) The plaque through $t = (0, k, s) \in K \times I_\delta$ identifies with:

$$P_t = \{u = (u_1, u_2, s) \in U \cong I_\epsilon \times V \times I_\delta \ ;\ u_2 \textit{ lies over } k\} ;$$

(c) For any $(u_1, u_2, s) \in U \cong I_\epsilon \times V \times I_\delta$ with u_2 lying over $k \in K$, we have:

$$\varphi(u_1, u_2, s) = \varphi(0, k, s) + \frac{u_1^3}{3} + u_1 s - \sum_{2 \leq j \leq i+1} \frac{u_j^2}{2} + \sum_{i+2 \leq j \leq p} \frac{u_j^2}{2}.$$

2. Proof of the foliated Morse inequalities

Denote by (M, F, Λ) a measured p-dimensional smooth foliation on a compact manifold M and let $\varphi \in C^\infty(M, \mathbb{R})$ be a generalized foliated Morse function. For simplicity, we shall assume that the holonomy groupoid of the foliation is Hausdorff, that Λ is absolutely continuous with respect to the Lebesgue class and that Λ-almost every leaf has no holonomy.

2.1. Twisted Laplacian

For $i = 0, 1, \ldots, p$ and any leaf L, denote by d^i_L the de Rham differential on forms of degree i on L. Let $\tau > 0$ be a scale parameter and consider the twisted de Rham differential $d^i_{\tau,L}$ defined on forms of degree i on L by the formula $d^i_{\tau,L}(\omega) = e^{-\tau\varphi} d^i_L(e^{\tau\varphi}\omega)$. For any Euclidean structure on F (we shall only consider Euclidean structures on F that are associated to metrics on M), the *twisted Laplacian* $\Delta^i_{\tau,L}$ is defined on $L^2(L, \Lambda^i T^\star F)$ by:

$$\Delta^i_{\tau,L} = d^{i-1}_{\tau,L}\left(d^{i-1}_{\tau,L}\right)^\star + \left(d^i_{\tau,L}\right)^\star d^i_{\tau,L}.$$

Since $\Delta^i_\tau = \left(\Delta^i_{\tau,L}\right)_L$ is a leafwise elliptic operator, we get $\dim_\Lambda(\mathrm{Ker}(\Delta^i_\tau)) < +\infty$ by [1].

Proposition 1. (*cf.* [2], *Lemma* 5.2). *For any $i = 0, 1, \ldots, p$, we have*

$$\beta_i = \dim_\Lambda(\mathrm{Ker}(\Delta^i_\tau)).$$

Proof. Consider the operator $T^i_\tau = \left(T^i_{\tau,L}\right)_L$ defined by $T^i_{\tau,L}(\omega) = e^{-\tau\varphi}\omega$. This operator belongs to the von Neumann algebra of the foliation and maps $\mathrm{Ker}(d^i) = \mathrm{Ker}(\Delta^i) \oplus \overline{\Im(d^{i-1})}$ into $\mathrm{Ker}(d^i_\tau) = \mathrm{Ker}(\Delta^i_\tau) \oplus \overline{\Im(d^{i-1}_\tau)}$ so that it decomposes in the form $T^i_\tau = \begin{pmatrix} U^i_\tau & 0 \\ * & V^i_\tau \end{pmatrix}$ with respect to these direct sums of Hilbert spaces. Since V^i_τ and T^i_τ are invertible, the field of operators $U^i_{\tau,L} : \mathrm{Ker}(\Delta^i_L) \to \mathrm{Ker}(\Delta^i_{\tau,L})$ is an isomorphism of square integrable fields of Hilbert spaces and hence $\beta_i = \dim_\Lambda(\mathrm{Ker}(\Delta^i)) = \dim_\Lambda(\mathrm{Ker}(\Delta^i_\tau))$. □

2.2. Twisted complex

Let $E > 0$ be a fixed "energy" and denote by $V^i_{\tau,L}$ the image in $L^2(L, \Lambda^i T^\star F)$ of the spectral projection of the operator $\Delta^i_{\tau,L}$ corresponding to the interval $]-\infty, E]$. The operator $d^i_{\tau,L}$ maps $V^i_{\tau,L}$ into $V^{i+1}_{\tau,L}$ so that its restriction $\delta^i_{\tau,L} = d^i_{\tau,L} \mid V^i_{\tau,L} : V^i_{\tau,L} \to V^{i+1}_{\tau,L}$ to $V^i_{\tau,L}$ defines a bounded field $\delta^i_\tau = \left(\delta^i_{\tau,L}\right)_L$ of operators from V^i_τ to V^{i+1}_τ. Since $\delta^{i+1}_{\tau,L} \circ \delta^i_{\tau,L} = 0$, we get a complex $\delta^i_\tau : V^i_\tau \to V^{i+1}_\tau$ called the *twisted complex.*

Proposition 2. (*cf.* [2], *Lemma* 11.2.4). *For $i = 0, 1, \ldots, p$, we have:*

$$\sum_{j=0}^{j=i}(-1)^j \beta_{i-j} \le \sum_{j=0}^{j=i}(-1)^j \dim_\Lambda(V^{i-j}_\tau) \text{ with equality for } i = p.$$

Proof. This follows from the relation:

$$(1)\ \sum_{j=0}^{j=i}(-1)^j\beta_{i-j} + \dim_\Lambda(\overline{\Im(\delta_\tau^i)}) = \sum_{j=0}^{j=i}(-1)^j \dim_\Lambda(V_\tau^{i-j})$$

for any $i = 0, 1, \ldots, p$, where $\delta_\tau^p = 0$. Relation (1) is proved by induction on i, from the orthogonal decomposition $V^i_{\tau,L} = \mathrm{Ker}(\Delta^i_{\tau,L}) \oplus \overline{\Im(\delta^{i-1}_{\tau,L})} \oplus \overline{\Im\left(\delta^i_{\tau,L}\right)^\star}$. This decomposition is obtained by classical arguments of Hodge's theory applied to the twisted complex. For $i = 0$, it reduces to $V^o_{\tau,L} = \mathrm{Ker}(\Delta^o_{\tau,L}) \oplus \overline{\Im\left(\delta^o_{\tau,L}\right)^\star}$ and gives immediately (1). For $i = 1$, this decomposition implies that $\dim_\Lambda(V^1_\tau) = \beta_1 + \dim_\Lambda(\overline{\Im(\delta^o_\tau)}) + \dim_\Lambda(\overline{\Im(\delta^1_\tau)^\star})$. Subtracting the equality $\dim_\Lambda(V^o_\tau) = \beta_o + \dim_\Lambda(\overline{\Im(\delta^o_\tau)^\star})$ from this relation, we get (1) for $i = 1$ since the initial support of the operator δ^o_τ has the same Murray-von Neumann dimension as its final support. By iterating this argument, we get (1). □

Proposition 2 shows that the proof of the foliated Morse inequalities reduces to the following result:

Theorem 1. *(cf. [2], Theorem 8.2). Let (M, F, Λ) and φ as above. For any fixed $E > 0$ and any $\epsilon > 0$, there exists a metric on M such that we have, for any $i = 0, 1, \ldots, p$ and τ large enough:*

$$\Lambda\left(A^i_1(\varphi)\right) - \epsilon \le \dim_\Lambda(V^i_\tau) \le \Lambda\left(A^i_1(\varphi)\right) + \epsilon.$$

To prove this theorem, fix $\eta > 0$ and choose an open neighborhood V of $A_2(\varphi)$ such that $\Lambda(V \cap C(\varphi)) \le \eta$. This is possible since the restriction of φ to Λ-almost leaf is a Morse function. We can choose V to be a finite union of foliated charts (called "birth-death charts") in which φ has the normal form quoted in section 1.3. For any $i = 0, 1, \ldots, p$, set $C^i_\eta = A^i_1(\varphi) \setminus \left(V \cap A^i_1(\varphi)\right)$. We thus define a compact subset C^i_η of $A^i_1(\varphi)$ such that $\Lambda(A^i_1(\varphi) \setminus C^i_\eta) \le \eta$. By the Morse lemma with parameters, we can choose a foliated neighborhood N^i_η of C^i_η disjoint from $A_2(\varphi)$ and a metric g_η on M with the following property: any $m \in C^i_\eta$ is contained in a foliation chart $U \cong \mathbb{R}^p \times \mathbb{R}^q$ with local coordinates $(u_1, \ldots, u_p, t_1, \ldots, t_q)$ such that:

(a) $C^i_\eta \cap U = \{(u,t) \in U;\ u = 0\}$;
(b) The restriction of the metric g_η to each plaque $t = C^{st}$ of U is the standard metric $(du_1)^2 + (du_2)^2 + \cdots + (du_p)^2$;
(c) We have $\varphi(u,t) = \varphi(0,t) - \sum_{1\le j\le i} \frac{u_j^2}{2} + \sum_{i+1\le j\le p} \frac{u_j^2}{2}$ for any $(u,t) \in U$ (in local coordinates).

Let us cover each C^i_η by a finite number of foliation's charts (called "Morse charts") satisfying the above conditions.

2.3. Proof of $\Lambda\left(\mathbf{A}_1^{\mathbf{i}}(\varphi)\right) - \eta \leq \dim_{\Lambda}(\mathbf{V}_\tau^{\mathbf{i}})$ for τ large enough

For any point $m \in C_\eta^i$, let U be a foliation chart containing m as above. From conditions (a), (b), (c), we get the following local expression for the twisted Laplacian:

$$\Delta_{\tau,L}^i(f du^I) = \sum_{j=1}^{j=p} \left\{ -\frac{\partial^2 f}{\partial u_j^2} + \tau^2 u_j^2 f + \tau \epsilon_j \epsilon_j^I f \right\} du^I$$

for each plaque L of U and any smooth compactly supported function f on L. Here, I is a multi-index such that $|I| = i$, ϵ_j satisfies $\epsilon_j = -1$ if $j \leq i$ and $\epsilon_j = 1$ if $j > i$, and $\epsilon_j^I = 1$ if $j \in I$ while $\epsilon_j^I = -1$ if $j \notin I$. Since $\Delta_{\tau,L}^i$ is a tensor product of harmonic oscillators, it is easy to show that

$$Sp(\Delta_{\tau,L}^i) = \{\tau\lambda_{I,N};\ |I| = i \text{ and } N = (N_1, \ldots, N_p) \in \mathbb{N}^p\},$$

where $\lambda_{I,N} = \sum_{j=1}^{j=p} \left[(1 + 2N_j) + \epsilon_j \epsilon_j^I\right]$. It follows that the first eigenvalue of $\Delta_{\tau,L}^i$ (corresponding to $I = (1, 2, \ldots, i)$ and $N = (0, 0, \ldots, 0)$) is equal to 0 while the other are larger than 2τ. The normalized eigenform ω_τ of degree i corresponding to the first eigenvalue writes in local coordinates $\omega_\tau(u) = C\tau^{\frac{p}{4}} \exp\left(-\frac{\tau}{2}\sum_{j=1}^{j=p} u_j^2\right) du^1 \wedge du^2 \wedge \cdots \wedge du^2$, where C is some normalizing constant. Let us multiply this form by a cut-off function of the form $u \to \theta(\frac{|u|^2}{\gamma})$, where $\theta : \mathbb{R} \to \mathbb{R}$ is a positive smooth function equal to 1 (resp. to 0) on $\left[-\frac{1}{2}, \frac{1}{2}\right]$ (resp. outside $[-1, 1]$) and $\gamma > 0$ a small parameter. We get a form $\omega_{\tau,\gamma}$ that we can plug on any plaque of the chart U for γ small enough, and which satisfies by direct computation:

$$\|\omega_{\tau,\gamma}\|_2^2 = 1 + O(e^{-\frac{\gamma}{8}\tau}) \text{ and } \langle \Delta_{\tau,L}^i(\omega_{\tau,\gamma}) \mid \omega_{\tau,\gamma}\rangle = O(\gamma^{\frac{p-3}{2}} \tau^{\frac{p+1}{2}} e^{-\gamma\tau})$$

for τ large enough. This allows to construct a measurable map $m \in C_\eta^i \to \omega_{\tau,\gamma}(m) \in V_{\tau,L_m}^i$ where $\omega_{\tau,\gamma}(m)$ is a smooth normalized form of degree i supported by a small neighborhood of m in the leaf L_m through m. We deduce that $(l^2(C_\eta^i \cap L))_L$ can be viewed as a subfield of the square integrable field $(V_{\tau,L}^i)_L$, and hence:

$$\Lambda\left(A_1^i(\varphi)\right) - \eta \leq \Lambda(C_\eta^i) = \dim_\Lambda((l^2(C_\eta^i \cap L))_L) \leq \dim_\Lambda(V_\tau^i).$$

2.4. Proof of $\dim_{\Lambda}(\mathbf{V}_\tau^{\mathbf{i}}) \leq \Lambda\left(\mathbf{A}_1^{\mathbf{i}}(\varphi)\right) + \mathbf{C}^{\mathbf{st}}\eta$ for τ large enough

To prove this inequality, we shall construct a random field $(H_{\tau,L})_L$ of Hilbert spaces, $H_{\tau,L} \subset L^2(L, \Lambda^i T^\star F)$, such that we have:

(i) $\dim_\Lambda(H_\tau) \leq \Lambda(A_1^i(\varphi)) + q\eta$ for $\tau \geq \tau(\eta)$,

(ii) $\langle \Delta_{\tau,L}^i(\omega_L) \mid \omega_L\rangle \geq C\tau^{\frac{2}{3}} \|\omega_L\|_2^2$ for any $\omega = (\omega_L)_L$ orthogonal to H_τ and Λ-almost leaf L.

Here, C is a constant, q a positive integer and $\tau(\eta) > 0$ a number depending only on $\eta > 0$. These conditions imply immediately that $\dim_\Lambda(V_\tau^i) \leq \Lambda(A_1^i(\varphi)) + q\eta$ for $\tau \geq \tau(\eta)$.

To construct this random field of Hilbert spaces, fix a leaf L without birth-death type leafwise singularities and recall that the critical manifold $C(\varphi)$ is covered by a finite number of Morse or birth-death foliation charts. Choose a smooth partition

$$1 = \sum_{\alpha \in A} \chi_{\alpha,\tau}^2 + \sum_{\beta \in B} \chi_\beta^2 + \chi_{\infty,\tau}^2$$

of the unit on L, where the smooth functions $\chi_{\alpha,\tau}^2$ (resp. χ_β^2) are compactly supported by the sections of L with the Morse charts $(U_\alpha)_{\alpha \in A}$ (resp. with the birth-death charts $(U_\beta)_{\beta \in B}$). Note that, despite the number of Morse or birth-death charts is finite, the leaf L may intersect each chart an infinite number of time. By using a formula of the type $\chi_{\alpha,\tau}(u) = \chi(\tau^{\frac{2}{5}}(u - c_\alpha))$ where $c_\alpha = \{L \cap U_\alpha \cap A_1(\varphi)\}$ and $\chi : \mathbb{R}^p \to R$ is a positive smooth function equal to 1 on $[-1, 1]^p$ and to 0 outside $[-2, 2]^p$, we may assume that $\left\| |\nabla_L(\chi_{\alpha,\tau})|^2 \right\|_\infty \leq C\tau^{\frac{4}{5}}$ for any $\alpha \in A$, where C is some fixed constant. In the same way, we may assume that $\left\| |\nabla_L(\chi_\beta)|^2 \right\|_\infty \leq C$ for any $\beta \in B$. This partition of unity allows us, by using the IMS formula:

$$f^2 \Delta_{\tau,L}^i + \Delta_{\tau,L}^i f^2 - 2f\Delta_{\tau,L}^i = -2\left|\nabla_L(f)\right|^2,$$

to decompose $\Delta_{\tau,L}^i$ in the following form:

$$\Delta_{\tau,L}^i = A_L + B_L + R_L + S_L,$$

where $A_L = \sum_{\alpha \in A} \chi_{\alpha,\tau} \Delta_{\tau,L}^i \chi_{\alpha,\tau}$, $B_L = \sum_{\beta \in B} \chi_\beta \Delta_{\tau,L}^i \chi_\beta$, $C_L = \chi_{\infty,\tau} \Delta_{\tau,L}^i \chi_{\infty,\tau}$ and D_L is the multiplication operator by the function:

$$\phi_L = -\left|\nabla_L(\chi_{\infty,\tau})\right|^2 - \sum_{\alpha \in A} \left|\nabla_L(\chi_{\alpha,\tau})\right|^2 - \sum_{\beta \in B} \left|\nabla_L(\chi_\beta)\right|.^2$$

From the controls imposed on $\left|\nabla_L(\chi_{\alpha,\tau})\right|^2$ and $\left|\nabla_L(\chi_\beta)\right|^2$, we get:

Claim 1. *There exists* $C > 0$ *such that we have, for any* $\omega \in L^2(L, \Lambda^i T^\star F)$:

$$\langle D_L \omega \mid \omega \rangle \geq -C\tau^{\frac{4}{5}} \sum_{\alpha \in A} \left\|\chi_{\alpha,\tau}\omega\right\|_2^2 - C \sum_{\beta \in B} \left\|\chi_\beta \omega\right\|_2^2 - C\tau^{\frac{4}{5}} \left\|\chi_{\infty,\tau}\omega\right\|_2^2.$$

On the other hand, the analysis of the bottom of the spectrum of $\Delta_{\tau,L}^i$ shows that we have $\chi_{\alpha,\tau}\Delta_{\tau,L}^i \chi_{\alpha,\tau} \geq C^{st}\tau\chi_{\alpha,\tau}^2$ if the index α corresponds to a point $c_\alpha \in A_1^j$ of Morse index $j \neq i$. The case $j = i$ corresponds to indices $\alpha \in A$ such that $c_\alpha \in C_\eta^i$. In this case, the eigenstate of zero energy of the twisted Laplacian $\Delta_{\tau,L}^i$ localized in U_α yield a Borel field $(\omega_{\alpha,\tau,L})_L$ of square integrable forms of degree i on the plaques of U_α. Since the second eigenvalue of $\Delta_{\tau,L}^i$ is larger than 2τ, there exists by the minimax principle a constant $C > 0$ such that we have

$\langle(\chi_{\alpha,\tau}\Delta^i_{\tau,L}\chi_{\alpha,\tau})\omega \mid \omega\rangle \geq C\tau \|\chi_{\alpha,\tau}\omega\|_2^2$ for any $\omega \in L^2(L, \Lambda^i T^\star F)$ orthogonal to $\omega_{\alpha,\tau,L}$. Denote by $H^1_\tau = (H^1_{\tau,L})_L$ the Borel field of Hilbert spaces generated by the $(\omega_{\alpha,\tau,L})_L$ where $\alpha \in A$ is such that $c_\alpha \in C^i_\eta$. We have:

Claim 2. $H^1_\tau = (H^1_{\tau,L})_L$ is a square integrable field of Hilbert spaces whose Murray-von Neumann dimension satisfies $\dim_\Lambda(H^1_\tau) \leq \Lambda(A^i_1(\varphi)) - \eta$ and such that

$$\langle A_L\omega \mid \omega\rangle \geq C^{st}\tau \sum_{\alpha\in A} \|\chi_{\alpha,\tau}\omega\|_2^2$$

for any $\omega \in L^2(L, \Lambda^i T^\star F)$ orthogonal to $H^1_{\tau,L}$.

On the other hand, consider the decomposition $\Delta^i_{\tau,L} = \Delta^i_L + \tau^2 |\nabla_L(\varphi)|^2 + \tau Q^i_L$ of $\Delta^i_{\tau,L}$ used by Witten's [11] to prove the classical Morse inequalities (see also [4]). Here, the operator Q^i_L is given by $Q^i_L = \sum_{k,l} \nabla^2_{k,l}(\varphi)\,[\mathrm{ext}(du_k), \mathrm{int}(du_l)]$ in local coordinates. Combining the fact that Q^i_L is lower bounded, the positivity of Δ^i_L and the lower bound $|\nabla_L(\varphi)(m)|^2 \geq C^{st} > 0$ for m away from the critical manifold $C(\varphi)$, we get:

Claim 3. There exists $C > 0$ such that we have $\langle C_L\omega \mid \omega\rangle \geq C\tau \|\chi_{\infty,\tau}\omega\|_2^2$ for any $\omega \in L^2(L, \Lambda^i T^\star F)$.

Finally, by using the local expression of φ in a "birth-death chart", we can show that there exists an integer $q > 0$ and, for any birth-death chart U_β, Borel fields $\omega_{\beta,k,\tau} = (\omega_{\beta,k,\tau,L})_L$ $(1 \leq k \leq q)$ of square integrable forms of degree i indexed by the points of $U_\beta \cap V \cap A^i_1(\varphi)$ and supported by the plaques of U_β, such that we have $\langle(\chi_\beta\Delta^i_{\tau,L}\chi_\beta)\omega \mid \omega\rangle \geq C^{st}\tau^{\frac{2}{3}} \|\chi_\beta\omega\|_2^2$ for any $\omega \in L^2(L, \Lambda^i T^\star F)$ orthogonal to the $\omega_{\beta,k,\tau}$'s. This assertion reduces in fact to the following theorem:

Theorem 2. *(cf. [2], Theorem 10.1). Let $\lambda, \tau > 0$ with $0 < \lambda < \frac{1}{4}$ and denote by $Q^\pm_{\tau,\lambda}$ the quadratic form on $H^1(]-1,1[)$ defined by*

$$Q^\pm_{\tau,\lambda}(f) = \int_{-1}^{1} \left\{|f'(u)|^2 + \left[\tau^2(u^2-\lambda)^2 \pm 2\tau u\right] |f(u)|^2\right\} du.$$

Then, there exists an integer $m > 0$, a real $\tau_o > 0$ and, for any $\tau \geq \tau_o$, m continuous functions $f_1, f_2, \ldots, f_m$ with compact support in $]-1,1[$ such that

$$Q^\pm_{\tau,\lambda}(f) \geq 10\tau^{\frac{2}{3}} \int_{-1}^{1} |f(u)|^2\, du$$

for any $f \in H^1(]-1,1[)$ orthogonal to the f_i's.

Denote by $H^2_\tau = (H^2_{\tau,L})_L$ the Borel field of Hilbert spaces generated by the $(\omega_{\beta,k,\tau,L})_L$ where $\beta \in B$, and recall that $\Lambda(V \cap A^i_1(\varphi)) \leq \eta$. Since the $\omega_{\beta,k,\tau}$ are indexed by k $(1 \leq k \leq q)$ and by the points of $U_\beta \cap V \cap A^i_1(\varphi)$, we have $\dim_\Lambda(H^2_\tau) \leq q\eta$.

This implies:

Claim 4. *The square integrable field* $H^2_\tau = (H^2_{\tau,L})_L$ *of Hilbert spaces satisfies:*

(i) $\dim_\Lambda(H^2_\tau) \leq q\eta$;

(ii) $\langle B_L\omega \mid \omega\rangle \geq C^{st}\tau^{\frac{2}{3}} \sum_{\beta\in B} \|\chi_\beta\omega\|_2^2$ *for any* $\omega \in L^2(L, \Lambda^i T^\star F)$ *orthogonal to* $H^2_{\tau,L}$.

Set $H_\tau = H^1_\tau \oplus H^2_\tau$. We thus define a random field $(H_{\tau,L})_L$ of Hilbert spaces such that $\dim_\Lambda(H_\tau) \leq \Lambda(A^i_1(\varphi)) + q\eta$ for τ large enough. By claims 1 to 4, we have $\langle\Delta^i_{\tau,L}(\omega_L) \mid \omega_L\rangle \geq C\tau^{\frac{2}{3}} \|\omega_L\|_2^2$ for any $\omega = (\omega_L)_L$ orthogonal to H_τ and the proof of 2.4 is complete. This achieves the proof of the Morse inequalities.

3. Applications

In analogy with the case of compact manifolds, we can use a *"foliated lacunary Morse principle"* (cf. [8], page 31) to compute the Betti numbers of several measured foliations. We can also use the foliated Morse inequalities to get the existence of compact leaves in 2-dimensional foliations. For instance, we have:

Corollary 1. *Let* (M, F, Λ) *be a 2-dimensional measured foliation on a compact manifold* M. *If there exists a foliated generalized Morse function* φ *on* M *such that* $\Lambda(A^1_1(\varphi)) < \Lambda(A^o_1(\varphi)) + \Lambda(A^2_1(\varphi))$, *the* Λ*-measure of the set of compact leaves is non zero.*

Proof. By the Morse inequalities, we get $2\beta_o - \beta_1 = \beta_o - \beta_1 + \beta_2 = c_o - c_1 + c_2 > 0$. If the set of compact leaves is Λ-negligible, we have $\beta_o = 0$ and hence $\beta_1 < 0$, a fact which is absurd. □

Corollary 2. *Let* (M, F, Λ) *be a 2-dimensional measured foliation on a compact manifold* M. *For any foliated generalized Morse function* φ *on* M *such that* $\Lambda(A^1_1(\varphi)) = 0$, *we have* $\Lambda(A^o_1(\varphi)) = \Lambda(A^2_1(\varphi))$. *Moreover, if this number is non zero, the foliation has a non* Λ*-negligible set of compact leaves.*

Proof. By the Morse inequalities, we get $\beta_1 \leq c_1 = 0$ and hence $\beta_1 = 0$. It follows that $\beta_o = c_o$ and hence $c_2 + c_o = \beta_2 - \beta_1 + \beta_o = 2\beta_o \leq 2c_o$. We deduce that $c_2 \leq c_o \leq \beta_o = \beta_2 \leq c_2$ and hence $c_o = c_2$. The conclusion immediately follows. □

References

[1] A. Connes, *Sur la théorie non commutative de l'intégration.* Lectures Notes in Math., Berlin, New-York, Springer, 1978.

[2] A. Connes and T. Fack, *Morse inequalities for measured foliations.* Preprint Université Claude-Bernard Lyon 1, 1991.

[3] Y.M. Eliashberg and N.M. Mishachev, *Wrinkling of smooth mappings. II Wrinkling of embeddings and K. Igusa's theorem.* Topology **39** (2000), 711–732.

[4] B. Helffer and J. Sjöstrand, *Puits multiples en mécanique semi-classique, IV, Etude du complexe de Witten.* Comm. in Partial Differential Equations **10** (1985), 245–340.

[5] L. Hörmander, *The Analysis of Linear Partial Differential operators, III.* Grundlehren der mathematischen Wissenschaften, **274**, Springer-Verlag, Berlin, Heidelberg, 1985.

[6] K. Igusa, *The stability theorem for smooth pseudoisotopies.* K-Theory **2** (1988), 1–355.

[7] K. Igusa, *C^1 local parametrized Morse theory.* Topology and its Applications **36** (1990), 209–231.

[8] J. Milnor, *Morse theory.* Annals of Math. Studies, **51**, Princeton University Press, 1969.

[9] R. Thom, *Généralisation de la théorie de Morse aux variétés feuilletées.* Ann. Inst. Fourier, Grenoble **14** (1964), 173–189.

[10] J.C. Tougeron, *Idéaux de fonctions différentiables.* Ann. Inst. Fourier, Grenoble **18** (1968), 177–240.

[11] E. Witten, *Supersymmetry and Morse Theory.* J. Diff. Geom. **17** (1982), 661–692.

Alain Connes
Institut des Hautes Etudes Scientifiques,
35 route de Chartes
F-91440 Bures sur Yvette, France
e-mail: `connes@ihes.fr`

Thierry Fack
Université Claude-Bernard Lyon 1

et

UMR CNRS 5208 (Institut Camille Jordan)
UFR de mathématiques,
Bâtiment 101 (Doyen Jean Braconnier),
43 boulevard du 11 novembre 1918
F-69622 Villeurbanne Cedex, France
e-mail: `thierry.fack@univ-lyon1.fr`

C^*-algebras and Elliptic Theory
Trends in Mathematics, 73–128

Index Theory for Generalized Dirac Operators on Open Manifolds

Jürgen Eichhorn

Abstract. In the first part of the paper, we give a short review of index theory on open manifolds. In the second part, we establish a general relative index theorem admitting compact topological perturbations and Sobolev perturbations of all other ingredients.

Mathematics Subject Classification (2000). Primary 58J20; Secondary 58J35.

Keywords. Relative index, heat kernel estimates.

1. Introduction

Let (M^n, g) be closed, oriented, $(E, h_E), (F, h_F) \longrightarrow M^n$ smooth vector bundles, $D : C^\infty(E) \longrightarrow C^\infty(F)$ an elliptic differential operator. Then $L_2(E) \supset \mathcal{D}_{\overline{D}} \xrightarrow{\overline{D}} L_2(F)$ is Fredholm, i.e., there exists $P : L_2(F) \longrightarrow L_2(E)$ s.t. $PD - \mathrm{id} = K_1$, $DP - \mathrm{id} = K_2$, K_i integral operators with C^∞ kernel $\mathcal{K}_i$ and hence compact. It follows $\dim\ker D, \dim\operatorname{coker} D < \infty$, $\mathrm{ind}_a D = \dim\ker D - \dim\operatorname{coker} D$ is well defined and there arise the question to calculate $\mathrm{ind}_a D$. The answer is given by the seminal Atiyah–Singer index theorem

Theorem 1.1.
$$\mathrm{ind}_a D = \mathrm{ind}_t D,$$
where
$$\mathrm{ind}_t D = \langle \mathrm{ch}\, \sigma(D) \mathcal{T}(M), [M] \rangle. \qquad \square$$

Assume now (M^n, g) open, E, F, D as above. K_1, K_2 still exist as operators with a smooth kernel where in good cases one can achieve that the support of $\mathcal{K}_i$ is located near the diagonal.

But there arise several troubles.

1) If K_i bounded is achieved then K_i must not be compact.
2) If K_i would be compact then $\mathrm{ind}_a D$ would be defined.
3) If $\mathrm{ind}_a D$ would be defined then $\mathrm{ind}_t D$ must not be defined.

4) If $\text{ind}_a D$, $\text{ind}_t D$ (as above) would be defined then they must not coincide. There are definite counterexamples.

There are 3 ways out from this difficult situation.

1) One could ask for special conditions in the open case under which an elliptic D is still Fredholm, then try to establish an index formula and finally present applications. These conditions could be conditions on D, on M and E or a combination of both. In [1] the author formulates an abstract (and very natural) condition for the Fredholmness of D and assumes nothing on the geometry. But in all substantial applications this condition can be assured by conditions on the geometry. The other extreme case is that discussed in [7], [10], [9], where the authors consider the L_2-index theorem for locally symmetric spaces. Under relatively restricting conditions concerning the geometry and topology at infinity the Fredholmness and an index theorem are proved in [5] and [6].
2) One could generalize the notion of Fredholmness (using other operator algebras) and then establish a meaningful index theory with applications. The discussion of these both approaches will be the content of the next paragraph.
3) Another approach will be relative index theory which is less restrictive concerning the geometrical situation (compared with the absolute case) but its outcome are only statements on the relative index, i.e., how much the analytical properties of D differ from those of D'. This approach will be discussed in detail in Section 3.
4) For open coverings $(\tilde{M}, \tilde{g})$ of closed manifolds (M^n, g) and lifted D there is an approach which goes back to Atiyah, (cf. [2]). This has been further elaborated by Cheeger, Gromov and others. The main point is that all considered (Hilbert-) modules are modules over a von Neumann algebra and one replaces the usual trace by a von Neumann trace. We will not dwell on this approach since there is a well established highly elaborated theory. Moreover special features of openness come not into. The openness is reflected by the fact that all modules under consideration are modules over the von Neumann algebra $\mathcal{N}(\pi)$, $\pi = \text{Deck}(\tilde{M} \longrightarrow M)$.

Section 2 presents a certain review of important absolute index theorems and Section 3 gives an outline including all essential proofs of the general relative index theory of the author.

2. Fredholmness, its generalization and group actions

This section is a brief review of absolute index theorems under additional strong assumptions. It shows that these approaches are successful only in special situations. In Section 3 we will establish very general relative index theorems.

We start with the first approach and with the question which elliptic operators over open manifolds are Fredholm in the classical sense above. Let (M^n, g) be open, oriented, complete, $(E, h) \longrightarrow (M^n, g)$ be a Hermitian vector bundle with involution $\tau \in \operatorname{End}(E)$, $E = E^+ \oplus E^-$, $D : C^\infty(E) \longrightarrow C^\infty(E)$ an essentially self-adjoint first-order elliptic operator satisfying $D\tau + \tau D = 0$. We denote $D^\pm = D|_{C^\infty(E^\pm)}$. Then we can write as usual

$$D = \begin{pmatrix} 0 & D^- \\ D^+ & 0 \end{pmatrix} : \begin{array}{c} C^\infty(E^+) \\ \oplus \\ C^\infty(E^-) \end{array} \longrightarrow \begin{array}{c} C^\infty(E^+) \\ \oplus \\ C^\infty(E^-) \end{array} . \tag{2.1}$$

The index $\operatorname{ind}_a D$ is defined as

$$\operatorname{ind}_a D := \operatorname{ind}_a D^+ := \dim \ker D^+ - \dim \operatorname{coker} D^+ = \dim \ker D^+ - \dim \ker D^- \tag{2.2}$$

if these numbers would be defined. Denote by $\Omega^{2,i}(E,D)$ the Sobolev space of order i of sections of E with D as generating differential operator. We essentially follow [1].

Proposition 2.1. *The following statements are equivalent*

a) *D is Fredholm.*

b) $\dim \ker D < \infty$ *and there is a constant $c > 0$ such that*

$$|D\varphi|_{L_2} \geq c \cdot |\varphi|_{L_2}, \quad \varphi \in (\ker D)^\perp \cap \Omega^{2,1}(E,D), \tag{2.3}$$

where $(\ker D)^\perp \equiv \mathcal{H}^\perp$ is the orthogonal complement of $\mathcal{H} = \ker D$ in $L_2(E)$.

c) *There exists a bounded non-negative operator $P : \Omega^{2,2}(E,D) \longrightarrow L_2(E)$ and bundle morphism $R \in C^\infty(\operatorname{End} E)$, R positive at infinity (i.e., there exists a compact $K \subset M$ and a $k > 0$ s. t. pointwise on $E|_{M\setminus K}$, $R \geq k$), such that on $\Omega^{2,2}(E,D)$*

$$D^2 = P + R. \tag{2.4}$$

d) *There exist a constant $c > 0$ and compact $K \subset M$ such that*

$$|D\varphi|_{L_2} \geq c \cdot |\varphi|, \quad \varphi \in \Omega^{2,1}(E,D), \quad \operatorname{supp}(\varphi) \cap K = \emptyset. \tag{2.5}$$

□

The main task now is to establish a meaningful index theorem. This has been performed in [1].

Theorem 2.2. *Let (M^n, g) be open, complete, oriented, $(E, h, \tau) = (E^+ \oplus E^-, h) \longrightarrow (M^n, g)$ a $\mathbb{Z}_2$-graded Hermitian vector bundle and $D : C_c^\infty(E) \longrightarrow C_c^\infty(E)$ first-order elliptic, essentially self-adjoint, compatible with the $\mathbb{Z}_2$-grading (i.e., supersymmetric), $D\tau + \tau D = 0$. Let $K \subset M$ be a compact subset such that 2.1a) for K is satisfied, and let $f \in C^\infty(M, \mathbb{R})$ be such that $f = 0$ on $U(K)$ and $f = 1$ outside a compact subset. Then there exists a volume density ω and a contribution I_ω such that*

$$\operatorname{ind}_a \overline{D}^+ = \int_M (\omega(1 - f(x)) \operatorname{dvol}_x(g) + I_\omega, \tag{2.6}$$

where ω has an expression locally depending on D and I_ω depends on D and f restricted to $\Omega = M \setminus K$. □

Until now the differential form $\omega\, \mathrm{dvol}_x(g)$ is mystery. One would like to express it by well-known canonical terms coming, e.g., from the Atiyah–Singer index form $\mathrm{ch}\, \sigma(D^+) \cup \mathcal{T}(M)$, where $\mathcal{T}(M)$ denotes the Todd genus of M. In fact this can be done.

Index Theorem 2.3 *Let (M^n, g) be open, oriented, complete, $(E, h, \tau) \longrightarrow (M^n, g)$ a $\mathbb{Z}_2$-graded Hermitian vector bundle, $D : C_c^\infty(E) \longrightarrow C_c^\infty(E)$ a first-order elliptic essentially self-adjoint supersymmetric differential operator, $D\tau + \tau D = 0$, which shall be assumed to be Fredholm. Let $K \subset M$ compact such that* 2.1d) *is satisfied. Then*

$$\mathrm{ind}_a D^+ = \int_K \mathrm{ch}\, \sigma(D^+) \cup \mathcal{T}(M) + I_\Omega, \tag{2.7}$$

where $\mathrm{ch}\, \sigma(D^+) \cup \mathcal{T}(M)$ is the Atiyah–Singer index form and I_Ω is a bounded contribution depending only on $D|_\Omega$, $\Omega = M \setminus K$. □

Remarks 2.4.

a) As we already mentioned, $\mathbb{Z}_2$-graded Clifford bundles and associated generalized Dirac operators D such that in $D^2 = \Delta^E + \mathcal{R}$, $\mathcal{R} \geq c \cdot \mathrm{id}$, $c > 0$, outside some compact $K \subset M$, yield examples for Theorem 2.3. A special case is the Dirac operator over a Riemannian spin manifold with scalar curvature $\geq c > 0$ outside $K \subset M$.
b) Much more general perturbations than compact ones will be considered in Section 3. □

The other case of a very special class of open manifolds are coverings $(\tilde{M}, \tilde{g})$ of a closed manifold (M^n, g). Let $E, F \longrightarrow (M^n, g)$ be Hermitian vector bundles over the closed manifold (M^n, g). $D : C^\infty(E) \longrightarrow C^\infty(F)$ be an elliptic operator, $(\tilde{M}, \tilde{g}) \longrightarrow (M, g)$ a Riemannian covering, $\tilde{D} : C_c^\infty(\tilde{E}) \longrightarrow C_c^\infty(\tilde{F})$ the corresponding lifting and $\Gamma = \mathrm{Deck}\, (\tilde{M}^n, \tilde{g}) \longrightarrow (M^n, g)$. The actions of Γ and $\tilde{D}$ commute. If $P : L_2(\tilde{M}, \tilde{E}) \longrightarrow \mathcal{H}$ is the orthogonal projection onto a closed subspace $\mathcal{H} \subset L_2(\tilde{M}, \tilde{E})$ then one defines the Γ-dimension $\dim_\Gamma \mathcal{H}$ of $\mathcal{H}$ as

$$\dim_\Gamma \mathcal{H} := \mathrm{tr}_\Gamma P,$$

where tr_Γ denotes the von Neumann trace and $\mathrm{tr}_\Gamma P$ can be any real number ≥ 0 or $= \infty$.

If one takes $\mathcal{H} = \mathcal{H}(\tilde{D}) = \ker \tilde{D} \subset L_2(\tilde{E})$, $\mathcal{H}^* = \mathcal{H}(\tilde{D}^*) = \ker(\tilde{D}^*) \subset L_2(\tilde{F})$ then one defines the Γ-index $\mathrm{ind}_\Gamma \tilde{D}$ as

$$\mathrm{ind}_\Gamma \tilde{D} := \dim_\Gamma \mathcal{H}(\tilde{D}) - \dim_\Gamma \mathcal{H}(\tilde{D}^*).$$

Atiyah proves in [2] the following main

Theorem 2.5. *Under the assumptions above there holds*

$$\mathrm{ind}_a D = \mathrm{ind}_\Gamma \tilde{D}.$$ □

It was this theorem which was the origin of the von Neumann analysis as a fastly growing area in geometry, topology and analysis. Moreover, the proof of Theorem 2.2 is strongly modeled by that of 2.5. Another very important special case which is related to the case above of coverings are locally symmetric spaces of finite volume. There is a vast number of profound contributions, e.g., [4], [7], [9], [10], [11]. We do not intend here to give a complete overview for reasons of space. But we will sketch the main features and main achievements of these approaches.

Let G be semisimple, noncompact, with finite center, $K \subset G$ maximal compact, $\tilde{X} = G/K$ a symmetric space of noncompact type, $\Gamma \subset G$ discrete, torsion free and vol $(\Gamma\backslash G) < \infty$. Then $X = \Gamma\backslash\tilde{X} = \Gamma\backslash G/K$ is a locally symmetric space of finite volume. If V_E, V_F are unitary K-modules then we obtain homogeneous vector bundles $\tilde{E} = G/K \times_K V_E \longrightarrow G/K = \tilde{X}$, $\tilde{F} = G/K \times_K V_F \longrightarrow G/K = \tilde{X}$, over $\tilde{X}$ and corresponding bundles $E, F \longrightarrow X$ over X. A G-invariant elliptic differential operator $\tilde{D} : C^\infty(\tilde{E}) \longrightarrow C^\infty(\tilde{F})$ descends to an elliptic operator $D : C^\infty(E) \longrightarrow C^\infty(F)$. There arise the following natural questions: to describe the $\tilde{D}$ in question, to establish a formula for the analytical index, to calculate the index via a topological index and an index theorem. We indicate (partial) answers given by Barbasch, Connes, Moscovici and Müller.

Denote by $R(k)$ the right regular representation $R(k)f(g) = f(gk)$, $\tau_E : K \longrightarrow U(V_E)$. Then $k \longrightarrow R(k) \otimes \tau_E(k)$ acts on $C^\infty(G) \otimes V_E$. We identify $C^\infty(\tilde{E})$ with $(C^\infty(G) \otimes V_E)^K$, similarly $L_2(\tilde{E})$ with $(L_2(G) \otimes V_E)^K$. If $\mathfrak{G}$ is the Lie algebra of G, $\mathfrak{G}_c$ its complexification, $\mathfrak{U}(\mathfrak{G})$ the universal enveloping algebra of $\mathfrak{G}$, $\tau_E : K \longrightarrow U(V_E)$, $\tau_F : K \longrightarrow U(V_F)$ are unitary representations then $(\mathfrak{U}(\mathfrak{G}) \otimes \operatorname{Hom}(V_E, V_F))^K$ shall denote the subspace of all elements in $\mathfrak{U}(\mathfrak{G}) \otimes \operatorname{Hom}(V_E, V_F)$ which are fixed under $k \longrightarrow Ad_G(k) \otimes \tau_E(k^{-1})^t \otimes \tau_F(k)$. Let $d = \sum_i X_i \otimes A_i \in (\mathfrak{A}(\mathfrak{G}) \otimes \operatorname{Hom}(V_E, V_F))^K$. Then $\tilde{D} = \sum_i R(X_i) \otimes A_i$ defines a differential operator $\tilde{D} : C^\infty(\tilde{E}) \longrightarrow C^\infty(\tilde{F})$ commuting with the action of G.

We state without proof the simple

Lemma 2.6.

a) *Any G-invariant differential operator $\tilde{D} : C^\infty(\tilde{E}) \longrightarrow C^\infty(\tilde{F})$ is of the form*

$$\tilde{D} = \sum_i R(X_i) \otimes A_i \tag{2.8}$$

above.

b) *The formal adjoint $\tilde{D}^*$ corresponds to*

$$d^* = \sum_i X_i^* \otimes A_i^* \in (\mathfrak{U}(\mathfrak{G}) \otimes \operatorname{Hom}(E, F))^K,$$

where $x \longrightarrow x^*$ denotes the conjugate-linear anti-automorphisms of $\mathfrak{U}(\mathfrak{G})$ such that $x^* = -\overline{x}$, $x \in \mathfrak{G}_c$. □

For a unitary representation $\pi : G \longrightarrow U(\mathcal{H}(\pi))$ and $d = \sum_i X_i \otimes A_i \in (\mathfrak{U}(\mathfrak{G}) \otimes \mathrm{Hom}\,(V_E, V_F))^K$ define $\pi(d) : \mathcal{H}(\pi)_\infty \otimes V_E \longrightarrow \mathcal{H}(\pi)_\infty \otimes V_F$ by

$$\pi(d) := \sum_i \pi(X_i) \otimes A_i.$$

Here $\mathcal{H}(\pi)_\infty$ denotes the space of C^∞-vectors of π. $\pi(d)$ induces an operator $d_\pi : (\mathcal{H}(\pi) \otimes V_E)^K \longrightarrow (\mathcal{H}(\pi) \otimes V_F)^K$.

Proposition 2.7. *Suppose that d is elliptic. Then*

$$\ker d_\pi = \{u \in (\mathrm{Hom}\,(\pi)_\infty \otimes V_E)^K \mid d_\pi u = 0\}$$

coincides with the orthogonal complement of

$$\mathrm{im}\; d_\pi^* = \{d_\pi^* v \mid v \in (\mathcal{H}(\pi)_\infty \otimes V_F)^K\}$$

in $(\mathcal{H}(\pi) \otimes V_E)^K$. □

Corollary 2.8.

a) $\ker d_\pi$ *is closed in* $(\mathcal{H}(\pi) \otimes E)^K$.

b) *The closure of* d_π^* *coincides with the Hilbert space adjoint of* d_π. □

Corollary 2.9. *Suppose that d is elliptic and*

$$\pi = \int_\Lambda^\oplus \pi_\lambda \, d\lambda, \quad \mathcal{H}(\pi) = \int_\Lambda^\oplus \mathcal{H}(\pi_\lambda)\, d\lambda$$

is an integral decomposition of π. *Then*

$$\ker d_\pi = \int_\Lambda^\oplus \ker d_{\pi_\lambda}\, d\lambda. \tag{2.9}$$

□

Now we come to the main part of our present discussions, the locally symmetric case. Identifying $L_2(E)$ with $(L_2(\Gamma\backslash G) \otimes V_E)^K$, and taking into consideration the decompositions

$$R^\Gamma = R_d^\Gamma \oplus R_c^\Gamma, \quad L_2(\Gamma\backslash G) = L_{2,d}(\Gamma\backslash G) \oplus L_{2,c}(\Gamma\backslash G)$$

of the right quasi-regular representation R^Γ of G on $L_2(\Gamma\backslash G)$, we obtain the decomposition

$$\begin{aligned} L_2(E) &= L_{2,d}(E) \oplus L_{2,c}(E), \\ L_{2,d}(E) &= (L_{2,d}(\Gamma\backslash G) \otimes V_E)^K, \\ L_{2,c}(E) &= (L_{2,c}(\Gamma\backslash G) \otimes V_E)^K, \end{aligned}$$

similarly for $F = \Gamma\backslash\tilde{F}$.

Consider now the operators $D = d_{R^\Gamma}$ and $D_d = d_{R_d^\Gamma} : C_c^\infty(E) \longrightarrow C_c^\infty(F)$.

Theorem 2.10. *Under the assumptions above (on G, K, Γ),*

$$\ker D = \ker D_d \tag{2.10}$$

and

$$\dim \ker D < \infty. \tag{2.11}$$

Denote by $\tilde{G}_d^\Gamma$ the set of all equivalence classes of irreducible unitary representations π of G whose multiplicity $m_\Gamma(\pi)$ in R_d^Γ is nonzero. In particular $L_{2,d}(\Gamma\backslash G) = \sum_{\pi\in\tilde{G}_d^\Gamma} m_\Gamma(\pi)\mathcal{H}(\pi)$.

Theorem 2.11. *Let $K \subset G$ be maximal compact, $\Gamma \in G$ discrete and torsion free, $\tau_E : K \longrightarrow V_E$, $\tau_F : K \longrightarrow V_F$ unitary representations, $\tilde{E} = G/K \times_K V_E$, $\tilde{F} = G/K \times_K V_F$, $E = \Gamma\backslash\tilde{E}$, $F = \Gamma\backslash\tilde{F}$ and $D = d_{R^\Gamma}$ a corresponding locally invariant elliptic differential operator acting between $L_2(E)$ and $L_2(F)$. Then*

$$\operatorname{ind}_a D = \dim \ker D - \dim \ker D^*$$

is well defined and

$$\operatorname{ind}_a D = \sum_{\pi\in\tilde{G}_d^\Gamma} m_\Gamma(\pi)(\dim(\mathcal{H}(\pi)\otimes E)^K - \dim(\mathcal{H}(\pi)\otimes F)^K). \tag{2.12}$$

□

Corollary 2.12. *Let $X = \Gamma\backslash G/K$ be a locally symmetric space of negative curvature with finite volume and $L_2(E) \supset \mathcal{D}_D \xrightarrow{D} L_2(F)$ a locally symmetric elliptic differential operator then* ind D *is defined and depends only on the K-modules $K \longrightarrow U(V_E), U(V_F)$ which define $\tilde{E}$, $\tilde{F}$, $E = \Gamma\backslash\tilde{E}$, $F = \Gamma\backslash\tilde{F}$.* □

The value of the formula in Theorem 2.11 is very limited since in general the $m_\Gamma(\pi)$ are not known. Hence there arises the task to find a meaningful expression for it. This has been done with great success, e.g., in [7] and [10], [11] where they essentially restrict to generalized Dirac operators. To be more precise, we must briefly recall what is a manifold with cusps. Here we densely follow [10]. Let G be a semisimple Lie group with finite center, $K \subset G$ a maximal compact subgroup. P_a split rank one parabolic subgroup of G with split component A, $P = UAM$ the corresponding Langlands decomposition, where U is the unipotent radical of P, A a $\mathbb{R}$-split torus of dimension one and M centralizes A. Set $S = UM$ and let Γ be a discrete uniform torsion free subgroup of S. Then $Y = \Gamma\backslash\tilde{Y} = \Gamma\backslash G/K$ is called a complete cusp of rank one. Put $K_M = M \cap K$, K_M is a maximal compact subgroup of M. If $X_M = M/K_M$ there is a canonical diffeomorphism $\tilde{\xi}$: $\mathbb{R}_+ \times U \times X_M \longrightarrow \tilde{Y}$. Set for $t \geq 0$ $\tilde{Y}_t = \tilde{\xi}([t,\infty[\times U \times X_M)$ and call $Y_t = \Gamma\backslash\tilde{Y}_t$ a cusp of rank one. Another, even more explicit description is given as follows. Let $\Gamma_M = M \cap (\cup\Gamma)$, $Z = S/S \cap K$. Then there is a canonical fibration P : $\Gamma\backslash Z \longrightarrow \Gamma_M\backslash X_M$ with fibre $\Gamma \cap U\backslash U$ a compact nilmanifold and a canonical diffeomorphism $\xi : [t,\infty[\times\Gamma\backslash Z \xrightarrow{\cong} Y_t$. The induced metric on $[t,\infty[\times\Gamma_2\backslash Z$ looks locally as $ds^2 = dr^2 + dx^2 + e^{-br}du_\lambda^2(x) + e^{-4br}du_{2\lambda}^2(x)$, where $|b| = \lambda$, dx^2 is the invariant metric on X_M induced by restriction of the Killing form.

Now a complete Riemannian manifold is called a manifold with cusps of rank one if X has a decomposition $X = X_0 \cup X_1 \cup \cdots \cup X_S$ such that X_0 is a compact manifold with boundary, for $i, j \geq 1$, $i \neq j$ holds $X_i \cap X_j = \emptyset$ and each X_j, $j \geq 1$, is a cusp of rank one. The first general statement for generalized Dirac operators on rank one cusps manifold is

Theorem 2.13. *Let X be a rank one cusp manifold, $(E, h, \nabla, \cdot) \longrightarrow (X, g_X)$ a Clifford bundle and D its corresponding generalized Dirac operator. Then D is essentially self-adjoint and*

$$\dim(\ker \overline{D}) < \infty. \tag{2.13}$$

The spectrum of $H = \overline{D}^2$ consists of a point spectrum and an absolutely continuous spectrum. If $L_2(E) = L_{2,d}(E) \oplus L_{2,c}(E)$ is the corresponding decomposition of $L_2(E)$ and $H_d = H|_{L_{2,d}(E)}$ then for $t > 0$

$$e^{-zH_d} \text{ is of trace class.} \tag{2.14}$$

□

As we mentioned after Corollary 2.12, the main task, main objective consists in the case of a $\mathbb{Z}_2$-grading to get an expression for $\text{ind}_a \overline{D}$. For the sake of simplicity we restrict to spaces $X = X_0 \cup Y_1$ as above with one cusp Y_1, $Y_0 \cup Y_1 = Y = \Gamma\backslash G/K$. Let $(E = E^+ \oplus E^-, h, \nabla, \cdot) \longrightarrow (Y, g)$ be a Z_2-graded Clifford bundle such that $E^\pm|_{Y_1} = \Gamma\backslash \tilde{E}^\pm$, where $\tilde{E}^\pm$ are homogeneous vector bundles over G/K and let $D^+ : C^\infty(Y, E_+^+) \longrightarrow C^\infty(Y, E^-)$ the corresponding generalized Dirac operator. We recall $K_M = M \cap K$, $X_M = M/K_M$. D^+ induces an elliptic differential operator $D_0^+ : C^\infty(\mathbb{R}_+ \times \Gamma_M\backslash X_M, E_M^+) \longrightarrow C^\infty(\mathbb{R}_+ \times \Gamma_M\backslash X_M, E_M^-)$, where $E_M^\pm$ are locally homogenous vector bundles over $\Gamma_M\backslash X_M$. From this come a self-adjoint differential operator $D_M : C^\infty(\Gamma_M\backslash X_M, E_M^+) \longrightarrow C^\infty(\Gamma_M\backslash X_M, E_M^-)$ and a bundle isomorphism $\beta : E_M^+ \longrightarrow E_M^-$ such that $D_0^+ = \beta\left(r\frac{\partial}{\partial r} + D_M\right)$. We set $\tilde{D}_M = D_M + \frac{m}{2}\text{id}$, $m = \dim u_\lambda|\lambda| + 2\dim u_{2\lambda}|\lambda|$, λ the unique simple root of the pair (P, A).

W. Müller then established in [10] the following general index theorem for a locally symmetric graded Dirac operator.

Theorem 2.14. *Assume $\ker \tilde{D}_M = \{0\}$, let $\eta(0)$ be the eta invariant of $\tilde{D}_M$ and ω_{D^+} the index form of D^+. Then*

$$\text{ind}_a D^+ = \int_X \omega_{D^+} + \mathcal{U} + \frac{1}{2}\eta(0), \tag{2.15}$$

where the term $\mathcal{U}$ is essentially given by the value of an L-series at zero and an expression in the scattering matrix at zero. □

Finally, application of an elaborated version of Theorem 2.14 allows to prove the famous Hirzebruch conjecture for Hilbert modular varieties. This has been done by W. Müller in [11].

There is another approach to Fredholmness by Gilles Carron, which relies on an inequality quite similar to 2.1 d).

Let $(E, h, \nabla, \cdot) \longrightarrow (M^n, g)$ be a Clifford bundle over the complete Riemannian manifold (M^n, g) and $D : C^\infty(E) \longrightarrow C^\infty(E)$ the associated generalized Dirac operator. D is called nonparabolic at infinity if there exists a compact set $K \subset M$ such that for any open and relative compact $U \subset M \setminus K$ there exists a constant $C(U) > 0$ such that

$$C(U)|\varphi|_{L_2(E|_U)} \leq |D\varphi|_{L_2(E|_{M\setminus K})} \text{ for all } \varphi \in C_c^\infty(E|_{M\setminus K}). \tag{2.16}$$

To exhibit the consequences of this inequality, we establish another characterization of it.

Proposition 2.15. *Let $(E, h, \nabla, \cdot) \longrightarrow (M^n, g)$ and D as above and let $W(E)$ be a Hilbert space of sections such that*

a) *$C_c^\infty(E)$ is dense in $W(E)$ and*

b) *the injection $C_c^\infty(E) \hookrightarrow \Omega_{\text{loc}}^{2,1}(E, D)$ extends continuously to $W(E) \longrightarrow \Omega_{\text{loc}}^{2,1}(E, D)$.*

Then $D : W(E) \longrightarrow L_2(E)$ is Fredholm if and only if there exist a compact $K \subset M$ and a constant $C(K) > 0$ such that

$$C(K) \cdot |\varphi|_W \leq |D\varphi|_{L_2(E|_{M\setminus K})} \text{ for all } \varphi \in C_c^\infty(E|_{M\setminus K}). \tag{2.17}$$

□

Remark 2.16. The norm $\mathcal{H}(\cdot)$ above is equivalent to the norm

$$\mathcal{N}_{\overline{U}(K)}(\cdot), \mathcal{N}_{\overline{U}(K)}(\varphi)^2 = |\varphi|^2_{L_2(E|_{\overline{U}(K)})} + |D\varphi|^2_{L_2(E)}. \tag{2.18}$$

Corollary 2.17. *$D : C^\infty(E) \longrightarrow C^\infty(E)$ is non-parabolic at infinity if and only if there exists a compact $K \subset M$ such that the completion of $C_c^\infty(E)$ w. r. t. $\mathcal{N}_K(\cdot)$,*

$$\mathcal{N}_K(\varphi)^2 = |\varphi|^2_{L_2(E|_K)} + |D\varphi|^2_{L_2} \tag{2.19}$$

yields a space $W(E)$ such that the injection $C_c^\infty(E) \longrightarrow \Omega_{\text{loc}}^{2,1}(E, D)$ continuously extends to $W(E)$. □

The point now is that we know if D is non-parabolic at infinity then $D : W(E) \longrightarrow L_2(E)$ is Fredholm. We emphasize, this does not mean $L_2(E) \supset \mathcal{D}_D \longrightarrow L_2(E)$ is Fredholm. We get a weaker Fredholmness, not the desired one. But in certain cases this can be helpful too.

Suppose again a $\mathbb{Z}_2$-grading of E and D, $D = \begin{pmatrix} 0 & D^- \\ D^+ & 0 \end{pmatrix}$, $L_2(E) = L_2(E^+) \oplus L_2(E^-)$, $W(E) = W(E^+) \oplus W(E^-)$. Following Gilles Carron, we now define the extended index $\text{ind}_e D^+$ as

$$\begin{aligned} \text{ind}_e D^+ &:= \dim \ker {}_W D^+ - \dim \ker {}_{L_2} D^- \\ &= \dim\{\varphi \in W(E^+) \mid D^+\varphi = 0\} - \\ &\quad - \dim\{\varphi \in L_2(E^-) \mid D^-\varphi = 0\}. \end{aligned} \tag{2.20}$$

If we denote $h_\infty(D^+) := \dim(\ker_W D^+ / \ker_{L_2} D^+)$ then we can (2.20) rewrite as

$$\text{ind}_e D^+ = h_\infty(D^+) + \text{ind}_{L_2} D^+ = h_\infty(D^+) + \dim\ker_{L_2} D^+ - \dim\ker_{L_2} D^- . \tag{2.21}$$

The most interesting question now are applications and examples. For $D =$ Gauß–Bonnet operator, there are in fact good examples (cf. [6]). For the general case it is not definitely clear, is non-parabolicity really a practical sufficient criterion for Fredholmness since in concrete cases it will be very difficult it to establish. In some well known standard cases which have been presented by Carron and which we will discuss now it is of great use.

Proposition 2.18. *Let $D : C^\infty(E) \longrightarrow C^\infty(E)$ be a generalized Dirac operator and assume that outside a compact $K \subset M$ the smallest eigenvalue $\lambda_{\min}(x)$ of $\mathcal{R}_x$ in $D^2 = \nabla^*\nabla + \mathcal{R}$ is ≥ 0. Then D is non-parabolic at infinity.* □

We obtain from Proposition 2.18

Corollary 2.19. *Assume the hypothesis of 2.18. Then $D : W_0(E) \longrightarrow L_2(E)$ is Fredholm.* □

Under certain additional assumptions the pointwise condition on $\lambda_{\min}(x)$ of $\mathcal{R}_x$ can be replaced by a (weaker) integral condition. Denote $\mathcal{R}_-(x) = \max\{0, -\lambda_{\min}(x)\}$, where $\lambda_{\min}(x)$ is the smallest eigenvalue of $\mathcal{R}_x$.

Theorem 2.20. *Suppose that for a $p > 2$ (M^n, g) satisfies the Sobolev inequality*

$$c_P(M) \left(\int\limits_M |u|^{\frac{2p}{p-2}}(x)\,\text{dvol}_x(g) \right)^{\frac{p-2}{2}} \leq \int\limits_M |du|^2(x)\,\text{dvol}_x(g) \text{ for all } u \in C_c^\infty(M) \tag{2.22}$$

and

$$\int\limits_M |\mathcal{R}_-|^{\frac{p}{2}}(x)\,\text{dvol}_x(g) < \infty.$$

Then $D : W_0(E) \longrightarrow L_2(E)$ is Fredholm. □

Another important example are manifolds with a cylindrical end which we already mentioned. In this case, there is a compact submanifold with boundary $K \subset M$ such that $(M \setminus K, g)$ is isometric to $(]0,\infty[\times \partial K, dr^2 + g_{\partial K})$. One assumes that $(E,h)|_{]0,\infty[\times\partial K}$ also has product structure and $D|_{M\setminus K} = \nu \cdot \left(\frac{\partial}{\partial r} + A\right)$, where $\nu\cdot$ is the Clifford multiplication with the exterior normal at $\{\gamma\} \times \partial K$ and A is first-order elliptic and self-adjoint on $E|_{\partial K}$.

Proposition 2.21. *D is non-parabolic at infinity.*

Proof. There are two proofs. The first one refers to [3]. According to proposition 2.5 of [3], there exists on $M \setminus K$ a parametrix $Q : L_2(E|_{M\setminus K}) \longrightarrow \Omega^{2,1}_{\text{loc}} E|_{M\setminus K}, D)$ such that $QD\varphi = \varphi$ for all $\varphi \in C_c^\infty(E|_{M\setminus K})$. Hence for $C_c^\infty(E|_{M\setminus K})$, $U \supset M \setminus K$ bounded,

$$|\varphi|_{L_2(E|_U)} = |QD\varphi|_{L_2(E|_U)} \leq |Q|_{L_2 \to \Omega^{2,1}} \cdot |D\varphi|_{L_2}.$$

The other proof is really elementary calculus. For $\varphi \in C_c^\infty(E|_{M\setminus K})$,

$$|\varphi(r,y)| = \left| \int_0^r \frac{\partial \varphi}{\partial r}\, dr \right| \leq \sqrt{r} \cdot \left| \frac{\partial \varphi}{\partial r} \right|_{L_2}.$$

Hence

$$|\varphi|^2_{L_2(E|_{]0,T[\times \partial K})} \leq \frac{T^2}{2} \left| \frac{\partial \varphi}{\partial r} \right|^2_{L_2} \leq \frac{T^2}{2} \left(\left| \frac{\partial \varphi}{\partial r} \right|^2_{L_2} + |A\varphi|^2_{L_2} \right) = \frac{T^2}{2} |D\varphi|^2_{L_2}. \qquad \square$$

The authors of [3] define extended L_2-sections of $E|_{]0,\infty[\times \partial K}$ as sections $\varphi \in L_{2,\mathrm{loc}}$, $\varphi(r,y) = \varphi_0(r,y) + \varphi_\infty(y)$, $\varphi_0 \in L_2$, $\varphi_\infty \in \ker A$.

Proposition 2.22. *The extended solutions of $D\varphi = 0$ are exactly the solutions of $D\varphi = 0$ in W.*

Proof. Let $\{\varphi_\lambda\}_{\lambda \in \sigma(A)}$ be a complete orthonormal system in $L_2(E|_{\partial K})$ consisting of the eigensections of A. Then we can a solution φ of $D\varphi = 0$ on $]0,\infty[\times \partial K$ decompose as

$$\varphi(r,y) = \sum_{\lambda \in \sigma(A)} c_\lambda e^{-\lambda r} \varphi_\lambda(y) \tag{2.23}$$

and $\varphi \in W$ if and only if $c_\lambda = 0$ for $\lambda < 0$. In this case

$$\varphi_0(r,y) = \sum_{\substack{\lambda \in \sigma(A) \\ \lambda > 0}} c_\lambda e^{-\lambda r} \varphi_\lambda(y), \quad \varphi_\infty(y) = \sum_{\lambda \in \sigma(A)} c_{0,i} \varphi_{0,i}(y). \qquad \square$$

This proposition can also be reformulated as

Proposition 2.23. *Denote by $P_{\leq 0}$ or $P_{<0}$ the spectral projection of A onto the sum of eigenspaces belonging to eigenvalues ≤ 0 or < 0, respectively. Then*

a) *φ is a solution in W of $D\varphi = 0$ if and only if*

$$D\varphi = 0 \text{ on } K \quad \text{and} \quad P_{<0}\varphi = 0 \text{ on } \partial K.$$

b) *φ is an L_2-solution of $D\varphi = 0$ if and only if*

$$D\varphi = 0 \text{ on } K \quad \text{and} \quad P_{\leq 0}\varphi = 0 \text{ on } \partial K. \qquad \square$$

There is a very general approach to index theory as established by Connes, Roe and others. The initial data are as follows: D an elliptic differential operator as above, $\mathfrak{B}$ an operator algebra, the K-theory $K_i(\mathfrak{B})$ of $\mathfrak{B}$, the cyclic cohomology $HC^*(\mathfrak{B})$ of $\mathfrak{B}$. Then one constructs the diagram

$$\begin{array}{ccc} D & \longrightarrow & \mathrm{Ind} D \in K_i(\mathfrak{B}) \\ \downarrow & & \downarrow \\ I_D \longrightarrow \langle I_D, m \rangle = \mathrm{ind}_t D & \stackrel{?}{=} & \mathrm{ind}_a D = \langle \mathrm{Ind} D, \zeta \rangle \end{array}$$

Here I_D is of cohomological nature, m a fundamental class, $\langle I_D, m\rangle$ a pairing, Ind$D$ comes from ellipticity and the 6 term exact sequence of K-theory, $\zeta \in HC^*(\mathfrak{B})$ and $\langle \mathrm{Ind}D, \zeta\rangle$ is the Connes' pairing.

Choice of $\mathfrak{B}, i, \zeta, m, \mathrm{Ind}_D$ yields a concrete index theory. We refer to [12], [13], [14], [15] for details. The classical index theory on closed manifolds is given by the choice $i = 0$, $\mathfrak{B}$ = ideal K of compact operators, $\mathrm{Ind}D \in K_0(K)$ = projectors – projectors, $HC^0 \ni \zeta$ = trace, $\mathrm{trInd}D = \mathrm{ind}_a D$, I_D = classical index from, $m = [M]$. The lack of all these (absolute) index theories for open manifolds is that they either refer to very special cases or there are not enough serious applications.

The next section is devoted to a general relative index theory.

3. Relative index theorems for generalized Dirac operators

Let (M^n, g) be closed, oriented $(E, h, \nabla, \cdot, \tau) \longrightarrow (M^n, g)$ a supersymmetric Clifford bundle with involution τ and $D = \begin{pmatrix} 0 & D^- \\ D^+ & 0 \end{pmatrix}$ the associated generalized Dirac operator. Then

$$\mathrm{ind}_a D^+ = \mathrm{tr}(\tau e^{-tD^2}).$$

Starting with this simple fact, one could attempt to define in the open case a relative index for a pair of generalized Dirac operators D, D' by

$$\mathrm{ind}(D, D') := \mathrm{tr}(\tau(e^{-tD^2} - e^{-tD'^2})).$$

With this intention in mind there arise immediately several problems.

1) One has to assure that D, D' are self-adjoint in the same Hilbert-space.
2) One has to assure that $e^{-tD^2} - e^{-tD'^2}$ is of trace class.
3) One has to assure that $\mathrm{tr}(\tau(e^{-tD^2} - e^{-tD'^2}))$ is independent of t.
4) Finally one has to present substantial applications.

The initial data for a fixed vector bundle $E \longrightarrow M$ and different Clifford structures are $(E, h, \nabla = \nabla^h, \cdot) \longrightarrow (M^n, g)$ and $(E, h', \nabla' = \nabla^{h'}, \cdot') \longrightarrow (M^n, g')$, respectively. These yield generalized Dirac operators $D = D(h, \nabla, \cdot, g)$ and $D' = D(h', \nabla', \cdot', g')$. D and D' act in different Hilbert spaces, i.e., $e^{-tD^2} - e^{-tD'^2}$ is not defined. But this can be repaired by two unitary transformations. Denote by D' already the result after performing this transformations.

To describe the possibly maximal perturbations $(h', \nabla', \cdot', g')$ of $(h, \nabla, \cdot, g)$, we introduced in [8] uniform structures of Clifford structures and defined generalized components. We indicate here briefly the main definitions. First of all, we restrict to manifolds and bundles $(E, h, \nabla) \longrightarrow (M^n, g)$ of bounded geometry of order k, i.e., we assume

$$r_{\mathrm{inj}}(M^n, g) > 0, \tag{I}$$

$$|(\nabla^g)^i R^g| \le C_i, \quad 0 \le i \le k, \tag{$B_k(M,g)$}$$

$$|\nabla^i R^E| \le D_i, \quad 0 \le i \le k. \tag{$B_k(E,\nabla)$}$$

For bounded geometry of order k, $k \geq r > \frac{n}{p} + 1$, we defined completed manifolds of diffeomorphisms $\mathcal{D}^{p,r}(M_1, M_2)$, $\mathcal{D}^{p,r}(E_1, E_2)$. This construction is very long and complicated. The smooth elements in $\mathcal{D}^{p,r}(\cdot,\cdot)$ shall be denoted by $\tilde{\mathcal{D}}^{p,r}(\cdot,\cdot)$. Here a diffeomorphism $f = (f_E, f_M) \in \tilde{\mathcal{D}}^{p,r}(E_1, E_2)$ if and only if $f_E = \exp_{\tilde{f}_E} X \circ \tilde{f}_E$, $\tilde{f}_E \in C^{\infty,r}(E_1, E_2)$ a diffeomorphism, $\tilde{f}_E^{-1} \in C^{\infty,r}(E_2, E_1)$ and X a smooth section of $f_E^* TE_2$, such that $|X|_{p,r} = \left(\int_M \sum_{i=0}^{r} |\nabla^i X|_x^p \, \mathrm{dvol}_x(g_1) \right)^{\frac{1}{p}} < \infty$. Here $\nabla = \tilde{f}_E^* \nabla^{TE_2}$, ∇^{TE_2} = Kaluza–Klein connection. $C^{\infty,r}(M,N)$ is the set of all smooth maps $f : (M^n, g_M) \longrightarrow (N^{n'}, g_N)$ such that $\sup_{x \in M} |\nabla^i df|_x < \infty$, $0 \leq i \leq r-1$. Here ∇ is the induced connection in $T^*M \otimes f^*TN$, $df \in C^\infty(T^*M \otimes f^*TN)$. f_M in (f_E, f_M) must satisfy analogous conditions.

Denote by $\mathrm{CL}\mathcal{B}^{N,n}(I, B_k)$ the set of (Clifford isometry classes) of all Clifford bundles $(E, h, \nabla^h, \cdot) \longrightarrow (M^n, g)$ of (module) rank N over n-manifolds, all with (I) and (B_k).

Lemma 3.1. *Let* $E_i = ((E_i, h_i, \nabla^{h_i}, \cdot_i) \longrightarrow (M_i^n, g_i)) \in \mathrm{CL}\mathcal{B}^{N,n}(I, B_k)$, $i = 1, 2$ *and* $f = (f_E, f_M) \in \tilde{\mathcal{D}}^{p,r+1}(E_1, E_2) \cap C^{\infty,k+1}(E_1, E_2)$ *be a vector bundle isomorphism between bundles of Clifford modules,* $f_E(X \cdot_1 \Phi) = (f_M)_* X \cdot_2 f_E \Phi$. *Then* $f^* E_2 := ((E_1, f_E^* h_2, f_E^* \nabla^{h_2}, f_E^* \cdot_2) \longrightarrow (M_1, f_M^* g_2)) \in \mathrm{CL}\mathcal{B}^{N,n}(I, B_k)$.

Proof. The definitions of $f_E^* h_2$, $f_E^* \nabla^{h_2}$, $f_M^* g_2$ are clear. $f_E^* \cdot_2$ is defined by $X(f_E^* \cdot_2)\Phi = f_E^{-1}(f_* X \cdot_2 f_E \Phi)$. It is now an easy calculation that $f^* E_2 \in \mathrm{CL}\mathcal{B}^{N,n}(I, B_k)$. □

In the sequel $|\cdot - \cdot|_{\ldots}$ denotes the Sobolev norm in the corresponding bundle. Let $k \geq r > \frac{n}{p} + 2$ and define for $E_1, E_2 \in \mathrm{CL}\mathcal{B}^{N,n}(I, B_k)$

$$
\begin{aligned}
d_{L,\mathrm{diff}}^{p,r}(E_1, E_2) \quad = \quad & \inf\{\max\{0, \log{}^b |df_E|\} + \max\{0, \log{}^b |df_E^{-1}|\} + \\
& \max\{0, \log{}^b |df_M|\} + \max\{0, \log{}^b |df_M^{-1}|\} + \\
& |g_1 - f_M^* g_2|_{g_1,p,r} + |h_1 - f_E^* h_2|_{g_1,h_1,\nabla^{h_1},p,r} + \\
& |\nabla^{h_1} - f_E^* \nabla^{h_2}|_{g_1,h_1,\nabla^{h_1},p,r} + |\cdot_1 - f_E^* \cdot_2|_{g_1,h_1,\nabla^{h_1},p,r} \\
& \mid f = (f_E, f_M) \in \tilde{\mathcal{D}}^{p,r}(E_1, E_2) \text{ is a } (k+1) - \text{bounded} \\
& \text{isomorphism of Clifford bundles}\}
\end{aligned}
$$

if $\{\ldots\} \neq \emptyset$ and $\inf\{\ldots\} < \infty$. In the other case set $d_{L,\mathrm{diff}}^{p,r}(E_1, E_2) = \infty$. $d_{L,\mathrm{diff}}^{p,r}$ is numerically not symmetric but nevertheless it defines a uniform structure which is by definition symmetric. Set for $\delta > 0$

$$V_\delta = \{(E_1, E_2) \in \mathrm{CL}\mathcal{B}^{N,n}(I, B_k))^2\} \mid d_{L,\mathrm{diff}}^{p,r}(E_1, E_2) < \delta\}.$$

Proposition 3.2. $\mathfrak{L} = \{V_\delta\}_{\delta > 0}$ *is a basis for a metrizable uniform structure* $\mathfrak{U}_{L,\mathrm{diff}}^{p,r}(\mathrm{CL}\mathcal{B}^{N,n}(I, B_k))$. □

Denote

$$\mathrm{CL}\mathcal{B}_{L,\mathrm{diff}\,,r}^{N,n,p}(I, B_k)$$

for the pair $(\mathrm{CL}\mathcal{B}^{N,n}(I,B_k),\mathcal{U}^{p,r})$ and

$$\mathrm{CL}\mathcal{B}^{N,n,p,r}_{L,\mathrm{diff}}(I,B_k)$$

for the completion. We introduce the generalized component $\operatorname{gen\,comp}(E) = \operatorname{gen\,comp}^{p,r}_{L,\mathrm{diff}}((E,h,\nabla^h)\longrightarrow(M,g))\subset \mathrm{CL}\mathcal{B}^{N,n,p,r}_{L,\mathrm{diff}}(I,B_k)$ by

$$\operatorname{gen\,comp}^{p,r}_{L,\mathrm{diff}}(E)=\{E'\in \mathrm{CL}\mathcal{B}^{N,n,p,r}_{L,\mathrm{diff}}(I,B_k)\mid d^{p,r}_{L,\mathrm{diff}}(E,E')<\infty\}.$$

$\operatorname{gen\,comp}(E)$ contains $\operatorname{arccomp}(E)$ and is endowed with a Sobolev topology induced from $\mathcal{U}^{p,r}_{L,\mathrm{diff}}$.

The last step in our uniform structures approach is the additional admission of compact topological perturbations. We assume $f=(f_E,f_M)|_{M_1\setminus K_1}$, $h=(h_E,h_M)|_{M_2\setminus K_2=f_M(M_1\setminus K_1)}$ vector bundle isomorphisms (not necessary Clifford isometric). Then we get $d^{p,r}_{L,\mathrm{diff}\,,\mathrm{rel}}(E_1,E_2)$, define V_δ, $\mathfrak{L}=\{v_\delta\}_{\delta>0}$, obtain the metrizable uniform structure $\mathfrak{U}^{p,r}_{L,\mathrm{diff}\,,\mathrm{rel}}(\mathrm{CL}\mathcal{B}^{N,n}(I,B_k))$ and finally the completion $\mathrm{CL}\mathcal{B}^{N,n,p,r}_{L,\mathrm{diff}\,,\mathrm{rel}}$. We set again

$$\begin{aligned}\operatorname{gen\,comp}(E) &= \operatorname{gen\,comp}^{p,r}_{L,\mathrm{diff}\,,\mathrm{rel}}(E)\\ &= \{E'\in \mathrm{CL}\mathcal{B}^{N,n,p,r}_{L,\mathrm{diff}\,,\mathrm{rel}}(I,B_k))\mid d^{p,r}_{L,\mathrm{diff}\,,\mathrm{rel}}(E,E')<\infty\}\end{aligned}$$

which contains the arc component and inherits a Sobolev topology from $\mathfrak{U}^{p,r}_{L,\mathrm{diff}\,,\mathrm{rel}}$.

The main goal is to prove that

$$e^{-tD^2}-e^{-t\tilde{D}'^2}$$

is for $t>0$ of trace class for where $\tilde{D}'$ is an appropriate transform of D'. We decompose the perturbation into several steps, 1) $\nabla\longrightarrow\nabla'$, all other fixed, 2) $h,\nabla\longrightarrow h',\nabla'$, $\cdot,g$ fixed, 3) $h,\nabla,\cdot\longrightarrow h',\nabla',\cdot'$, g fixed and finally 4) $h,\nabla,\cdot,g\longrightarrow h',\nabla',\cdot',g'$. The last step consists in even admitting compact topological perturbations. The first (and simplest) step is settled by

Theorem 3.3. *Assume* $(E,\nabla)\longrightarrow(M^n,g)$ *with* $(I),(B_k)$, (E,∇) *with* (B_k), $k\geq r>n+2$, $n\geq 2$, $\nabla'\in\operatorname{comp}(\nabla)\cap\mathcal{C}_E(B_k)\subset\mathcal{C}^{1,r}_E(B_k)$, $D=D(g,\nabla)$, $D'=D(g,\nabla')$ *generalized Dirac operators. Then*

$$e^{-tD^2}-e^{-tD'^2}\quad and\quad De^{-tD^2}-D'e^{-tD'^2}$$

are trace class operators for $t>0$ *and their trace norm is uniformly bounded on compact* t*-intervals* $[a_0,a_1]$, $a_0>0$. □

Here $\nabla'\in\operatorname{comp}^{1,r}(\nabla)$ means $|\nabla-\nabla'|_{\nabla,1,r}<\infty$ and both connections satisfy $(B_k(E))$.

The first step in the proof is Duhamel's principle. We remark that

$$\mathcal{D}_D=\mathcal{D}_{D'},\quad \mathcal{D}_{D^2}=\mathcal{D}_{D'^2}.$$

Lemma 3.4. *Assume $t > 0$. Then*

$$e^{-tD^2} - e^{-tD'^2} = \int_0^t e^{-sD^2}(D'^2 - D^2)e^{-(t-s)D'^2}\, ds. \qquad \square$$

Proof. The assertion means at heat kernel level

$$W(t,m,p) - W'(t,m,p) = -\int_0^t \int_M (W(s,m,q), (D^2 - D'^2)W'(t-s,q,p))_q \, dq\, ds,$$

where $(,)_q$ means the fibrewise scalar product at q and $dq = \mathrm{dvol}_q(g)$. Hence we have to prove this equation, which is an immediate consequence of Duhamel's principle. We present the proof, which is the last of the following 7 facts and implications.

1) For $t > 0$ is $W(t,m,p) \in L_2(M,E,dp) \cap \mathcal{D}_D^2$
2) If $\Phi, \Psi \in \mathcal{D}_D^2$ then $\int (D^2\Phi, \Psi) - (\Phi, D^2\Psi)$ dvol $= 0$ (Greens formula).
3) $((D^2 + \frac{\partial}{\partial \tau})\Phi(\tau,g)\Psi(t-\tau,q))_q - (\Phi(\tau,g),(D^2 + \frac{\partial}{\partial t})\Psi(t-\tau,q))_q$
   $= (D^2(\Phi(\tau,q),\Psi(t-\tau,q))_q - (\Phi(\tau,q),D^2\Psi(t-\tau,q))_q + \frac{\partial}{\partial \tau}(\Phi(\tau,g),\Psi(t-\tau,q))_q$
   .
4) $\int_\alpha^\beta \int_M ((D^2 + \frac{\partial}{\partial \tau})\Phi(\tau,q),\Psi(t-\tau,q))_q - (\Phi(\tau,q),(D^2 + \frac{\partial}{\partial t})\Psi(t-\tau,q))_q \, dq\, d\tau$
   $= \int_M [(\Phi(\beta,q),\Psi(t-\beta,q)_q - (\Phi(\alpha,q),\Psi(t-\alpha,q))_q]\, dq.$
5) $\Phi(t,q) = W(t,m,q), \Psi(t,q) = W'(t,q,p)$ yields
   $-\int_\alpha^\beta \int_M (W(\tau,m,q),(D^2 + \frac{\partial}{\partial t})W'(t-\tau,q,p)\, dq\, d\tau$
   $= \int_M [(W(\beta,m,q),W'(t-\beta,q,p))_q - (W(\alpha,m,q),W'(t-\alpha,q,p))_q]\, dq\,.$
6) Performing $\alpha \to 0^+, \beta \to t^-$ in 5) yields
   $-\int_0^t \int_M (W(s,m,q),(D^2 + \frac{\partial}{\partial t})W'(t-s,q,p))_q\, dq\, ds = W(t,m,p) - W'(t,m,p).$
7) Finally, using $D^2 + \frac{\partial}{\partial t} = D^2 - D'^2 + D'^2 + \frac{\partial}{\partial t}$ and $(D'^2 + \frac{\partial}{\partial t})W' = 0$ we obtain
   $W(t,m,p) - W'(t,m,p) = -\int_0^t \int_M (W(s,m,q),(D^2 - D'^2)W'(t-s,q,p))_q\, dq\, ds$

which is the assertion. $\square$

If we write $D^2 - D'^2 = D(D-D') + (D-D')D'$ then

$$e^{-tD^2} - e^{-tD'^2} = -\int_0^t e^{-sD^2}(D^2 - D'^2)e^{-(t-s)D'^2}\, ds$$

$$= -\int_0^t e^{-sD^2} D(D-D')e^{-(t-s)D'^2}\, ds$$

$$- \int_0^t e^{-sD^2} (D-D')D'e^{-(t-s)D'^2}\, ds$$

$$= \int_0^t e^{-sD^2} D\eta e^{-(t-s)D'^2}\, ds$$

$$+ \int_0^t e^{-sD^2} \eta D' e^{-(t-s)D'^2}\, ds,$$

where $\eta = \eta^{\mathrm{op}}$ in the sense of Section 3, $\eta^{\mathrm{op}}\,(\Psi)|_x = \sum_{i=1}^n e_i \eta_{e_i}(\Psi)$ and $|\eta^{\mathrm{op}}\,|_{op,x} \le C \cdot |\eta|_x$, C independent of x. We split $\int_0^t = \int_0^{\frac{t}{2}} + \int_{\frac{t}{2}}^t$,

$$e^{-tD^2} - e^{-tD'^2} = \int_0^{\frac{t}{2}} e^{-sD^2} D\eta e^{-(t-s)D'^2}\, ds \qquad (\mathrm{I}_1)$$

$$+ \int_0^{\frac{t}{2}} e^{-sD^2} \eta D' e^{-(t-s)D'^2}\, ds \qquad (\mathrm{I}_2)$$

$$+ \int_{\frac{t}{2}}^t e^{-sD^2} D\eta e^{-(t-s)D'^2}\, ds \qquad (\mathrm{I}_3)$$

$$+ \int_{\frac{t}{2}}^t e^{-sD^2} \eta D' e^{-(t-s)D'^2}\, ds. \qquad (\mathrm{I}_4)$$

We want to show that each integral $(\mathrm{I}_1) - (\mathrm{I}_4)$ is a product of Hilbert–Schmidt operators and to estimate their Hilbert–Schmidt norm. Consider the integrand of (I_4),

$$(e^{-sD^2}\eta)(D'e^{-(t-s)D'^2}).$$

There holds

$$\begin{aligned} |e^{-(t-s)D'^2}|_{L_2 \to H^1} &\le C \cdot (t-s)^{-\frac{1}{2}} \\ |D'e^{-(t-s)D'^2}|_{L_2 \to L_2} &\le |D'|_{H^1 \to L_2} \cdot |e^{-(t-s)D'^2}|_{L_2 \to H^1} \\ &\le C \cdot (t-s)^{-\frac{1}{2}}. \end{aligned}$$

Write

$$(e^{-sD^2}\eta)(D'e^{-(t-s)D'^2}) = (e^{-\frac{s}{2}D^2}f)(f^{-1}e^{-\frac{s}{2}D^2}\eta)(D'e^{-(t-s)D'^2}).$$

Here f shall be a scalar function which acts by multiplication. The main point is the right choice of f. $e^{-\frac{s}{2}D^2}f$ has the integral kernel

$$W(\frac{s}{2}, m, p)f(p) \tag{3.1}$$

and $f^{-1}e^{-\frac{s}{2}D^2}\eta$ has the kernel

$$f^{-1}(m)W(\frac{s}{2}, m, p)\eta(p). \tag{3.2}$$

We have to make a choice such that (3.1), (3.2) are square integrable over $M \times M$ and that their L_2-norm is on compact t-intervals uniformly bounded.

We decompose the L_2-norm of (3.1) as

$$\int_M \int_M |W(\frac{s}{2}, m, p)|^2 |f(m)|^2 \; dm \; dp = \tag{3.3}$$

$$\int_M \int_{dist(m,p) \geq c} |W(\frac{s}{2}, m, p)|^2 |f(m)|^2 \; dp \; dm = \tag{3.4}$$

$$\int_M \int_{dist(m,p) < c} |W(\frac{s}{2}, m, p)|^2 |f(m)|^2 \; dp \; dm \tag{3.5}$$

We use the fact that for any $T > 0$ and sufficiently small $\varepsilon > 0$ there exists $C > 0$ such that

$$|W(t, m, p)| \leq e^{-(t-\varepsilon)\inf \sigma(D^2)} \cdot C \cdot C(m) \cdot C(p)$$

for all $t \in]T, \infty[$ and obtain for $s \in]\frac{t}{2}, t[$

$$(3.5) \leq \int_M C_1 |f(m)|^2 \text{vol } B_c(m) \; dm \leq C_2 \int_M |f(m)|^2 \; dm$$

Moreover, for any $\varepsilon > 0$, $T > 0$, $\delta > 0$ there exists $C > 0$ such that for $r > 0$, $m \in M$, $T > t > 0$ holds

$$\int_{M \setminus B_r(m)} |W(t, m, p)|^2 dp \leq C \cdot C(m) \cdot e^{-\frac{(r-\varepsilon)^2}{(4+\delta)t}},$$

which yields

$$\int_M \int_{dist(m,p) \geq c} |W(\frac{s}{2}, m, p)|^2 |f(m)|^2 \; dp \; dm \leq \int_M C_3 e^{-\frac{-(c-\varepsilon)^2}{4+\delta}\frac{s}{2}} |f(m)|^2 \; dm \; \leq$$

$$\leq C_3 \cdot e^{-\frac{-(c-\varepsilon)^2}{4+\delta}\frac{s}{2}} \int_M |f(m)|^2 \; dm, \quad c > \varepsilon. \tag{3.6}$$

Hence the estimate of $\int\limits_M\int\limits_M |W(\frac{s}{2},m,p)|^2|f(m)|^2 dpdm$ for $s\in[\frac{t}{2},t]$ is done if

$$\int\limits_M |f(m)|^2\, dm < \infty$$

and then $|e^{-\frac{s}{2}D^2}f|_2 \le C_4\cdot|f|_{L_2}$, where $C_4 = C_4(t)$ contains a factor e^{-at}, $a>0$, if $\inf\sigma(D^2)>0$.

For (3.2) we have to estimate

$$\int\limits_M\int\limits_M |f(m)|^{-2}|(W(\frac{s}{2},m,p),\eta^{\mathrm{op}}\,(p)\cdot)_p|^2\, dp\, dm \tag{3.7}$$

We recall a simple fact about Hilbert spaces. Let X be a Hilbert space, $x\in X, x\neq 0$. Then $|x| = \sup\limits_{|y|=1}|\langle x,y\rangle|$,

$$|x|^2 = \Big(\sup_{|y|=1}|\langle x,y\rangle|\Big)^2. \tag{3.8}$$

This follows from $|\langle x,y\rangle|\le|x|\cdot|y|$ and equality for $y=\frac{x}{|x|}$. We apply this to $E\to M$, $X = L_2(M,E,dp)$, $x = x(m) = W(t,m,p),\eta^{\mathrm{op}}\,(p)\cdot)_p = W(t,m,p)\circ\eta^{\mathrm{op}}\,(p)$ and have to estimate

$$\sup_{\substack{\Phi\in C_c^\infty(E)\\ |\Phi|_{L_2}=1}} N(\Phi) = \sup_{\substack{\Phi\in C_c^\infty(E)\\ |\Phi|_{L_2}=1}} |\langle\delta(m), e^{-tD^2}\eta^{\mathrm{op}}\,\Phi\rangle|_{L_2} \tag{3.9}$$

The heat kernel is of Sobolev class,

$$W(t,m,\cdot)\in H^{\frac{r}{2}}(E),\quad |W(t,m,\cdot)|_{H^{\frac{r}{2}}}\le C_5(t). \tag{3.10}$$

Hence we have can restrict in (3.9) to

$$\sup_{\substack{\Phi\in C_c^\infty(E)\\ |\Phi|_{L_2}=1\\ |\Phi|_{H^{\frac{r}{2}}}\le C_5}} N(\Phi) \tag{3.11}$$

In the sequel we estimate (3.11). For doing this, we recall some simple facts concerning the wave equation

$$\frac{\partial\Phi_s}{\partial s} = iD\Phi_s,\quad \Phi_0=\Phi,\quad \Phi\ C^1 \text{ with compact support.} \tag{3.12}$$

It is well known that (3.12) has a unique solution Φ_s which is given by

$$\Phi_s = e^{isD}\Phi \tag{3.13}$$

and

$$\operatorname{supp}\,\Phi_s\subset U_{|s|}\ (\operatorname{supp}\,\Phi) \tag{3.14}$$

$U_{|s|} = |s|$-neighborhood. Moreover,

$$|\Phi_s|_{L_2} = |\Phi|_{L_2},\quad |\Phi_s|_{H^{\frac{r}{2}}} = |\Phi|_{H^{\frac{r}{2}}}. \tag{3.15}$$

We fix a uniformly locally finite cover $\mathcal{U} = \{U_\nu\}_\nu = \{B_d(x_\nu)\}_\nu$ by normal charts of radius $d < r_{inj}(M,g)$ and associated decomposition of unity $\{\varphi_\nu\}_\nu$ satisfying

$$|\nabla^i \varphi_\nu| \le C \text{ for all } \nu, \ 0 \le i \le k+2 \tag{3.16}$$

Write

$$\begin{aligned} N(\Phi) &= |\langle \delta(m), e^{-tD^2} \eta^{\mathrm{op}}\, \Phi \rangle| = \frac{1}{\sqrt{4\pi t}} \left| \langle \delta(m), \int\limits_{-\infty}^{+\infty} e^{\frac{-s^2}{4t}} e^{isD} (\eta^{\mathrm{op}}\, \Phi)\, ds \rangle \right|_{L_2(dp)} \\ &= \frac{1}{\sqrt{4\pi t}} \left| \int\limits_{-\infty}^{+\infty} e^{\frac{-s^2}{4t}} (e^{isD} \eta^{\mathrm{op}}\, \Phi)(m)\, ds \right|_{L_2(dp)}. \end{aligned} \tag{3.17}$$

We decompose

$$\eta^{\mathrm{op}}\, (\Phi) = \sum_\nu \varphi_\nu \eta^{\mathrm{op}}\, \Phi. \tag{3.18}$$

(3.18) is a locally finite sum, (3.12) linear. Hence

$$(\eta^{\mathrm{op}}\, (\Phi))_s = \sum_\nu (\varphi_\nu \eta^{\mathrm{op}}\, \Phi)_s. \tag{3.19}$$

Denote as above

$$|\ |_{p,i} \equiv |\ |_{W^{p,i}},$$

in particular

$$|\ |_{2,i} \equiv |\ |_{W^{2,i}} \sim |\ |_{H^i}, \quad i \le k. \tag{3.20}$$

Then we obtain from (3.15), (3.16) and an Sobolev embedding theorem

$$\begin{aligned} |(\varphi_\nu \eta^{\mathrm{op}}\, \Phi)_s|_{H^{\frac{r}{2}}} = |\varphi_\nu \eta^{\mathrm{op}}\, \Phi|_{H^{\frac{r}{2}}} &\le C_6 |\varphi_\nu \eta^{\mathrm{op}}\, \Phi|_{2,\frac{r}{2}} \le C_7 |\eta^{\mathrm{op}}\, \Phi|_{2,\frac{r}{2},U_\nu} \\ &\le C_8 |\eta|_{2,\frac{r}{2},U_\nu} \le C_9 |\eta|_{1,r-1,U_\nu} \end{aligned} \tag{3.21}$$

since $r - 1 - \frac{n}{i} \ge \frac{r}{2} - \frac{n}{2}, r - 1 \ge \frac{r}{2}, 2 \ge i$ for $r > n+2$ and $|\Phi|_{H^{\frac{r}{2}}} \le C_5$. This yields together with the Sobolev embedding the estimate

$$\begin{aligned} |(\eta^{\mathrm{op}}\, \Phi)_s(m)| &\le C_{10} \cdot \sum_{\substack{\nu \\ m \in U_s(U_\nu)}} |(\varphi_\nu \eta^{\mathrm{op}}\, \Phi)_s|_{2,\frac{r}{2}} \\ &\le C_{11} \cdot \sum_{\substack{\nu \\ m \in U_s(U_\nu)}} |\eta|_{1,r-1,U_\nu} \le C_{12} \cdot |\eta|_{1,r-1,B_{2d+|s|}(m)} \\ &= C_{12} \cdot \mathrm{vol}\, (B_{2d+|s|}(m)) \cdot \left(\frac{1}{\mathrm{vol}\, B_{2d+|s|}(m)} \cdot |\eta|_{1,r-1,B_{2d+|s|}(m)} \right). \end{aligned} \tag{3.22}$$

There exist constants A and B, independent of m s. t.

$$\mathrm{vol}\, (B_{2d+|s|}(m)) \le A \cdot e^{B_{|s|}}.$$

Write

$$e^{-\frac{s^2}{4t}} \cdot \mathrm{vol}\, (B_{2d+|s|}(m)) \le C_{13} \cdot e^{-\frac{9}{10}\frac{s^2}{4t}}, \quad C_{13} = A \cdot e^{10 B^2 t}, \tag{3.23}$$

thus obtaining

$$N(\Phi) \le C_{14} \int_0^\infty e^{-\frac{9}{10}\frac{s^2}{4t}} \left(\frac{1}{\text{vol } B_{2d+|s|}(m)} \cdot |\eta|_{1,r-1,B_{2d+|s|}(m)} \right) ds,$$

$$C_{14} = C_{12} \cdot C_{13} = C_{12} \cdot A \cdot e^{10B^2 t}.$$

Now we apply Buser/Hebey's inequality

$$\int_M |u - \overline{u}_c| \, \text{dvol}_x(g) \le C \cdot c \int_M |\nabla u| \, \text{dvol}_x(g)$$

for $u \in W^{1,1}(M) \sim C^\infty(M)$, $c \in]0, R[$, $\text{Ric}\,(g) \ge k$, $C = C(n,k,R)$ and

$$\overline{u}_c(x) := \frac{1}{\text{vol } B_c(x)} \int_{B_c(x)} u(y) \, \text{dvol}_y$$

with $R = 3d + s$ and infer

$$\begin{aligned} &\int_M \frac{1}{\text{vol } B_{2d+|s|}(m)} \cdot |\eta|_{1,r-1,B_{2d+|s|}(m)} \, dm \\ &\le |\eta|_{1,r-1} + C(3d+s) \cdot (2d+s) |\nabla \eta|_{1,r-1} \\ &\le |\eta|_{1,r-1} + C(3d+s) \cdot (2d+s) |\eta|_{1,r-1}. \end{aligned} \tag{3.24}$$

$C(3d+s)$ depends on $3d+s$ at most linearly exponentially, i.e.,

$$C(3d+s) \cdot (2d+s) \le A_1 e^{B_1 s}.$$

This implies

$$\int_0^\infty e^{-\frac{9}{10}\frac{s^2}{4t}} \int_M \frac{1}{\text{vol } B_{2d+|s|}(m)} \cdot |\eta|_{1,r-1,B_{2d+|s|}(m)} \, dm \, ds \tag{3.25}$$

$$\begin{aligned} &\le \; = \int_0^\infty e^{-\frac{9}{10}\frac{s^2}{4t}} (|\eta|_{1,r-1} + C(3d+s) \cdot (2d+s) |\eta|_{1,r-1}) \, ds \\ &\le \int_0^\infty e^{-\frac{8}{10}\frac{s^2}{4t}} \, ds (|\eta|_{1,r-1} + A_1 e^{10 B_1^2 t} |\eta|_{1,r} \\ &= \sqrt{t} \cdot \frac{1}{2} \sqrt{5\pi} (|\eta|_{1,r-1} + A_1 e^{10 B_1^2 t} |\eta|_{1,r}) \;<\; \infty. \end{aligned}$$

The function $\mathbb{R}_+ \times M \to \mathbb{R}$,

$$(s,m) \to e^{-\frac{9}{10}\frac{s^2}{4t}} \left(\frac{1}{\text{vol } B_{2d+|s|}(m)} \cdot |\eta|_{1,r-1,B_{2d+|s|}(m)} \right)$$

is measurable, nonnegative, the integrals (3.24), (3.25) exist, hence according to the principle of Tonelli, this function is 1-summable, the Fubini theorem is applicable and

$$\tilde{\eta} := C_{10} \cdot \int_0^\infty e^{-\frac{9}{10}\frac{s^2}{4t}} \left(\frac{1}{\operatorname{vol} B_{2d+|s|}(m)} \cdot |\eta|_{1,r-1,B_{2d+|s|}(m)} \right) ds$$

is (for $\eta \not\equiv 0$) everywhere $\neq 0$ and i-summable. We proved

$$\int |(W(t,m,p), \eta^{\mathrm{op}} \cdot)_p|^2 \leq \tilde{\eta}(m)^2. \tag{3.26}$$

Now we set

$$f(m) = (\tilde{\eta}(m))^{\frac{1}{2}} \tag{3.27}$$

and infer $f(m) \neq 0$ everywhere, $f \in L_2$ and

$$\begin{aligned} |f^{-1} e^{-\frac{s}{2}D^2} \circ \eta|_{L_2}^2 &= \int_M \int_M f(m)^{-2} |((W(\frac{s}{2}, m, p), \eta^{\mathrm{op}}\ )_p|^2 \, dp \, dm \\ &\leq \int_M \frac{1}{\tilde{\eta}(m)} \tilde{\eta}(m)^2 \, dm = \int_M \tilde{\eta}(m) \, dm \\ &\leq C_{12} \cdot A \cdot e^{10B^2 s} \sqrt{s} \cdot \frac{1}{2} \sqrt{5\pi} (|\eta|_{1,r-1} + A_1 e^{10 B_1^2 s} |\eta|_{1,r}) \\ &\leq C_{15} \sqrt{s} e^{10B^2 s} |\eta|_{1,r}, \end{aligned} \tag{3.28}$$

i.e.,

$$|f^{-1} e^{-\frac{s}{2}D^2} \circ \eta|_2 \leq C_{15}^{\frac{1}{2}} \cdot s^{\frac{1}{4}} \cdot e^{5B^2 s} \cdot |\eta|_{1,r}^{\frac{1}{2}}. \tag{3.29}$$

Here according to the term $A_1 e^{10 B_1^2 s}$, C_{15} still depends on s.

We obtain

$$\begin{aligned} &|e^{-\frac{s}{2}D^2} \circ f|_{L_2} \cdot |f^{-1} \circ e^{-\frac{s}{2}D^2} \circ \eta| \leq C_4 |f|_{L_2} \cdot C_{15}^{\frac{1}{2}} \cdot s^{\frac{1}{4}} \cdot e^{5B^2 s} \cdot |\eta|_{1,r}^{\frac{1}{2}} \\ &\leq C_4 \cdot C_{15} \sqrt{s} e^{10B^2 s} |\eta|_{1,r} = C_{16} \cdot \sqrt{s} \cdot e^{10B^2 s} |\eta|_{1,r}. \end{aligned} \tag{3.30}$$

This yields $e^{-sD^2} \circ \eta$ is of trace class,

$$|e^{-sD^2} \eta|_1 \leq |e^{-\frac{s}{2}D^2} \circ f|_2 \cdot |f^{-1} e^{-\frac{s}{2}D^2} \eta|_2 \leq C_{16} \sqrt{s} e^{10B^2 s} |\eta|_{1,r}, \tag{3.31}$$

$e^{-sD^2} \circ \eta \circ D' \circ e^{-(t-s)D'^2}$ is of trace class,

$$\begin{aligned} |e^{-sD^2} \circ \eta \circ D' \circ e^{-(t-s)D'^2}|_1 &\leq |e^{-sD^2} \eta|_1 \cdot |D' e^{-(t-s)D'^2}|_{\mathrm{op}} \\ &\leq C_{16} \sqrt{s} e^{10B^2 s} |\eta|_{1,r} \cdot C' \cdot \frac{1}{\sqrt{t-s}}, \end{aligned} \tag{3.32}$$

$$\left| \int_{\frac{t}{2}}^{t} (e^{-sD^2} \circ \eta \circ D' \circ e^{-(t-s)D'^2} ds \right|_1 \leq \int_{\frac{t}{2}}^{t} |e^{-sD^2} \eta \circ D' e^{-(t-s)D'^2}|_1 \, ds$$

$$\leq C_{16} \cdot C' \cdot e^{10B^2 t} |\eta|_{1,r} \cdot \int_{\frac{t}{2}}^{t} \left(\frac{s}{t-s} \right)^{\frac{1}{2}} ds, \tag{3.33}$$

$$\int_{\frac{t}{2}}^{t} \left(\frac{s}{t-s} \right)^{\frac{1}{2}} ds = [\sqrt{s(t-s)} + \frac{t}{2} \arcsin \frac{2s-t}{t}]_{\frac{t}{2}}^{t} = -\frac{t}{2} + \frac{t}{2}\frac{\pi}{2} = \frac{t}{2}(\frac{\pi}{2} - 1),$$

$$\begin{aligned} \left| \int_{\frac{t}{2}}^{t} (e^{-sD^2} \circ \eta \circ D' \circ e^{-(t-s)D'^2} ds \right|_1 &\leq C_{16} \cdot C' \cdot e^{10B^2 t} \cdot (\frac{\pi}{2} - 1) \cdot \frac{t}{2} |\eta|_{1,r} \\ &= C_{17} e^{10B^2 t} \cdot t \cdot |\eta|_{1,r}. \end{aligned} \tag{3.34}$$

Here $C_{17} = C_{17}(t)$ and $C_{17}(t)$ can grow exponentially in t if the volume grows exponentially. (3.34) expresses the fact that (I_4) is of trace class and its trace norm is uniformly bounded on any t-interval $[a_0, a_1]$, $a_0 > 0$. The treatment of (I_1)–(I_3) is quite parallel to that of (I_4). Write the integrand of (I_3), (I_2) or (I_1) as

$$(De^{-\frac{s}{2}D^2})[(e^{-\frac{s}{4}D^2} f)(f^{-1} e^{-\frac{s}{4}D^2} \eta)] e^{-(t-s)D'^2} \tag{3.35}$$

or

$$(e^{-sD^2})[(\eta e^{-\frac{(t-s)}{4}D'^2} f^{-1})(f e^{-\frac{(t-s)}{4}D'^2})] D' e^{-\frac{t-s}{2}D'^2} \tag{3.36}$$

or

$$(e^{-sD^2} D)[(\eta e^{-\frac{(t-s)}{2}D'^2} f^{-1})(f e^{-\frac{(t-s)}{2}D'^2})], \tag{3.37}$$

respectively. Then in the considered intervals the expression $[\dots]$ are of trace class which can literally be proved as for (I_4). The main point in (I_4) was the estimate of $f^{-1}e^{-\tau D^2}\eta$. In (3.36), (3.37) we have to estimate expressions $\eta e^{-\tau D'^2} f^{-1}$. Here we use the fact that $\eta = \eta^{\mathrm{op}}$ is symmetric with respect to the fibre metric h: the endomorphism $\eta_{e_i}(\cdot)$ is skew symmetric as the Clifford multiplication $e_i\cdot$ which yields together that η^{op} is symmetric. Then the L_2-estimate of $(\eta^{\mathrm{op}} \cdot W'(\tau, m, p), \cdot)$ is the same as that of $W'(\tau, m, p), \eta^{\mathrm{op}}(p)\cdot)$ and we can perform the same procedure as that starting with (3.6). The only distinction are other constants. Here essentially enters the equivalence of the D- and D'-Sobolev spaces, i.e., the symmetry of our uniform structure. The factors outside $[\dots]$ produce $\frac{1}{\sqrt{s}}$ on $[\frac{t}{2}, t]$, $\frac{1}{\sqrt{t-s}}$ and $\frac{1}{\sqrt{s}}$ on $[0, \frac{t}{2}]$ (up to constants). Hence (I_1) – (I_3) are of trace class with uniformly bounded trace norm on any t-interval $[a_0, a_1]$, $a_0 > 0$. This finishes the proof of the first part of Theorem 3.3.

We must still prove the trace class property of

$$e^{-tD^2} D - e^{-tD'^2} D'. \tag{3.38}$$

Consider the decomposition

$$e^{-tD^2}D - e^{-tD'^2}D' = e^{-\frac{t}{2}D^2}D(e^{-\frac{t}{2}D^2} - e^{-\frac{t}{2}D'^2}) \tag{3.39}$$

$$+ \quad (e^{-\frac{t}{2}D^2}D - e^{-\frac{t}{2}D'^2}D')e^{-\frac{t}{2}D'^2}. \tag{3.40}$$

According to the first part, $e^{-\frac{t}{2}D^2} - e^{-\frac{t}{2}D'^2}$ is for $t > 0$ of trace class. Moreover, $e^{-\frac{t}{2}D^2}D = De^{-\frac{t}{2}D^2}$ is for $t > 0$ bounded, its operator norm is $\leq \frac{C}{\sqrt{t}}$. Hence their product is for $t > 0$ of trace class and has bounded trace norm for $t \in [a_0, a_1]$, $a_0 > 0$. (3.39) is done. We can write (3.40) as

$$\begin{aligned}
&(e^{-\frac{t}{2}D^2}D - e^{-\frac{t}{2}D'^2}D')e^{-\frac{t}{2}D'^2} \\
&= [e^{-\frac{t}{2}D^2}(D - D') + (e^{-\frac{t}{2}D^2} - e^{-\frac{t}{2}D'^2})D'] \cdot e^{-\frac{t}{2}D'^2} \\
&= [-e^{\frac{t}{2}D^2}\eta]e^{-\frac{t}{2}D'^2} + [\int_0^{\frac{t}{2}} e^{-sD^2}D\eta e^{-(\frac{t}{2}-s)D'^s}\, ds \\
&+ \int_0^{\frac{t}{2}} e^{-sD^2}\eta D'e^{-(\frac{t}{2}-s)D'^2}\, ds](D'e^{-\frac{t}{2}D'^2}).
\end{aligned} \tag{3.41}$$

Now

$$[e^{-\frac{t}{2}D^2}\eta] \cdot e^{-\frac{t}{2}D'^2} = [(e^{-\frac{t}{4}D^2}f)(f^{-1}e^{-\frac{t}{4}D^2}\eta)]e^{-\frac{t}{2}D'^2}. \tag{3.42}$$

(3.42) is of trace class and its trace norm is uniformly bounded on any $[a_0, a_1]$, $a_0 > 0$, according the proof of the first part. If we decompose $\int_0^{\frac{t}{2}} = \int_0^{\frac{t}{4}} + \int_{\frac{t}{4}}^{\frac{t}{2}}$ then we obtain back from the integrals in (3.41) the integrals $(I_1) - (I_4)$, replacing $t \to \frac{t}{2}$. These are done. $D'e^{-\frac{t}{2}D'^2}$ generates $C/\sqrt{t}$ in the estimate of the trace norm. Hence we are done. □

Our procedure is to admit much more general perturbations than those of $\nabla = \nabla^h$ only. Nevertheless, the discussion of more general perturbations is modelled by the case of ∇-perturbation. In this next step, we admit perturbations of $g, \nabla^h, \cdot$, fixing h, the topology and vector bundle structure of $E \longrightarrow M$. The next main result shall be formulated as follows.

Theorem 3.5. *Let* $E = (E, h, \nabla = \nabla^h, \cdot) \longrightarrow (M^n, g)$ *be a Clifford bundle with* (I), $(B_k(M, g))$, $(B_k(E, \nabla))$, $k \geq r + 1 > n + 3$, $E' = (E, h, \nabla' = \nabla'^h, \cdot') \longrightarrow (M^n, g') \in \operatorname{gen\,comp}^{1,r+1}_{L,\text{diff}\ ,F}(E) \cap \mathrm{CL}\mathcal{B}^{N,n}(I, B_k)$, $D = D(g, h, \nabla = \nabla^h, \cdot)$, $D' = D(g', h, \nabla' = \nabla'^h, \cdot')$ *the associated generalized Dirac operators. Then for* $t > 0$

$$e^{-tD^2} - e^{-tD'^2_{L_2}} \tag{3.43}$$

is of trace class and the trace norm is uniformly bounded on compact t*-intervals* $[a_0, a_1]$, $a_0 > 0$.

Here $D'^2_{L_2}$ is the unitary transformation of D'^2 to $L_2 = L_2((M,E),g,h)$. 3.5 needs some explanations. D acts in $L_2 = L_2((M,E),g,h)$, D' in $L'_2 = L_2((M,E),g',h)$. L_2 and L'_2 are quasi isometric Hilbert spaces. As vector spaces they coincide, their scalar products can be quite different but must be mutually bounded at the diagonal after multiplication by constants. D is self-adjoint on $\mathcal{D}_D$ in L_2, D' is self-adjoint on $\mathcal{D}_{D'}$ in L'_2 but not necessarily in L_2. Hence $e^{-tD'^2}$ and $e^{-tD^2} - e^{-tD'^2}$ are not defined in L_2. One has to graft D^2 or D'^2. Write $\mathrm{dvol}_q(g) \equiv dq(g) = \alpha(q) \cdot dq(g') \equiv \mathrm{dvol}_q(g')$. Then

$$\begin{aligned}&0 < c_1 \leq \alpha(q) \leq c_2, \alpha, \alpha^{-1} \text{ are } (g, \nabla^g)\text{- and } (g', \nabla^{g'})\text{-bounded}\\ &\text{up to order } 3, |\alpha - 1|_{g,1,r+1}, |\alpha - 1|_{g',1,r+1} < \infty,\end{aligned} \tag{3.44}$$

since $g' \in \mathrm{comp}^{1,r+1}(g)$. Define $U : L_2 \longrightarrow L'_2$, $U\Phi = \alpha^{\frac{1}{2}}\Phi$. Then U is a unitary equivalence between L_2 and L'_2, $U^* = U^{-1}$. $D'_{L_2} := U^*D'U$ acts in L_2, is self-adjoint on $U^{-1}(\mathcal{D}_{D'})$, since U is a unitary equivalence. The same holds for $D'^2_{L_2} = U^*D'^2U = (U^*D'U)^2$. It follows from the definition of the spectral measure, the spectral integral and the spectral representations $D'^2 = \int \lambda^2\, dE'_\lambda$, $e^{-tD'^2} = \int e^{-t\lambda^2}\, dE'_\lambda$ that $D'^2_{L_2} = U^*D'^2U = U^* \int \lambda^2\, dE'_\lambda U = \int \lambda^2\, d(U^*E'_\lambda U)$ and

$$e^{-tD'^2_{L_2}} = \int e^{-t\lambda^2}\, d(U^*E'_\lambda U) = U^*(\int e^{-t\lambda^2}\, dE'_\lambda)U = U^*e^{-tD'^2}U. \tag{3.45}$$

In (3.43) $e^{-tD'^2_{L_2}}$ means $e^{-tD'^2_{L_2}} = e^{-t(U^*D'U)^2} = U^*e^{-tD'^2}U$. We obtain from $g' \in \mathrm{comp}^{1,r+1}(g)$, $\nabla'^h \in \mathrm{comp}^{1,r+1}(\nabla^h g)$, $\cdot' \in \mathrm{comp}^{1,r+1}(\cdot)$, $D - \alpha^{-\frac{1}{2}}D'\alpha^{\frac{1}{2}} = D - D' - \frac{grad'\alpha \cdot'}{2\alpha}$ and (3.44) the following lemma concerning the equivalence of Sobolev spaces.

Lemma 3.6. *$W^{1,i}(E,g,h,\nabla^h) = W^{1,i}(E,g',h,\nabla'^h)$ as equivalent Banach spaces, $0 \leq i \leq r+1$.* □

Corollary 3.7. *$W^{2,i}(E,g,h,\nabla^h) = W^{2,i}(E,g',h,\nabla'^h)$ as equivalent Hilbert spaces, $0 \leq j \leq \frac{r+1}{2}$.* □

Corollary 3.8. *$H^j(E,D) \cong H^jK(E,D')$, $0 \leq j \leq \frac{r+1}{2}$.* □

3.6 has a parallel version for the endomorphism bundle $\mathrm{End}\, E$.

Lemma 3.9. *$\Omega^{1,1,i}(\mathrm{End}\, E, g, h, \nabla^h) \cong \Omega^{1,1,i}(\mathrm{End}\, E, g', h, \nabla'^h)$ $0 \leq i \leq r+1$.* □

Lemma 3.10. *$\Omega^{1,2,j}(\mathrm{End}\, E, g, h, \nabla^h) \cong \Omega^{1,2,j}(\mathrm{End}\, E, g', h, \nabla'^h)$ $0 \leq j \leq \frac{r+1}{2}$.* □

$e^{-tD'^2_{L_2}} : L_2 \longrightarrow L_2$ has evidently the heat kernel

$$W'_{L_2}(t,m,p) = \alpha^{-\frac{1}{2}}(m)W'(t,m,p)\alpha^{\frac{1}{2}}(p)$$

$W' \equiv W_{L'_2}$. Our next task is to obtain an explicit expression for $e^{-tD^2} - e^{-tD'^2_{L_2}}$. For this we apply again Duhamel's principle. The steps 1)–4) in the proof of 3.4

remain. Then we set $\Phi(t,q) = W(t,m,q)$, $\Psi(t,q) = W'_{L_2}(t,m,q)$ and obtain

$$-\int_{\alpha}^{\beta}\int_{M} h_q(W(\tau,m,q),(D^2+\frac{\partial}{\partial t})W'_{L_2}(t-\tau,q,p))\,dq(g)\,d\tau$$
$$= \int_{M} [h_q(W(\beta,m,q),W'_{L_2}(t-\beta,q,p) - h_q(W(\alpha,m,q),W'_{L_2}(t-\alpha,q,p)]\,dq(g).$$

Performing $\alpha \longrightarrow 0^+, \beta \longrightarrow t$ and using $dq(g) = \alpha(q)dq(g')$ yields

$$-\int_{0}^{t}\int_{M} h_q(W(s,m,q),(D^2+\frac{\partial}{\partial t})W'(t-s,q,p))\,dq(g)\,ds =$$
$$= -\int_{0}^{t}\int_{M} [h_q(W(s,m,q),(D^2-D'^2_{L_2})W'_{L_2}(t-s,q,p)\,dq(g)\,ds$$
$$= W(t,m,p)\alpha(p) - W'_{L_2}(t,m,p). \tag{3.46}$$

(3.46) expresses the operator equation

$$e^{-tD^2}\alpha - e^{-tD'^2_{L_2}} = -\int_0^t e^{-sD^2}(D^2-D'^2_{L_2})e^{-(t-s)D'^2_{L_2}}\,ds.$$
$$e^{-tD^2}\alpha - e^{-tD'^2_{L_2}} = e^{-tD^2}(\alpha-1) + e^{-tD^2} - e^{-tD'^2_{L_2}},$$

hence

$$e^{-tD^2} - e^{-tD'^2_{L_2}} = -e^{-tD^2}(\alpha-1) - \int_0^t e^{-sD^2}(D^2-D'^2_{L_2})e^{-(t-s)D'^2_{L_2}}\,ds. \tag{3.47}$$

As we mentioned in (3.44), $(\alpha-1) = \frac{dq(g)}{dq(g')} - 1 = \frac{\sqrt{\det g}}{\sqrt{\det g'}} - 1 \in \Omega^{0,1,r+1}$ since $g \in \mathrm{comp}^{1,r+1}(g)$. We write $e^{-tD^2}(\alpha-1) = (e^{-\frac{t}{2}D^2}f)(f^{-1}e^{-\frac{t}{2}D^2}(\alpha-1))$, determine f as in the proof of Theorem 3.3 from $\eta_\alpha = \alpha-1$ and obtain $e^{-tD^2}(\alpha-1)$ is of trace class with trace norm uniformly bounded on any t-interval $[a_0,a_1]$, $a_0>0$. Decompose $D^2 - D'^2_{L_2} = D(D-D'_{L_2}) + (D-D'_{L_2})D'_{L_2}$. We need explicit analytic expressions for this. $D(D-D'_{L_2}) = D(D-\alpha^{-\frac{1}{2}}D'\alpha^{\frac{1}{2}}) = D(D-D') - D\frac{\mathrm{grad}\,'\alpha\cdot'}{2\alpha}$, $(D-D'_{L_2})D'_{L_2} = ((D-D') - \frac{\mathrm{grad}\,'\alpha\cdot'}{2\alpha})\alpha^{-\frac{1}{2}}D'\alpha^{\frac{1}{2}}$. If we set again $D-D' = -\eta$ then we have to consider as before with $\frac{\operatorname{grad}'\alpha}{2\alpha} = \frac{\operatorname{grad}'\alpha\cdot'}{2\alpha}$ where $\operatorname{grad}' \equiv \operatorname{grad}_{g'}$

$$\int_0^{\frac{t}{2}} e^{-sD^2} D(\eta - \frac{\operatorname{grad}'\alpha}{2\alpha}) e^{-(t-s)D'^2_{L_2}}\, ds$$

$$+ \int_0^{\frac{t}{2}} e^{-sD^2} (\eta - \frac{\operatorname{grad}'\alpha}{2\alpha}) D'_{L_2} e^{-(t-s)D'^2_{L_2}}\, ds$$

$$+ \int_{\frac{t}{2}}^{t} e^{-sD^2} D(\eta - \frac{\operatorname{grad}'\alpha}{2\alpha}) e^{-(t-s)D'^2_{L_2}}\, ds$$

$$+ \int_{\frac{t}{2}}^{t} e^{-sD^2} (\eta - \frac{\operatorname{grad}'\alpha}{2\alpha}) D'_{L_2} e^{-(t-s)D'^2_{L_2}}\, ds.$$

It follows immediately from $g' \in \operatorname{comp}^{1,r+1}(g)$ that the vector field $\frac{\operatorname{grad}'\alpha}{\alpha} \in \Omega^{0,1,r}(TM)$. If we write $\eta_0^{\mathrm{op}} = -\frac{\operatorname{grad}'\alpha\cdot'}{\alpha}$ then η_0^{op} is a zero-order operator, $|\eta_0|_r < \infty$ and we literally repeat the procedure for $(I_1) - (I_4)$ as before, inserting $\eta_0 = -\frac{\operatorname{grad}'\alpha\cdot'}{\alpha}$ for η there. Hence there remains to discuss the integrals

$$\int_0^t e^{-sD^2} D\eta e^{-(t-s)D'^2_{L_2}}\, ds + \int_0^t e^{-sD^2} \eta D'_{L_2} e^{-(t-s)D'^2_{L_2}}\, ds. \tag{3.48}$$

The next main step is to insert explicit expressions for $D - D'$.

Let $m_0 \in M$, $U = U(m_0)$ a manifold and bundle coordinate neighborhood with coordinates $x^1, \dots, x^n$ and local bundle basis $\Phi_1, \dots, \Phi_n : U \longrightarrow E|_U$. Setting $\nabla_{\frac{\partial}{\partial x_i}} \Phi_\alpha \equiv \nabla_i \Phi_\alpha = \Gamma^\beta_{i\alpha} \Phi_\beta$, $\nabla \Phi_\alpha = dx^i \otimes \Gamma^\beta_{i\alpha} \Phi_\beta$, we can write $D\Phi_\alpha = \Gamma^\beta_{i\alpha} g^{ik} \frac{\partial}{\partial x^k} \cdot \Phi_\beta$, $D'\Phi_\alpha = \Gamma'^\beta_{i\alpha} g'^{ik} \frac{\partial}{\partial x^k} \cdot' \Phi_\beta$, or for a local section Φ

$$D\Phi = g^{ik} \frac{\partial}{\partial x^k} \cdot \nabla_i \Phi, \quad D'\Phi = g'^{ik} \frac{\partial}{\partial x^k} \cdot' \nabla'_i \Phi. \tag{3.49}$$

This yields

$$\begin{aligned} -(D - D')\Phi &= g^{ik} \frac{\partial}{\partial x^k} \cdot \nabla_i \Phi - g'^{ik} \frac{\partial}{\partial x^k} \cdot' \nabla'_i \Phi \\ &= [(g^{ik} - g'^{ik}) \frac{\partial}{\partial x^k} \cdot \nabla_i + g'^{ik} \frac{\partial}{\partial x^k} \cdot (\nabla_i - \nabla'_i) \\ &\quad + g'^{ik} \frac{\partial}{\partial x^k} (\cdot - \cdot') \nabla'_i] \Phi, \end{aligned} \tag{3.50}$$

i.e., we can write

$$-(D - D')\Phi = (\eta_1^{\mathrm{op}} + \eta_2^{\mathrm{op}} + \eta_3^{\mathrm{op}})\Phi, \tag{3.51}$$

where locally

$$\eta_1^{\text{op}}\,\Phi \;=\; (g^{ik} - g'^{ik})\frac{\partial}{\partial x^k}\cdot \nabla_i \Phi, \tag{3.52}$$

$$\eta_2^{\text{op}}\,\Phi \;=\; g'^{ik}\frac{\partial}{\partial x^k}\cdot (\nabla_i - \nabla_i')\Phi, \tag{3.53}$$

$$\eta_3^{\text{op}}\,\Phi \;=\; g'^{ik}\frac{\partial}{\partial x^k}(\cdot - \cdot')\nabla_i'\Phi. \tag{3.54}$$

Here $(g'^{ik}) = (g'_{jl})^{-1}$. We simply write η_ν instead η_ν^{op} , hence

$$(3.48) \;=\; \int_0^t e^{-sD^2} D(\eta_1 + \eta_2 + \eta_3) e^{-(t-s)D'^2_{L_2}}\, ds \tag{3.55}$$

$$+\; \int_0^t e^{-sD^2}(\eta_1 + \eta_2 + \eta_3) D'_{L_2} e^{-(t-s)D'^2_{L_2}}\, ds. \tag{3.56}$$

We have to estimate

$$\int_0^t e^{-sD^2} D\eta_\nu e^{-(t-s)D'^2_{L_2}}\, ds \tag{3.57}$$

and

$$\int_0^t e^{-sD^2} \eta_\nu D'_{L_2} e^{-(t-s)D'^2_{L_2}}\, ds. \tag{3.58}$$

Decompose $\int\limits_0^t = \int\limits_0^{\frac{t}{2}} + \int\limits_{\frac{t}{2}}^t$ which yields

$$\int_0^{\frac{t}{2}} e^{-sD^2} D\eta_\nu e^{-(t-s)D'^2_{L_2}}\, ds, \tag{I$_{\nu,1}$}$$

$$\int_0^{\frac{t}{2}} e^{-sD^2} \eta_\nu D'_{L_2} e^{-(t-s)D'^2_{L_2}}\, ds, \tag{I$_{\nu,2}$}$$

$$\int_{\frac{t}{2}}^t e^{-sD^2} D\eta_\nu e^{-(t-s)D'^2_{L_2}}\, ds, \tag{I$_{\nu,3}$}$$

$$\int_{\frac{t}{2}}^t e^{-sD^2} \eta_\nu D'_{L_2} e^{-(t-s)D'^2_{L_2}}\, ds. \tag{I$_{\nu,4}$}$$

$(I_{\nu,1})$–$(I_{\nu,4})$ look as (I_1)–(I_4) as before. But in distinction to that, not all $\eta_\nu = \eta_\nu^{\text{op}}$ are operators of order zero. Only η_2 is a zero-order operator, generated by an

End$E$-valued 1-form η_2. η_1 and η_3 are first-order operators. We start with $\nu = 2$, $\eta_2 \cdot |\eta_2|_{1,r} < \infty$ is a consequence of $E' \in \mathrm{comp}^{1,r+1}_{L,\mathrm{diff}}(E)$ and we are from an analytical point of view exactly in the situation as before. $(I_{2,1})$–$(I_{2,4})$ can be estimated quite parallel to (I_1)–(I_4) and we are done. There remains to estimate $(I_{\nu,j})$, $\nu \neq 2$, $j = 1, \ldots, 4$. We start with $\nu = 1$, $j = 3$ and write

$$e^{-sD^2} D\eta_1 e^{-(t-s)D'^2} = (De^{-\frac{s}{2}D^2})(e^{-\frac{s}{4}D^2} \cdot f)(f^{-1}e^{-\frac{s}{4}D^2}\eta_1)(e^{-(t-s)D'^2}). \quad (3.59)$$

$De^{-\frac{s}{2}D^2}$ and $e^{-(t-s)D'^2}$ are bounded in $[\frac{t}{2}, t]$ and we perform their estimate as before. $e^{-\frac{s}{4}D^2} \cdot f$ is Hilbert–Schmidt if $f \in L_2$. There remains to show that for appropriate f

$$f^{-1}e^{-\frac{s}{4}D^2}\eta_1$$

is Hilbert–Schmidt. Recall $r + 1 > n + 3$, $n \geq 2$, which implies $\frac{r}{2} > \frac{n}{2} + 1$, $r - 1 - n \geq \frac{r}{2} - \frac{n}{2}$, $r - 1 \geq \frac{r}{2}$, $2 \geq i$. If we write in the sequel pointwise or Sobolev norms we should always write $|\Psi|_{g',h,m'}$, $|\Psi|_{H^\nu(E,D')}$, $|\Psi|_{g',h,\nabla',2,\frac{r}{2}}$, $|g - g'|_{g',m}$, $|g - g'|_{g',1,r}$ etc. or the same with respect to g, h, ∇, D, depending on the situation. But we often omit the reference to $g', h, \nabla', D, m, g, h \ldots$ in the notation. The justification for doing this in the Sobolev case is the symmetry of our uniform structure.

Now

$$\begin{aligned}
(\eta_1\Phi)(m) &= ((g^{ik} - g'^{ik})\frac{\partial}{\partial x^k} \cdot \nabla_i\Phi)|_m, \quad (3.60)\\
|\eta_1\Phi|_m &= |\eta_1\Phi|_{g,h,m}\\
&\leq C_1 \cdot |g - g'|_{g,m} \cdot \left(\sum_{k=1}^n \left|\frac{\partial}{\partial x^k}\right|^2_{g,m}\right)^{\frac{1}{2}} \cdot \left(\sum_{i=1}^n |\nabla_i\Phi|^2_{h,m}\right)^{\frac{1}{2}}.
\end{aligned}$$

To estimate $\sum\limits_{k=1}^n \left|\frac{\partial}{\partial x^k}\right|^2_{g,m}$ more concretely we assume that $x^1, \ldots, x^n$ are normal coordinates with respect to g, i.e., we assume a (uniformly locally finite) cover of M by normal charts of fixed radius $\leq r_{inj}(M, g)$. Then $\left|\frac{\partial}{\partial x^k}\right|^2_{g,m} = g\left(\frac{\partial}{\partial x^k}, \frac{\partial}{\partial x^k}\right) = g_{kk}(m)$, and there is a constant $C_2 = C_2(R, r_{inj}(M, g))$ s. t. $\left(\sum\limits_{i=1}^n |\nabla_i\Phi|^2_{h,m}\right)^{\frac{1}{2}} \leq C_2$. Using finally $|\nabla_X\Phi| \leq |X| \cdot |\nabla\Phi|$, we obtain

$$|\eta_1\Phi|_m \leq C \cdot |g - g'|_g \cdot |\nabla\Phi|_{h,m}. \quad (3.61)$$

(3.61) extends by the Leibniz rule to higher derivatives $|\nabla^k\eta_1\Phi|_m$, where the polynomials on the right hand side are integrable by the module structure theorem (this is just the content of this theorem). (3.60), (3.61) also hold (with other constants) if we perform some of the replacements $g \longrightarrow g'$, $\nabla \longrightarrow \nabla'$: We remark that the expressions $D(g, h, \nabla^h, \cdot$, $D(g', h, \nabla^h, \cdot)$ are invariantly defined, hence

$$[D(g, h, \nabla^h, \cdot) - D(g, h, \nabla^h, \cdot)](\Phi|_U) = ((g^{ik} - g'^{ik})\partial_k) \cdot \nabla_i(\Phi|_U). \quad (3.62)$$

We have to estimate the kernel of

$$h_p(W(t,m,p), \eta_1^{\text{op}} \cdot) \tag{3.63}$$

in $L_2((M,E),g,h)$ and to show that this represents the product of two Hilbert–Schmidt operators in $L_2 = L_2((M,E),g,h)$. We cannot immediately apply the procedure as before since η_1^{op} is not of zero order but we would be done if we could write (3.63) as

$$(\eta_{1,1}^{\text{op}}(p)W(t,m,p), \eta_{1,0}^{\text{op}} \cdot), \tag{3.64}$$

$\eta_{1,1}^{\text{op}}$ of first order, $\eta_{1,0}^{\text{op}}$ of zeroth order.

Then we would replace W by $\eta_{1,1}^{\text{op}}(p)W(t,m,p)$, apply $k \geq r+1 > n+3$, and obtain

$$\eta_{1,1}^{\text{op}} W(t,m,\cdot) \in H^{\frac{r}{2}}(E), \quad |W(t,m,\cdot)|_{H^{\frac{r}{2}}} \leq C(t) \tag{3.65}$$

and would then literally proceed as before.

Let $\Phi \in C_c^\infty(U)$. Then

$$\begin{aligned}
&\int (W(t,m,p), \eta_1^{\text{op}}(p)\Phi(p))_p \, dvol_p(g) \\
&= \int (((g^{ik} - g'^{ik})\partial_k) \cdot \nabla_i W, \Phi)_p \, dvol_p(g) \\
&\quad - \int (W, (\nabla_i (g^{ik} - g'^{ik})\partial_k) \cdot \Phi) \, dvol_p(g) \\
&= - \int (\nabla_i W, (g^{ik} - g'^{ik})\partial_k \cdot \Phi)_p \, dvol_p(g) \\
&\quad - \int (W, (\nabla_i ((g^{ik} - g'^{ik})\partial_k)) \cdot \Phi)_p \, dvol_p(g).
\end{aligned}$$

This can easily be globalized by introducing a u. l. f. cover by normal charts $\{U_\alpha\}_\alpha$ of fixed radius, an associated decomposition of unity $\{\varphi_\alpha\}_\alpha$ as follows:

$$\begin{aligned}
&\int (W, \eta_1^{\text{op}} (\sum \varphi_\alpha \Phi)) = \sum_\alpha \int (W, \eta_1^{\text{op}} (\varphi_\alpha \Phi)) \\
&= \sum_\alpha \int (\nabla_{\alpha,i} W, ((g_\alpha^{ik} - g'^{ik}_\alpha)\partial_{\alpha,k} \cdot \varphi_\alpha \Phi) \\
&\quad - \sum_\alpha \int (W, (\nabla_{\alpha,i}((g_\alpha^{ik} - g'^{ik}_\alpha)\partial_k)) \cdot \varphi_\alpha \Phi)
\end{aligned}$$

$$= - \int (\sum_\alpha \nabla_{\alpha,i} W, \varphi_\alpha ((g_\alpha^{ik} - g'^{ik}_\alpha)\partial_{\alpha,k}) \cdot \Phi) \tag{3.66}$$

$$- \int (W, \sum_\alpha \varphi_\alpha (\nabla_{\alpha,i}((g_\alpha^{ik} - g'^{ik}_\alpha)\partial_k)) \cdot \Phi). \tag{3.67}$$

Using (3.66), (3.67), we write

$$\begin{aligned} N(\Phi) &= |\langle \delta(m), e^{tD^2} \eta_1^{\mathrm{op}} \, \Phi\rangle|_{L_2(M,E,dp)} \\ &= |(W(t,m,p), \eta_1^{\mathrm{op}} \, \Phi)_p|_{L_2(M,E,dp)} \\ &= |(\eta_{1,1}^{\mathrm{op}}(p) W(t,m,p), \eta_{1,0}^{\mathrm{op}} \Phi)_p \\ &\quad + (W(t,m,p,\eta_{1,0,0}^{\mathrm{op}} \Phi))_p|_{L_2(M,E,dp)}. \end{aligned} \tag{3.68}$$

Now we use $|\nabla_X \chi| \leq |X| \cdot |\nabla \chi|$, that the cover is u.l.f. and $|\nabla W| \leq C_1 \cdot (|DW| + W)$ (since we have bounded geometry) and obtain

$$\begin{aligned} N(\Phi) &\leq C \cdot (|(DW(t,m,p), \eta_{1,0}^{\mathrm{op}} \Phi|_{L_2(M,E,dp)} + |W(t,m,p,\eta_{1,0,0}^{\mathrm{op}} \Phi|_{L_2(dp)} \\ &\equiv C \cdot (N_1(\Phi) + N_2(\Phi)). \end{aligned} \tag{3.69}$$

Hence we have to estimate

$$\sup_{\substack{\Phi \in C_c^\infty(E) \\ |\Phi|_{L_2}=1}} N_1(\Phi) = \sup_{\substack{\Phi \in C_c^\infty(E) \\ |\Phi|_{L_2}=1}} |\langle \delta(m), (De^{-tD^2}) \eta_{1,0}^{\mathrm{op}} \Phi\rangle|_{L_2 dp} \tag{3.70}$$

and

$$\sup_{\substack{\Phi \in C_c^\infty(E) \\ |\Phi|_{L_2}=1}} N_2(\Phi) = \sup_{\substack{\Phi \in C_c^\infty(E) \\ |\Phi|_{L_2}=1}} |\langle \delta(m), (e^{-tD^2}) \eta_{1,0,0}^{\mathrm{op}} \Phi\rangle|_{L_2 dp}. \tag{3.71}$$

According to $k > r+1 > n+3$,

$$\begin{aligned} &D(W(t,m,\cdot), W(t,m,\cdot) \in H^{\frac{r}{2}}(E), \\ &|(D(W(t,m,\cdot)|_{H^{\frac{r}{2}}}, |W(t,m,\cdot)|_{H^{\frac{r}{2}}} \leq C_1(t) \end{aligned} \tag{3.72}$$

and we can restrict in (3.70), (3.71) to

$$\sup_{\substack{\Phi \in C_c^\infty(E) \\ |\Phi|_{L_2}=1 \\ |\Phi|_{H^{\frac{r}{2}}} \leq C_1(t)}} N_i(\Phi). \tag{3.73}$$

$\eta_{1,0}^{\mathrm{op}}$, $\eta_{1,0,0}^{\mathrm{op}}$ are of order zero and we estimate them by

$$C \cdot |g - g'|_{g,2,\frac{r}{2}} \leq C'|g - g'|_{g,1,r-1} \tag{3.74}$$

$$\begin{aligned} \text{and} \quad D \cdot |\nabla(g - g')|_{g,2,\frac{r}{2}} &\leq D'|\nabla(g - g')|_{g,1,r-1} \\ &\leq D''|g - g'|_{g,1,r} \end{aligned} \tag{3.75}$$

respectively. As we have seen already, into the estimate (3.75) enters $|\nabla \eta|_{1,r-1}$, i.e., in our case $|\nabla^2(g - g')|_{r-1} \sim |g - g'|_{r+1}$. For this reason we assumed $E' \in \mathrm{comp}_{L,\mathrm{diff}\,,F}^{1,r+1}(E)$. In the expression for $N_1(\Phi)$ there is now a slight deviation,

$$N_1(\Phi) = \frac{1}{\sqrt{4\pi t}} \frac{1}{2t} \left| \int_{-\infty}^{+\infty} s \cdot e^{-\frac{s^2}{4t}} e^{isD} \eta_{1,0}^{\mathrm{op}} \Phi(m) \, ds \right|. \tag{3.76}$$

We estimate in (3.76) $s \cdot e^{-\frac{1}{18}\frac{s^2}{4t}}$ by a constant and write

$$e^{-\frac{17}{18}\frac{s^2}{4t}} \cdot \mathrm{vol}\,(B_{2d+s}(m)) \leq C \cdot e^{-\frac{9}{10}\frac{s^2}{4t}}$$

and proceed now for $N_1(\Phi)$, $N_2(\Phi)$ literally as before. Hence (3.59) is of trace class, its trace norm is uniformly bounded on any t-interval $[a_0, a_1]$, $a_0 > 0$. $(I_{1,3})$ is done. $(I_{1,4})$ is absolutely parallel to $(I_{1,3})$, even better, since the left hand factor D is missing. $|D'_{L_2} e^{-(t-s)D'^2_{L_2}}|_{\text{op}}$ now produces the factor $\frac{1}{\sqrt{t-s}}$ which is integrable over $[\frac{t}{2}, t]$. Write the integrand of $(I_{1,1})$ as

$$(De^{-sD^2})(\eta_1 e^{-\frac{(t-s)}{2}D'^2_{L_2}} f^{-1})(f e^{-\frac{(t-s)}{2}D'^2_{L_2}}). \tag{3.77}$$

We proceed with (3.77) as before. Here η_1 already stands at the right place, we must not perform partial integration. Into the estimate enters again the first derivative of W'. De^{-sD^2} generates the factor $\frac{1}{\sqrt{s}}$ which is integrable on $[0, \frac{t}{2}]$. We write $(I_{1,2})$ as

$$\int_0^{\frac{t}{2}} e^{-sD^2}[(\eta_1 e^{-\frac{(t-s)}{4}D'^2_{L_2}} f^{-1})(f e^{-\frac{(t-s)}{4}D'^2_{L_2}})] e^{-\frac{(t-s)}{2}D'^2_{L_2}} D'^2_{L_2}\, ds \tag{3.78}$$

and proceed as before.

Consider finally the case $\nu = 3$, locally

$$\eta_3^{\text{op}}\,\Phi = g'^{ik}\frac{\partial}{\partial x^k}(\cdot - \cdot')\nabla'_i\Phi.$$

The first step in this procedure is quite similar as in the case $\nu = 1$ to shift the derivation to the left of W and to shift all zero-order terms to the right.

Let X be a tangent vector field and Φ a section.

Lemma 3.11. $X(\cdot - \cdot')\nabla'_i\Phi = \nabla'_i(X(\cdot - \cdot')\Phi) +$ *zero-order terms.*

Proof. $X(\cdot - \cdot')\nabla'_i\Phi = [X(\cdot - \cdot')\nabla'_i\Phi - \nabla'_i(X(\cdot - \cdot')\Phi)] + \nabla'_i(X(\cdot - \cdot')\Phi)$. We are done if $[\dots]$ on the right-hand side contains no derivatives of Φ. But an easy calculation yields

$$\begin{aligned}
&[X(\cdot - \cdot')\nabla'_i\Phi - \nabla'_i(X(\cdot - \cdot')\Phi)] \\
&= X\cdot(\nabla'_i - \nabla_i)\Phi - (\nabla'_i - \nabla_i)(X\cdot\Phi) \\
&\quad +(\nabla'_i - \nabla_i)X \cdot'\Phi + (\nabla_i X)(\cdot' - \cdot)\Phi.
\end{aligned} \tag{3.79}$$

□

Hence for $\Phi, \Psi \in C_c^\infty(U)$

$$\begin{aligned}
&\int h(\Psi, g'^{ik}\frac{\partial}{\partial x^k}(\cdot - \cdot')\nabla'_i\Phi)_p dp(g) \\
&= \int h(\Psi, \nabla'_i(g'^{ik}\frac{\partial}{\partial x^k}(\cdot - \cdot')\Phi)_p\, dp(g) \qquad (3.80) \\
&+ \int h(\Psi, g'^{ik}\frac{\partial}{\partial x^k}(\nabla'_i - \nabla_i)\Phi - (\nabla'_i - \nabla_i)g'^{ik}\frac{\partial}{\partial x^k}\cdot\Phi)_p \\
&+(\nabla'_i - \nabla_i)X\cdot'\Phi + \left(\nabla_i g'^{ik}\frac{\partial}{\partial x^k}\right)(\cdot' - \cdot)\Phi)_p\, dp(g). \qquad (3.81)
\end{aligned}$$

(3.79) equals to

$$\int h(\nabla_i'^{*}\Psi, g'^{ik}\frac{\partial}{\partial x^k}(\cdot - \cdot')\Phi)_p\, dp(g). \tag{3.82}$$

If Φ is Sobolev and $\Psi = W$ then we obtain again by a u.l.f. cover by normal charts $\{U_\alpha\}_\alpha$ and an associated decomposition of unity $\{\varphi_\alpha\}_\alpha$

$$\begin{aligned}
&\int h(W, \eta_3^{\mathrm{op}}\, \Phi)_p\, dp(g) \\
&= \int h(W, \sum_\alpha g'^{ik}_{\ \alpha}\frac{\partial}{\partial x^k}(\cdot - \cdot')\nabla'_{\alpha,i}(\varphi_\alpha\Phi))_p\, dp(g) \\
&= \int h(\nabla'^{\,*}_{\alpha,i}W, \sum_\alpha \varphi_\alpha g'^{ik}_{\ \alpha}\frac{\partial}{\partial x^k_\alpha}(\cdot - \cdot')\Phi)_p\, dp(g) \qquad (3.83)\\
&+ \int h(W, \eta_{3,0}^{\mathrm{op}}\,\Phi)_p\, dp(g), \qquad (3.84)
\end{aligned}$$

where $\eta_{3,0}^{\mathrm{op}}\,\Phi$ is the right component in $h(\cdot,\cdot)$ under the integral (7.41), multiplied with φ_α and summed up over α.

Now we proceed literally as before. Start with

$$\begin{aligned}
(I_{3,3}) &= \int_{\frac{t}{2}}^{t} e^{-sD^2} D\eta_3^{\mathrm{op}}\, e^{-(t-s)D'^2_{L_2}}\, ds \\
&= \int_{\frac{t}{2}}^{t} (De^{-\frac{s}{2}D^2})[(e^{-\frac{s}{4}D^2}f)(f^{-1}e^{-\frac{s}{4}D^2}\eta_3^{\mathrm{op}}\,)]e^{-(t-s)D'^2_{L_2}}\, ds. \qquad (3.85)
\end{aligned}$$

We want that for suitable $f \in L_2$, $f^{-1}e^{\frac{s}{4}D^2}\eta_3^{\mathrm{op}}$ is Hilbert–Schmidt. For this we have to estimate $h(W(t,m,p), \eta_3^{\mathrm{op}}\,\cdot)_p$ and to show it defines an integral operator with finite $L_2((M,E),dp)$-norm. We estimate

$$\begin{aligned}
N(\Phi) &= |\langle\delta(m), e^{-tD^2}\eta_3^{\mathrm{op}}\,\Phi\rangle|_{L_2((M,E),dp)} \qquad (3.86)\\
&= |h(W(t,m,p),\eta_3^{\mathrm{op}}\,\Phi)_p|_{L_2((M,E),dp)}. \qquad (3.87)
\end{aligned}$$

Using (3.83) and (3.84), we write

$$\begin{aligned}
N(\Phi) &= |h(W(t,m,p),\eta_3^{\mathrm{op}}\,\Phi)_p|_{L_2(dp)} \\
&= |h(\eta_{3,1}^{\mathrm{op}}\,W(t,m,p),\eta_{3,0}^{\mathrm{op}}\,\Phi)_p \\
&\quad + h(W(t,m,p),\eta_{3,0,0}^{\mathrm{op}}\Phi)_p|_{L_2(dp)}. \qquad (3.88)
\end{aligned}$$

Now we use $|\nabla'^{\,*}_X\chi| \le C_1|\nabla'_X\chi| \le C_2|X|\cdot|\nabla'\chi| \le C_3|X|(|\nabla\chi| + |\chi|)$, that the cover is u.l.f. and $|\nabla W| \le C_4(|DW| + |W|)$ and obtain

$$\begin{aligned}
N(\Phi) &\le C(|hDW(t,m,p),\eta_{3,0}^{\mathrm{op}}\Phi)_p|_{L_2(dp)} + |h(W(t,m,p),\eta_{3,0,0}^{\mathrm{op}}\Phi)_p|_{L_2(dp)} = \\
&= C(N_1(\Phi) + N_2(\Phi)).
\end{aligned}$$

Here we again essentially use the bounded geometry.

Hence we have to estimate

$$\sup_{\substack{\Phi\in C_c^\infty(E)\\ |\Phi|_{L_2}=1}} N_1(\Phi) = \sup_{\substack{\Phi\in C_c^\infty(E)\\ |\Phi|_{L_2}=1}} |\langle\delta(m),(De^{-tD^2})\eta_{3,0}^{\mathrm{op}}\Phi\rangle|_{L_2(dp)} \tag{3.89}$$

and

$$\sup_{\substack{\Phi\in C_c^\infty(E)\\ |\Phi|_{L_2}=1}} N_2(\Phi) = \sup_{\substack{\Phi\in C_c^\infty(E)\\ |\Phi|_{L_2}=1}} |\langle\delta(m),e^{-tD^2}\eta_{3,0,0}^{\mathrm{op}}\Phi\rangle|_{L_2(dp)}. \tag{3.90}$$

According to $k > r+1 > n+3$,

$$\begin{gathered} DW(t,m,\cdot), W(t,m,\cdot) \in H^{\frac{r}{2}(E)} \\ |DW(t,m,\cdot)|_{H^{\frac{r}{2}}}, |W(t,m,\cdot)|_{H^{\frac{r}{2}}} \le C_1(t) \end{gathered} \tag{3.91}$$

and we can restrict in (3.90), (3.91) to

$$\sup_{\substack{\Phi\in C_c^\infty(E)\\ |\Phi|_{L_2}=1\\ |\Phi|_{H^{\frac{r}{2}}}\le C_1(t)}} \tag{3.92}$$

$\eta_{3,0}^{\mathrm{op}}$, $\eta_{3,0,0}^{\mathrm{op}}$ are of order zero and can be estimated by

$$C_0|\cdot - \cdot'|_{2,\frac{r}{2}} \le C_1|\cdot - \cdot'|_{1,r-1} \tag{3.93}$$

and

$$D_0\cdot(|\nabla-\nabla'|_{2,\frac{r}{2}} + |\cdot - \cdot'|_{2,\frac{r}{2}} \le D_1\cdot(|\nabla-\nabla'|_{2,r-1} + |\cdot - \cdot'|_{1,r-1} \tag{3.94}$$

respectively.

Now we proceed literally as for $(I_{1,3})$, replacing (3.76) by

$$N_1(\Phi) = \frac{1}{\sqrt{4\pi t}}\frac{1}{2t}\left|\int_{-\infty}^{+\infty} se^{-\frac{s^2}{4t}}e^{isD}\eta_{3,0}^{\mathrm{op}}\Phi(m)\,ds\right|. \tag{3.95}$$

$(I_{3,3})$ is done, $(I_{3,4})$, $(I_{3,1})$, $(I_{3,2})$ are absolutely parallel to the case $\nu = 1$.

This finishes the proof of 3.5. □

Theorem 3.12. *Suppose the hypotheses of* 3.5. *Then*

$$De^{-tD^2} - D'_{L_2}e^{-tD'^2_{L_2}}$$

is of trace class and the trace norm is uniformly bounded on compact t-intervals $[a_0,a_1]$, $a_0>0$.

Proof. The proof is a simple combination of the proofs of 3.3 and 3.5. □

Now we additionally admit perturbation of the fibre metric h. Before the formulation of the theorem we must give some explanations. Consider the Hilbert spaces $L_2(g,h) = L_2((M,E),g,h)$, $L_2(g',h) = L_2((M,E),g',h)$, $L_2(g',h') = L_2((M,E),g',h') \equiv L_2'$ and the maps

$$i_{(g',h),(g',h')} : L_2(g',h) \longrightarrow L_2(g',h'), \quad i_{(g',h),(g',h')}\Phi = \Phi$$
$$U_{(g,h),(g',h)} : L_2(g,h) \longrightarrow L_2(g',h), \quad U_{(g,h),(g',h)}\Phi = \alpha^{\frac{1}{2}}\Phi$$

where $dp(g) = \alpha(p)dp(g')$. Then we set

$$\begin{aligned} D'_{L_2(g,h)} &= D'_{L_2} := U^*_{(g,h),(g',h)} i^*_{(g',h),(g',h')} D' i_{(g',h),(g',h')} U_{(g,h),(g',h)} \\ &\equiv U^* i^* D' i U. \end{aligned} \tag{3.96}$$

Here i^* is even locally defined (since g' is fixed) and $i^*_p = \text{dual}\,{}_h^{-1} \circ i' \circ dual_{h'}$, where $\text{dual}\,{}_h(\Phi(p)) = h_p(\cdot, \Phi(p))$. In a local basis field $\Phi_1, \ldots, \Phi_N$, $\Phi(p) = \xi^i(p)\Phi_i(p)$,

$$i^*_p\Phi(p) = h^{kl}h'_{ik}\xi^i\Phi_l(p). \tag{3.97}$$

It follows from (3.97) that for $h' \in \text{comp}^{1,r+1}(h)$ i^*, i^{*-1} are bounded up to order k,

$$\begin{aligned} & i^* - 1, i^{*-1} - 1 \in \Omega^{0,1,r+1}(\text{Hom}\,((E,h',\nabla^{h'}) \\ & \longrightarrow (M,g'),(E,h,\nabla^h) \longrightarrow (M,g'))) \end{aligned} \tag{3.98}$$

and

$$\begin{aligned} & i^* - 1, i^{*-1} - 1 \in \Omega^{0,2,\frac{r+1}{2}}(\text{Hom}\,((E,h',\nabla^{h'}) \\ & \longrightarrow (M,g'),(E,h,\nabla^h) \longrightarrow (M,g'))). \end{aligned} \tag{3.99}$$

$D' \equiv \overline{D'}$ is self-adjoint on $D_{\overline{D'}} = \overline{C_c^\infty(E)}^{|\,|_{D'}}$, where $|\Phi|^2_{D'} = |\Phi|^2_{L_2'} + |D'\Phi|^2_{L_2'}$. $i : L_2(g',h) \longrightarrow L_2(g',h') \equiv L_2'$ and $i^* : L_2(g',h') \longrightarrow L_2(g',h)$ are for $h' \in \text{comp}^{1,r+1}(h)$ quasi isometries with bounded derivatives, they map $C_c^\infty(E)$ 1–1 onto $C_c^\infty(E)$ and $i^*D'i$ is self-adjoint on $\overline{C_c^\infty(E)}^{|\,|_{i^*D'i}} = D_{i^*D'i} \subset L_2((M,E),g',h) \equiv L_2(g',h)$. We obtain as a consequence that $e^{-t(i^*D'i)^2}$ is defined and self-adjoint in $L_2((M,E),g',h) = L_2(g',h)$, maps for $t > 0$ and $i,j \in \mathbb{Z}$ $H^i(E,i^*D'i)$ continuously into $H^j(E,i^*D'i)$ and has the heat kernel $W'_{g',h}(t,m,p) = \langle\delta(m), e^{-t(i^*D'i)^2}\delta(p)\rangle$, $W'(t,m,p)$ satisfies the same general estimates as $W(t,m,p)$. By exactly the same arguments we obtain that $e^{-tU^*(i^*D'i)^2U} = e^{-t(U^*i^*D'iU)^2} = U^*e^{-t(i^*D'i)^2}U$ is defined in $L_2 = L_2((M,E),g,h)$, self-adjoint and has the heat kernel $W'_{L_2}(t,m,p) = W'_{g,h}(t,m,p) = \alpha^{-\frac{1}{2}}(m)W'_{g',h}(t,m,p)\alpha(p)^{\frac{1}{2}}$. Here we assume $g' \in \text{comp}^{1,r+1}(g)$. Now we are able to formulate our main theorem.

Theorem 3.13. *Let* $E = ((E,h,\nabla = \nabla^h,\cdot) \longrightarrow (M^n,g))$ *be a Clifford bundle with* (I), $(B_k(M,g))$, $(B_k(E,\nabla))$, $k \geq r+1 > n+3$, $E' = ((E,h,\nabla' = \nabla^{h'},\cdot') \longrightarrow (M^n,g)) \in \text{gen}\,\text{comp}^{1,r+1}_{L,\text{diff}}(E) \cap \text{CL}\mathcal{B}^{N,n}(I,B_k)$, $D = D(g,h,\nabla = \nabla^h,\cdot)$, $D' = D(g',h,\nabla' = \nabla^{h'},\cdot')$ *the associated generalized Dirac operators,*

$dp(g) = \alpha(p)dp(g')$, $U = \alpha^{\frac{1}{2}}$. *Then for* $t > 0$

$$e^{-tD^2} - U^* e^{-t(i^* D' i)^2} U \tag{3.100}$$

is of trace class and the trace norm is uniformly bounded on compact t*-intervals* $[a_0, a_1]$, $a_0 > 0$.

Proof. We are done if we could prove the assertions for

$$e^{-t(UD'U^*)^2} - e^{-t(i^* D' i)^2} = Ue^{-tD^2}U^* - e^{-t(i^* D' i)^2} \tag{3.101}$$

since $U^*(3.101)U = (3.100)$. To get a better explicit expression for (3.101), we apply again Duhamel's principle. This holds since Greens formula for UD^2U^* holds,

$$\int h_q(UD^2U^*\Phi, \Psi) - h(\Phi, UD^2U^*\Psi)\; dq(g') = 0.$$

We obtain

$$\begin{aligned}
&-\int_0^t \int_M h_q(\alpha^{\frac{1}{2}}(m)W(s,m,q)\alpha^{-\frac{1}{2}}(q), \\
&\qquad \left(UD^2U^* + \frac{\partial}{\partial t}\right) W'_{g',h}(t-s,q,p)\; dq(g')\; ds \\
&= -\int_0^t \int_M h_q(\alpha^{\frac{1}{2}}(m)W(s,m,q)\alpha^{-\frac{1}{2}}(q), \\
&\qquad (UD^2U^* - (i^* D' i)^2) W'_{g',h}(t-s,q,p)\; dq(g')\; ds \\
&= \alpha^{\frac{1}{2}}(m)W(s,m,q)\alpha^{-\frac{1}{2}}(q) - W'_{g',h'}(t,m,p) \\
&= W_{g',h}(t,m,p) - W'_{g',h}(t,m,p).
\end{aligned} \tag{3.102}$$

(3.102) expresses the operator equation

$$\begin{aligned}
&e^{-t(UDU^*)^2} - e^{-t(i^* D' i)^2} \\
&= -\int_0^t e^{-s(U^* DU)^2}((UDU^*)^2 - (i^* D' i)^2)e^{-(t-s)(i^* D' i)^2}\; ds \\
&= -\int_0^t e^{-s(UDU^*)^2} UDU^*(UDU^* - i^* D' i)e^{-(t-s)(i^* D' i)^2}\; ds \\
&\quad - \int_0^t e^{-s(UDU^*)^2}(UDU^* - i^* D' i)(i^* D' i)e^{-(t-s)(i^* D' i)^2}\; ds.
\end{aligned}$$

We write (3.104) as

$$-\int_0^t \alpha^{\frac{1}{2}} e^{-sD^2} D\alpha^{-\frac{1}{2}}(\alpha^{\frac{1}{2}} D\alpha^{-\frac{1}{2}} - i^* D' i) e^{-(t-s)(i^* D' i)^2}\, ds$$

$$= -\int_0^t \alpha^{\frac{1}{2}} e^{-sD^2} D\alpha^{-\frac{1}{2}}(D - i^* D' i - \frac{\operatorname{grad} \alpha \cdot}{2\alpha}) e^{-(t-s)(i^* D' i)^2}\, ds$$

$$= -\int_0^t \alpha^{\frac{1}{2}} e^{-sD^2} D\alpha^{-\frac{1}{2}} i^*((i^{*-1} - 1)D + (D - D') - i^{*-1}\frac{\operatorname{grad} \alpha \cdot}{2\alpha})$$
$$\cdot e^{-(t-s)(i^* D' i)^2}\, ds$$

$$= \int_0^t \alpha^{\frac{1}{2}} e^{-sD^2} D(\eta_0 + \eta_1 + \eta_2 + \eta_3 + \eta_4) e^{-(t-s)(i^* D' i)^2}\, ds,$$

$$\eta_0 = \frac{\operatorname{grad} \alpha \cdot}{2\alpha^{\frac{3}{2}}}, \quad \eta_i = -\alpha^{-\frac{1}{2}} i^* \eta_i(7), \quad i = 1, 2, 3,$$

$$\eta_1(7) = (3.52), \quad \eta_2(7) = (3.53), \quad \eta_3(7) = (3.54), \quad \eta_4 = \alpha^{-\frac{1}{2}} i^{*-1}(i^* - 1)D.$$

Here η_0 and η_2 are of zeroth order. η_1 and η_3 can be discussed as in (3.60)–(3.95). η_4 can be discussed analogous to η_1, η_3 as before, i.e., η_4 will be shifted via partial integration to the left (up to zero-order terms) and $\alpha^{-\frac{1}{2}} i^*(i^* - 1)$ thereafter again to the right. In the estimates one has to replace W by DW and nothing essentially changes as we exhibited in (3.77). We perform in (3.104) the same decomposition and have to estimate 20 integrals,

$$\int_0^{\frac{t}{2}} \alpha^{\frac{1}{2}} e^{-sD^2} D\eta_\nu e^{-(t-s)(i^* D' i)^2}\, ds, \tag{I$_{\nu,1}$}$$

$$\int_0^{\frac{t}{2}} \alpha^{\frac{1}{2}} e^{-sD^2} \eta_\nu (i^* D' i) e^{-(t-s)(i^* D' i)^2}\, ds, \tag{I$_{\nu,2}$}$$

$$\int_{\frac{t}{2}}^t \alpha^{\frac{1}{2}} e^{-sD^2} D\eta_\nu e^{-(t-s)(i^* D' i)^2}\, ds, \tag{I$_{\nu,3}$}$$

$$\int_{\frac{t}{2}}^t \alpha^{\frac{1}{2}} e^{-sD^2} \eta_\nu (i^* D' i) e^{-(t-s)(i^* D' i)^2}\, ds, \tag{I$_{\nu,4}$}$$

$\nu = 0, \ldots, 4$ and to show that these are products of Hilbert–Schmidt operators and have uniformly bounded trace norm on compact t-intervals. This has been completely modelled in the proof of 3.5. □

Finally we obtain

Theorem 3.14. *Assume the hypotheses of 3.13. Then for* $t > 0$

$$e^{tD^2}D - U^* e^{-t(i^*D'i)^2}(i^*D'i)U$$

is of trace class and its trace norm is uniformly bounded on compact t-intervals $[a_0, a_1]$, $a_0 > 0$.

□

The operators $i^*D'^2 i$ and $(i^*D'i)^2$ are different in general. We should still compare $e^{-ti^*D'^2 i}$ and $e^{-t(i^*D'i)^2}$.

Theorem 3.15. *Assume the hypotheses of* 3.13. *Then for* $t > 0$

$$e^{-t(i^*D'^2 i)} - e^{-t(i^*D'i)^2}$$

is of trace class and the trace norm is uniformly bounded on compact t-intervals $[a_0, a_1]$, $a_0 > 0$.

Proof. We obtain again immediately from Duhamel's principle

$$e^{-ti^*D'^2 i} - e^{-t(i^*D'i)^2} = -\int_0^t e^{-s(i^*D'^2 i)}(i^*D'^2 i - (i^*D'i)^2)e^{-(t-s)(i^*D'i)^2}\, ds$$

$$= -\int_0^t e^{-s(i^*D'^2 i)} i^*D'(1 - ii^*)D'ie^{-(t-s)(i^*D'i)^2}\, ds$$

$$= -\int_0^t e^{-s(i^*D'^2 i)}(i^*D'i)i^{-1}(1 - ii^*)i^{*-1}(i^*D'i)e^{-(t-s)(i^*D'i)^2}\, ds. \tag{3.105}$$

In $[\frac{t}{2}, t]$ we shift $i^*D'i$ again to the left of the kernel $W'_{e^{-s(i^*D^2 i)}}$ via partial integration and estimate

$$(i^*D'ie^{-\frac{s}{2}(i^*D'^2 i)})[(e^{-\frac{s}{4}(i^*D'^2 i)})f)(f^{-1}e^{-\frac{s}{4}(i^*D'^2 i)}i^{-1}(1 - ii^*)i^{*-1})]$$
$$((i^*D'i)e^{-(t-s)(i^*D'i)^2})$$

as before. In $[0, \frac{t}{2}]$ we write the integrand of (3.105) as

$$(e^{-s(i^*D'^2 i)}i^*D'i)[((i^*i)^{-1}e^{-\frac{t-s}{4}(i^*D'i)^2}f^{-1})(fe^{-\frac{t-s}{4}(i^*D'i)^2})]$$
$$(e^{-\frac{t-s}{2}(i^*D'i)^2}(i^*D'i))$$

and proceed as in the corresponding cases. □

Theorem 3.16. *Assume the hypotheses of* 3.13. *Then for* $t > 0$

$$e^{-tD^2} - e^{-t(U^*i^*D^2 iU)} \equiv e^{-tD^2} - U^*e^{-t(i^*D^2 i)}U$$

is of trace class and the trace norm is uniformly bounded on any t-interval $[a_0, a_1]$, $a_0 > 0$.

Proof. This immediately follows from 3.13 and 3.15. □

Finally the last class of admitted perturbations are compact topological perturbations which will be studied now.

Let $E = ((E, h, \nabla^h) \longrightarrow (M^n, g)) \in \mathrm{CL}\mathcal{B}^{N,n}(I, B_k)$ be a Clifford bundle, $k \geq r+1 > n+3$, $E' = ((E, h', \nabla^{h'}) \longrightarrow (M'^n, g')) \in \mathrm{comp}^{1,r+1}_{L,\mathrm{diff},\mathrm{rel}}(E) \cap \mathrm{CL}\mathcal{B}^{N,n}(I, B_k)$. Then there exist $K \subset M$, $K' \subset M'$ and a vector bundle isomorphism (not necessarily an isometry) $f = (f_E, f_M) \in \tilde{\mathcal{D}}^{1,r+2}(E|_{M\setminus K}, E'|_{M'\setminus K'})$ s.t.

$$g|_{M\setminus K} \text{ and } f_M^* g'|_{M\setminus K} \text{ are quasi isometric,} \tag{3.106}$$

$$h|_{E|_{M\setminus K}} \text{ and } f_E^* h'|_{E|_{M\setminus K}} \text{ are quasi isometric,} \tag{3.107}$$

$$|g|_{M\setminus K} - f_M^* g'|_{M\setminus K}|_{g,1,r+1} < \infty, \tag{3.108}$$

$$|h|_{E_{M\setminus K}} - f_E^* h'|_{E|M\setminus K}|_{g,h,\nabla^h,1,r+1} < \infty, \tag{3.109}$$

$$|\nabla^h|_{E|_{M\setminus K}} - f_E^* \nabla^{h'}|_{E|_{M\setminus K}}|_{g,h,\nabla^h,1,r+1} < \infty, \tag{3.110}$$

$$|\cdot|_{M\setminus K} - f_E^* \cdot'|_{M\setminus K}|_{g,h,\nabla^h,1,r+1} < \infty. \tag{3.111}$$

(3.106)–(3.111) also hold if we replace f by f^{-1}, $M\setminus K$ by $M'\setminus K'$ and $g, h, \nabla^h, \cdot$ by $g', h', \nabla^{h'}, \cdot'$. If we consider the complete pull back $f_E^*(E'|_{M'\setminus K'})$, i.e., the pull back together with all Clifford data, then we have on $M\setminus K$ two Clifford bundles, $E|_{M\setminus K}$, $f_E^*(E'|_{M'\setminus K'}$ which are as vector bundles isomorphic and we denote $f_E^*(E'|_{M'\setminus K'}$ again by E' on $M\setminus K$, i.e., $g'_{new} = (f_M|_{M\setminus K})^* g'_{old}$ etc. (3.106)–(3.111) and the symmetry of our uniform structure $\mathfrak{U}^{1,r+1}_{L,\mathrm{diff},\mathrm{rel}}$ imply

$$\begin{aligned} W^{1,i}(E|_{M\setminus K}) &\cong W^{1,i}(E'|_{M\setminus K}), \quad 0 \leq i \leq r+1, \\ W^{2,j}(E|_{M\setminus K}) &\cong W^{2,j}(E'|_{M\setminus K}), \quad 0 \leq j \leq \frac{r+1}{2}, \\ H^j(E|_{M\setminus K}, D) &\cong H^j(E'|_{M\setminus K}, D'), \quad 0 \leq j \leq \frac{r+1}{2}, \end{aligned} \tag{3.112}$$

$$\begin{aligned} \Omega^{1,1,i}(End(E|_{M\setminus K})) &\cong \Omega^{1,1,i}(End(E'|_{M\setminus K})), \quad 0 \leq i \leq r+1, \\ \Omega^{1,2,j}(End(E|_{M\setminus K})) &\cong \Omega^{1,2,j}(End(E'|_{M\setminus K})), \quad 0 \leq j \leq \frac{r+1}{2}. \end{aligned}$$

Here the Sobolev spaces are defined by restriction of corresponding Sobolev sections.

We now fix our set up for compact topological perturbations. Set $\mathcal{H} = L_2((K, E|_K), g, h) \oplus L_2((K', E'|_{K'}), g', h') \oplus L_2((M\setminus K, E), g, h)$ and consider the following maps

$$\begin{aligned} &i_{L_2,K'} : L_2((K', E'|_{K'}), g', h') \longrightarrow \mathcal{H}, \\ &i_{L_2,K'}(\Phi) = \Phi, \\ &i^{-1} : L_2((M'\setminus K', E'|_{M'\setminus K'}), g', h') \longrightarrow L_2((M'\setminus K', E'|_{M'\setminus K'}), g', h), \\ &i^{-1}\Phi = \Phi, \end{aligned}$$

$$U^* : L_2((M' \setminus K', E'|_{M' \setminus K'}), g', h) \longrightarrow L_2((M' \setminus K', E'|_{M' \setminus K'}), g, h),$$
$$U^*\Phi = \alpha^{-\frac{1}{2}},$$

where $dq(g) = \alpha(q)dq(g')$. We identify $M \setminus K$ and $M' \setminus K'$ as manifolds and $E'|_{M' \setminus K'}$ and $E|_{M \setminus K}$ as vector bundles. Then we have natural embeddings

$$i_{L_2,M} : L_2((M, E), g, h) \longrightarrow \mathcal{H},$$
$$i_{L_2,K'} \oplus U^*i^{-1} : L_2((M', E'), g', h') \longrightarrow \mathcal{H},$$
$$(i_{L_2,K'} \oplus U^*i^{-1})\Phi = i_{L_2,K'}\chi_{K'}\Phi + U^*i^{-1}\chi_{M' \setminus K'}\Phi.$$

The images of these two embeddings are closed subspaces of $\mathcal{H}$. Denote by P and P' the projection onto these closed subspaces. D is defined on $\mathcal{D}_D \subset \text{im } P$. We extend it onto $(\text{im } P)^\perp$ as zero operator. The definition of (the shifted) D' is a little more complicated. For the sake of simplicity of notation we write $U^*i^{-1} \equiv i_{L_2,K'} \oplus U^*i^{-1} = id \oplus U^*i^{-1}$, keeping in mind that $i_{L_2,K'}$ fixes χ_K, Φ and the scalar product. Moreover we set also $iU\chi_{K'}\Phi = U^*i^*\chi_{K'}\Phi = \chi_{K'}\Phi$. Let $\Phi \in \mathcal{D}_{D'}$, $\chi_{K'}\Phi + U^*i^{-1}\chi_{M' \setminus K'}\Phi$ its image in $\mathcal{H}$. Then $(U^*i^*D'iU)(\chi_{K'}\Phi + U^*i^{-1}\chi_{M' \setminus K'}, \Phi) = U^*i^*D'\Phi = \chi_{K'}D'\Phi + U^*i^*\chi_{M' \setminus K'}D'\Phi$. Now we set as $\mathcal{D}_{U^*i^*D'iU} \subset H$

$$\mathcal{D}_{U^*i^*D'iU} = \{\chi_{K'}\Phi + U^*i^{-1}\chi_{M' \setminus K'}\Phi | \Phi \in \mathcal{D}_{D'}\} \oplus (\text{im } P')^\perp. \tag{3.113}$$

It follows very easy from the selfadjointness of D' on $\mathcal{D}_{D'}$ and (8.7) that $U^*i^*D'iU$ is self-adjoint on $\mathcal{D}_{U^*i^*D'iU}$, if we additionally set $U^*i^*D'iU = 0$ on $(\text{im } P')^\perp$.

Remark 3.17. *If g and h do not vary then we can spare the whole $i - U$-procedure, $i = U = id$. Nevertheless this case still includes interesting perturbations. Namely perturbations of $\nabla, \cdot$ and compact topological perturbations.* □

We set for the sake of simplicity $\tilde{D'} = U^*i^*D'iU$. The first main result of this section is the following

Theorem 3.18. *Let $E = ((E, h, \nabla^h) \longrightarrow (M^n, g)) \in \text{CL}\mathcal{B}^{N,n}(I, B_k)$, $k \geq r + 1 > n + 3$, $E' \in \text{gen comp}^{1,r+1}_{L,\text{diff },\text{rel}}(E) \cap \text{CL}\mathcal{B}^{N,n}(I, B_k)$. Then for $t > 0$*

$$e^{-tD^2}P - e^{-t\tilde{D'}^2}P' \tag{3.114}$$

and

$$e^{-tD^2}D - e^{-t\tilde{D'}^2}\tilde{D'} \tag{3.115}$$

are of trace class and their trace norms are uniformly bounded on any t-interval $[a_0, a_1]$, $a_0 > 0$.

For the proof we make the following construction. Let $V \subset M \setminus K$ be open, $\overline{M \setminus K} \setminus V$ compact, $\text{dist}(V, \overline{M \setminus K} \setminus (M \setminus K)) \geq 1$ and denote by $B \in L(\mathcal{H})$ the multiplication operator $B = \chi_\nu$. The proof of 3.18 consists of two steps. First we prove 3.18 for the restriction of (3.114), (3.115) to V, i.e., for $B(3.114)B$, thereafter for $(1 - B)(3.114)B$, $B(3.114)(1 - B)$ and the same for (3.115).

Theorem 3.19. *Assume the hypotheses of* 3.18. *Then*

$$B(e^{-tD^2}P - e^{-t\tilde{D}'^2}P')B, \tag{3.116}$$

$$B(e^{-tD^2}D - e^{-t\tilde{D}'^2}\tilde{D}')B, \tag{3.117}$$

$$B(e^{-tD^2}P - e^{-t\tilde{D}'^2}P')(1-B), \tag{3.118}$$

$$(1-B)(e^{-tD^2}P - e^{-t\tilde{D}'^2}P')B, \tag{3.119}$$

$$B(e^{-tD^2}D - e^{-t\tilde{D}'^2}\tilde{D}')(1-B), \tag{3.120}$$

$$(1-B)(e^{-tD^2}D - e^{-t\tilde{D}'^2}\tilde{D}')B, \tag{3.121}$$

$$(1-B)(e^{-tD^2}P - e^{-t\tilde{D}'^2}P')(1-B), \tag{3.122}$$

$$(1-B)(e^{-tD^2}D - e^{-t\tilde{D}'^2}\tilde{D}')(1-B) \tag{3.123}$$

are of trace class and their trace norms are uniformly bounded on any t-interval $[a_0, a_1]$, $a_0 > 0$.

3.18 immediately follows from 3.19. We start with the assertion for (3.116). Introduce functions $\varphi, \psi, \gamma \in C^\infty(M, [0,1])$ with the following properties.

1) $\text{supp}\,\varphi \subset M \setminus K$, $(1-\varphi) \in C_c^\infty(M \setminus K)$, $\varphi|_V = 1$.
2) ψ with the same properties as φ and additionally $\psi = 1$ on $\text{supp}\,\varphi$, i.e., $\text{supp}\,(1-\psi) \cap \text{supp}\,\varphi = 0$.
3) $\gamma \in C_c^\infty(M)$, $\gamma = 1$ on $\text{supp}\,(1-\varphi)$, $\gamma|_V = 0$.

Define now an approximate heat kernel $E(t,m,p)$ on M by

$$E(t,m,p) := \gamma(m)W(t,m,p)(1-\varphi(p)) + \psi(m)\tilde{W}'(t,m,p)\varphi(p).$$

Applying Duhamel's principle yields

$$-\int_\alpha^\beta \int_M h_q\left(W(s,m,q), \left(\frac{\partial}{\partial t} + D^2\right) E(t-s,q,p)\right)\chi_\nu(p)\, dq(g)\, ds$$
$$= \int_M [h_q(W(\beta,m,q), E(t-\beta,q,p)) - h_q(W(\alpha,m,q),$$
$$E(t-\alpha,q,p))]\chi_\nu(p)\, dq(g). \tag{3.124}$$

Performing $\alpha \longrightarrow 0^+$, $\beta \longrightarrow t^-$ in (3.124), we obtain

$$-\int_\alpha^\beta \int_M h_q\left(W(s,m,q), \left(D^2 + \frac{\partial}{\partial t}\right) E(t-s,q,p)\right)\chi_\nu(p)\, dq(g)\, ds$$
$$= \lim_{\beta \to t^-} \int_M [h_q(W(\beta,m,q), E(t-\beta,q,p))\chi_\nu(p)\, dq(g)$$
$$-E(t,m,p)\chi_\nu(p). \tag{3.125}$$

Now we use

$$\chi_V(p)(1-\varphi(p)) = 0 \tag{3.126}$$

and obtain

$$\begin{aligned}
&\lim_{\beta\to t^-} \int_M [h_q(W(\beta,m,q), E(t-\beta,q,p))\chi_\nu(p)\, dq(g) \\
&= \lim_{\beta\to t^-} \int_M [h_q(W(\beta,m,q), \psi(q)\tilde{W}'(t-\beta,q,p))\varphi(p)\chi_\nu(p)\, dq(g) \\
&= W(t,m,p)
\end{aligned}$$

since $\tilde{W}'(\tau,q,p)$ is the heat kernel of $e^{-\tau\tilde{D}'^2}$. This yields

$$\begin{aligned}
&-\int_0^t \int_M h_q(W(s,m,q), \left(D^2 + \frac{\partial}{\partial t}\right) E(t-s,q,p))\chi_V(p)\, dq(g)\, ds \\
&= -\int_0^t \int_M h_q(W(s,m,q), (D^2\psi(q) - \psi(q)\tilde{D}'^2)\tilde{W}'(t-s,q,p))\chi_V(p)\, dq(g)\, ds \\
&= [W(t,m,p) - \tilde{W}'(t,m,p)] \cdot \chi_V(p).
\end{aligned} \tag{3.127}$$

(3.127) expresses the operator equation

$$(e^{-tD^2}P - e^{-t\tilde{D}'^2}P')B = -\int_0^t e^{-sD^2}(D^2\psi - \psi\tilde{D}'^2)e^{-(t-s)\tilde{D}'^2}B\, ds \tag{3.128}$$

in $\mathcal{H}$ at kernel level.

We rewrite (3.128) as in the foregoing cases.

(3.128)

$$= -\int_0^t e^{-sD^2}(D(D-\tilde{D}')\psi + (D-\tilde{D}')\tilde{D}'\psi + \tilde{D}'^2\psi - \psi\tilde{D}'^2)e^{-(t-s)\tilde{D}'^2}\, ds$$

$$= -\left[\int_0^{\frac{t}{2}} e^{-sD^2}(D(D-\tilde{D}')\psi e^{-(t-s)\tilde{D}'^2}\, ds\right. \tag{3.129}$$

$$+ \int_0^{\frac{t}{2}} e^{-sD^2}(D-\tilde{D}')\tilde{D}'\psi e^{-(t-s)\tilde{D}'^2}\, ds \tag{3.130}$$

$$+ \int_0^{\frac{t}{2}} e^{-sD^2}(\tilde{D}'^2\psi - \psi\tilde{D}'^2)\, ds \tag{3.131}$$

$$+\int_{\frac{t}{2}}^{t} e^{-sD^2} D(D-\tilde{D}')\psi e^{-(t-s)\tilde{D}'^2}\, ds \tag{3.132}$$

$$+\int_{\frac{t}{2}}^{t} e^{-sD^2} (D-\tilde{D}')\tilde{D}'\psi e^{-(t-s)\tilde{D}'^2}\, ds \tag{3.133}$$

$$\left.+\int_{\frac{t}{2}}^{t} e^{-sD^2} (\tilde{D}'^2\psi-\psi\tilde{D}'^2) e^{-(t-s)\tilde{D}'^2}\, ds\right] . \tag{3.134}$$

Write the integrand of (3.132) as

$$(e^{-\frac{s}{2}D^2}D)[(e^{-\frac{s}{4}D^2}f)(f^{-1}e^{-\frac{s}{4}D^2}(D-\tilde{D}')\psi)]e^{-(t-s)}\tilde{D}'^2,$$

$|e^{-\frac{s}{2}D^2}D|_{\text{op}} \le \frac{C}{\sqrt{s}}$, $|e^{-(t-s)\tilde{D}'^2}|_{\text{op}} \le C'$ and $[\ldots]$ is the product of two Hilbert–Schmidt operators if f can be chosen $\in L_2$ and such that $f^{-1}e^{-\frac{s}{4}D^2}(D-\tilde{D}')\psi$ is Hilbert–Schmidt. We know from the preceding considerations, sufficient for this is that $(D-\tilde{D}')\psi$ has Sobolev coefficients of order $r+1$ (and $p=1$).

$$\begin{aligned}
(D-\tilde{D}')\psi &= (D-\alpha^{-\frac{1}{2}}i^*D'i\alpha^{\frac{1}{2}})\psi \\
&= \left(D - i^*\frac{\text{grad}\,'\alpha}{2\alpha}\cdot' - i^*D'\right)\psi \\
&= i^*\left((i^{*-1}-1)D + (D-D') - \frac{\text{grad}\,'\alpha\cdot'}{2\alpha}\right)\psi \\
&= i^*\Big[i^{*-1}(\text{ grad }\psi\cdot + \psi D) + \text{ grad }\Psi(\cdot - \cdot') \\
&\quad +(\text{ grad }\psi - \text{ grad }'\psi)\cdot' + \psi(D-D') - \frac{\text{grad}\,'\alpha\cdot'}{2\alpha}\psi\Big].
\end{aligned}$$

i^* is bounded up to order k, $i^{*-1}-1$ is $(r+1)$-Sobolev, grad ψ, grad $'\psi$ have compact support, $0 \le \psi \le 1$, $\frac{\text{grad}\,\alpha\cdot'}{2\alpha}$ is $(r+1)$-Sobolev and $\psi(D-D')$ is completely discussed in (3.51)–(3.95). Hence (3.132) is completely done.

Write the integrand of (3.133) as

$$[(e^{\frac{s}{2}D^2}f)(f^{-1}e^{\frac{s}{2}D^2}(D-D'))](D'e^{-(t-s)\tilde{D}'^2}).$$

$[\ldots]$ is the product of two Hilbert–Schmidt operators with bounded trace norm on t-intervals $[a_0,a_1]$, $a_0>0$. An easy calculation yields

$$\tilde{D}'\psi = \alpha^{-\frac{1}{2}}i^*D'i\alpha^{\frac{1}{2}}\psi = i^*\text{ grad }\psi\cdot' + \psi\tilde{D}',$$

hence

$$|\tilde{D}'\psi e^{-(t-s)\tilde{D}'^2}|_{\text{op}} = |(i^*\text{ grad }'\psi\cdot' + \psi\tilde{D}')e^{-(t-s)\tilde{D}'^2}|_{\text{op}} \le C + \frac{C'}{\sqrt{t-s}},$$

(3.133) is done.

Rewrite finally the integrands of (3.129), (3.130) as

$$\begin{aligned}
&(e^{-sD^2}D)[((D-\tilde{D}')\psi e^{-\frac{t-s}{2}\tilde{D}'^2}f^{-1})(fe^{-\frac{t-s}{2}\tilde{D}'^2})]\\
&= e^{-sD^2}Di^*[(((i^{*-1}-1)(\text{ grad }\psi\cdot+\psi D)\\
&\quad+\text{ grad }\psi(\cdot-\cdot')+(\text{ grad }\psi-\text{ grad }'\psi)\cdot'\\
&\quad+\psi(D-D')-\frac{\text{grad }\alpha}{2\alpha}\cdot'\psi)e^{-\frac{t-s}{2}\tilde{D}'^2}f^{-1})(fe^{-\frac{t-s}{2}\tilde{D}'^2})]
\end{aligned}$$

and

$$\begin{aligned}
&e^{-sD^2}i^*[((D-\tilde{D}')\psi e^{-\frac{t-s}{4}\tilde{D}'^2}f^{-1})(fe^{-\frac{t-s}{4}\tilde{D}'^2})](\tilde{D}'e^{-\frac{t-s}{2}\tilde{D}'^2})\\
&= e^{-sD^2}i^*[(((i^{*-1}-1)(\text{ grad }\psi\cdot+\psi D)+\text{ grad }\psi(\cdot-\cdot')\\
&\quad+(\text{ grad }\psi-\text{ grad }'\psi)\cdot'+\psi(D-D')-\frac{\text{grad }\alpha}{2\alpha}\cdot'\psi)\\
&\quad e^{-\frac{t-s}{4}\tilde{D}'^2}f^{-1})(fe^{-\frac{t-s}{4}\tilde{D}'^2})](\tilde{D}'e^{-\frac{t-s}{2}\tilde{D}'^2}),
\end{aligned}$$

respectively, and (3.129), (3.130) are done. The remaining integrals are (3.131) and (3.134). We have to find an appropriate expression for $\tilde{D}'^2\psi-\psi\tilde{D}'^2$.

$$\begin{aligned}
\tilde{D}'^2 &= (\alpha^{-\frac12}i^*D'i\alpha^{\frac12})(\alpha^{-\frac12}i^*D'i\alpha^{\frac12}) &(3.135)\\
&= \alpha^{-\frac12}i^*D'i^*\left(\frac{\text{grad }'\alpha}{2\alpha^{\frac12}}\cdot'+\alpha^{\frac12}D'\right)\\
&= i^*\left(D'\alpha^{-\frac12}+\frac{\text{grad }'\alpha}{2\alpha^{\frac32}}\cdot'\right)i^*\left(\frac{\text{grad }'\alpha}{2\alpha^{\frac12}}\cdot'+\alpha^{\frac12}D'\right)\\
&= i^*D'i^*D'+i^*D'i^*\frac{\text{grad }'\alpha}{2\alpha}\cdot'+i^*\frac{\text{grad }'\alpha}{2\alpha}\cdot'i^*\frac{\text{grad }'\alpha}{2\alpha}\cdot'\\
&\quad+i^*\frac{\text{grad }'\alpha}{2\alpha}\cdot'i^*D'. &(3.136)
\end{aligned}$$

Hence

$$\begin{aligned}
&\tilde{D}'^2\psi-\psi\tilde{D}'^2 = i^*D'i^*D'\psi-\psi i^*D'i^*D'\\
&+i^*D'i^*\frac{\text{grad }'\alpha}{2\alpha}\cdot'\psi-\psi i^*D'i^*\frac{\text{grad }'\alpha}{2\alpha}\cdot' &(3.137)\\
&+i^*\frac{\text{grad }'\alpha}{2\alpha}\cdot'i^*\frac{\text{grad }'\alpha}{2\alpha}\cdot'\psi-\psi i^*\frac{\text{grad }'\alpha}{2\alpha}\cdot'i^*\frac{\text{grad }'\alpha}{2\alpha}\cdot'\\
&+i^*\frac{\text{grad }'\alpha}{2\alpha}\cdot'i^*D'\psi-\psi i^*\frac{\text{grad }'\alpha}{2\alpha}\cdot'i^*D'\\
&= i^*D'i^*\text{ grad }'\psi\cdot'+i^*\text{ grad }'\psi\cdot'i^*D'+\psi i^*D'i^*D'-\psi i^*D'i^*D'\\
&+i^*(\text{ grad }'\psi\cdot'+\psi D')i^*\frac{\text{grad }'\alpha}{2\alpha}\cdot'-\psi i^*D'i^*\frac{\text{grad }'\alpha}{2\alpha}\cdot' &(3.138)\\
&+i^*\frac{\text{grad }'\alpha}{2\alpha}\cdot'i^*(\text{ grad }'\psi\cdot'+\psi D')-\psi i^*\frac{\text{grad }'\alpha}{2\alpha}\cdot'i^*D'
\end{aligned}$$

$$= i^* D' i^* \text{ grad }'\psi \cdot' + i^* \text{ grad }'\psi \cdot' i^* D' \tag{3.139}$$

$$+ i^* \text{ grad }'\psi \cdot' i^* \frac{\text{grad }'\alpha}{2\alpha} \cdot' \tag{3.140}$$

$$+ i^* \frac{\text{grad }'\alpha}{2\alpha} \cdot' i^* \text{ grad }'\psi \cdot' . \tag{3.141}$$

The terms in (3.139) are first-order operators but grad $'\psi$ has compact support and we are done. The terms in (3.140), (3.141) are zero-order operators and we are also done since grad $'\psi$ has compact support.

Hence $(e^{-tD^2}P - e^{-t\tilde{D}'^2}P')B$, $B(e^{-tD^2}P - e^{-t\tilde{D}'^2}P')B$ are of trace class and the trace norm in uniformly bounded on any compact t-interval $[a_0, a_1]$, $a_0 > 0$. The assertions for (3.116) are done.

Next we study the operator

$$(e^{\frac{t}{2}D^2}P - e^{-\frac{t}{2}\tilde{D}'^2}P')(1-B). \tag{3.142}$$

Denote by M_ε the multiplication operator with $\exp(-\varepsilon \text{dist}(m,K)^2)$. We state that for ε small enough

$$M_\varepsilon e^{-tD^2}B, \quad M_\varepsilon e^{-tD'^2}B \tag{3.143}$$

and

$$M_\varepsilon^{-1} e^{-tD^2}\chi_G, \quad M_\varepsilon^{-1} e^{-tD'^2}\chi_G \tag{3.144}$$

are Hilbert–Schmidt for every compact $G \subset M$ or $G' \subset M'$. Write

$$(e^{\frac{t}{2}D^2}P - e^{-\frac{t}{2}\tilde{D}'^2}P')(1-B)$$

$$= [e^{-\frac{t}{2}D^2}PM_\varepsilon] \cdot [M_\varepsilon^{-1}(e^{-\frac{t}{2}D^2}P - e^{-\frac{t}{2}\tilde{D}'^2}P')(1-B)] \tag{3.145}$$

$$+[e^{-\frac{t}{2}D^2}P - e^{-\frac{t}{2}\tilde{D}'^2}P')M_\varepsilon] \cdot [M_\varepsilon^{-1}e^{-\frac{t}{2}\tilde{D}'^2}P'(1-B)]. \tag{3.146}$$

According to (3.1)–(3.5) and (3.143), each of the factors $[\cdots]$ in (3.145), (3.146) is Hilbert–Schmidt and we obtain that (3.139) is of trace class and has uniformly bounded trace norm in any t-interval $[a_0, a_1]$, $a_0 > 0$. The same holds for

$$B(e^{\frac{t}{2}D^2}P - e^{-\frac{t}{2}\tilde{D}'^2}P')(1-B) \tag{3.147}$$

$$(1-B)(e^{\frac{t}{2}D^2}P - e^{-\frac{t}{2}\tilde{D}'^2}P')B \tag{3.148}$$

$$(1-B)(e^{\frac{t}{2}D^2}P - e^{-\frac{t}{2}\tilde{D}'^2}P')(1-B) \tag{3.149}$$

by multiplication of (3.142) from the left by B etc., i.e., the assertions for (3.118), (3.119), (3.122) are done. Write now

$$(e^{-\frac{t}{2}D^2}D - e^{-\frac{t}{2}\tilde{D}'^2}D')B$$

$$= (e^{-\frac{t}{2}D^2}D)(e^{\frac{t}{2}D^2}P - e^{-\frac{t}{2}\tilde{D}'^2}P')B \tag{3.150}$$

$$+(e^{-\frac{t}{2}D^2}D - e^{-\frac{t}{2}\tilde{D}'^2}D')(e^{-\frac{t}{2}D'^2}P)B. \tag{3.151}$$

(3.150) is done already by (3.128) and $|e^{-\frac{t}{2}D^2}D|_{\text{op}} \le \frac{C}{\sqrt{t}}$. Decompose (3.151) as the sum of

$$e^{-\frac{t}{2}D^2}P(D-\tilde{D}')\cdot(e^{-\frac{t}{2}\tilde{D}'^2}\tilde{D}') = [e^{-\frac{t}{2}D^2}P(-\eta)]\cdot(e^{-\frac{t}{2}\tilde{D}'^2}\tilde{D}')B \tag{3.152}$$

and

$$(e^{-\frac{t}{2}D^2}P - e^{-\frac{t}{2}\tilde{D}'^2}P')(e^{-\frac{t}{2}\tilde{D}'^2}\tilde{D}')B \tag{3.153}$$

[...] in (3.152) is done. Rewrite $e^{-\frac{t}{2}D^2}P - e^{-\frac{t}{2}\tilde{D}'^2}P'$ as

$$(e^{-\frac{t}{2}D^2}P - e^{-\frac{t}{2}\tilde{D}'^2}P)B \tag{3.154}$$

$$+(e^{\frac{t}{2}D^2}P - e^{-\frac{t}{2}\tilde{D}'^2}P')(1-B). \tag{3.155}$$

(3.154), (3.155) are done already, hence (3.153) and hence $(e^{\frac{t}{2}D^2}P - e^{-\frac{t}{2}\tilde{D}'^2}P')B$, (3.117), (3.120), (3.121), (3.123). This finishes the proof of 3.19. □

The proof of Theorem 3.18 now follows from 3.19 by adding up the four terms containing $e^{\frac{t}{2}D^2}P - e^{-\frac{t}{2}\tilde{D}'^2}P'$ or $e^{\frac{t}{2}D^2}D - e^{-\frac{t}{2}\tilde{D}'^2}\tilde{D}'$, respectively. □

Remark 3.20. *We could perform the proof of* 3.18, 3.19 *also along the lines of* (3.101)–(3.104), *performing first a unitary transformation, proving the trace class property and performing the back transformation, as we indicate in* (3.101). *This procedure is completely equivalent to the proof of* 3.18, 3.19 *presented above.* □

The operators $U^*i^*D'^2iU$ and $(U^*i^*D'^2iU)^2$ are distinct in general and we have still to compare $e^{-t(U^*i^*D'^2iU)}P'$ and $e^{-t(U^*i^*D'^2iU)^2}P'$. According to our remark above, it is sufficient to prove the trace class property of

$$e^{-t(i^*D'^2i)}P - e^{-t(i^*D'i)^2}P' \tag{3.156}$$

in

$$\mathcal{H}' = L_2((K,E),g,h) \oplus L_2((K',E'),g',h') \oplus L_2((M\setminus K,E),g',h).$$

Here we have an embedding

$$\begin{aligned} &i_{L_2,K'} \oplus i^{-1} : L_2((M',E'),g',h') \longrightarrow \mathcal{H}' \\ &(i_{L_2,K'} \oplus i^{-1})\Phi = i_{L_2,K'}\chi_{K'}\Phi + i^{-1}\chi_{M'\setminus K'}\Phi, \end{aligned} \tag{3.157}$$

where

$$\begin{aligned} &i^{-1} : L_2((M'\setminus K', E'|_{M'\setminus K'}),g',h') \longrightarrow L_2((M'\setminus K', E'|_{M'\setminus K'}),g'h), \\ &i^{-1}\Phi = \Phi, \end{aligned}$$

and

$$i^*D'i(\chi_K,\Phi + i^{-1}\chi_{M'\setminus K'}\Phi) := i^*D'\Phi = \chi_{K'}D'\Phi + i^*\chi_{M'\setminus K'}D'\Phi,$$

$i^*D'^2i$ similar, all with the canonical domains of definition analogous to (3.113). P' is here the projection onto im $(i_{L_2,K'} \oplus i^{-1})$. We define $i^*D'^2i$, $(i^*D'i)^2$ to be zero on im $P'^{\perp}$.

Remark 3.21. *Quite similar we could embed* $L_2((M,E),g,h)$ *into* $\mathcal{H}'$*, define* P*,* UDU^* *and the assertion* 3.18 *would be equivalent to the assertion for*

$$e^{-t(UDU^*)^2}P - e^{-t(i^*D'i)^2}P'. \tag{3.158}$$

Applying the (extended) U^* *from the right,* U *from the left, yields just the expression* (3.114). □

Theorem 3.22. *Assume the hypotheses of* 3.18. *Then*

$$e^{-t(i^*D'^2i)}P' - e^{-t(i^*D'i)^2}P' \tag{3.159}$$

is of trace class and its trace norm is uniformly bounded on compact t*-intervals* $[a_0, a_1]$, $a_0 > 0$.

Proof. We prove this by establishing the assertion for the four cases arising from multiplication by B, $1-B$. Start with (3.159). B. Duhamel's principle again yields

$$\begin{aligned} &(e^{-t(i^*D'^2i)}P' - e^{-t(i^*D'i)^2}P')B \\ &= -\int_0^t e^{-s(i^*D'^2i)}((i^*D'^2i) - (i^*D'i)^2)e^{-(t-s)(i^*D'i)^2}\,ds. \end{aligned} \tag{3.160}$$

An easy calculation yields

$$\begin{aligned} (i^*D'^2i)\psi - \psi(i^*D'i)^2 &= i^*D'^2\psi - \psi i^*D'i^*D' \\ &= i^*D' \text{ grad }'\psi \cdot' + i^* \text{ grad }'\psi \cdot' D + \psi i^*D'^2 \\ &\quad -(\psi i^*D'^2 + \psi i^*D'(i^*-1)D') \\ &= i^*D' \text{ grad }'\psi \cdot' + i^* \text{ grad }'\psi \cdot' D' \qquad (3.161) \\ &\quad -\psi i^*D'(i^*-1)D'. \qquad (3.162) \end{aligned}$$

The first-order operators in (3.161) contain the compact support factor $\text{grad }'\psi$ and we are done. Here i^*D' (coming from the first term or from $\text{grad }'\psi \cdot' D' = \text{grad }'\psi \cdot' i^{*-1}i^*D$) will be connected with $e^{-s(i^*D'^2i)}$ or $e^{-(t-s)(i^*D'i)^2}$, depending on the interval $[\frac{t}{2}, t]$ or $[0, \frac{t}{2}]$. The (D')'s of the second-order operator (3.162) can be distributed analogous to the proof of 3.15. The remaining main point is $0 \le \psi \le 1$ and $i^* - 1$ Sobolev of order $r+1$, i.e., $i^* - 1 \in \Omega^{0,1,r+1}(\text{Hom}((E'|_{M'\setminus K'}, g', h'), (E|_{M\setminus K}, g', h)))$.

The assertion for (3.159)$\cdot B$ is done. Quite analogously (and parallel to the proofs of (3.116), (3.118), (3.119), (3.122)) one discusses the other 3 cases. □

We obtain as a corollary from 3.18 and 3.22

Theorem 3.23. *Assume the hypothesis of* 3.18. *Then for* $t > 0$

$$e^{-tD^2}P - e^{-t(U^*i^*D'^2iU)}P'$$

is of trace class in $\mathcal{H}'$ *and the trace norm is uniformly bounded on compact* t*-intervals* $[a_0, a_1]$, $a_0 > 0$. □

We proved that after fixing $E \in \mathrm{CL}\mathcal{B}^{N,n}(I, B_k)$, $k \geq r+1 > n+3$, we can attach to any $E' \in \mathrm{gen\,comp}^{1,r+1}_{L,\mathrm{diff}\,,\mathrm{rel}}(E)$ two number-valued invariants, namely

$$E' \longrightarrow \mathrm{tr}(e^{-tD^2}P - e^{-t(U^*i^*D'iU)^2}P') \tag{3.163}$$

and

$$E' \longrightarrow \mathrm{tr}(e^{-tD^2}P - e^{-tU^*i^*D'^2iU}P'). \tag{3.164}$$

This is a contribution to the classification inside a component but still unsatisfactory insofar as it

1) could depend on t,
2) will depend on the $K \subset M$, $K' \subset M'$ in question,
3) is not yet clear the meaning of this invariant.

We are in a much more comfortable situation if we additionally assume that the Clifford bundles under consideration are endowed with an involution $\tau : E \longrightarrow E$, s.t.

$$\tau^2 = 1, \quad \tau^* = \tau \tag{3.165}$$

$$[\tau, X]_+ = 0 \text{ for } X \in TM \tag{3.166}$$

$$[\nabla, \tau] = 0 \tag{3.167}$$

Then $L_2((M,E), g, h) = L_2(M, E^+) \oplus L_2(M, E^-)$

$$D = \begin{pmatrix} 0 & D^- \\ D^+ & 0 \end{pmatrix}$$

and $D^- = (D^+)^*$. If M^n is compact then as usual

$$\mathrm{ind}D := \mathrm{ind}D^+ := \dim\ker D^+ - \dim\ker D^- \equiv \mathrm{tr}(\tau e^{-tD^2}), \tag{3.168}$$

where we understand τ as

$$\tau = \begin{pmatrix} I & 0 \\ 0 & -I \end{pmatrix}.$$

For open M^n $\mathrm{ind}D$ in general is not defined since τe^{-tD^2} is not of trace class. The appropriate approach on open manifolds is relative index theory for pairs of operators D, D'. If D, D' are self-adjoint in the same Hilbert space and $e^{tD^2} - e^{-tD'^2}$ would be of trace class then

$$\mathrm{ind}(D, D') := \mathrm{tr}(\tau(e^{-tD^2} - e^{-tD'^2})) \tag{3.169}$$

makes sense, but at the first glance (3.169) should depend on t.

If we restrict to Clifford bundles $E \in \mathrm{CL}\mathcal{B}^{N,n}(I, B_k)$ with involution τ then we assume that the maps entering in the definition of $\mathrm{comp}^{1,r+1}_{L,\mathrm{diff}\,,F}(E)$ or $\mathrm{gen\,comp}^{1,r+1}_{L,\mathrm{diff}\,,\mathrm{rel}}(E)$ are τ-compatible, i.e., after identification of $E|_{M\setminus K}$ and $f_E^* E'|_{M'\setminus K}$ holds

$$[f_E^* \nabla^{h'}, \tau] = 0, \quad [f^* \cdot', \tau]_+ = 0. \tag{3.170}$$

We call $E|_{M\setminus K}$ and $E'|_{M'\setminus K'}$ τ-compatible. Then, according to the preceding theorems,

$$\operatorname{tr}(\tau(e^{-tD^2}P - e^{-t(U^*i^*D'iU)^2}P')) \tag{3.171}$$

makes sense.

Theorem 3.24. *Let* $((E,h,\nabla^h) \longrightarrow (M^n,g),\tau) \in \mathrm{CL}\mathcal{B}^{N,n}(I,B_k)$ *be a graded Clifford bundle,* $k \geq r > n+2$.

a) *If* $\nabla'^h \in \operatorname{comp}^{1,r}(\nabla) \subset C_E^{1,r}(B_k)$, ∇' *τ-compatible, i.e.,* $[\nabla',\tau] = 0$ *then*

$$\operatorname{tr}(\tau(e^{-tD^2} - e^{-tD'^2}))$$

is independent of t.

b) *If* $E' \in \operatorname{gen\,comp}^{1,r+1}_{L,\mathrm{diff},\mathrm{rel}}(E)$ *is τ-compatible with* E*, i.e.,* $[\tau, X\cdot']_+ = 0$ *for* $X \in TM$ *and* $[\nabla',\tau] = 0$*, then*

$$\operatorname{tr}(\tau(e^{-tD^2}P - e^{-t(U^*i^*D'iU)^2}P'))$$

is independent of t.

Proof. a) follows from our 3.3. b) follows from our 3.25. □

Proposition 3.25. *If* $E' \in \operatorname{gen\,comp}^{1,r+1}_{L,\mathrm{diff},\mathrm{rel}}(E)$ *and*

$$\tau(e^{-tD^2}P - e^{-t(U^*i^*D'iU)^2}P')$$
$$\tau(e^{-tD^2}D - e^{-t(U^*i^*D'iU)^2}(U^*i^*D'iU))$$

are for $t > 0$ *of trace class and the trace norm of*

$$\tau(e^{-tD^2}D - e^{-t(U^*i^*D'iU)^2}(U^*i^*D'iU))$$

is uniformly bounded on compact t-intervals $[a_0,a_1]$, $a_0 > 0$*, then*

$$\operatorname{tr}(\tau(e^{-tD^2}p - e^{-t(U^*i^*D'iU)^2}(U^*i^*D'iU)))$$

is independent of t.

Proof. Let $(\varphi_i)_i$ be a sequence of smooth functions $\in C_c^\infty(\overline{M\setminus K})$, satisfying $\sup|d\varphi_i| \underset{i\to\infty}{\longrightarrow} 0$, $0 \leq \varphi_i \leq \varphi_{i+1}$ and $\varphi_i \underset{i\to\infty}{\longrightarrow} 1$. Denote by M_i the multiplication operator with φ_i on $L_2((M\setminus K, E|_{M\setminus K}), g, h)$. We extend M_i by 1 to the complement of $L_2((M\setminus K, E), g, h)$ in H. We have to show

$$\frac{d}{dt}\operatorname{tr}\tau(e^{-tD^2}P - e^{-t(U^*i^*D'iU)^2}P') = 0.$$

$e^{-tD^2}P - e^{-t(U^*i^*D'iU)^2}P'$ is of trace class, hence

$$\operatorname{tr}\tau(e^{-tD^2}P - e^{-t(U^*i^*D'iU)^2}P') = \lim_{j\to\infty} \operatorname{tr}\tau M_j(e^{-tD^2}P - e^{-t(U^*i^*D'iU)^2}P')M_j.$$

M_j restricts to compact sets and we can differentiate under the trace and we obtain

$$\begin{aligned}
&\frac{d}{dt}\mathrm{tr}\tau M_j(e^{-tD^2}P - e^{-t(U^*i^*D'iU)^2}P')M_j \\
&= \frac{d}{dt}\mathrm{tr}\tau(M_jU^*(e^{-t(UDU^*)^2}P - e^{-t(i^*D'i)^2}P')UM_j \\
&= -\mathrm{tr}\tau(U^*M_j(e^{-t(UDU^*)^2}(UDU^*)^2 - e^{-t(i^*D'i)^2}(i^*D'i)^2)M_jU).
\end{aligned}$$

Consider $\mathrm{tr}\tau(U^*M_j(e^{-t(UDU^*)^2}(UDU^*)^2M_jU) = \mathrm{tr}\tau M_je^{-tD^2}D^2M_j$. There holds $\mathrm{tr}\tau(M_je^{-tD^2}D^2M_j) = \mathrm{tr}M_j \text{ grad } \varphi_i \cdot \tau De^{-tD^2}$. Quite similar

$$\begin{aligned}
&\mathrm{tr}\tau(M_j(e^{-t(i^*D'i)^2}(i^*D'i)^2)M_j) \\
&= \mathrm{tr}\tau\varphi_je^{-\frac{t}{2}(i^*D'i)^2}(i^*Di)(i^*D'i)e^{-\frac{t}{2}(i^*D'i)^2}\varphi_j \\
&= \mathrm{tr}(i^*Di)e^{-\frac{t}{2}(i^*D'i)^2}\varphi_j\tau\varphi_je^{-\frac{t}{2}(i^*D'i)^2}(i^*D'i) \\
&= \mathrm{tr}e^{-\frac{t}{2}(i^*D'i)^2}(i^*D'i)\varphi_j^2\tau e^{-\frac{t}{2}(i^*D'i)^2}(i^*D'i) \\
&= \mathrm{tr}e^{-\frac{t}{2}(i^*D'i)^2}i^*(2\varphi_j \text{ grad }'\varphi_j \cdot' + \varphi_j^2D')i\tau e^{-\frac{t}{2}(i^*D'i)^2}(i^*D'i) = \\
&= \mathrm{tr}2i^*M_j \text{ grad }'\varphi_j \cdot' i\tau(i^*D'i)e^{-t(i^*D'i)^2} - \mathrm{tr}\tau M_je^{-t(i^*D'i)^2}(i^*D'i)^2M_j,
\end{aligned}$$

hence

$$\mathrm{tr}\tau(M_je^{-t(i^*D'i)^2}(i^*D'i)^2M_j) = \mathrm{tr}M_ji^* \text{ grad }'\varphi_j \cdot' i\tau(i^*D'i)e^{-t(i^*D'i)^2}$$

and finally

$$\begin{aligned}
&\frac{d}{dt}\mathrm{tr}\tau M_j(e^{-tD^2}P - e^{-t(U^*i^*D'iU)^2}P')M_j \\
&= \mathrm{tr}\tau M_j[\text{ grad } \varphi_j \cdot e^{-tD^2}D - \text{ grad }'\varphi_j \cdot' e^{-t(U^*i^*D'iU)^2}(U^*i^*D'iU)] \\
&= \mathrm{tr}\tau M_j[(\text{ grad } \varphi_j - \text{ grad }'\varphi_j) \cdot e^{-tD^2} + \text{ grad }'\varphi_j(\cdot - \cdot')e^{-tD^2} \\
&+ \text{ grad }'\varphi_j \cdot' (e^{-tD^2} - e^{-t(U^*i^*D'iU)^2}(U^*i^*D'iU))].
\end{aligned}$$

But this tends to zero uniformly for t in compact intervals since grad φ_j, grad $'\varphi_j$ do so. □

We denote $Q^\pm = D^\pm$

$$Q = \begin{pmatrix} 0 & Q^+ \\ Q^- & 0 \end{pmatrix}, \quad H = \begin{pmatrix} H^+ & 0 \\ 0 & H^- \end{pmatrix} = \begin{pmatrix} Q^-Q^+ & 0 \\ 0 & Q^+Q^- \end{pmatrix} = Q^2, \tag{3.172}$$

$Q'^\pm = U^*i^*D'^\pm iU = (U^*i^*D'iU)^\pm$, Q', H' analogous, assuming (3.165)–(3.167) as before and $\cdot'$, ∇' τ-compatible. H, H' form by definition a supersymmetric scattering system if the wave operators

$$W^\mp(H,H') := \lim_{t\to\mp\infty} e^{itH}e^{-tH'} \cdot P_{ac}(H') \text{ exist and are complete} \tag{3.173}$$

and

$$QW^\mp(H,H') = W^\mp(H,H')H' \text{ on } \mathcal{D}_{H'} \cap \mathcal{H}'_{ac}(H'). \tag{3.174}$$

Here $P_{ac}(H')$ denotes the projection on the absolutely continuous subspace $\mathcal{H}'_{ac}(H') \subset \mathcal{H}$ of H'.

A well-known sufficient criterion for forming a supersymmetric scattering system is given by

Proposition 3.26. *Assume for the graded operators Q, Q' (= supercharges)*

$$e^{-tH} - e^{-tH'} \quad \text{and} \quad e^{-tH}Q - e^{-tH'}Q$$

are for $t > 0$ of trace class. Then they form a supersymmetric scattering system. □

Corollary 3.27. *Assume the hypotheses of* 3.24*. Then D, D' or $D, U^*i^*D'iU$ form a supersymmetric scattering system, respectively. In particular, the restriction of D, D' or $D, U^*i^*D'iU$ to their absolutely continuous spectral subspaces are unitarily equivalent, respectively.* □

Until now we have seen that under the hypotheses of 3.24

$$\operatorname{ind}(D, \tilde{D}') = \operatorname{tr}_{\mathcal{T}}(e^{-tD^2}P - e^{-t\tilde{D}'^2}P'), \tag{3.175}$$

$\tilde{D}' = D'$ or $\tilde{D}' = U^*i^*D'iU$, is a well defined number, independent of $t > 0$ and hence yields an invariant of the pair (E, E'), still depending on K, K'. Hence we should sometimes better write

$$\operatorname{ind}(D, \tilde{D}', K, K'). \tag{3.176}$$

We want to express in some good cases $\operatorname{ind}(D, \tilde{D}', K, K')$ by other relevant numbers. Consider the abstract setting (9.127). If $\inf \sigma_e(H) > 0$ then $\operatorname{ind}D := \operatorname{ind}D^+$ is well defined.

Lemma 3.28. *If $e^{-tH}P - e^{-tH'}P'$ is of trace class for all $t > 0$ and $\inf \sigma_e(H)$, $\inf \sigma_e(H') > 0$ then*

$$\lim_{t \to \infty} \operatorname{tr}_{\mathcal{T}}(e^{-tH}P - e^{-tH'}P') = \operatorname{ind}Q^+ - \operatorname{ind}Q^-. \tag{3.177}$$

□

We infer from this

Theorem 3.29. *Assume the hypotheses of* 3.24 *and $\inf \sigma_e(D^2) > 0$. Then $\inf \sigma_e(D'^2), \inf \sigma_e(U^*i^*D'iU)^2 > 0$ and for each $t > 0$*

$$\operatorname{tr}_{\mathcal{T}}(e^{-tD^2} - e^{-t\tilde{D}'^2}) = \operatorname{ind}D^+ - \operatorname{ind}D'^+. \tag{3.178}$$

Proof. In the case 3.24.a, $\inf \sigma_e(D'^2) > 0$ follows from a standard fact and (3.178) then follows from 3.28. Consider the case 3.24.b. We can replace the comparison of $\sigma_e(D^2)$ and $\sigma_e((U^*i^*D'iU)^2)$ by that of $\sigma_e(UD^2U^*)$ and $\sigma_e((i^*D'i)^2)$. Moreover, for self-adjoint A, $0 \notin \sigma_e(A)$ if and only if $\inf \sigma_e(A^2) > 0$. Assume $0 \notin \sigma_e(UDU^*)$ and $0 \in \sigma_e(i^*D'i)$. We must derive a contradiction. Let $(\Phi_\nu)_\nu$ be a Weyl sequence for $0 \in \sigma_e(i^*D'i)$ satisfying additionally $|\Phi_\nu|_{L_2} = 1$, supp $\Phi_\nu \subseteq M \setminus K = M' \setminus K'$ and for any compact $L \subset M \setminus K = M' \setminus K'$

$$|\Phi_\nu|_{L_2(M\setminus L} \underset{\nu\to\infty}{\longrightarrow} 0. \tag{3.179}$$

We have $\lim\limits_{\nu\to\infty} i^*D'i\Phi_\nu = 0$. Then also $\lim\limits_{\nu\to\infty} D'\Phi_\nu = 0$. We use in the sequel the following simple fact. If β is an L_2-function, in particular if β is even Sobolev, then

$$|\beta \cdot \Phi_\nu|_{L_2} \longrightarrow 0. \tag{3.180}$$

Now $(UDU^*)\Phi_\nu = (UDU^* - D')\Phi_\nu + D'\Phi_\nu$. Here $D'\Phi_\nu \underset{\nu\to\infty}{\longrightarrow} 0$. Consider $(UDU^* - D')\Phi_\nu = (\alpha^{\frac{1}{2}} D \alpha^{-\frac{1}{2}} - D')\Phi_\nu = \left(-\frac{\text{grad }\alpha}{2\alpha} \cdot + D - D'\right)\Phi_\nu$. Assume $\alpha \not\equiv 1$. Then $\beta = \left|\frac{\text{grad }\alpha}{2\alpha}\right| \in \Omega^{0,2,\frac{r}{2}}(T(M \setminus K))$ satisfies the assumptions above and

$$\lim_{\nu\to\infty} \left| -\frac{\text{grad }\alpha}{2\alpha} \cdot \Phi_\nu \right|_{L_2} = 0. \tag{3.181}$$

If $\alpha \equiv 1$ this term does not appear. Write, according to (3.51)–(3.54),

$$(D - D')\Phi_\nu = \eta_1^{\text{op}}\, \Phi_\nu + \eta_2^{\text{op}}\, \Phi_\nu + \eta_3^{\text{op}}\, \Phi_\nu. \tag{3.182}$$

$\left|g'^{ik} \frac{\partial}{\partial x^k} \cdot\right|$ is bounded (we use a uniformly locally finite cover by normal charts, an associated bounded decomposition of unity etc.). $\beta = |\nabla - \nabla'|$ is Sobolev hence L_2 and by (3.180)

$$|\eta_2^{\text{op}}\, \Phi_\nu|_{L_2} \underset{\nu\to\infty}{\longrightarrow} 0. \tag{3.183}$$

Now $|\nabla\Phi_\nu|_{L_2} \le C_1(|\Phi_\nu|_{L_2} + |D\Phi_\nu|_{L_2}) \le C_2(|\Phi_\nu|_{L_2} + |D'\Phi_\nu|_{L_2})$. $g - g'$ is Sobolev, hence, according to (3.180) with $\beta = |g - g'|$, $||g - g'| \cdot \Phi_\nu|_{L_2} \underset{\nu\to\infty}{\longrightarrow} 0$ and finally $||g - g'| \cdot D'\Phi_\nu|_{L_2} \underset{\nu\to\infty}{\longrightarrow} 0$. This yields

$$|\eta_1^{\text{op}}\, \Phi_\nu|_{L_2} \underset{\nu\to\infty}{\longrightarrow} 0. \tag{3.184}$$

We conclude in the same manner from $\cdot - \cdot'$ Sobolev and $|\nabla'\Phi_\nu|_{L_2} \le C_3(|\Phi_\nu|_{L_2} + |D'\Phi_\nu|_{L_2})$ that

$$|\eta_3^{\text{op}}\, \Phi_\nu|_{L_2} \underset{\nu\to\infty}{\longrightarrow} 0. \tag{3.185}$$

(3.181)–(3.185) yield $(UDU^*)\Phi_\nu \longrightarrow 0$, $0 \in \sigma_e(UDU^*)$, $\inf \sigma_e(D^2) = 0$, a contradiction, hence $\inf \sigma_e((U^*i^*D'iU)^2) > 0$, $\inf \sigma_e((i^*D'i)^2) > 0$, $0 \notin \sigma_e(i^*D'i)$, $0 \notin \sigma_e(D')$, $\inf \sigma_e(D'^2) > 0$. We infer from 3.25 and 3.28 that for $t > 0$

$$\text{tr}\tau e^{-tD^2} - e^{-t(U^*i^*D'iU)^2} = \text{ind}D^+ - \text{ind}(U^*i^*D'iU)^+. \tag{3.186}$$

We are done if we can show

$$\text{ind}(U^*i^*D'iU)^+ = \text{ind}D'^+. \tag{3.187}$$

$\Phi \in \ker(U^*i^*D'iU)^+$ means

$$\begin{aligned}(U^*i^*D'iU)^+\Phi &= (U^*i^*D'iU)(\chi_{K'}\Phi + U^*i^{-1}\chi_{M'\setminus K'}\Phi)\\ &= \chi_{K'}D'^+\Phi + U^*i^*\chi_{M'\setminus K'}D'^+\Phi\\ &= 0.\end{aligned}$$

But this is equivalent to $D'^+\Phi = 0$. Similar for D'^-. (3.187) holds and hence (3.178). □

It would be desirable to express $\mathrm{ind}(D, \tilde{D}', K, K')$ by geometric topological terms. In particular, this would be nice in the case $\inf \sigma_e(D^2) > 0$. In the compact case, one sets $\mathrm{ind}_a D := \mathrm{ind}_a D^+ = \dim\ker D^+ - \dim\ker(D^+)^* = \dim\ker D^+ - \dim\ker D^- = \lim\limits_{t\to\infty} \mathrm{tr}\tau e^{-tD^2}$. On the other hand, for $t \to 0^+$ there exists the well-known asymptotic expansion for the kernel of τe^{-tD^2}. Its integral at the diagonal yields the trace. If $\mathrm{tr}\tau e^{-tD^2}$ is independent of t (as in the compact case), we get the index theorem where the integrand appearing in the L_2-trace consists only of the t-free term of the asymptotic expansion. Here one would like to express things in the asymptotic expansion of the heat kernel of $e^{-tD'^2}$ instead of $e^{-t(U^*i^*D'iU)^2}$. For this reason we restrict in the definition of the topological index to the case $E' \in \mathrm{comp}^{1,r+1}_{L,\mathrm{diff}\,,F}(E)$ or $E' \in \mathrm{comp}^{1,r+1}_{L,\mathrm{diff}\,,F,\mathrm{rel}}(E)$, i.e., we admit Sobolev perturbation of $g, \nabla^h, \cdot$ but the fibre metric h should remain fixed. Then for $D' = D(g', h, \nabla'^h, \cdot')$ in $L_2((M,E), g, h)$ the heat kernel of $e^{-t(U^*D'U)^2} = U^*e^{-tD'^2}U$ equals to $\alpha(q)^{-\frac{1}{2}}W'(t,q,p)\alpha(p)^{\frac{1}{2}}$. At the diagonal this equals to $W'(t,m,m)$, i.e., the asymptotic expansion at the diagonal of the original $e^{-tD'^2}$ and the transformed to $L_2((M,E), g, h)$ coincide.

Consider

$$\mathrm{tr}\tau W(t,m,m) \underset{t\to 0^+}{\sim} t^{-\frac{n}{2}} b_{-\frac{n}{2}}(D,m) + \cdots + b_0(D,m) + \cdots \tag{3.188}$$

and

$$\mathrm{tr}\tau W'(t,m,m) \underset{t\to 0^+}{\sim} t^{-\frac{n}{2}} b_{-\frac{n}{2}}(D',m) + \cdots + b_0(D',m) + \cdots . \tag{3.189}$$

We state without proof

Lemma 3.30.

$$b_i(D,m) - b_i(D',m) \in L_1, \quad -\frac{n}{2} \le i \le 1. \tag{3.190}$$

□

Define for $E' \in \mathrm{gen\,comp}^{1,r+1}_{L,\mathrm{diff}\,,F}(E)$

$$\mathrm{ind}_{\mathrm{top}}\,(D, D') := \int\limits_M b_0(D,m) - b_0(D',m). \tag{3.191}$$

According to (3.190), $\mathrm{ind}_{\mathrm{top}}\,(D, D')$ is well defined.

Theorem 3.31. *Assume* $E' \in \operatorname{gen\,comp}^{1,r+1}_{L,\mathrm{diff}\,,F,\mathrm{rel}}(E)$

a) *Then*

$$\operatorname{ind}(D, D', K, K') = \int_K b_0(D,m) - \int_{K'} b_0(D',m) \tag{3.192}$$

$$+ \int_{M\setminus K = M'\setminus K'} b_0(D,m) - b_0(D',m). \tag{3.193}$$

b) *If* $E' \in \operatorname{gen\,comp}^{1,r+1}_{L,\mathrm{diff}\,,F}(E)$ *then*

$$\operatorname{ind}(D, D') = \operatorname{ind}_{\mathrm{top}}(D, D'). \tag{3.194}$$

c) *If* $E' \in \operatorname{gen\,comp}^{1,r+1}_{L,\mathrm{diff}\,,F}(E)$ *and* $\inf \sigma_e(D^2) > 0$ *then*

$$\operatorname{ind}_{\mathrm{top}}(D, D') = \operatorname{ind}_a D - \operatorname{ind}_a D'. \tag{3.195}$$

Proof. All this follows from 3.24, the asymptotic expansion, (3.190) and the fact that the L_2-trace of a trace class integral operator equals to the integral over the trace of the kernel. □

Remarks 3.32.

1) If $E' \in \operatorname{gen\,comp}^{1,r+1}_{L,\mathrm{diff}\,,\mathrm{rel}}(E)$, g and g', ∇^h and ∇'^h, $\cdot$ and $\cdot'$ coincide in $V = M \setminus L = M' \setminus L'$, $L \supseteq K$, $L' \supseteq K'$, then in (3.47)–(3.95) $\alpha - 1$ and the η's have compact support and we conclude from (3.143), (3.144) and the standard heat kernel estimates that

$$\int_V |W(t,m,m) - W'(t,m,m)|\, dm \le C \cdot e^{-\frac{d}{t}} \tag{3.196}$$

and obtain

$$\operatorname{ind}(D, D', L, L') = \int_L b_o(D,m) - \int_{L'} b_0(D',m). \tag{3.197}$$

This follows immediately from 3.31. a).

2) The point here is that we admit much more general perturbations than in preceding approaches to prove relative index theorems.

3) $\inf \sigma_e(D^2) > 0$ is an invariant of $\operatorname{gen\,comp}^{1,r+1}_{L,\mathrm{diff}\,,F}(E)$. If we fix E, D as reference point in $\operatorname{gen\,comp}^{1,r+1}_{L,\mathrm{diff}\,,F}(E)$ then 3.31 c) enables us to calculate the analytical index for all other D's in the component from $\operatorname{ind} D$ and a pure integration.

4) $\inf \sigma_e(D^2) > 0$ is satisfied, e.g., if in $D^2 = \nabla^*\nabla + \mathcal{R}$ the operator $\mathcal{R}$ satisfies outside a compact K the condition

$$\mathcal{R} \ge \kappa_0 \cdot \mathrm{id}, \kappa_0 > 0. \tag{3.198}$$

(3.198) is an invariant of $\operatorname{gen\,comp}^{1,r+1}_{L,\mathrm{diff}\,,F}(E)$ (with possibly different K, κ_0). □

It is possible that ind$D$, indD' are defined even if $0 \in \sigma_e$. For the corresponding relative index theorem we need the scattering index.

To define the scattering index and in the next section relative ζ-functions, we must introduce the spectral shift function of Birman/Krein/Yafaev. Let A, A' be bounded self-adjoint operators, $V = A - A'$ of trace class, $R'(z) = (A' - z)^{-1}$. Then the spectral shift function

$$\xi(\lambda) = \xi(\lambda, A, A') := \pi^{-1} \lim_{\varepsilon \to 0} \arg \det(1 + VR'(\lambda + i\varepsilon)) \tag{3.199}$$

exists for a.e. $\lambda \in \mathbb{R}$. $\xi(\lambda)$ is real valued, $\in L_1(\mathbb{R})$ and

$$\operatorname{tr}(A - A') = \int_{\mathbb{R}} \xi(\lambda)\, d\lambda, \quad |\xi|_{L_1} \leq |A - A'|_1. \tag{3.200}$$

If $I(A, A')$ is the smallest interval containing $\sigma(A) \cup \sigma(A')$ then $\xi(\lambda) = 0$ for $\lambda \notin I(A, A')$.

Let

$$\mathcal{G} = \{f : \mathbb{R} \longrightarrow \mathbb{R} \mid f \in L_1 \quad \text{and} \quad \int_{\mathbb{R}} |\widehat{f}(p)|(1 + |p|)\, dp < \infty\}.$$

Then for $\varphi \in \mathcal{G}$, $\varphi(A) - \varphi(A')$ is of trace class and

$$\operatorname{tr}(\varphi(A) - \varphi(A')) = \int_{\mathbb{R}} \varphi'(\lambda)\xi(\lambda)\, d\lambda. \tag{3.201}$$

We state without proof

Lemma 3.33. *Let $H, H' \geq 0$, self-adjoint in $\mathcal{H}$, $e^{-tH} - e^{-tH'}$ for $t > 0$ of trace class. Then there exist a unique function $\xi = \xi(\lambda) = \xi(\lambda, H, H') \in L_{1,\text{loc}}(\mathbb{R})$ such that for > 0, $e^{-t\lambda}\xi(\lambda) \in L_1(\mathbb{R})$ and the following holds.*

a) $\operatorname{tr}(e^{-tH} - e^{-tH'}) = -t \int_0^\infty e^{-t\lambda}\xi(\lambda)\, d\lambda$.

b) *For every $\varphi \in \mathcal{G}$, $\varphi(H) - \varphi(H')$ is of trace class and*

$$\operatorname{tr}(\varphi(H) - \varphi(H')) = \int_{\mathbb{R}} \varphi'(\lambda)\xi(\lambda)\, d\lambda.$$

c) $\xi(\lambda) = 0$ *for* $\lambda < 0$. □

We apply this to our case $E' \in \text{gen comp}^{1,r+1}_{L,\text{diff},\text{rel}}(E)$. According to 9.4, D and $U^*i^*D'iU$ form a supersymmetric scattering system, $H = D^2$, $H' = (U^*i^*D'iU)^2$. In this case

$$e^{2\pi i\xi(\lambda,H,H')} = \det S(\lambda),$$

where $S = (W^+)^*W^- = \int S(\lambda)\, dE'(\lambda)$ and $H'_{ac} = \int \lambda\, dE'(\lambda)$.

Let $P_d(D)$, $P_d(U^*i^*D'iU)$ be the projector on the discrete subspace in $\mathcal{H}$, respectively and $P_c = 1 - P_d$ the projector onto the continuous subspace.

Moreover we write

$$D^2 = \begin{pmatrix} H^+ & 0 \\ 0 & H^- \end{pmatrix}, \quad (U^*i^*D'iU)^2 = \begin{pmatrix} H'^+ & 0 \\ 0 & H'^- \end{pmatrix}. \tag{3.202}$$

We make the following additional assumption.

$$e^{-tD^2} P_d(D), e^{-t(U^*i^*D'iU)^2} P_d(U^*i^*D'iU) \quad \text{are for } t > 0 \text{ of trace class.} \tag{3.203}$$

Then for $t > 0$

$$e^{-tD^2} P_c(D) - e^{-t(U^*i^*D'iU)^2} P_c(U^*i^*D'iU)$$

is of trace class and we can in complete analogy to (3.199) define

$$\begin{aligned} \xi^c(\lambda, H^\pm, H'^\pm) \quad := \quad & -\pi \lim_{\varepsilon \to 0^+} \arg\det[1 + (e^{-tH^\pm} P_c(H^\pm) - e^{-tH'^\pm} P_c(H'^\pm)) \\ & (e^{-tH'^\pm} P_c(H'^\pm) - e^{-\lambda t} - i\varepsilon)^{-1}] \end{aligned} \tag{3.204}$$

According to (3.200),

$$\operatorname{tr}(e^{-tH^\pm} P_c(H^\pm) - e^{-tH'^\pm} P_c(H'^\pm)) = -t \int_0^\infty \xi^c(\lambda, H^\pm, H'^\pm) e^{-t\lambda}\, d\lambda. \tag{3.205}$$

We denote as after (3.175) $\tilde{D}' = D'$ in the case $\nabla' \in \operatorname{comp}^{1,r}(\nabla)$ and $\tilde{D}' = U^*i^*D'iU$ in the case $E' \in \operatorname{gen\,comp}^{1,r+1}_{L,\mathrm{diff}\,,\mathrm{rel}}(E)$. The assumption (3.203) in particular implies that for the restriction of D and $\tilde{D}'$ to their discrete subspace the analytical index is well defined and we write $\operatorname{ind}_{a,d}(D, \tilde{D}') = \operatorname{ind}_{a,d}(D) - \operatorname{ind}_{a,d}(\tilde{D}')$ for it. Set

$$n^c(\lambda, D, \tilde{D}') := -\xi^c(\lambda, H^+, H'^+) + \xi^c(\lambda, H^-, H'^-). \tag{3.206}$$

Theorem 3.34. *Assume the hypotheses of* 3.24 *and* (3.203). *Then*

$$n^c(\lambda, D, \tilde{D}') = n^c(D, \tilde{D}')$$

is constant and

$$\operatorname{ind}(D, \tilde{D}') - \operatorname{ind}_{a,d}(D, \tilde{D}') = n^c(D, \tilde{D}'). \tag{3.207}$$

Proof.

$$\begin{aligned} \operatorname{ind}(D, \tilde{D}') &= \operatorname{tr}_\tau(e^{-tD^2} P - e^{-t\tilde{D}'^2} P') \\ &= \operatorname{tr}_\tau e^{-tD^2} P_d(D) P - \operatorname{tr}_\tau e^{-t\tilde{D}'^2} P_d(\tilde{D}') P' \\ &\quad + \operatorname{tr}_\tau(e^{-tD^2} P_c(D) - e^{-t\tilde{D}'^2} P_c(\tilde{D}')) \\ &= \operatorname{ind}_{a,d}(D, \tilde{D}') + t \int_0^\infty e^{-t\lambda} n^c(\lambda, D, \tilde{D}')\, d\lambda. \end{aligned}$$

According to 3.24, $\operatorname{ind}(D, \tilde{D}')$ is independent of t. The same holds for $\operatorname{ind}_{a,d}(D, \tilde{D}')$. Hence $t \int_0^\infty e^{-t\lambda} n^c(\lambda, D, \tilde{D}')\, d\lambda$ is independent of t.

This is possible only if

$$\int_0^\infty e^{-t\lambda} n^c(\lambda, D, D')\, d\lambda = \frac{1}{t} \quad \text{or} \quad n^c(\lambda, D, \tilde{D}') \quad \text{is independent of } \lambda. \qquad \square$$

Corollary 3.35. *Assume the hypotheses of* 3.34 *and additionally* $\inf \sigma_e(D^2|_{(\ker D^2)^\perp}) > 0$. *Then* $n^c(D, \tilde{D}') = 0$.

Proof. In this case $\mathrm{ind}_{a,d}(D, \tilde{D}') = \mathrm{ind} D - \mathrm{ind} \tilde{D}' = \mathrm{ind}(D, \tilde{D}')$, hence $n^c = 0$. $\square$

This finishes the outline of our relative index theory.

References

[1] N. Anghel, *An abstract index theorem on noncompact Riemannian manifolds.* Houston J. Math. **19** (1993), 223–237.

[2] M.F. Atiyah, *Elliptic operators, discrete groups and von Neumann algebras.* Soc. Math. France, Astérisque **32–33** (1976), 43–72.

[3] M.F. Atiyah, V.K. Patodi, I.M. Singer, *Spectral asymmetry and Riemannian geometry I.* Math. Proc. Comb. Philos. Soc. **77** (1975), 43–69.

[4] D. Barbasch, H. Moscovici, *L_2-index and the Selberg trace formula.* J. Funct. Analysis **53** (1983), 151–201.

[5] G. Carron, *Index theorems for noncompact manifolds.* J. Reine Angew. Mathematik **541** (2001), 81–115.

[6] G. Carron, *L_2-cohomology and Sobolev inequalities.* Math. Ann. **314** (1999), 613–639.

[7] A. Connes, H. Moscovici, *The L_2-index theorem for homogeneous spaces of Lie groups.* Ann. of Math. **115** (1982), 291–330.

[8] J. Eichhorn, *Relative zeta functions, determinants, torsion, index theorems and invariants for open manifolds.* preprint 24/2001 Greifswald University, 110 pp., to appear in LNM.

[9] H. Moscovici, *L_2-index of elliptic operators on locally symmetric spaces of finite volume.* Contemp. Math. **10** (1982), 129–138.

[10] W. Müller, *L_2-index theory, eta invariants and values of L-functions.* Contemp. Math. **105** (1990), 145–189.

[11] W. Müller, *Signature defects of cusps of Hilbert modular varieties and values of L-series at $s = 1$.* J. Diff. Geom. **20** (1984), 55–119.

[12] J. Roe, *An index theorem on open manifolds I.* J. Diff. Geom. **27** (1988), 87–113.

[13] J. Roe, *An index theorem on open manifolds II.* J. Diff. Geom. **27** (1988), 115–136.

[14] J. Roe, *Index theory, coarse geometry and topology of open manifolds.* CBMS Reg. Conf. Series in Math. vol. 90, Providence, 1996.

[15] G. Yu, *K-theoretic indices of Dirac type operators on complete manifolds and the Roe algebra.* K-Theory **11** (1997), 1–15.

Jürgen Eichhorn
Institut für Mathematik und Informatik, Universität Greifswald
D-17489 Greifswald, Germany
e-mail: `eichhorn@uni-greifswald.de`

C^*-algebras and Elliptic Theory
Trends in Mathematics, 129–150

Semiclassical Asymptotics and Spectral Gaps for Periodic Magnetic Schrödinger Operators on Covering Manifolds

Yuri A. Kordyukov

Abstract. We survey a method to prove the existence of gaps in the spectrum of periodic second-order elliptic partial differential operators, which was suggested by Kordyukov, Mathai and Shubin, and describe applications of this method to periodic magnetic Schrödinger operators on a Riemannian manifold, which is the universal covering of a compact manifold. We prove the existence of arbitrarily large number of gaps in the spectrum of these operators in the asymptotic limits of the strong electric field or the strong magnetic field under Morse type assumptions on the electromagnetic potential. We work on the level of spectral projections (and not just their traces) and obtain an asymptotic information about classes of these projections in K-theory. An important corollary is a vanishing theorem for the higher traces in cyclic cohomology for the spectral projections. This result is then applied to the quantum Hall effect.

1. Introduction

The problem of finding examples of periodic second-order elliptic partial differential operators, which have gaps in their spectrum, is of great importance, for instance, in heat conduction, acoustics, and propagation of electro-magnetic waves in photonic crystals and was studied recently (see, for instance, [5, 6, 7, 22, 23, 24, 26, 27, 28, 30, 33] and a recent survey [25] and references therein). In this paper, we will focus on this problem for periodic magnetic Schrödinger operators on covering spaces of compact manifolds. Strictly speaking, these operators are not periodic, even if the electric and magnetic fields are, but they are invariant under a projective action of the fundamental group, as will be explained in Section 3.

Supported by Russian Foundation of Basic Research (grant no. 04-01-00190).

So let (M,g) be a closed Riemannian oriented manifold of dimension $n \geq 2$, $\widetilde{M}$ be its universal cover and $\widetilde{g}$ be the lift of g to $\widetilde{M}$ so that $\widetilde{g}$ is a Γ-invariant Riemannian metric on $\widetilde{M}$ where Γ denotes the fundamental group of M acting on $\widetilde{M}$ by the deck transformations. (Actually, the results, which we will describe below, apply to any covering $\widetilde{M}$ of M such that its first Betti number $b_1(\widetilde{M})$ vanishes.) Let $\mathbf{B}$ be a real-valued Γ-invariant closed 2-form on $\widetilde{M}$. We assume that $\mathbf{B}$ is exact. Choose a real-valued 1-form $\mathbf{A}$ on $\widetilde{M}$ such that $d\mathbf{A} = \mathbf{B}$. $\mathbf{A}$ defines a Hermitian connection $\nabla_{\mathbf{A}} = d + i\mathbf{A}$ on the trivial line bundle $\mathcal{L}$ over $\widetilde{M}$, whose curvature is $i\mathbf{B}$. Physically we can think of $\mathbf{A}$ as the electromagnetic vector potential for a magnetic field $\mathbf{B}$. Suppose that E is a Hermitian vector bundle on M and $\widetilde{E}$ the lift of E to the universal cover $\widetilde{M}$. Let $\widetilde{\nabla}^E$ denote a Γ-invariant Hermitian connection on $\widetilde{E}$. Then consider the Hermitian connection $\nabla = \widetilde{\nabla}^E \otimes \mathrm{id} + \mathrm{id} \otimes \nabla_{\mathbf{A}}$ on $\widetilde{E} \otimes \mathcal{L} = \widetilde{E}$. Let V be a Γ-invariant self-adjoint endomorphism of the bundle $\widetilde{E}$. A periodic magnetic Schrödinger operator is a second-order elliptic differential operator

$$H_{\mathbf{A},V} = \nabla^*\nabla + V,$$

acting on the Hilbert space $L^2(\widetilde{M}, \widetilde{E})$.

A well-known method to produce operators with spectral gaps is to study differential operators with high contrast in (some of) the coefficients. We consider two cases of high contrast in the coefficients of periodic magnetic Schrödinger operators.

1. Strong electric field limit. This means the study of the asymptotic behavior of the spectrum of the operator

$$H_{\mathbf{A},\mu^{-2}V} = \nabla^*\nabla + \mu^{-2}V,$$

where $V \geq 0$ and the coupling constant μ tends to zero. In this case, there is a periodic array of electric wells which are separated from one another by electric barriers. The wells get deeper as μ approaches zero, which makes tunnelling from one well into any other well more and more difficult, and wells gets (asymptotically) isolated. Therefore, the spectrum of $H_{\mathbf{A},\mu^{-2}V}$ as $\mu \to 0$ concentrates on the spectrum of the Hamiltonian for the crystal with perfectly isolated atoms (called a model operator below). The spectrum of the model operator is a discrete set, consisting of eigenvalues of infinite multiplicity. Hence, this spectral concentration produces arbitrarily large numbers of gaps in the spectrum of $H_{\mathbf{A},\mu^{-2}V}$ if μ is sufficiently small.

2. Strong magnetic field limit. This means the study of the asymptotic behavior of the spectrum of the operator

$$H_{\lambda\mathbf{A},0} = (d + i\lambda\mathbf{A})^*(d + i\lambda\mathbf{A}),$$

when the coupling constant λ tends to $+\infty$ or, equivalently, the asymptotic behavior of the spectrum of the operator

$$H^h = h^2 H_{-h^{-1}\mathbf{A},0} = (ih\,d + \mathbf{A})^*(ih\,d + \mathbf{A}),$$

when the semiclassical parameter $h > 0$ tends to zero. In this case, the spectral gaps are produced by a periodic array of wells created by the magnetic field.

There are several methods to prove the existence of spectral gaps. Some of them are essentially based on one-dimensional calculations and separation of variables. Another, simple, but powerful method to produce examples of periodic elliptic operators with spectral gaps was suggested by Hempel and Herbst [22]. It is based on a well-known fact that norm resolvent convergence of self-adjoint operators implies their spectral convergence on any compact interval of the real line. As a consequence, we get that if a sequence T_n of self-adjoint operators converges to T in norm resolvent sense and $(a,b)\cap\sigma(T)=\emptyset$ (where $\sigma(T)$ denotes the spectrum of T), then for any $\varepsilon>0$, $(a+\varepsilon,b-\varepsilon)\cap\sigma(T_n)=\emptyset$ if n is sufficiently large. In particular, if $\sigma(T)$ is a discrete set with each point in the spectrum an eigenvalue of infinite multiplicity, then the spectrum of T_n concentrates at a discrete set of points and has an arbitrarily large number of spectral gaps as $n\to\infty$.

This method was applied in [22] to the study of the strong electric field limit for the periodic Schrödinger operator in the case when $\widetilde{M}=\mathbb{R}^n$, $\widetilde{E}$ is the trivial line bundle and $\Gamma=\mathbb{Z}^n$. Consider a closed subset S of $\mathbb{R}^n$ such that the interior of S is non-empty and S can be represented as $S=\cup_{j\in\mathbb{Z}^n}S_j$ (up to a set of measure zero) where the S_j are pairwise disjoint compact sets with $S_j=S_0+j$. Put $\Omega=\mathbb{R}^n\setminus S$. It is shown that the operator $-\Delta+\mu^{-2}\chi_\Omega$ converges in norm resolvent sense to the Dirichlet Laplacian $-\Delta_S$ on the closed set S. Note that $-\Delta_S$ is a countable direct sum of copies of $-\Delta_{S_0}$. Therefore, the spectrum of $-\Delta_S$ is a discrete set with each point in the spectrum an eigenvalue of infinite multiplicity. It follows that the spectrum of $-\Delta+\mu^{-2}\chi_\Omega$ concentrates at a discrete set of points and has an arbitrarily large number of spectral gaps as $\mu\to 0$. The same arguments work for any operator $H_{0,\mu^{-2}V}=-\Delta+\mu^{-2}V$ with a $\mathbb{Z}^n$-periodic real-valued (continuous or measurable and bounded) potential $V\geq 0$ in $\mathbb{R}^n$ such that the set $S=\{x\in\mathbb{R}^n : V(x)=0\}$ satisfies the above conditions.

Hempel and Herbst also studied in [22] the strong magnetic field limit for the periodic Schrödinger operator in the case when $\widetilde{M}=\mathbb{R}^n$, $\widetilde{E}$ is the trivial line bundle and $\Gamma=\mathbb{Z}^n$. Let $S=\{x\in\mathbb{R}^n : \mathbf{B}(x)=0\}$ and $S_{\mathbf{A}}=\{x\in\mathbb{R}^n : \mathbf{A}(x)=0\}$. Assume that the set $S\setminus S_{\mathbf{A}}$ has measure zero, the interior of S is non-empty and S can be represented as $S=\cup_{j\in\mathbb{Z}^n}S_j$ (up to a set of measure zero) where the S_j are pairwise disjoint compact sets with $S_j=S_0+j$. It is shown that, as $\lambda\to\infty$, $H_{\lambda\mathbf{A},0}$ converges in norm resolvent sense to the Dirichlet Laplacian $-\Delta_S$ on the closed set S. Therefore, as $\lambda\to\infty$, the spectrum of $H_{\lambda\mathbf{A},0}$ concentrates around the eigenvalues of $-\Delta_S$ and gaps opens up in the spectrum of $H_{\lambda\mathbf{A},0}$.

On the other hand, Hempel and Herbst also proved in [22] that, if $S_{\mathbf{A}}$ has measure zero, then, as $\lambda\to\infty$, $H_{\lambda\mathbf{A},0}$ converges in strong resolvent sense to the zero operator in $L^2(\mathbb{R}^n)$. So, in this case, the method does not work.

In this paper, we describe a new method to prove the existence of gaps for magnetic Schrödinger operators, which works in the case when the bottom S of electric or magnetic wells has measure zero and the electromagnetic potential has regular behavior near the bottom. This method was suggested in [27] for the study of the strong electric field limit and applied in [28] for the study of the strong magnetic field limit. Let us formulate the main results obtained by this method.

1. Strong electric field limit. In this case, we consider a more general self-adjoint elliptic second-order differential operator given by

$$H(\mu) = \mu\nabla^*\nabla + B + \mu^{-1}V = \mu H_{\mathbf{A},\mu^{-2}V} + B,$$

where B is a Γ-invariant self-adjoint endomorphism of the bundle $\widetilde{E}$. Assume that V satisfies in addition the following *Morse type condition:*

(E1) For all $x \in \widetilde{M}$, $V(x) \geq 0$, and V has at least one zero point.

(E2) If the matrix $V(x_0)$ is degenerate for some x_0 in $\widetilde{M}$, then $V(x_0) = 0$ and there is a positive constant c such that

$$V(x) \geq c|x - x_0|^2 I$$

for all x in a neighborhood of x_0, where I denotes the identity endomorphism of $\widetilde{E}$.

We remark that all functions $V = |df|^2$, where $|df|$ denotes the pointwise norm of the differential of a Γ-invariant Morse function f on $\widetilde{M}$, are examples of Morse type potentials.

Theorem 1 ([27]). *Under the assumptions* (E1)–(E2), *there exists an increasing sequence* $\{\lambda_m, m \in \mathbb{N}\}$, *satisfying* $\lambda_m \to \infty$ *as* $m \to \infty$, *such that for any* a *and* b, *satisfying* $\lambda_m < a < b < \lambda_{m+1}$ *with some* m, *there exists* $\mu_0 > 0$ *such that* $[a, b] \cap \sigma(H(\mu)) = \emptyset$ *for all* $\mu \in (0, \mu_0)$. *In particular, there exists arbitrarily large number of gaps in the spectrum of* $H(\mu)$ *provided the coupling constant* μ *is sufficiently small.*

The special case of Theorem 1 in the absence of a magnetic field was established in [33], using variational method, and the special case of this result in the presence of a magnetic field but in the scalar case was established in [30] using the same method as in [33].

2. Strong magnetic field limit. Assume that $\widetilde{E}$ is the trivial line bundle. For any $x \in \widetilde{M}$, denote by $B(x)$ the anti-symmetric linear operator on the tangent space $T_x\widetilde{M}$ associated with the 2-form $\mathbf{B}$:

$$\tilde{g}_x(B(x)u, v) = \mathbf{B}_x(u, v), \quad u, v \in T_x\widetilde{M}.$$

The trace-norm $|B(x)|$ of $B(x)$ is given by the formula

$$|B(x)| = [\mathrm{Tr}(B^*(x) \cdot B(x))]^{1/2}.$$

We will assume that:

(M1) There exists at least one zero of B.

(M2) There exists an integer $k > 0$ such that, if $B(x_0) = 0$, then there exists a positive constant C such that

$$C^{-1}|x - x_0|^k \leq |B(x)| \leq C|x - x_0|^k$$

for all x in some neighborhood of x_0.

Theorem 2 ([28]). *Under the assumptions* (M1) *and* (M2), *there exists an increasing sequence* $\{\lambda_m, m \in \mathbb{N}\}$, *satisfying* $\lambda_m \to \infty$ *as* $m \to \infty$, *such that for any* a *and* b, *satisfying* $\lambda_m < a < b < \lambda_{m+1}$ *with some* m, $[ah^{\frac{2k+2}{k+2}}, bh^{\frac{2k+2}{k+2}}] \cap \sigma(H^h) = \emptyset$ *for any* $h > 0$ *small enough. In particular, there exists arbitrarily large number of gaps in the spectrum of* H^h *provided the parameter* h *is sufficiently small.*

The method of the proof of Theorems 1 and 2 developed in [27, 28] completely replaces a variational method mentioned above which was used in [33] and [30] to prove Theorem 1. First of all, this method also uses the idea of model operator suggested in [33]. The construction of the model operator in the strong electric field limit was given in [33] and in the strong magnetic field limit was given in [28], following the ideas of [17]. It is supplemented by a direct construction of an intertwining operator between two spectral projections: of the original operator and the model operator. This construction uses cut-off functions, the polar decomposition and closed image technique and can be formulated in a pure functional analytic setting. In particular, it allows to treat these two operators in a symmetric way unlike the treatment in [33, 30] where the proofs of the upper and lower estimates for the spectrum distribution functions required separate and very different proofs.

Note that similar ideas were used earlier. First of all, one should mention [33], which uses variational methods and cut-off functions and where the idea of the model operator was crystallized for the first time (see also [30]). Variational methods and cut-off functions were also used in [3, Proposition 5.2] to establish existence of a gap near zero in the spectrum of the Witten deformation for the periodic Laplacians on forms, and in [2] (see Section 4, in particular, Lemmas 4.3 and 4.10) to prove vanishing of the relative index term in the gluing formula for the η-invariant in the adiabatic limit. In [9], Helffer and Sjöstrand used cut-off functions and perturbation arguments based on the Riesz projection formula (see [9, Theorem 2.4 and Proposition 2.5]) together with Agmon type weighted estimates to study the tunneling effect for Schrödinger operators with electric wells. These methods were extended to magnetic Schrödinger operators on compact manifolds in [10, 11, 13, 14, 15, 16]. It is quite possible that the technique developed in these papers can also be applied to the problems discussed here.

We believe, however, that our use of the polar decomposition and closed image technique to establish C^*-algebra equivalence of spectral projections and the symmetric use of the original and the model operators are new and may lead to further important results.

As mentioned above, the essential part of our method can be formulated in a pure functional analytic setting. These results are described in Section 2. Section 3 contains a necessary information on the magnetic Schrödinger operators and related operator algebras. Section 4 and Section 5 describe applications of the general technique to the study of the strong electric field limit and the strong magnetic field limit accordingly and outline the proofs of Theorems 1 and 2 (we refer the reader to [27, 28] for more details). In Section 4, we also describe (without proofs) extensions of Theorem 1 and its applications to the quantum Hall effect.

2. An abstract setting

In this section we describe a general functional-analytic setting suggested in [27], where we can state general results on spectral concentration and the existence of spectral gaps.

Consider Hilbert spaces $\mathcal{H}_1$ and $\mathcal{H}_2$ equipped with inner products $(\cdot,\cdot)_1$ and $(\cdot,\cdot)_2$ and semi-bounded from below self-adjoint operators A_1 in $\mathcal{H}_1$ and A_2 in $\mathcal{H}_2$ with the domains $\mathrm{Dom}(A_1)$ and $\mathrm{Dom}(A_2)$ respectively. So we have with some $\lambda_{01}, \lambda_{02} \leq 0$:

$$(A_l u, u)_l \geq \lambda_{0l} \|u\|_l^2, \quad u \in \mathrm{Dom}(A_l), \quad l = 1, 2. \tag{1}$$

It is convenient to fix unitary isomorphisms $\mathcal{V}_1 : \mathcal{H}_1 \to \mathcal{H}$ and $\mathcal{V}_2 : \mathcal{H}_2 \to \mathcal{H}$, where $\mathcal{H}$ is a fixed Hilbert space.

We will assume that the operators A_1 and A_2 have symmetries. This is expressed by an assumption that there exists a C^*-algebra $\mathfrak{A}$ equipped with a faithful $*$-representation $\pi : \mathfrak{A} \to \mathcal{B}(\mathcal{H})$ in $\mathcal{H}$ such that, if we denote by π_1 and π_2 the corresponding representations of $\mathfrak{A}$ in $\mathcal{H}_1$ and $\mathcal{H}_2$ accordingly:

$$\pi_l(a) = \mathcal{V}_l^{-1} \circ \pi(a) \circ \mathcal{V}_l, \quad l = 1, 2, \quad a \in \mathfrak{A},$$

then:

Assumption 1. For any $t > 0$ and for any $l = 1, 2$, the operator e^{-tA_l} belongs to $\pi_l(\mathfrak{A})$. Moreover, there exists a faithful, normal, semi-finite trace τ on the von Neumann algebra $\pi(\mathfrak{A})''$ such that, for any $t > 0$ and for any $l = 1, 2$, $\tau(\mathcal{V}_l e^{-tA_l} \mathcal{V}_l^{-1}) < \infty$.

Denote by $E_l(\lambda)$ $(l = 1, 2)$ the spectral projection of the operator A_l, corresponding to the semi-axis $(-\infty, \lambda]$. By standard arguments, it follows that, for any λ, $\mathcal{V}_l E_l(\lambda) \mathcal{V}_l^{-1} \in \pi(\mathfrak{A})''$, and $\tau(\mathcal{V}_l E_l(\lambda) \mathcal{V}_l^{-1}) < \infty$ $(l = 1, 2)$.

We will assume that the quadratic forms of A_1 and A_2 are close on some common part in $\mathcal{H}_1$ and $\mathcal{H}_2$ and sufficiently large on the complements. To formulate these assumptions more precisely, first, let us fix a Hilbert space $\mathcal{H}_0$ equipped with injective bounded linear maps $i_1 : \mathcal{H}_0 \to \mathcal{H}_1$ and $i_2 : \mathcal{H}_0 \to \mathcal{H}_2$. We suppose that there are given bounded linear maps $p_1 : \mathcal{H}_1 \to \mathcal{H}_0$ and $p_2 : \mathcal{H}_2 \to \mathcal{H}_0$ such that $p_1 \circ i_1 = \mathrm{id}_{\mathcal{H}_0}$ and $p_2 \circ i_2 = \mathrm{id}_{\mathcal{H}_0}$. The whole picture can be represented by the following diagram (note that this diagram is not commutative).

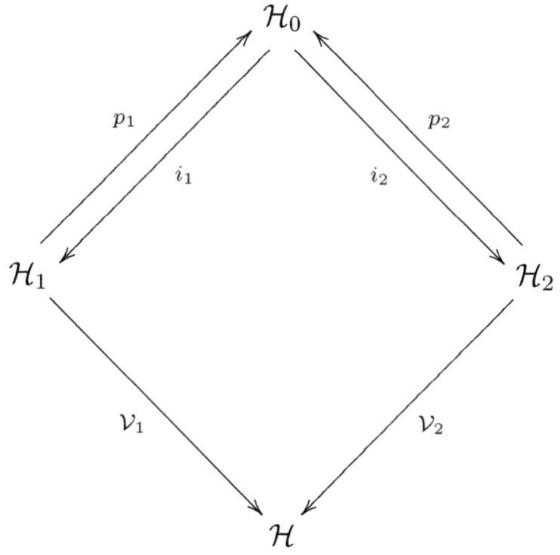

To have a possibility to restrict the quadratic forms of the operators A_l to $i_l(\mathcal{H}_0)$, we introduce a self-adjoint bounded operator J in $\mathcal{H}_0$ (a cut-off operator) such that:

Assumption 2. The operator $\mathcal{V}_2 i_2 J p_1 \mathcal{V}_1^{-1}$ belongs to $\pi(\mathfrak{A})''$, $(i_2 J p_1)^* = i_1 J p_2$, and the operator $\pi(a)\mathcal{V}_2(i_2 J p_1)\mathcal{V}_1^{-1}$ belongs to $\pi(\mathfrak{A})$ for any $a \in \mathfrak{A}$.

Since the operators $i_l : \mathcal{H}_0 \to \mathcal{H}_l, l = 1, 2$, are bounded and have bounded left-inverse operators p_l, they are topological monomorphisms, i.e., they have closed image and the maps $i_l : \mathcal{H}_0 \to \operatorname{Im} i_l$ are topological isomorphisms. Therefore, we can assume that the estimate

$$\rho^{-1}\|i_2 J u\|_2 \le \|i_1 J u\|_1 \le \rho\|i_2 J u\|_2, \quad u \in \mathcal{H}_0, \tag{2}$$

holds with some $\rho > 1$ (depending on J).

Assumption 3. For $u \in \mathcal{H}_0$, we have

$$i_1 J u \in \operatorname{Dom}(A_1) \Longleftrightarrow i_2 J u \in \operatorname{Dom}(A_2) \left(\overset{\text{def}}{\Longleftrightarrow} u \in D \right).$$

Introduce the corresponding cut-off operators J_l in $\mathcal{H}_l, l = 1, 2$, by the formula $J_l = i_l J p_l$.

Assumption 4. For $l = 1, 2$, J_l is a self-adjoint bounded operator in $\mathcal{H}_l$ such that J_l maps the domain of A_l to itself and $0 \le J_l \le \mathrm{id}_{\mathcal{H}_l}$.

Introduce a self-adjoint positive bounded linear operator J_l' in $\mathcal{H}_l$ by the formula $J_l^2 + J_l'^2 = \mathrm{id}_{\mathcal{H}_l}$.

Assumption 5. For $l = 1, 2$, the operator J_l' maps the domain of A_l to itself.

Assumption 6. For $l = 1, 2$, the operators $[J_l, [J_l, A_l]]$ and $[J_l', [J_l', A_l]]$ extend to bounded operators in $\mathcal{H}_l$.

The operator partition of unity $\{J_l, J_l'\}$ allows to decompose the quadratic forms of the operator A_l, using a so-called IMS-localization formula:

$$\begin{aligned}(A_l u, u)_l =&(A_l J_l u, J_l u)_l + (A_l J_l' u, J_l' u)_l + \frac{1}{2}([J_l,[J_l,A_l]]u,u)_l \\ &+\frac{1}{2}([J_l',[J_l',A_l]]u,u)_l, \quad u\in \mathrm{Dom}(A_l), \quad l=1,2.\end{aligned}$$

We need an estimate for the error of this localization formula:

$$\max(\|[J_l,[J_l,A_l]]\|_l, \|[J_l',[J_l',A_l]]\|_l) \leq \gamma_l, \quad l=1,2. \tag{3}$$

Finally, we assume that the quadratic forms of the operators A_1 and A_2 are close on D: for some $\beta_1, \beta_2 \geq 1$ and $\varepsilon_1, \varepsilon_2 > 0$, we have

$$(A_2 i_2 Ju, i_2 Ju)_2 \leq \beta_1 (A_1 i_1 Ju, i_1 Ju)_1 + \varepsilon_1 \|i_1 Ju\|_1^2, \quad u\in D, \tag{4}$$

$$(A_1 i_1 Ju, i_1 Ju)_1 \leq \beta_2 (A_2 i_2 Ju, i_2 Ju)_2 + \varepsilon_2 \|i_2 Ju\|_2^2, \quad u\in D, \tag{5}$$

and are large enough on the complement of D: for some $\alpha_l > 0$, we have

$$(A_l J_l' u, J_l' u)_l \geq \alpha_l \|J_l' u\|_l^2, \quad u \in \mathrm{Dom}(A_l), \quad l=1,2. \tag{6}$$

Recall that two orthogonal projections P and Q in a unital $*$-algebra $\mathcal{A}$ are said to be *Murray-von Neumann equivalent* in $\mathcal{A}$ if there is an element $V \in \mathcal{A}$ such that $P = V^*V$ and $Q = VV^*$.

Theorem 3. *Under current assumptions, let* $b_1 > a_1$ *and*

$$a_2 = \rho\left[\beta_1\left(a_1+\gamma_1+\frac{(a_1+\gamma_1-\lambda_{01})^2}{\alpha_1-a_1-\gamma_1}\right)+\varepsilon_1\right], \tag{7}$$

$$b_2 = \frac{\beta_2^{-1}(b_1\rho^{-1}-\varepsilon_2)(\alpha_2-\gamma_2)-\alpha_2\gamma_2+2\lambda_{02}\gamma_2-\lambda_{02}^2}{\alpha_2-2\lambda_{02}+\beta_2^{-1}(b_1\rho^{-1}-\varepsilon_2)}. \tag{8}$$

Suppose that $\alpha_1 > a_1+\gamma_1$, $\alpha_2 > b_2+\gamma_2$ *and* $b_2 > a_2$. *If* $(a_1,b_1)\cap\sigma(A_1)=\emptyset$, *then:*

(1) $(a_2,b_2)\cap\sigma(A_2)=\emptyset$;

(2) *for any* $\lambda_1\in(a_1,b_1)$ *and* $\lambda_2\in(a_2,b_2)$, *the projections* $\mathcal{V}_1E_1(\lambda_1)\mathcal{V}_1^{-1}$ *and* $\mathcal{V}_2E_2(\lambda_2)\mathcal{V}_2^{-1}$ *belong to* $\mathfrak{A}$ *and are Murray-von Neumann equivalent in* $\mathfrak{A}$.

Remark. Since $\rho > 1, \beta_1 \geq 1, \gamma_1 > 0$ and $\varepsilon_1 > 0$, we, clearly, have $a_2 > a_1$. The formula (8) is equivalent to the formula

$$b_1 = \rho\left[\beta_2\left(b_2+\gamma_2+\frac{(b_2+\gamma_2-\lambda_{02})^2}{\alpha_2-b_2-\gamma_2}\right)+\varepsilon_2\right],$$

which is obtained from (7), if we replace $\alpha_1, \beta_1, \gamma_1, \varepsilon_1, \lambda_{01}$ by α_2, β_2, γ_2, ε_2, λ_{02} accordingly and a_1 and a_2 by b_2 and b_1 accordingly. In particular, this implies that $b_1 > b_2$.

Remark. In applications of Theorem 3, all the data depend on a positive parameter, and one can check the assumptions of the theorem provided the parameter is sufficiently small.

Proof of Theorem 3 (*outline*). Take arbitrary $\lambda_1 \in (a_1, b_1)$ and $\lambda_2 \in (a_2, b_2)$. Consider the bounded operator $T = \mathcal{V}_2 E_2(\lambda_2) i_2 J p_1 E_1(\lambda_1) \mathcal{V}_1^{-1}$ in $\mathcal{H}$. By assumption, T belongs to the von Neumann algebra $\pi(\mathfrak{A})''$.

The key step in the proof is to show the estimates

$$\begin{aligned} \|Tu\| &\geq \varepsilon \|u\|, \quad u \in \mathcal{V}_1(\operatorname{Im} E_1(\lambda_1)), \\ \|T^* u\| &\geq \varepsilon \|u\|, \quad u \in \mathcal{V}_2(\operatorname{Im} E_2(\lambda_2)), \end{aligned}$$

that can be done, using the assumptions and some elementary facts from the operator theory. By these estimates, it follows that the operators $T : \mathcal{V}_1(\operatorname{Im} E_1(\lambda_1)) \to \mathcal{V}_2(\operatorname{Im} E_2(\lambda_2))$ and $T^* : \mathcal{V}_2(\operatorname{Im} E_2(\lambda_2)) \to \mathcal{V}_1(\operatorname{Im} E_1(\lambda_1))$ are injective and have closed image, and, therefore, are bijective.

Let $T = US, U, S \in \pi(\mathfrak{A})''$, be the polar decomposition of T. Since $\operatorname{Ker} T = \mathcal{V}_1(\operatorname{Im} E_1(\lambda_1)) = \operatorname{Im} \mathcal{V}_1 E_1(\lambda_1) \mathcal{V}_1^{-1}$ and $\operatorname{Im} T = \mathcal{V}_2(\operatorname{Im} E_2(\lambda_2)) = \operatorname{Im} \mathcal{V}_2 E_2(\lambda_2) \mathcal{V}_2^{-1}$, U is a partial isometry that performs the Murray-von Neumann equivalence of the projections $\mathcal{V}_1 E_1(\lambda_1) \mathcal{V}_1^{-1}$ and $\mathcal{V}_2 E_2(\lambda_2) \mathcal{V}_2^{-1}$ in the von Neumann algebra $\pi(\mathfrak{A})''$.

Since $(a_1, b_1) \cap \sigma(A_1) = \emptyset$, the spectral density function $\tau(\mathcal{V}_1 E_1(\lambda_1) \mathcal{V}_1^{-1})$ is constant for any $\lambda_1 \in (a_1, b_1)$. Using the Murray-von Neumann equivalence of $\mathcal{V}_1 E_1(\lambda_1) \mathcal{V}_1^{-1}$ and $\mathcal{V}_2 E_2(\lambda_2) \mathcal{V}_2^{-1}$ and the tracial property, we conclude that the spectral density function $\tau(\mathcal{V}_2 E_2(\lambda_2) \mathcal{V}_2^{-1})$ is constant for any $\lambda_2 \in (a_2, b_2)$. Since the trace τ is faithful, we get $(a_2, b_2) \cap \sigma(A_2) = \emptyset$, that completes the proof of the first part of Theorem 3.

Note that $E_1(\lambda_1) = \chi_{[e^{-t\lambda_1}, \infty)}(e^{-tA_1})$. Using the fact that λ_1 belongs to a gap in the spectrum A_1 and $e^{-tA_1} \in \pi_1(\mathfrak{A})$, one can replace $\chi_{[e^{-t\lambda_1}, \infty)}$ by a continuous function and obtain that $E_1(\lambda_1) \in \pi_1(\mathfrak{A})$ for any $\lambda_1 \in (a_1, b_1)$. Similarly, $E_2(\lambda_2) \in \pi_2(\mathfrak{A})$ for any $\lambda_2 \in (a_2, b_2)$. By assumption, it follows that T belongs to the C^*-algebra $\pi(\mathfrak{A})$. It remains to apply the following lemma to show that the partial isometry U belongs to the C^*-algebra $\pi(\mathfrak{A})$.

Lemma 4. *Let $\mathfrak{A}$ be a C^*-algebra, $\mathcal{H}$ a Hilbert space equipped with a faithful $*$-representation of $\mathfrak{A}$, $\pi : \mathfrak{A} \to \mathcal{B}(\mathcal{H})$. If $P \in \pi(\mathfrak{A})$ has closed image and $P = US$ is its polar decomposition, then $U, S \in \pi(\mathfrak{A})$.*

□

3. Magnetic translations and related operator algebras

The proofs of Theorems 1 and 2 are given by application of Theorem 3 in some concrete situations. In both cases, the operator A_2 will be essentially the periodic magnetic operator $H_{\mathbf{A},V}$ and the operator A_1 will be the model operator, which is obtained as an approximation of A_2 near the bottoms of wells (electric or magnetic). In particular, in both cases the spectrum of A_1 is a discrete set of eigenvalues of infinite multiplicities, which is precisely the sequence $\{\lambda_m\}$, entering in the formulations of the theorems. This Section contains necessary information on properties of the periodic magnetic Schrödinger operator $H_{\mathbf{A},V}$ and related notions.

As above, let M be a compact connected Riemannian manifold, Γ be its fundamental group and $p : \widetilde{M} \to M$ be its universal cover. Let $\mathbf{B}$ be a closed Γ-invariant real-valued 2-form on M, which is exact. So $\mathbf{B} = d\mathbf{A}$ where $\mathbf{A}$ is a real-valued 1-form on $\widetilde{M}$. Let V be a Γ-invariant electric potential. The corresponding Hermitian connection $\nabla_{\mathbf{A}}$ is no longer Γ-invariant, but $\nabla_{\mathbf{A}}$ turns out to be invariant under a projective representation T of the group Γ on $L^2(\widetilde{M}, \widetilde{E})$ (see, for instance, [30, 27] and references therein for more details). The operators $T_\gamma, \gamma \in \Gamma$, of this representation, called magnetic translations, are unitary operators in $L^2(\widetilde{M}, \widetilde{E})$, satisfying

$$T_e = \mathrm{id}, \quad T_{\gamma_1}T_{\gamma_2} = \sigma(\gamma_1,\gamma_2)T_{\gamma_1\gamma_2}, \quad \gamma_1,\gamma_2 \in \Gamma.$$

Here σ is a 2-cocycle on Γ i.e. $\sigma : \Gamma \times \Gamma \to U(1)$ satisfies

$$\begin{aligned} \sigma(\gamma, e) =& \sigma(e,\gamma) = 1, \quad \gamma \in \Gamma; \\ \sigma(\gamma_1,\gamma_2)\sigma(\gamma_1\gamma_2,\gamma_3) =& \sigma(\gamma_1,\gamma_2\gamma_3)\sigma(\gamma_2,\gamma_3), \quad \gamma_1,\gamma_2,\gamma_3 \in \Gamma. \end{aligned}$$

In other words, T is a projective (Γ,σ)-unitary representation (or a (Γ,σ)-action) on $L^2(\widetilde{M}, \widetilde{E})$, It is easy to see that the periodic magnetic operator $H_{\mathbf{A},V}$ also commutes with the (Γ,σ)-action T on $L^2(\widetilde{M}, \widetilde{E})$.

Denote by $\ell^2(\Gamma)$ the standard Hilbert space of complex-valued L^2-functions on the discrete group Γ. An orthonormal basis of $\ell^2(\Gamma)$ is formed by δ-functions $\{\delta_\gamma\}_{\gamma\in\Gamma}$, $\delta_\gamma(\gamma') = 1$ if $\gamma = \gamma'$ and 0 otherwise.

For any $\gamma \in \Gamma$, define a unitary operator T^L_γ in $\ell^2(\Gamma)$ by

$$T^L_\gamma f(\gamma') = f(\gamma^{-1}\gamma')\bar{\sigma}(\gamma,\gamma^{-1}\gamma'), \quad \gamma' \in \Gamma, \quad f \in \ell^2(\Gamma).$$

The operators T^L_γ define a $(\Gamma,\bar{\sigma})$-unitary representation in $\ell^2(\Gamma)$.

Define a twisted group algebra $\mathbb{C}(\Gamma,\bar{\sigma})$ which consists of complex-valued functions with finite support on Γ, with the twisted convolution operation

$$(f * g)(\gamma) = \sum_{\gamma_1,\gamma_2:\gamma_1\gamma_2=\gamma} f(\gamma_1)g(\gamma_2)\bar{\sigma}(\gamma_1,\gamma_2),$$

and with the involution

$$f^*(\gamma) = \sigma(\gamma,\gamma^{-1})\overline{f(\gamma^{-1})}.$$

Associativity of the multiplication is equivalent to the cocycle condition.

The correspondence $f \in \mathbb{C}(\Gamma,\bar{\sigma}) \mapsto T^L(f) \in \mathcal{B}(\ell^2(\Gamma))$, where $T^L(f)u = f * u, u \in \ell^2(\Gamma)$, defines a $*$-representation of the twisted group algebra $\mathbb{C}(\Gamma,\bar{\sigma})$ in $\ell^2(\Gamma)$. The weak closure of the image of $\mathbb{C}(\Gamma,\bar{\sigma})$ in this representation coincides with the *(left) twisted group von Neumann algebra* $\mathcal{A}^L(\Gamma,\bar{\sigma})$. The corresponding norm closure is the *reduced twisted group C^*-algebra* which is denoted $C^*_r(\Gamma,\bar{\sigma})$.

The von Neumann algebra $\mathcal{A}^L(\Gamma,\bar{\sigma})$ can be described in terms of the matrix elements. For any $A \in \mathcal{B}(\ell^2(\Gamma))$, its matrix elements are defined as $A_{x,y} =$

$(A\delta_y, \delta_x), x, y \in \Gamma$. Then, for any $A \in \mathcal{B}(\ell^2(\Gamma))$, the inclusion $A \in \mathcal{A}^L(\Gamma, \bar{\sigma})$ is equivalent to the relations

$$A_{x\gamma, y\gamma} = \bar{\sigma}(x, \gamma)\sigma(y, \gamma)A_{x,y}\,, \quad x, y, \gamma \in \Gamma.$$

A finite *von Neumann trace* $\mathrm{tr}_{\Gamma,\bar{\sigma}} : \mathcal{A}^L(\Gamma, \bar{\sigma}) \to \mathbb{C}$ is defined by the formula

$$\mathrm{tr}_{\Gamma,\bar{\sigma}} A = (A\delta_e, \delta_e).$$

We can also write $\mathrm{tr}_{\Gamma,\bar{\sigma}} A = A_{\gamma,\gamma} = (A\delta_\gamma, \delta_\gamma)$ for any $\gamma \in \Gamma$ because the right-hand side does not depend on γ.

Let $\mathcal{F} \subset \widetilde{M}$ be a connected fundamental domain for the action of Γ. This allows us to define a (Γ, σ)-equivariant isometry $\mathbf{U} : L^2(\widetilde{M}, \widetilde{E}) \cong \ell^2(\Gamma) \otimes L^2(\mathcal{F}, \widetilde{E}|_{\mathcal{F}})$ by the formula

$$\mathbf{U}(\phi) = \sum_{\gamma \in \Gamma} \delta_\gamma \otimes i^*(T_\gamma \phi), \quad \phi \in L^2(\widetilde{M}, \widetilde{E}), \tag{9}$$

where $i : \mathcal{F} \to \widetilde{M}$ denotes the inclusion map.

If $\mathcal{H}$ is a Hilbert space, then let $\mathcal{K}(\mathcal{H})$ denote the algebra of compact operators in $\mathcal{H}$, and $\mathcal{K} = \mathcal{K}(\ell^2(\mathbb{N}))$, where $\mathbb{N} = \{1, 2, 3, \ldots\}$. Consider the C^* algebra $\mathfrak{A} = C^*_r(\Gamma, \bar{\sigma}) \otimes \mathcal{K}$. Let $\mathcal{H}$ be the Hilbert space $\ell^2(\Gamma) \otimes \ell^2(\mathbb{N})$ and $\mathcal{H}_2$ be the Hilbert space $L^2(\widetilde{M}, \widetilde{E})$. There is a natural representation π of the algebra $\mathfrak{A}$ in $\mathcal{H}$ given by the tensor product of the representation T^L of $C^*_r(\Gamma, \bar{\sigma})$ in $\ell^2(\Gamma)$ and the natural representation of $\mathcal{K}$ in $\ell^2(\mathbb{N})$. We have $\pi(C^*_r(\Gamma, \bar{\sigma}) \otimes \mathcal{K})'' = \mathcal{A}^L(\Gamma, \bar{\sigma}) \otimes \mathcal{B}(\ell^2(\mathbb{N}))$. Choose an arbitrary unitary isomorphism $V_2 : L^2(\mathcal{F}, \widetilde{E}|_{\mathcal{F}}) \to \ell^2(\mathbb{N})$ and define a unitary operator $\mathcal{V}_2 : \mathcal{H}_2 \to \mathcal{H}$ as $\mathcal{V}_2 = (\mathrm{id} \otimes V_2) \circ \mathbf{U}$. Let π_2 be the corresponding representation of $\mathfrak{A}$ in $\mathcal{H}_2$.

As shown in [27], for any $t > 0$, the operator $e^{-tH_{\mathbf{A},V}}$ belongs to $\pi_2(\mathfrak{A})$. Moreover, if we define a trace τ on $\mathcal{A}^L(\Gamma, \bar{\sigma}) \otimes \mathcal{B}(\ell^2(\mathbb{N}))$ as the tensor product of the finite von Neumann trace $\mathrm{tr}_{\Gamma,\bar{\sigma}}$ on $\mathcal{A}^L(\Gamma, \bar{\sigma})$ and the standard trace on $\mathcal{B}(\ell^2(\mathbb{N}))$, then, for any $t > 0$, $\tau(\mathcal{V}_2 e^{-tH_{\mathbf{A},V}} \mathcal{V}_2^{-1}) < \infty$.

4. Strong electric fields

4.1. The model operator

The construction of the model operator in the case of strong electric field was given in [33]. Let us use the notation of Theorem 1. Choose a fundamental domain $\mathcal{F} \subset \widetilde{M}$ so that there are no zeros of V on the boundary of $\mathcal{F}$. This is equivalent to saying that the translations $\{\gamma\mathcal{F},\ \gamma \in \Gamma\}$ cover the set $V^{-1}(0)$ (the set of all zeros of V). Let $V^{-1}(0) \cap \mathcal{F} = \{\bar{x}_j | \, j = 1, \ldots, N\}$ be the set of all zeros of V in $\mathcal{F}$; $\bar{x}_i \neq \bar{x}_j$ if $i \neq j$.

The *model operator* K is an operator in $L^2(\mathbb{R}^n, \mathbb{C}^k)^N$ defined as a direct sum

$$K = \bigoplus_{1 \le j \le N} K_j,$$

where K_j is a self-adjoint second-order differential operator in $L^2(\mathbb{R}^n, \mathbb{C}^k)$ (a harmonic oscillator) which corresponds to the zero $\bar{x}_j$ and is obtained as the quadratic part of $H(1)$ near $\bar{x}_j$. Fix local coordinates on $\widetilde{M}$ and trivialization of the bundle $\widetilde{E}$ in a small neighborhood $B(\bar{x}_j, r)$ of $\bar{x}_j$ for every $j = 1, \dots, N$. We assume that $\bar{x}_j$ becomes zero in these local coordinates. Then K_j has the form

$$K_j = H_j^{(2)} + \bar{B}_j + V_j^{(2)},$$

where all the components are obtained from H as follows. The second-order term $H_j^{(2)}$ is a homogeneous second-order differential operator with constant coefficients (without lower-order terms) given by

$$H_j^{(2)} = -\sum_{i,k=1}^{n} g^{ik}(\bar{x}_j) \frac{\partial^2}{\partial x_i \partial x_k},$$

where (g^{ik}) is the inverse matrix to the matrix of the Riemannian metric (g_{ik}). (Note that $H_j^{(2)}$ does not depend on $\mathbf{A}$.) The zeroth-order term $V_j^{(2)}$ is obtained by taking the quadratic part of V in the chosen coordinates near $\bar{x}_j$:

$$V_j^{(2)} = \frac{1}{2} \sum_{i,k=1}^{n} \frac{\partial^2 V}{\partial x_i \partial x_k}(\bar{x}_j) x_i x_k.$$

Finally, $\bar{B}_j$ is an endomorphism of the fiber of the bundle $\widetilde{E}$ over the point $\bar{x}_j$ given by

$$\bar{B}_j = B(\bar{x}_j), \quad j = 1, \dots, N.$$

We will also need the operator $K(\mu) = \oplus_{1 \leq j \leq N} K_j(\mu)$, where

$$K_j(\mu) = \mu H_j^{(2)} + \bar{B}_j + \mu^{-1} V_j^{(2)}, \quad \mu > 0.$$

Note that $K(\mu)$ is obtained by a simple scaling from the operator $K = K(1)$ and has a discrete spectrum independent of μ. If we take a direct sum of all harmonic oscillators $K_j(\mu)$ over all zeros of V in $\widetilde{M}$ (and not only in a fundamental domain) then we will get another version of the model operator, the operator $\mathrm{id} \otimes K(\mu)$, which acts in $\ell^2(\Gamma) \otimes L^2(\mathbb{R}^n, \mathbb{C}^k)^N$ and has the same spectrum, which is a discrete set, consisting of eigenvalues of infinite multiplicity.

4.2. Proof of Theorem 1

For the proof, we apply Theorem 3 in the following setting. As in Section 3, let $\mathfrak{A}$ be the C^* algebra $C_r^*(\Gamma, \bar{\sigma}) \otimes \mathcal{K}$, let $\mathcal{H}$ be the Hilbert space $\ell^2(\Gamma) \otimes \ell^2(\mathbb{N})$ equipped with the representation π of the algebra $\mathfrak{A}$ and let $\mathcal{H}_2 = L^2(\widetilde{M}, \widetilde{E})$. Let $\mathcal{V}_2 : \mathcal{H}_2 \to \mathcal{H}$ be a unitary operator defined as $\mathcal{V}_2 = (\mathrm{id} \otimes V_2) \circ \mathbf{U}$, where $V_2 : L^2(\mathcal{F}, \widetilde{E}|_{\mathcal{F}}) \to \ell^2(\mathbb{N})$ is an arbitrary unitary isomorphism. Let π_2 be the corresponding representation of $\mathfrak{A}$ in $\mathcal{H}_2$.

We will use the notation of Section 4.1. Put $\mathcal{H}_1 = \ell^2(\Gamma) \otimes L^2(\mathbb{R}^n, \mathbb{C}^k)^N$. Choose an arbitrary unitary isomorphism $V_1 : L^2(\mathbb{R}^n, \mathbb{C}^k)^N \to \ell^2(\mathbb{N})$. Define a unitary operator $\mathcal{V}_1 : \mathcal{H}_1 \to \mathcal{H}$ as $\mathcal{V}_1 = \mathrm{id} \otimes V_1$. Let π_1 the corresponding representation $\mathfrak{A}$ in $\mathcal{H}_1$.

Consider self-adjoint, semi-bounded from below operators A_1 in $\mathcal{H}_1$ and A_2 in $\mathcal{H}_2$:

$$A_1 = \mathrm{id} \otimes K(\mu), \quad A_2 = H(\mu).$$

Assumption 1 is clear for the operator A_1. It holds for the operator A_2 by the results mentioned at the end of Section 3.

Let $\mathcal{H}_0 = \ell^2(\Gamma) \otimes \left(\oplus_{j=1}^N L^2(B(\bar{x}_j, r), \widetilde{E}|_{B(\bar{x}_j,r)})\right)$. An inclusion $i_1 : \mathcal{H}_0 \to \mathcal{H}_1$ is defined as $i_1 = \mathrm{id} \otimes j_1$, where j_1 is the inclusion

$$\bigoplus_{j=1}^{N} L^2(B(\bar{x}_j, r), \widetilde{E}|_{B(\bar{x}_j,r)}) \to L^2(\mathbb{R}^n, \mathbb{C}^k)^N$$

given by the chosen local coordinates and trivializations of the vector bundle $\widetilde{E}$. An inclusion $i_2 : \mathcal{H}_0 \to \mathcal{H}_2$ is defined as $i_2 = \mathbf{U}^* \circ (\mathrm{id} \otimes j_2)$, where j_2 is the natural inclusion

$$\bigoplus_{j=1}^{N} L^2(B(\bar{x}_j, r), \widetilde{E}|_{B(\bar{x}_j,r)}) \to L^2(\mathcal{F}, \widetilde{E}|_{\mathcal{F}}).$$

The operator $p_1 : \mathcal{H}_1 \to \mathcal{H}_0$ is defined as $p_1 = \mathrm{id} \otimes r_1$, where r_1 is the restriction operator

$$L^2(\mathbb{R}^n, \mathbb{C}^k)^N \to \bigoplus_{j=1}^{N} L^2(B(\bar{x}_j, r), \widetilde{E}|_{B(\bar{x}_j,r)}).$$

The operator $p_2 : \mathcal{H}_1 \to \mathcal{H}_0$ is defined as $p_2 = (\mathrm{id} \otimes r_2) \circ \mathbf{U}$, where r_2 is the restriction operator

$$L^2(\mathcal{F}, \widetilde{E}|_{\mathcal{F}}) \to \bigoplus_{j=1}^{N} L^2(B(\bar{x}_j, r), \widetilde{E}|_{B(\bar{x}_j,r)}).$$

Fix a function $\phi \in C_0^\infty(\mathbb{R}^n)$ such that $0 \le \phi \le 1$, $\phi(x) = 1$ if $|x| \le 1$, $\phi(x) = 0$ if $|x| \ge 2$. Fix a number κ, $0 < \kappa < 1/2$, which we shall choose later. For any $\mu > 0$ define $\phi^{(\mu)}(x) = \phi(\mu^{-\kappa} x)$. For any $\mu > 0$ small enough, let $\phi_j = \phi^{(\mu)} \in C_c^\infty(B(\bar{x}_j, r))$ in the fixed coordinates near $\bar{x}_j$. Denote also $\phi_{j,\gamma} = (\gamma^{-1})^* \phi_j$. (This function is supported near $\gamma \bar{x}_j$.) We will always take $\mu \in (0, \mu_0)$ where μ_0 is sufficiently small, so in particular the supports of all functions $\phi_{j,\gamma}$ are disjoint.

Let $\Phi = \oplus_{j=1}^N \phi_j \in \oplus_{j=1}^N C_c^\infty(B(\bar{x}_j, r)) \subset C^\infty(\mathcal{F})$. Consider a (Γ, σ)-equivariant, self-adjoint, bounded operator J in $\mathcal{H}_0$ defined as $J = \mathrm{id} \otimes \Phi$, where Φ denotes the multiplication operator by the function Φ in $\oplus_{j=1}^N L^2(B(\bar{x}_j, r), \widetilde{E}|_{B(\bar{x}_j,r)})$. Assumptions 2, 3, 4 and 5 can be easily checked.

We will use local coordinates near $\bar{x}_j$ such that the Riemannian volume element at the point $\bar{x}_j$ coincides with the Euclidean volume element given by the chosen local coordinates. Similarly we will fix a trivialization of the bundle $\widetilde{E}$ near $\bar{x}_j$ such that the Hermitian metric becomes trivial in this trivialization. Then the

estimate (2) holds with
$$\rho = 1 + O(\mu^\kappa). \tag{10}$$

If A is a second-order differential operator in $C^\infty(\mathbb{R}^n, \mathbb{C}^k)$ with the principal symbol $a_2(x,\xi)$, which is a matrix-valued function on $T^*\mathbb{R}^n$, and J is the multiplication operator by a function $\phi \in C_c^\infty(\mathbb{R}^n)$, then the operator $[J,[J,A]]$ is the multiplication operator by the function $-a_2(x, d\phi(x))$. The principal symbols $a_{1,j}^{(2)} \in C^\infty(T^*\mathbb{R}^n)$ of $K_j(\mu), j = 1,2,\dots,N$, and $a_2^{(2)} \in C^\infty(T^*\widetilde{M})$ of $H(\mu)$ are given respectively by

$$a_{1,j}^{(2)}(x,\xi) = \mu \sum_{i,k=1}^{n} g^{ik}(\bar{x}_j)\xi_i\xi_k, \quad (x,\xi) \in T^*\mathbb{R}^n;$$

$$a_2^{(2)}(x,\xi) = \mu \sum_{i,k=1}^{n} g^{ik}(x)\xi_i\xi_k, \quad (x,\xi) \in T^*\widetilde{M}.$$

Using these facts and that $d\phi = O(\mu^{-\kappa})$, one can easily check Assumption 6 and the estimate (3) with
$$\gamma_l = O(\mu^{1-2\kappa}), \quad l = 1,2. \tag{11}$$

Since $V_j^{(2)} \geq c_j\mu^{2\kappa}, j = 1,2,\dots,N$, outside any neighborhood of 0 with some $c_j > 0$ and $V \geq c_0\mu^{2\kappa}$ outside any neighborhood of the set $\{\bar{x}_j | j = 1,\dots,N\}$, with some $c_0 > 0$, the estimates (6) hold with
$$\alpha_l = c\mu^{-1+2\kappa}, \quad l = 1,2. \tag{12}$$

It is easy to see that the constants λ_{0l} can be chosen independently of μ:
$$\lambda_{0l} = \text{const}, \quad l = 1,2. \tag{13}$$

Finally, using the fact that the corresponding coefficients of A_1 and A_2 are close near the points $\bar{x}_j$, one can easily show that the estimates (4) and (5) hold with
$$\beta_l = 1 + O(\mu^\kappa), \quad \varepsilon_l = O(\mu^{3\kappa-1}). \tag{14}$$

To complete the proof of Theorem 1, let $\{\lambda_m : m \in \mathbb{N}\}$, $\lambda_1 < \lambda_2 < \lambda_3 < \dots$, be the spectrum (without taking into account multiplicities) of the operator $K(\mu)$, which is independent of μ. Take any a and b such that $\lambda_m < a < b < \lambda_{m+1}$ with some m. Clearly, the spectrum of the operator A_1 coincides with the spectrum of the operator $K(\mu)$. Therefore, $[a,b] \cap \sigma(A_1) = \emptyset$. There exists an open interval (a_1,b_1) that contains $[a,b]$ and does not intersect with the spectrum of A_1. Using the formulas (10), (11), (12), (13) and (14), one can see that, for a_2 and b_2 given by (7) and (8), we have

$$a_2 = a_1 + O(\mu^s), \quad b_2 = b_1 + O(\mu^s), \quad \mu \to 0, \tag{15}$$

where $s = \min\{3\kappa - 1, 1 - 2\kappa\}$. The best possible value of s which is

$$s = \max_\kappa \min\{3\kappa - 1, 1 - 2\kappa\} = \frac{1}{5}$$

is attained when $\kappa = 2/5$.

Hence, if $\mu > 0$ is small enough, we have $\alpha_1 > a_1 + \gamma_1$, $\alpha_2 > b_2 + \gamma_2$, $b_2 > a_2$ and the interval (a_2, b_2) contains $[a, b]$. By Theorem 3, we conclude that $(a_2, b_2) \cap \sigma(A_2) = \emptyset$, that completes the proof of Theorem 1.

4.3. Semiclassical approximation of spectral projections

Theorem 3 also provides some information about the K-theory classes of the spectral projections of the magnetic Schrödinger operator in the strong electric field limit.

For a unital $*$-algebra $\mathcal{A}$, denote by $K_0(\mathcal{A})$ the K-group of $\mathcal{A}$ and by $\tilde{K}_0(\mathcal{A})$ the reduced K-group of $\mathcal{A}$. By definition, $\tilde{K}_0(\mathcal{A}) = \operatorname{coker} \pi_* \cong K_0(\mathcal{A})/\mathbb{Z}$, where $\pi_* : K_0(\mathbb{C}) \cong \mathbb{Z} \to K_0(\mathcal{A})$ is induced by the homomorphism $\pi : \mathbb{C} \to \mathcal{A} : \lambda \mapsto \lambda \cdot 1_A$, and the K-group $K_0(\mathcal{A})$ of a non-unital $*$-algebra $\mathcal{A}$ is the reduced K-group $\tilde{K}_0(\widetilde{\mathcal{A}})$ of the algebra $\widetilde{\mathcal{A}}$ obtained from $\mathcal{A}$ by adjoining a unit.

Recall that for a C^*-algebra $\mathcal{A}$, the Morita invariance of K-theory asserts that there is a natural isomorphism $K_0(\mathcal{A}) \cong K_0(\mathcal{A} \otimes \mathcal{K})$.

Theorem 5. *Under assumptions of Theorem 1, assume that $\lambda \in \mathbb{R}$ does not coincide with λ_k for any k. Let $E(\lambda) = \chi_{(-\infty,\lambda]}(H(\mu))$ and $E^0(\lambda) = \chi_{(-\infty,\lambda]}(K(\mu))$ denote the spectral projections. There exists a (Γ, σ)-equivariant isometry $U : L^2(\widetilde{M}, \widetilde{E}) \to \ell^2(\Gamma) \otimes L^2(\mathbb{R}^n, \mathbb{C}^k)^N$ and a constant $\mu_0 > 0$ such that for all $\mu \in (0, \mu_0)$, the projections $UE(\lambda)U^*$ and $\mathrm{id} \otimes E^0(\lambda)$ are in $C_r^*(\Gamma, \bar{\sigma}) \otimes \mathcal{K}(L^2(\mathbb{R}^n, \mathbb{C}^k)^N)$ and are Murray-von Neumann equivalent in $C_r^*(\Gamma, \bar{\sigma}) \otimes \mathcal{K}(L^2(\mathbb{R}^n, \mathbb{C}^k)^N)$. In particular,*

$$\begin{aligned}[UE(\lambda)U^*] =&[\mathrm{id} \otimes E^0(\lambda)] \\ &\in K_0(C_r^*(\Gamma, \bar{\sigma}) \otimes \mathcal{K}(L^2(\mathbb{R}^n, \mathbb{C}^k)^N)) \cong K_0(C_r^*(\Gamma, \bar{\sigma})); \\ [E(\lambda)] =&0 \in \tilde{K}_0(C_r^*(\Gamma, \bar{\sigma})).\end{aligned}$$

Denote by Tr_Γ the trace on the algebra $C_r^*(\Gamma, \bar{\sigma}) \otimes \mathcal{K}(L^2(\mathbb{R}^n, \mathbb{C}^k)^N)$, which is the tensor product of the trace $\mathrm{tr}_{\Gamma,\bar{\sigma}}$ on $C_r^*(\Gamma, \bar{\sigma})$ and the standard trace Tr on $\mathcal{K}(L^2(\mathbb{R}^n, \mathbb{C}^k)^N)$. As an immediate consequence of Theorem 5, we get the following

Corollary 6. *In the notation of Theorem 5 one has,*

$$\mathrm{Tr}_\Gamma(UE(\lambda)U^*) = \operatorname{rank}(E^0(\lambda)) \qquad \text{for all} \quad \mu \in (0, \mu_0).$$

4.4. Higher cocycles and equivalence of spectral projections in smooth subalgebras

As a consequence of Theorem 5, one can also show a vanishing theorem for the pairing of the spectral projections $E(\lambda)$ of the operator $H(\mu)$ with higher cyclic cocycles as $\mu \to 0$. One of the main motivation for these results is an application to the quantum Hall effect described in the next section.

It should be noted that higher cyclic cocycles are usually defined not on the whole C^*-algebra $C_r^*(\Gamma, \bar{\sigma}) \otimes \mathcal{K}(L^2(\mathbb{R}^n, \mathbb{C}^k)^N)$, but on some dense subalgebra (better, on a smooth one). Recall that a $*$-subalgebra $\mathfrak{A}_0$ of a C^*-algebra $\mathfrak{A}$ is said to be a smooth subalgebra if $\mathfrak{A}_0$ is a dense $*$-subalgebra of $\mathfrak{A}$, stable under the holomorphic functional calculus. A useful property of smooth subalgebras is the following. If $\mathfrak{A}_0$ is a smooth subalgebra of a C^*-algebra $\mathfrak{A}$, then the inclusion map $\mathfrak{A}_0 \to \mathfrak{A}$ induces an isomorphism in K-theory.

Theorem 7. *In the notation of Theorem 5, assume that $\lambda \in \mathbb{R}$ does not coincide with λ_k for any k. There is a smooth subalgebra $\mathcal{B}(\Gamma, \sigma)$ of $C^*_r(\Gamma, \bar{\sigma}) \otimes \mathcal{K}(L^2(\mathbb{R}^n, \mathbb{C}^k)^N)$ such that the spectral projections $UE(\lambda)U^*$ and $\mathrm{id} \otimes E^0(\lambda)$ are in $\mathcal{B}(\Gamma, \sigma)$ and are also Murray-von Neumann equivalent in $\mathcal{B}(\Gamma, \sigma)$. That is, for all $\mu \in (0, \mu_0)$, one has*

$$[UE(\lambda)U^*] = [\mathrm{id} \otimes E^0(\lambda)] \in K_0(\mathcal{B}(\Gamma, \sigma)).$$

The following corollary uses in addition the Rapid Decay property (RD) for discrete groups. This property is related with the Haagerup inequality, which estimates the convolution norm in terms of the word lengths. Groups that are either virtually nilpotent or word hyperbolic have property (RD). For these groups, it is also known that every group cohomology class can be represented by a group cocycle $c \in Z^j(\Gamma, \mathbb{R})$ that is of polynomial growth, cf. [8].

Corollary 8. *Let Γ be a discrete group that has property (RD). Let $c \in Z^j(\Gamma, \mathbb{R})$ (j even > 0) be a normalised group cocycle that is of polynomial growth, and τ_c the induced cyclic cocycle on the twisted group algebra $\mathbb{C}(\Gamma, \bar{\sigma})$. Then the tensor product cocycle $\tau_c \# \mathrm{Tr}$ extends continuously to $\mathcal{B}(\Gamma, \sigma)$, and in the notation of Theorem 5 one has, for all $\mu \in (0, \mu_0)$*

$$\tau_c \# \mathrm{Tr}(UE(\lambda)U^*, \ldots, UE(\lambda)U^*) = \tau_c \# \mathrm{Tr}(\mathrm{id} \otimes E^0(\lambda), \ldots, \mathrm{id} \otimes E^0(\lambda)) = 0.$$

4.5. Applications to the quantum Hall effect

The Kubo formula for the Hall conductance both in the usual model of the integer quantum Hall effect on the Euclidean plane and in the model of the fractional quantum Hall effect on the hyperbolic plane can be naturally interpreted as a (densely defined) cyclic 2-cocycle tr_K on the algebra $\mathcal{B}(\Gamma, \sigma)$, [1, 4, 29]. The Hall conductance cocycle tr_K can also be shown to be given by a quadratically bounded group cocycle. Moreover, it is well-known that $\mathbb{Z}^2$ and cocompact Fuchsian groups have property (RD). Therefore we have the following consequence of Corollary 8.

Corollary 9. *Let $\widetilde{M}$ be either the Euclidean plane $\mathbb{R}^2$ or the hyperbolic plane $\mathbb{H}$, $\widetilde{E}$ the trivial line bundle, V a Morse type potential. In the notation of Theorem 1, assume that $\lambda \in \mathbb{R}$ does not coincide with λ_k for any k. Let $P_\lambda = \chi_{(-\infty, \mu^{-1}\lambda]}(H_{\mathbf{A}, \mu^{-2}V})$ denote the spectral projection. Then for all sufficiently small values of the coupling constant μ, the Hall conductance vanishes,*

$$\sigma_\lambda = \mathrm{tr}_K(P_\lambda, P_\lambda, P_\lambda) = 0.$$

That is, the low energy bands do not contribute to the Hall conductance.

In the case of the Euclidean plane and when the magnetic field is uniform, this result was established by a different method in [31].

5. Strong magnetic fields

5.1. The model operator

The construction of the model operator in the case of the strong magnetic field was given in [28], using the ideas of [17]. We will use the notation of Theorem 2. Choose a fundamental domain $\mathcal{F} \subset \widetilde{M}$ so that there are no zeros of B on the boundary of $\mathcal{F}$. This is equivalent to saying that the translations $\{\gamma\mathcal{F},\ \gamma \in \Gamma\}$ cover the set of all zeros of B. Let $\{\bar{x}_j \,|\, j = 1, \ldots, N\}$ denote all the zeros of B in $\mathcal{F}$; $\bar{x}_i \neq \bar{x}_j$ if $i \neq j$.

The *model operator* K^h associated with H^h is an operator in $L^2(\mathbb{R}^n)^N$ given by

$$K^h = \bigoplus_{1 \le j \le N} K_j^h,$$

where K_j^h is a self-adjoint second-order differential operator in $L^2(\mathbb{R}^n)$ which corresponds to the zero $\bar{x}_j$. Let us fix local coordinates $f_j : U(\bar{x}_j) \to \mathbb{R}^n$ on $\widetilde{M}$ defined in a small neighborhood $U(\bar{x}_j)$ of $\bar{x}_j$ for every $j = 1, \ldots, N$. We assume that $f_j(\bar{x}_j) = 0$ and the image $f_j(U(\bar{x}_j))$ is a fixed ball $B = B(0, r) \subset \mathbb{R}^n$ centered at the origin.

Write $\mathbf{B}$ in the local coordinates f_j as a 2-form $\mathbf{B}_j(X)$ on $B(0, r)$ and $\mathbf{A}$ as a 1-form $\mathbf{A}_j$ on $B(0, r)$. By [12], there exists a real-valued function $\theta_j \in C^\infty(B(0, r))$ such that

$$|\mathbf{A}_j(X) - d\theta_j(X)| \le C|X|^{k+1}, \quad X \in B(0, r).$$

Write the 1-form $\mathbf{A}_j - d\theta_j$ as

$$\mathbf{A}_j(X) - d\theta_j(X) = \sum_{l=1}^{n} a_l(X)\, dX_l, \quad X \in B(0, r).$$

Let $\mathbf{A}_{1,j}$ be a 1-form on $\mathbb{R}^n$ with polynomial coefficients given by

$$\mathbf{A}_{1,j}(X) = \sum_{l=1}^{n} \sum_{|\alpha|=k+1} \frac{X^\alpha}{\alpha!} \frac{\partial^\alpha a_l}{\partial X^\alpha}(0)\, dX_l, \quad X \in \mathbb{R}^n.$$

Take any extension of the function θ_j to a smooth, compactly supported function in $\mathbb{R}^n$ denoted also by θ_j and put

$$\mathbf{A}_j^0(X) = \mathbf{A}_{1,j}(X) + d\theta_j(X), \quad X \in \mathbb{R}^n.$$

Then we have

$$d\mathbf{A}_j^0(X) = d\mathbf{A}_{1,j}(X) = \mathbf{B}_j^0(X), \quad X \in \mathbb{R}^n,$$

where $\mathbf{B}_j^0$ is a closed 2-form on $\mathbb{R}^n$ with polynomial coefficients. Moreover, we have

$$\begin{aligned} |\mathbf{B}_j(X) - \mathbf{B}_j^0(X)| &\le C|X|^{k+1}, \quad X \in B(0, r), \\ |\mathbf{A}_j(X) - \mathbf{A}_j^0(X)| &\le C|X|^{k+2}, \quad X \in B(0, r). \end{aligned} \tag{16}$$

By definition, K_j^h is the self-adjoint differential operator with asymptotically polynomial coefficients in $L^2(\mathbb{R}^n)$ given by

$$K_j^h = (ih\,d + \mathbf{A}_j^0)^*(ih\,d + \mathbf{A}_j^0),$$

where the adjoint is taken with respect to the Hilbert structure in $L^2(\mathbb{R}^n)$ given by the flat Riemannian metric $(g_{lm}(0))$ in $\mathbb{R}^n$. The operator K_j^h has discrete spectrum (cf., for instance, [21, 12]). By gauge invariance, the operator K_j^h is unitarily equivalent to the Schrödinger operator

$$H_j^h = (ih\,d + \mathbf{A}_{1,j})^*(ih\,d + \mathbf{A}_{1,j}),$$

associated with the homogeneous 1-form $\mathbf{A}_{1,j}$. Using a simple scaling $X \mapsto h^{\frac{1}{k+2}}X$, it can be shown that the operator H_j^h is unitarily equivalent to the operator $h^{\frac{2k+2}{k+2}}H_j^1$. So we conclude that the operator $h^{-\frac{2k+2}{k+2}}K^h$ has discrete spectrum independent of h, which is denoted by $\{\lambda_m : m \in \mathbb{N}\}$, $\lambda_1 < \lambda_2 < \lambda_3 < \dots$ (not taking into account multiplicities). This fact explains the appearance of a scaling factor $h^{\frac{2k+2}{k+2}}$ in Theorem 2. The sequence $\{\lambda_m : m \in \mathbb{N}\}$ is precisely the sequence, which we need for the proof of Theorem 2.

5.2. Proof of Theorem 2

For the proof, we apply Theorem 3 in the following setting. As in Section 3, let $\mathfrak{A}$ be the C^* algebra $C_r^*(\Gamma, \bar{\sigma}) \otimes \mathcal{K}$, let $\mathcal{H}$ be the Hilbert space $\ell^2(\Gamma) \otimes \ell^2(\mathbb{N})$ equipped with the representation π of the algebra $\mathfrak{A}$ and $\mathcal{H}_2 = L^2(\widetilde{M})$. Let $\mathcal{V}_2 : \mathcal{H}_2 \to \mathcal{H}$ be a unitary operator defined as $\mathcal{V}_2 = (\mathrm{id} \otimes V_2) \circ \mathbf{U}$, where $V_2 : L^2(\mathcal{F}) \to \ell^2(\mathbb{N})$ is an arbitrary unitary isomorphism. Let π_2 be the corresponding representation of $\mathfrak{A}$ in $\mathcal{H}_2$.

We will use the notation of Section 5.1. Put $\mathcal{H}_1 = \ell^2(\Gamma) \otimes L^2(\mathbb{R}^n)^N$. Choose an arbitrary unitary isomorphism $V_1 : L^2(\mathbb{R}^n)^N \to \ell^2(\mathbb{N})$ and define an unitary operator $\mathcal{V}_1 : \mathcal{H}_1 \to \mathcal{H}$ as $\mathcal{V}_1 = \mathrm{id} \otimes V_1$. Let π_1 be the corresponding representation of $\mathfrak{A}$ in $\mathcal{H}_1$.

Consider self-adjoint, semi-bounded from below operators A_1 in $\mathcal{H}_1$ and A_2 in $\mathcal{H}_2$:

$$A_1 = \mathrm{id} \otimes h^{-\frac{2k+2}{k+2}}K^h, \quad A_2 = h^{-\frac{2k+2}{k+2}}H^h.$$

Assumption 1 is clear for the operator A_1. It holds for the operator A_2 by the results mentioned at the end of Section 3.

Let $\mathcal{H}_0 = \ell^2(\Gamma) \bigotimes \left(\bigoplus_{j=1}^N L^2(U(\bar{x}_j))\right)$. An inclusion $i_1 : \mathcal{H}_0 \to \mathcal{H}_1$ is defined as $i_1 = \mathrm{id} \otimes j_1$, where j_1 is the inclusion

$$\bigoplus_{j=1}^{N} L^2(U(\bar{x}_j)) \cong L^2(B(0,r))^N \hookrightarrow L^2(\mathbb{R}^n)^N$$

given by the chosen local coordinates. An inclusion $i_2 : \mathcal{H}_0 \to \mathcal{H}_2$ is defined as $i_2 = \mathbf{U}^* \circ (\mathrm{id} \otimes j_2)$, where j_2 is the natural inclusion

$$\bigoplus_{j=1}^{N} L^2(U(\bar{x}_j)) \hookrightarrow L^2(\mathcal{F}).$$

The operator $p_1 : \mathcal{H}_1 \to \mathcal{H}_0$ is defined as $p_1 = \mathrm{id} \otimes r_1$, where r_1 is the restriction operator

$$L^2(\mathbb{R}^n)^N \to L^2(B(0,r))^N \cong \bigoplus_{j=1}^{N} L^2(U(\bar{x}_j)).$$

The operator $p_2 : \mathcal{H}_1 \to \mathcal{H}_0$ is defined as $p_2 = (\mathrm{id} \otimes r_2) \circ \mathbf{U}$, where r_2 is the restriction operator

$$L^2(\mathcal{F}) \to \bigoplus_{j=1}^{N} L^2(U(\bar{x}_j)).$$

Fix a function $\phi \in C_c^\infty(\mathbb{R}^n)$ such that $0 \le \phi \le 1$, $\phi(x) = 1$ if $|x| \le 1$, $\phi(x) = 0$ if $|x| \ge 2$. Fix a number $\kappa > 0$, which we shall choose later. For any $h > 0$ define $\phi^{(h)}(x) = \phi(h^{-\kappa}x)$. For any $h > 0$ small enough, let $\phi_j = \phi^{(h)} \in C_c^\infty(U(\bar{x}_j))$ in the fixed coordinates near $\bar{x}_j$. Denote also $\phi_{j,\gamma} = (\gamma^{-1})^*\phi_j$. (This function is supported in the neighborhood $U(\gamma\bar{x}_j) = \gamma(U(\bar{x}_j))$ of $\gamma\bar{x}_j$.) We will always take $h \in (0, h_0)$ where h_0 is sufficiently small, so in particular the supports of all functions $\phi_{j,\gamma}$ are disjoint.

Let $\Phi \in C^\infty(\bigcup_{j=1}^N U(\bar{x}_j))$ be equal to ϕ_j on $U(\bar{x}_j)$, $j = 1, 2, \ldots, N$. Consider a (Γ, σ)-equivariant, self-adjoint, bounded operator J in $\mathcal{H}_0$ defined as $J = \mathrm{id} \otimes \Phi$, where Φ denotes the multiplication operator by the function Φ in the space $\oplus_{j=1}^N L^2(U(\bar{x}_j))$. Assumptions 2, 3, 4 and 5 can be easily checked.

We will use local coordinates near $\bar{x}_j$ such that the Riemannian volume element at the point $\bar{x}_j$ coincides with the Euclidean volume element given by the chosen local coordinates. Then the estimate (2) holds with

$$\rho = 1 + O(h^\kappa). \tag{17}$$

The principal symbols $a_{1,j}^{(2)} \in C^\infty(T^*\mathbb{R}^n)$ of $K_j^h, j = 1, 2, \ldots, N$, and $a_2^{(2)} \in C^\infty(T^*\widetilde{M})$ of H^h are given respectively by

$$a_{1,j}^{(2)}(x,\xi) = h^2 \sum_{i,k=1}^{n} g^{ik}(\bar{x}_j)\xi_i\xi_k, \quad (x,\xi) \in T^*\mathbb{R}^n;$$

$$a_2^{(2)}(x,\xi) = h^2 \sum_{i,k=1}^{n} g^{ik}(x)\xi_i\xi_k, \quad (x,\xi) \in T^*\widetilde{M}.$$

As in Section 4.2, one can check Assumption 6 and the estimates (3) with

$$\gamma_l = O(h^{-\frac{2k+2}{k+2}+2-2\kappa}), \quad l = 1, 2. \tag{18}$$

Using the asymptotic lower bounds for the quadratic forms associated to the operators K_j^h and H^h given by [17, Theorems 4.4 and 4.5], it can be proved that the estimates (6) hold with

$$\alpha_l = O(h^{-\frac{2k+2}{k+2}+k\kappa+1}), \quad l = 1, 2. \tag{19}$$

The constants $\lambda_{0l}, l = 1, 2$, can be chosen to be independent of h:

$$\lambda_{01} = \lambda_{02} = 0. \tag{20}$$

Finally, using (16), it can be easily verified that the estimates (4) and (5) hold with

$$\beta_l = 1 + O(h^\kappa), \quad \varepsilon_l = O(h^{2\kappa(k+2)-\kappa-\frac{2k+2}{k+2}}), \quad l = 1, 2. \tag{21}$$

Now we complete the proof of Theorem 2. As above, let $\{\lambda_m : m \in \mathbb{N}\}$, $\lambda_1 < \lambda_2 < \lambda_2 < \ldots$, be the spectrum (without taking into account multiplicities) of the operator $h^{-\frac{2k+2}{k+2}} K^h$, which is independent of h. Take any a and b such that $\lambda_m < a < b < \lambda_{m+1}$ with some m. Clearly, the spectrum of the operator A_1 coincides with the spectrum of the operator $h^{-\frac{2k+2}{k+2}} K^h$. Therefore, $[a, b] \cap \sigma(A_1) = \emptyset$. Take any open interval (a_1, b_1) that contains $[a, b]$ and does not intersect with the spectrum of A_1. Using the estimates (17), (18), (19), (20) and (21), one can see that, for a_2 and b_2 given by (7) and (8), we have

$$a_2 = a_1 + O(h^s), \quad b_2 = b_1 + O(h^s), \quad h \to 0, \tag{22}$$

where $s = \min\{(2k+3)\kappa - \frac{2k+2}{k+2}, -\frac{2k+2}{k+2} + 2 - 2\kappa\}$. The best possible value of s which is

$$s = \max_\kappa \min\{(2k+3)\kappa - \frac{2k+2}{k+2}, -\frac{2k+2}{k+2} + 2 - 2\kappa\} = \frac{2}{(2k+5)(k+2)}$$

is attained when $\kappa = \frac{2}{2k+5}$.

Hence, if $h > 0$ is small enough, we have $\alpha_1 > a_1 + \gamma_1$, $\alpha_2 > b_2 + \gamma_2$, $b_2 > a_2$ and the interval (a_2, b_2) contains $[a, b]$. By Theorem 3, we conclude that $(a_2, b_2) \cap \sigma(A_2) = \emptyset$, that completes the proof of Theorem 2.

5.3. Semiclassical approximation of spectral projections

Like in the case of strong electric field, Theorem 3 also provides some information about the K-theory class of the spectral projection of the magnetic Schrödinger operator in the limit of strong magnetic field.

Theorem 10. *In the notation of Theorem 2, assume that $\lambda \in \mathbb{R}$ does not coincide with λ_k for any k. Let $E^h(\lambda) = \chi_{(-\infty,\lambda]}(H^h)$ and $E^0(\lambda) = \chi_{(-\infty,\lambda]}(K^h)$ denote the spectral projections. There exists a (Γ, σ)-equivariant isometry $U : L^2(\widetilde{M}) \to \ell^2(\Gamma) \otimes L^2(\mathbb{R}^n)^N$ and a constant $h_0 > 0$ such that for all $h \in (0, h_0)$, the spectral projections $UE(h^{\frac{2k+2}{k+2}}\lambda)U^*$ and $\mathrm{id} \otimes E^0(h^{\frac{2k+2}{k+2}}\lambda)$ are in $C_r^*(\Gamma, \bar\sigma) \otimes \mathcal{K}(L^2(\mathbb{R}^n)^N)$ and Murray-von Neumann equivalent in $C_r^*(\Gamma, \bar\sigma) \otimes \mathcal{K}(L^2(\mathbb{R}^n)^N)$.*

References

[1] J. Bellissard, A. van Elst, H. Schulz-Baldes, *The non-commutative geometry of the quantum Hall effect.* J. Math. Phys. **35** (1994), 5373–5451.

[2] U. Bunke, *On the gluing problem for the η-invariant.* J. Differential Geom. **41** (1995), 397–448.

[3] D. Burghelea, L. Friedlander, T. Kappeler, P. McDonald, *Analytic and Reidemeister torsion for representations in finite type Hilbert modules.* Geom. Funct. Anal. **6** (1996), 751–859.

[4] A. Carey, K. Hannabus, V. Mathai, P. McCann, *Quantum Hall Effect on the hyperbolic plane.* Comm. Math. Phys. **190** (1998), 629–673.

[5] A. Figotin, P. Kuchment, *Band-Gap Structure of Spectra of Periodic Dielectric and Acoustic Media. I. Scalar model.* SIAM J. Appl. Math. **56** (1996), 68–88. *II. Two-dimensional photonic crystals.* SIAM J. Appl. Math. **56** (1996), 1561–1620.

[6] A. Figotin, P. Kuchment, *Spectral properties of classical waves in high-contrast periodic media.* SIAM J. Appl. Math. **58** (1998), 683–702.

[7] L. Friedlander, *On the density of states of periodic media in the large coupling limit.* Comm. Partial Differential Equations **27** (2002), 355–380.

[8] M. Gromov, *Volume and bounded cohomology.* Inst. Hautes Études Sci. Publ. Math. **56** (1982), 5–99.

[9] B. Helffer, J. Sjöstrand, *Multiple wells in the semiclassical limit. I.* Comm. Partial Differential Equations **9** (1984), 337–408.

[10] B. Helffer, J. Sjöstrand, *Effet tunnel pour l'équation de Schrödinger avec champ magnétique.* Ann. Scuola Norm. Sup. Pisa Cl. Sci. (4) **14** (1987), 625–657.

[11] B. Helffer, J. Sjöstrand, *Analyse semi-classique pour l'équation de Harper (avec application à l'équation de Schrödinger avec champ magnétique).* Mém. Soc. Math. France (N.S.) **34** (1988).

[12] B. Helffer, A. Mohamed, *Caractérisation du spectre essential de l'opérateur de Schrödinger avec un champ magnétique.* Ann. Inst. Fourier (Grenoble) **38** (1988), 95–112.

[13] B. Helffer, J. Sjöstrand, *Analyse semi-classique pour l'équation de Harper. II. Comportement semi-classique près d'un rationnel.* Mém. Soc. Math. France (N.S.) **40** (1990).

[14] B. Helffer, J. Sjöstrand, *Semiclassical analysis for Harper's equation. III. Cantor structure of the spectrum.* Mém. Soc. Math. France (N.S.) **39** (1989), 1–124.

[15] B. Helffer, J. Sjöstrand, *Équation de Schrödinger avec champ magnétique et équation de Harper.* In *Schrödinger operators (Sønderborg, 1988)*, Lecture Notes in Phys., **345**, Springer, 1989, pp. 118–197.

[16] B. Helffer, J. Sjöstrand, *On diamagnetism and de Haas-van Alphen effect.* Ann. Inst. H. Poincaré Phys. Théor. **52** (1990), 303–375.

[17] B. Helffer, A. Mohamed, *Semiclassical analysis for the ground state energy of a Schrödinger operator with magnetic wells.* J. Funct. Anal. **138** (1996), 40–81.

[18] B. Helffer, A. Morame, *Magnetic bottles in connection with superconductivity,* J. Funct. Anal. **185** (2001), 604–680.

[19] B. Helffer, A. Morame, *Magnetic bottles for the Neumann problem: the case of dimension* 3. Spectral and inverse spectral theory (Goa, 2000) Proc. Indian Acad. Sci. (Math. Sci.) **112** (2002), 71–84.

[20] B. Helffer, A. Morame, *Magnetic bottles for the Neumann problem: Curvature effects in the case of dimension 3.* Ann. Sci. Ec. Norm. Sup. 4 série **37** (2004), 105–170.

[21] B. Helffer, J. Nourrigat, *Hypoellipticité maximale pour des opérateurs polynômes de champs de vecteurs.* Birkhäuser, 1985.

[22] R. Hempel, I. Herbst, *Strong magnetic fields, Dirichlet boundaries, and spectral gaps.* Comm. Math. Phys. **169** (1995), 237–259.

[23] R. Hempel, I. Herbst, *Bands and gaps for periodic magnetic Hamiltonians.* In: *Operator Theory: Advances and Applications, Vol. 78.* Birkhäuser, 1995, pp. 175–184.

[24] R. Hempel, K. Lienau, *Spectral properties of periodic media in the large coupling limit.* Comm. Partial Differential Equations **25** (2000), 1445–1470.

[25] R. Hempel, O. Post, *Spectral gaps for periodic elliptic operators with high contrast: an overview.* In: *Progress in analysis, Vol. I, II (Berlin, 2001).* World Sci. Publishing, 2003, pp. 577–587.

[26] I. Herbst, S. Nakamura, *Schrödinger operators with strong magnetic fields: Quasi-periodicity of spectral orbits and topology.* In: *Differential operators and spectral theory.* Amer. Math. Soc. Transl. Ser. 2, v. 189, Amer. Math. Soc., 1999, pp. 105–123.

[27] Yu. A. Kordyukov, V. Mathai, M. Shubin, *Equivalence of projections in semiclassical limit and a vanishing theorem for higher traces in K-theory.* J. Reine Angew. Math. **581** (2005), 193–236.

[28] Yu. A. Kordyukov, *Spectral gaps for periodic Schrödinger operators with strong magnetic fields.* Comm. Math. Phys. **253** (2005), 371–384.

[29] M. Marcolli, V. Mathai, *Twisted index theory for good orbifolds, II: fractional quantum numbers.* Comm. Math. Phys. **217** (2001), 55–87.

[30] V. Mathai, M. Shubin, *Semiclassical asymptotics and gaps in the spectra of magnetic Schrödinger operators.* Geom. Dedicata **91** (2002), 155–173.

[31] S. Nakamura, J. Bellissard, *Low energy bands do not contribute to the quantum Hall effect.* Comm. Math. Phys. **131** (1990), 283–305.

[32] O. Post, *Periodic manifolds with spectral gaps.* J. Differential Equations **187** (2003), 23–45.

[33] M. Shubin, *Semiclassical asymptotics on covering manifolds and Morse Inequalities.* Geom. Funct. Anal. **6** (1996), 370–409.

Yuri A. Kordyukov
Institute of Mathematics,
Russian Academy of Sciences,
112 Chernyshevsky str.
Ufa 450077, Russia
e-mail: yuri@imat.rb.ru

C^*-algebras and Elliptic Theory
Trends in Mathematics, 151–156

The Group of Unital C^*-extensions

Vladimir Manuilov and Klaus Thomsen

Abstract. Let A and B be separable C^*-algebras, A unital and B stable. It is shown that there is a natural six-terms exact sequence which relates the group which arises by considering *all* semi-split extensions of A by B to the group which arises by restricting the attention to *unital* semi-split extensions of A by B. The six-terms exact sequence is an unpublished result of G. Skandalis.

Mathematics Subject Classification (2000). Primary 46L05; Secondary 46L80.
Keywords. C^*-algebra, extension, unital extension, absorbing homomorphism.

Let A, B be separable C^*-algebras, B stable. As is well known the C^*-algebra extensions of A by B can be identified with $\mathrm{Hom}(A.Q(B))$, the set of $*$-homomorphisms $A \to Q(B)$ where $Q(B) = M(B)/B$ is the generalized Calkin algebra. Two extensions $\varphi, \psi : A \to Q(B)$ are *unitarily equivalent* when there is a unitary $u \in M(B)$ such that $\mathrm{Ad}\, q(u) \circ \psi = \varphi$, where $q : M(B) \to Q(B)$ is the quotient map. The unitary equivalence classes of extensions of A by B have an abelian semi-group structure thanks to the stability of B: Choose isometries $V_1, V_2 \in M(B)$ such that $V_1V_1^* + V_2V_2^* = 1$, and define the sum $\varphi \oplus \psi : A \to Q(B)$ of $\varphi, \psi \in \mathrm{Hom}(A, Q(B))$ by

$$(\psi \oplus \varphi)(a) = \mathrm{Ad}\, q(V_1) \circ \psi(a) + \mathrm{Ad}\, q(V_2) \circ \varphi(a). \tag{1}$$

The isometries, V_1 and V_2, are fixed in the following. An extension $\varphi : A \to Q(B)$ is *split* when there is a $*$-homomorphisms $\pi : A \to M(B)$ such that $\varphi = q \circ \pi$. To trivialize the split extensions we declare two extensions $\varphi, \psi : A \to Q(B)$ to be *stably equivalent* when there there is a split extension π such that $\psi \oplus \pi$ and $\varphi \oplus \pi$ are unitarily equivalent. This is an .equivalence relation because the sum (1) of two split extensions is itself split. We denote by $\mathrm{Ext}(A, B)$ the semigroup of stable equivalence classes of extensions of A by B. It was proved in [5], as a generalization of results of Kasparov, that there exists an *absorbing split extension* $\pi_0 : A \to Q(B)$, i.e., a split extension with the property that $\pi_0 \oplus \pi$ is unitarily equivalent to π_0 for every split extension π. Thus two extensions φ, ψ are stably equivalent if and only if $\varphi \oplus \pi_0$ and $\psi \oplus \pi_0$ are unitarily equivalent. The classes of stably equivalent extensions of A by B is an abelian semigroup $\mathrm{Ext}(A, B)$ in which

The first named author was partially supported by RFFI grant 05-01-00923.

any split extension (like 0) represents the neutral element. As is well documented the semi-group is generally not a group, and we denote by

$$\mathrm{Ext}^{-1}(A,B)$$

the abelian group of invertible elements in $\mathrm{Ext}(A,B)$. It is also well known that this group is one way of describing the KK-groups of Kasparov. Specifically, $\mathrm{Ext}^{-1}(A,B) = KK(SA,B) = KK(A,SB)$.

Assume now that A is unital. It is then possible, and sometimes even advantageous, to restrict attention to unital extensions of A by B, i.e., to short exact sequences

$$0 \longrightarrow B \longrightarrow E \longrightarrow A \longrightarrow 0$$

of C^*-algebras with E is unital, or equivalently to $*$-homomorphisms $A \to Q(B)$ that are unital. The preceding definitions are all amenable to such a restriction, if done consistently. Specifically, we say that a unital extension $\varphi : A \to Q(B)$ is *unitally split* when there is a unital $*$-homomorphism $\pi : A \to M(B)$ such that $\varphi = q \circ \pi$. The sum (1) of two unital extensions is again unital, and we say that two unital extensions $\varphi, \psi : A \to Q(B)$ are *unitally stably equivalent* when there is a unital split extension π such that $\psi \oplus \pi$ and $\varphi \oplus \pi$ are unitarily equivalent. It was proved in [5] that there always exists a *unitally absorbing split extension* $\pi_0 : A \to Q(B)$, i.e., a unitally split extension with the property that $\pi_0 \oplus \pi$ is unitarily equivalent to π_0 for every unitally split extension π. Thus two unital extensions φ, ψ are unitally stably equivalent if and only if $\varphi \oplus \pi_0$ and $\psi \oplus \pi_0$ are unitarily equivalent. The classes of unitally stably equivalent extensions of A by B is an abelian semi-group which we denote by $\mathrm{Ext}_{\mathrm{unital}}(A,B)$. The unitally absorbing split extension π_0, or any other unitally split extension, represents the neutral element of $\mathrm{Ext}_{\mathrm{unital}}(A,B)$, and we denote by

$$\mathrm{Ext}^{-1}_{\mathrm{unital}}(A,B)$$

the abelian group of invertible elements in $\mathrm{Ext}_{\mathrm{unital}}(A,B)$. As we shall see there is a difference between $\mathrm{Ext}^{-1}_{\mathrm{unital}}(A,B)$ and $\mathrm{Ext}^{-1}(A,B)$ arising from the fact that while the class in $\mathrm{Ext}^{-1}(A,B)$ of a unital extension $A \to Q(B)$ can not be changed by conjugating it with a unitary from $Q(B)$, its class in $\mathrm{Ext}^{-1}_{\mathrm{unital}}(A,B)$ can. In a sense the main result of this note is that this is the only way in which the two groups differ.

Note that there is a group homomorphism

$$\mathrm{Ext}^{-1}_{\mathrm{unital}}(A,B) \to \mathrm{Ext}^{-1}(A,B),$$

obtained by forgetting the word 'unital'. It will be shown that this forgetful map fits into a six-terms exact sequence

$$\begin{array}{ccccc} K_0(B) & \xrightarrow{u_0} & \mathrm{Ext}^{-1}_{\mathrm{unital}}(A,B) & \longrightarrow & \mathrm{Ext}^{-1}(A,B) \\ {\scriptstyle i_0^*}\uparrow & & & & \downarrow{\scriptstyle i_1^*} \\ \mathrm{Ext}^{-1}(A,SB) & \longleftarrow & \mathrm{Ext}^{-1}_{\mathrm{unital}}(A,SB) & \xleftarrow[u_1]{} & K_1(B) \end{array}$$

where SB is the suspension of B, i.e., $SB = C_0(0,1) \otimes B$, and the maps u_k and $i_k^*, k = 0, 1$, will be defined shortly. This six-terms exact sequence is mentioned in 10.11 of [4], but the proof was never published.

Fix a unitally absorbing $*$-homomorphism $\alpha_0 : A \to M(B)$, which exists by Theorem 2.4 of [5]. It follows then from Theorem 2.1 of [5] that $\alpha = q \circ \alpha_0$ is a unitally absorbing split extension as defined above.

Lemma 1. *The $*$-homomorphisms* $\operatorname{Ad} V_1 \circ \alpha_0 : A \to M(B)$ *and* $\left(\begin{smallmatrix} \alpha_0 & \\ & 0 \end{smallmatrix}\right) : A \to M(M_2(B))$ *are both absorbing.*

Proof. There is a $*$-isomorphism $M(M_2(B)) = M_2(M(B)) \to M(B)$ given by

$$\left(\begin{smallmatrix} m_{11} & m_{12} \\ m_{21} & m_{22} \end{smallmatrix}\right) \mapsto V_1 m_{11} V_1^* + V_1 m_{12} V_2^* + V_2 m_{21} V_1^* + V_2 m_{22} V_2^*,$$

which sends $M_2(B)$ to B and $\left(\begin{smallmatrix} \alpha_0 & \\ & 0 \end{smallmatrix}\right)$ to $\operatorname{Ad} V_1 \circ \alpha_0$, so it suffices to show that the latter is an absorbing $*$-homomorphism. By definition, cf. Definition 2.6 of [5], we must show that the unital $*$-homomorphism $A \oplus \mathbb{C} \ni (a, \lambda) \mapsto V_1 \alpha_0(a) V_1^* + \lambda V_2 V_2^*$ is unitally absorbing. For this we check that it has property 1) of Theorem 2.1 of [5]. So let $\varphi : A \oplus \mathbb{C} \to B$ be a completely positive contraction. Since α_0 has property 1), there is a sequence $\{W_n\}$ in $M(B)$ such that $\lim_{n\to\infty} W_n^* b = 0$ for all $b \in B$ and $\lim_{n\to\infty} W_n^* \alpha_0(a) W_n = \varphi(a)$ for all $a \in A$. Since B is stable there is a sequence $\{S_n\}$ of isometries in $M(B)$ such that $\lim_{n\to\infty} S_n^* b = 0$ for all $b \in B$. Set

$$T_n = V_1 W_n + V_2 S_n \varphi(0,1)^{\frac{1}{2}}.$$

Then $\lim_{n\to\infty} T_n b = 0$ for all $b \in B$, and

$$T_n^* \left(V_1 \alpha_0(a) V_1^* + \lambda V_2 V_2^*\right) T_n = W_n^* \alpha_0(a) W_n + \varphi(0, \lambda)$$

for all n. Since the last expression converges to $\varphi(a, \lambda)$ as n tends to infinity, the proof is complete. □

Set

$$C_\alpha = \left\{ m \in M_2(M(B)) : \; m \left(\begin{smallmatrix} \alpha_0(a) & \\ & 0 \end{smallmatrix}\right) - \left(\begin{smallmatrix} \alpha_0(a) & \\ & 0 \end{smallmatrix}\right) m \in M_2(B) \ \forall a \in A \right\}$$

and

$$A_\alpha = \left\{ m \in C_\alpha : \; m \left(\begin{smallmatrix} \alpha_0(a) & \\ & 0 \end{smallmatrix}\right) \in M_2(B) \ \forall a \in A \right\}.$$

We can define a $*$-homomorphism $C_\alpha \to \alpha(A)' \cap Q(B)$ such that

$$\left(\begin{smallmatrix} m_{11} & m_{12} \\ m_{21} & m_{22} \end{smallmatrix}\right) \mapsto q(m_{11}).$$

Then kernel is then A_α, so we have a $*$-isomorphism $C_\alpha / A_\alpha \simeq \alpha(A)' \cap Q(B)$. By Lemma 1, $\left(\begin{smallmatrix} \alpha_0 & \\ & 0 \end{smallmatrix}\right)$ is an absorbing $*$-homomorphism, so we conclude from Theorem 3.2 of [5] that there is an isomorphism

$$K_1\left(\alpha(A)' \cap Q(B)\right) \simeq KK(A, B). \tag{2}$$

Since the unital $*$-homorphism $\mathbb{C} \to M(B)$ is unitally absorbing, this gives us also the well-known isomorphism

$$K_1\left(Q(B)\right) \simeq KK(\mathbb{C}, B). \tag{3}$$

Let $i : \mathbb{C} \to A$ be the unital $*$-homorphism. For convenience we denote the map $K_1(\alpha(A)' \cap Q(B)) \to K_1(Q(B))$ induced by the inclusion $\alpha(A)' \cap Q(B) \subseteq Q(B)$ by i^*. It is then easy to check that the isomorphisms (2) and (3) match up to make the diagram

$$\begin{array}{ccc} K_1(\alpha(A)' \cap Q(B)) & \xrightarrow{i^*} & K_1(Q(B)) \\ \downarrow & & \downarrow \\ KK(A,B) & \xrightarrow{i^*} & KK(\mathbb{C},B) \end{array} \tag{4}$$

commute.

Let v be a unitary in $M_n(Q(B))$. By composing the $*$-homomorphism $\operatorname{Ad} v \circ (1_n \otimes \alpha) : A \to M_n(Q(B))$ with isomorphism $M_n(Q(B)) \simeq Q(B)$ which is canonical in the sense that it arises from an isomorphism $M_n(B) \simeq B$, we obtain a unital extension $e(v) : A \to Q(B)$ of A by B. By use of a unitary lift of $\begin{pmatrix} v & \\ & v^* \end{pmatrix}$ one sees that $e(v) \oplus e(v^*)$ is split, proving that $e(v)$ represents an element in $\operatorname{Ext}^{-1}_{\text{unital}}(A,B)$. If $v_t, t \in [0,1]$, is a norm-continuous path of unitaries in $M_n(Q(B))$ there is a partition $0 = t_0 < t_1 < t_2 < \cdots < t_N = 1$ of $[0,1]$ such that $v_{t_i} v^*_{t_{i+1}}$ is in the connected component of 1 in the unitary group of $M_n(Q(B))$ and hence has a unitary lift to $M_n(M(B))$. It follows that $e(v_0) = e(v_1)$, and it is then clear that the construction gives us a group homomorphism

$$u : K_1(Q(B)) \to \operatorname{Ext}^{-1}_{\text{unital}}(A,B).$$

Lemma 2. *The sequence*

$$\begin{array}{ccccc} K_1(Q(B)) & \xrightarrow{u} & \operatorname{Ext}^{-1}_{\text{unital}}(A,B) & \longrightarrow & \operatorname{Ext}^{-1}(A,B) \\ \uparrow{\scriptstyle i^*} & & & & \downarrow{\scriptstyle i^*} \\ K_1(\alpha(A)' \cap Q(B)) & & & & \operatorname{Ext}^{-1}(\mathbb{C},B) \end{array}$$

is exact.

Proof. **Exactness at $K_1(Q(B))$:** If v is a unitary in $M_n(\alpha(A)' \cap Q(B))$, the extension $\operatorname{Ad} v \circ (1_n \otimes \alpha) = 1_n \otimes \alpha$ (of A by $M_n(B)$) is split, proving that $u \circ i^* = 0$. To show that $\ker u \subseteq \operatorname{im} i^*$, let $v \in Q(B)$ be a unitary such that $u[v] = 0$. Then $\operatorname{Ad} v \circ \alpha \oplus \alpha$ is unitarily equivalent to $\alpha \oplus \alpha$, which means that there is a unitary $S \in M(M_2(B))$ such that

$$\operatorname{Ad}\left(\left(\operatorname{id}_{M_2(\mathbb{C})} \otimes q\right)(S) \begin{pmatrix} v & \\ & 1 \end{pmatrix}\right) \begin{pmatrix} \alpha(a) & \\ & \alpha(a) \end{pmatrix} = \begin{pmatrix} \alpha(a) & \\ & \alpha(a) \end{pmatrix} \tag{5}$$

for all $a \in A$. Since the unitary group of $M(M_2(B))$ is normconnected by [3] or [2], the unitary $\begin{pmatrix} v & \\ & 1 \end{pmatrix}$ is homotopic to $\left(\operatorname{id}_{M_2(\mathbb{C})} \otimes q\right)(S) \begin{pmatrix} v & \\ & 1 \end{pmatrix}$ which is in $M_2(\alpha(A)' \cap Q(B))$ by (5). This implies that $[v] \in \operatorname{im} i^*$. The same argument works when v is a unitary in $M_n(Q(B))$ for some $n \geq 2$.

Exactness at $\operatorname{Ext}^{-1}_{\text{unital}}(A,B)$: For any unitary $v \in Q(B)$,

$$(\operatorname{Ad} v \circ \alpha) \oplus 0 = \operatorname{Ad}\left(\operatorname{id}_{M_2(\mathbb{C})} \otimes q\right)(T) \circ (\alpha \oplus 0),$$

where $T \in M_2(M(B))$ is a unitary lift of $\begin{pmatrix} v & \\ & v^* \end{pmatrix}$. Hence $[\mathrm{Ad}\, v \circ \alpha] = 0$ in $\mathrm{Ext}^{-1}(A, B)$. The same argument works when v is a unitary $M_n(Q(B))$ for some $n \geq 2$, and we conclude that the composition

$$K_1(Q(B)) \longrightarrow \mathrm{Ext}^{-1}_{\mathrm{unital}}(A, B) \longrightarrow \mathrm{Ext}^{-1}(A, B)$$

is zero. Let $\varphi : A \to Q(B)$ be a unital extension such that $[\varphi] = 0$ in $\mathrm{Ext}^{-1}(A, B)$. By Lemma 1, this means that there is a unitary $T \in M(M_3(B))$ such that

$$\mathrm{Ad}\left(\mathrm{id}_{M_3(\mathbb{C})} \otimes q\right)(T) \circ \begin{pmatrix} \varphi & & \\ & \alpha & \\ & & 0 \end{pmatrix} = \begin{pmatrix} \alpha & & \\ & \alpha & \\ & & 0 \end{pmatrix}.$$

It follows that $\left(\mathrm{id}_{M_3(\mathbb{C})} \otimes q\right)(T) = \begin{pmatrix} V & \\ & r \end{pmatrix}$ for some unitaries $V \in M_2(Q(B))$ and $r \in Q(B)$. Hence

$$\begin{pmatrix} \varphi & \\ & \alpha \end{pmatrix} = \mathrm{Ad}\, V^* \circ \begin{pmatrix} \alpha & \\ & \alpha \end{pmatrix}.$$

Thus $[\varphi] = u[V^*]$.

Exactness at $\mathrm{Ext}^{-1}(A, B)$**:** It is obvious that i^* kills the image of $\mathrm{Ext}^{-1}_{\mathrm{unital}}(A, B)$, so consider an invertible extension $\varphi : A \to Q(B)$ such that $[\varphi \circ i] = 0$ in $\mathrm{Ext}^{-1}(\mathbb{C}, B)$. By Lemma 1, applied with $A = \mathbb{C}$, this means that there is a unitary $T \in M_3(M(B))$ such that

$$\left(\mathrm{id}_{M_3(\mathbb{C})} \otimes q\right)(T) \begin{pmatrix} \varphi(1) & & \\ & 1 & \\ & & 0 \end{pmatrix} \left(\mathrm{id}_{M_3(\mathbb{C})} \otimes q\right)(T^*) = \begin{pmatrix} 0 & & \\ & 1 & \\ & & 0 \end{pmatrix}. \quad (6)$$

Set $\psi = \varphi \oplus \alpha \oplus 0$. It follows from (6) that there are isometries $W_1, W_2, W_3 \in M(B)$ and a unitary $u \in M(B)$ such that $W_i^* W_j = 0, i \neq j$, $W_1 W_1^* + W_2 W_2^* + W_3 W_3^* = 1$ and $\mathrm{Ad}\, q(u) \circ \psi(1) = q(W_2 W_2^*)$. Then $\mathrm{Ad}\, q(u) \circ \psi + \mathrm{Ad}\, q(W_1) \circ \alpha + \mathrm{Ad}\, q(W_3) \circ \alpha$ is a unital extension which is invertible because it admits a completely positive contractive lifting to $M(B)$ since ψ does, cf. [1]. As it represents the same class in $\mathrm{Ext}^{-1}(A, B)$ as φ, the proof is complete. □

In order to complete the sequence of Lemma 2, let $i_1^* : \mathrm{Ext}^{-1}(A, B) \to K_1(B)$ be the composition

$$\mathrm{Ext}^{-1}(A, B) \longrightarrow K_1\left(\alpha(A)' \cap Q(B)\right) \xrightarrow{i^*} K_1\left(Q(SB)\right) \longrightarrow K_1(B), \quad (7)$$

where the first map is the isomorphism (2) and the last is the well-known isomorphism. Let $u_1 : K_1(B) \to \mathrm{Ext}^{-1}_{\mathrm{unital}}(A, SB)$ be the composition

$$K_1(B) \longrightarrow K_1\left(Q(SB)\right) \xrightarrow{u} \mathrm{Ext}^{-1}_{\mathrm{unital}}(A, SB),$$

where the first map is the well-known isomorphism (the inverse of the one used in (7)) and second is the u-map as defined above, but with SB in place of B. Let $i_0^* : \mathrm{Ext}^{-1}(A, SB) \to K_0(B)$ be the composition

$$\mathrm{Ext}^{-1}(A, SB) \xrightarrow{i^*} \mathrm{Ext}^{-1}(\mathbb{C}, SB) \longrightarrow K_0(B),$$

where the second map is the well-known isomorphism. Finally, let $u_0 : K_0(B) \to \mathrm{Ext}^{-1}_{\mathrm{unital}}(A, B)$ be the composition

$$K_0(B) \longrightarrow K_1(Q(B)) \xrightarrow{u} \mathrm{Ext}^{-1}_{\mathrm{unital}}(A, B),$$

where the first map is the well-known isomorphism. We have now all the ingredients to prove

Theorem 3. *The sequence*

$$\begin{array}{ccccc}
K_0(B) & \xrightarrow{u_0} & \mathrm{Ext}^{-1}_{\mathrm{unital}}(A, B) & \longrightarrow & \mathrm{Ext}^{-1}(A, B) \\
{\scriptstyle i_0^*}\Big\uparrow & & & & \Big\downarrow{\scriptstyle i_1^*} \\
\mathrm{Ext}^{-1}(A, SB) & \longleftarrow & \mathrm{Ext}^{-1}_{\mathrm{unital}}(A, SB) & \xleftarrow[u_1]{} & K_1(B)
\end{array}$$

is exact.

Proof. If we apply Lemma 2 with B replaced by SB we find that the sequence

$$\begin{array}{ccccc}
\mathrm{Ext}^{-1}(\mathbb{C}, SB) & & & & K_1(\alpha(A)' \cap Q(SB)) \\
{\scriptstyle i^*}\Big\uparrow & & & & \Big\downarrow{\scriptstyle i^*} \\
\mathrm{Ext}^{-1}(A, SB) & \longleftarrow & \mathrm{Ext}^{-1}_{\mathrm{unital}}(A, SB) & \xleftarrow[u]{} & K_1(Q(SB))
\end{array}$$

is exact. Thanks to the commuting diagram (4) we can patch this sequence together with the sequence from Lemma 2 with the stated result. □

References

[1] W. Arveson, *Notes on extensions of C^*-algebras.* Duke Math. J. **44** (1977), 329–355.

[2] J. Cuntz and N. Higson, *Kuiper's Theorem for Hilbert Modules.* Contemporary Mathematics **62**, 429–435, AMS, Providence, 1987.

[3] J.A. Mingo, *On the contractibility of the unitary group of the Hilbert space over a C^*-algebra.* Integral Eq. Operator Theory **5** (1982), 888–891.

[4] G. Skandalis, *Kasparov's bivariant K-theory and applications.* Exposition. Math. **9** (1991), no. 3, 193–250.

[5] K. Thomsen, *On absorbing extensions.* Proc. Amer. Math. Soc. **129** (2001), 1409–1417.

Vladimir Manuilov
Dept. of Mechanics and Mathematics, Moscow State University
Leninskie Gory, Moscow, 119992 Russia
e-mail: manuilov@mech.math.msu.su

Klaus Thomsen
IMF, Department of Mathematics
Ny Munkegade, DK-8000 Aarhus C, Denmark
e-mail: matkt@imf.au.dk

C^*-algebras and Elliptic Theory
Trends in Mathematics, 157–186

Lefschetz Theory on Manifolds with Singularities

Vladimir Nazaikinskii and Boris Sternin

Abstract. The semiclassical method in Lefschetz theory is presented and applied to the computation of Lefschetz numbers of endomorphisms of elliptic complexes on manifolds with singularities. Two distinct cases are considered, one in which the endomorphism is geometric and the other in which the endomorphism is specified by Fourier integral operators associated with a canonical transformation. In the latter case, the problem includes a small parameter and the formulas are (semiclassically) asymptotic. In the first case, the parameter is introduced artificially and the semiclassical method gives exact answers. In both cases, the Lefschetz number is the sum of contributions of interior fixed points given (in the case of geometric endomorphisms) by standard formulas plus the contribution of fixed singular points. The latter is expressed as a sum of residues in the lower or upper half-plane of a meromorphic operator expression constructed from the conormal symbols of the operators involved in the problem.

Mathematics Subject Classification (2000). Primary 58J20; Secondary 32S50.

Keywords. Lefschetz number, singular manifold, elliptic operator, Fourier integral operator, semiclassical method.

Introduction

Lefschetz theory has been intensively developing starting from the fundamental paper [1] by Atiyah and Bott, who substantially generalized Lefschetz's original result [2], and is now an important branch of elliptic theory. Numerous papers dealing with the computation of the Lefschetz number in various cases have been published in more than thirty years since the appearance of [1]. In particular, recently a number of results have been proved concerning the *Lefschetz numbers of*

A preliminary version of the paper was published as a preprint at Chalmers University of Technology and supported by a grant from the Swedish Royal Academy of Sciences. We also acknowledge support from the RFBR under grants Nos.05-01-00466 and 05-01-00982.

endomorphisms of elliptic complexes on manifolds with singularities. (In particular, we note the papers [3–5].) These results are largely due to a new powerful method for the direct computation of the Lefschetz number on the basis of semiclassical asymptotics [6–8]. Being combined with the localization principle, the method not only permits one to give a short straightforward proof of Atiyah–Bott–Lefschetz theorems in known and new cases but also extends the theory from *geometric endomorphisms* to *endomorphisms associated with arbitrary quantized canonical transformations* (Fourier–Maslov integral operators; see [9–13]).

This paper is essentially a review of results pertaining to Lefschetz theory on manifolds with singularities. We start by recalling the definition of the Lefschetz number and the main results due to Atiyah and Bott [1] (see Subsection 1.1). Next, in Subsection 1.2 we very briefly describe the above-mentioned semiclassical method, which occurs intrinsically in the proofs. The main new results obtained by this method on smooth manifolds are presented in Subsection 1.3. (Formally, these results are special cases of the corresponding theorems for manifolds with singularities under the assumption that the set of singular points is empty.) Then we proceed to results concerning manifolds with singularities. Section 2 deals with the Atiyah–Bott–Lefschetz theorem for geometric endomorphisms, and in Section 3 we consider the same theorem for endomorphisms given by quantized canonical transformations. Although geometric endomorphisms are a special case of endomorphisms given by general quantized canonical transformations, the separation of the material into two sections is quite natural: the answers in the general case are asymptotic rather than exact, and different, more cumbersome conditions are imposed on the operators forming the complex.

1. Preliminaries

1.1. The Atiyah–Bott–Lefschetz Theorem

Let

$$0 \longrightarrow E_0 \xrightarrow{D_0} E_1 \xrightarrow{D_1} \cdots \xrightarrow{D_{m-1}} E_m \longrightarrow 0 \tag{1}$$

be a complex of vector spaces over $\mathbb{C}$ with finite-dimensional cohomology, and let

$$T = \{T_j : E_j \to E_j\}\big|_{j=0,\dots,m} \tag{2}$$

be an endomorphism of the complex (1), i.e., a set of linear mappings such that the diagram

$$\begin{array}{ccccccccccc} 0 & \longrightarrow & E_0 & \xrightarrow{D_0} & E_1 & \xrightarrow{D_1} & \cdots & \xrightarrow{D_{m-1}} & E_m & \longrightarrow & 0 \\ & & \downarrow{\scriptstyle T_0} & & \downarrow{\scriptstyle T_1} & & & & \downarrow{\scriptstyle T_m} & & \\ 0 & \longrightarrow & E_0 & \xrightarrow{D_0} & E_1 & \xrightarrow{D_1} & \cdots & \xrightarrow{D_{m-1}} & E_m & \longrightarrow & 0 \end{array} \tag{3}$$

commutes. Then

$$T_j(\operatorname{Ker} D_j) \subset \operatorname{Ker} D_j, \quad T_j(\operatorname{Im} D_{j-1}) \subset \operatorname{Im} D_{j-1}, \quad j = 0, \ldots, m,$$

where $\operatorname{Ker} A$ and $\operatorname{Im} A$ are the kernel and the range, respectively, of an operator A and the mappings D_{-1} and D_m are zero by convention. It follows that the endomorphism (2) induces mappings

$$\widetilde{T}_j : H^j(E) \to H^j(E), \quad j = 0, \ldots, m, \tag{4}$$

of the cohomology spaces $H^j(E) = \operatorname{Ker} D_j / \operatorname{Im} D_{j-1}$ of the complex (1). The *Lefschetz number* of the endomorphism (2) is defined as the alternating sum

$$\mathcal{L} = \sum_{j=0}^{m} (-1)^j \operatorname{Trace} \widetilde{T}_j \tag{5}$$

of traces of the finite-dimensional operators (4).

The classical Atiyah–Bott–Lefschetz theorem [1] deals with the case in which (1) is an elliptic complex of differential operators on a smooth compact manifold M without boundary and (2) is a geometric endomorphism associated with a smooth mapping $f : M \to M$:

$$\begin{aligned} E_j &= C^\infty(M, F_j), && \text{where } F_j \text{ is a vector bundle over } M, \\ T_j\varphi(x) &= A_j(x)\varphi(f(x)), && \text{where } A_j(x) : F_{jf(x)} \to F_{jx} \text{ is a homomorphism.} \end{aligned} \tag{6}$$

Here F_{jx} is the fiber of F_j over a point x.

Suppose that the fixed points of f are nondegenerate in the sense that

$$\det\left(1 - \frac{\partial f}{\partial x}(x)\right) \neq 0, \quad x \in \operatorname{fix}(f). \tag{7}$$

(Here $\operatorname{fix}(f)$ is the set of fixed points of f.) Then they are isolated, and the Atiyah–Bott–Lefschetz theorem states that the Lefschetz number can be expressed by the formula

$$\mathcal{L} = \sum_{x \in \operatorname{fix}(f)} \sum_{j=0}^{m} \frac{(-1)^j \operatorname{Trace} A_j(x)}{|\det(1 - \partial f(x)/\partial x)|}. \tag{8}$$

Thus the Lefschetz number of the endomorphism (2) is expressed in classical terms. We also note that the operators D_j themselves do not occur in (8); one only requires the diagram (3) to commute.

The *Lefschetz fixed point theorem* is the special case of formula (8) in which (1) is the de Rham complex on a smooth compact oriented m-dimensional manifold M and T is the endomorphism induced on differential forms by a smooth mapping $f : M \to M$. In other words,

$$E_k = \Lambda^k(M)$$

is the space of differential k-forms on M,

$$D_k = d : \Lambda^k(M) \to \Lambda^{k+1}(M)$$

is the exterior differential, and

$$T_k = f^* : \Lambda^k(M) \to \Lambda^k(M)$$

is the induced mapping of differential forms.

The Lefschetz fixed point theorem states that in this case one has

$$\mathcal{L} \equiv \mathcal{L}(f) = \sum_{x \in \operatorname{fix}(f)} \operatorname{sgn}\det\left(\frac{\partial f}{\partial x}(x) - 1\right)$$

under the assumption that all fixed points are nondegenerate.

1.2. The semiclassical method

The class of geometric endomorphisms is not a natural framework for the problem on the Lefschetz number if one does not restrict oneself to complexes of differential operators but has in mind also pseudodifferential operators. As differential operators form a subclass of the more general class of pseudodifferential operators, so geometric endomorphisms form a subclass of the class of Fourier integral operators, and hence one can naturally try to obtain a Lefschetz type formula for the case in which (1) is an elliptic complex of *pseudodifferential* operators and the endomorphism (2) is given by a set of Fourier integral operators. In a special case, this formula was obtained by Fedosov [14], who considered endomorphisms given by evolution operators for the Schrödinger equation. The general theory (in the case of smooth compact manifolds) was developed by Sternin and Shatalov [6–8, 15]. It turns out that once we pass to endomorphisms associated with mappings of the phase space, the theory necessarily becomes *asymptotic*. To obtain meaningful formulas, one must introduce a small parameter $h \in (0, 1]$ and consider semiclassical pseudodifferential operators (or $1/h$-pseudodifferential operators; e.g., see [9–11]) and Fourier–Maslov integral operators associated with a canonical transformation

$$g : T^*M \to T^*M.$$

(Thus it is *symplectic* rather than contact geometry that underlies the Lefschetz formula.) Then the Lefschetz number depends on h, and under appropriate assumptions about the fixed points of g one obtains an expression for the asymptotics of the Lefschetz number as $h \to 0$ by applying the stationary phase method to the trace integrals representing this number.

Namely, the Lefschetz number of the endomorphism (2) (with $m = 1$ for simplicity) is given by the formula (e.g., see, [14])

$$\mathcal{L}(D, T) = \operatorname{Trace}[T_0(1 - RD_0)] - \operatorname{Trace}[T_1(1 - D_0R)], \tag{9}$$

where R is an arbitrary almost inverse of D_0 modulo trace class operators. (By definition, this means that the operators $1 - D_0R$ and $1 - RD_0$ are trace class, so that the traces in (9) are well defined.) Now if D_0 and R are pseudodifferential operators and T_0 and T_1 are Fourier–Maslov integral operators, then $T_0(1 - RD_0)$ and $T_1(1 - D_0R)$ are also Fourier–Maslov integral operators associated with the same canonical transformation as T_0 and T_1. Hence the problem is reduced to the evaluation of *traces of Fourier–Maslov integral operators*. This is carried out with the help of the stationary phase method; only fixed points of g give a nonzero contribution to the asymptotics of these traces as $h \to 0$. As usual in the stationary phase method, the contribution of each isolated component of the set of fixed points can be treated separately (*localization*, or, more precisely, *microlocalization*), which permits one to separate the contributions of interior and singular fixed points in applications to manifolds with singularities.

1.3. Semiclassical Lefschetz formulas for smooth manifolds

We see that the semiclassical method provides a straightforward computation of the Lefschetz number. Now let us give the corresponding results in more detail. First, we state the theorem about the trace of a Fourier integral operator in the simplest form.

Let M be a smooth closed manifold of dimension n. Suppose that M is oriented and equipped with a positive volume form dx. For a canonical transformation

$$g : T^*M \to T^*M \tag{10}$$

and a smooth function φ on the cotangent space T^*M satisfying appropriate conditions at infinity in the fibers, by $T(g, \varphi)$ we denote the Fourier–Maslov integral operator (i.e., the Fourier integral operator with a small parameter $h \in (0, 1]$) with amplitude φ associated with the graph of the transformation (10). We assume that the graph is a quantized Lagrangian submanifold of $T^*M \times T^*M$. (A detailed definition of Fourier–Maslov integral operators can be found in [11], and precise conditions guaranteeing that the operator $T(g, \varphi)$ is well defined are given in [7].)

Now assume that the conditions in [7] guaranteeing that $T(g, \varphi)$ is trace class are satisfied. Then the following theorem holds [7].

Theorem 1.1. *Suppose that the transformation g has finitely many fixed points $\alpha_1, \dots, \alpha_k \in T^*M$ and these fixed points are nondegenerate in the sense that $\det(1 - g_*) \neq 0$ at any of them, where*

$$g_* : T_{\alpha_i}T^*M \to T_{\alpha_i}T^*M$$

is the induced mapping of tangent spaces. Then the trace $\operatorname{Trace} T(g, \varphi)$ *has the following asymptotic expansion as $h \to 0$:*

$$\operatorname{Trace} T(g, \varphi) \equiv \sum_{j=1}^{k} \exp\left\{\frac{i}{h} S_j\right\} \frac{\varphi(\alpha_j)}{\sqrt{\det(1 - g_*(\alpha_j))}} + O(h).$$

Here S_j is the value of the generating function of g at the point α_j (the choice of this function occurs in the definition of the Fourier–Maslov integral operator $T(g, \varphi)$), and the branch of the square root is chosen in a special way (see [7]).

Remark 1.2. A similar theorem is proved in [7] for the case in which the canonical transformation can have manifolds of fixed points.

Now we are in a position to present the Lefschetz formulas obtained in [6–8, 15] by the semiclassical method for endomorphisms of elliptic complexes on smooth manifolds. Consider the commutative diagram

$$\begin{array}{ccccccc} 0 & \longrightarrow & C^\infty(M,F_1) & \xrightarrow{\widehat{D}} & C^\infty(M,F_2) & \longrightarrow & 0 \\ & & \Big\downarrow{\scriptstyle \widehat{T}_1} & & \Big\downarrow{\scriptstyle \widehat{T}_2} & & \\ 0 & \longrightarrow & C^\infty(M,F_1) & \xrightarrow{\widehat{D}} & C^\infty(M,F_2) & \longrightarrow & 0, \end{array} \tag{11}$$

where F_1 and F_2 are vector bundles on M, the $\widehat{T}_j = T(g,\varphi_j)$, $j=1,2$, are Fourier–Maslov integral operators associated with the canonical transformation (10), and $\widehat{D}$ is an elliptic $1/h$-pseudodifferential operator on M. For $1/h$-pseudodifferential operators, ellipticity means the existence of a $1/h$-pseudodifferential operator $\widehat{R}$ on M such that the operators $1-\widehat{D}\widehat{R}$ and $1-\widehat{R}\widehat{D}$ belong to Hörmander's class $L^{-\infty}(M)$ uniformly with respect to the parameter $h \in (0,1]$. The following theorem was proved in [6–8, 15].

Theorem 1.3. *Suppose that g has only nondegenerate fixed points $\alpha_1,\dots,\alpha_N$. Then the Lefschetz number of the diagram (11) has the asymptotics*

$$\mathcal{L} = \sum_{k=1}^{N} \exp\left(\frac{i}{h}S_k\right) \frac{\operatorname{Trace}\varphi_1(\alpha_k) - \operatorname{Trace}\varphi_2(\alpha_k)}{\sqrt{\det(1-g_*(\alpha_k))}} + O(h), \tag{12}$$

where φ_1 and φ_2 are the amplitudes of the endomorphisms $\widehat{T}_1$ and $\widehat{T}_2$, respectively, S_k is the value of the generating function of g at the point α_k, and the branch of the square root is chosen in a special way (see [7]).

We point out that the classical result due to Atiyah and Bott (for the case in which the mapping f is a diffeomorphism) can be derived from this asymptotic Lefschetz formula by a scaling procedure. Namely, by multiplying the differential operator D_j by h^{k_j}, where $k_j = \operatorname{ord} D_j$, we make it a semiclassical pseudodifferential operator; the geometric endomorphisms are Fourier–Maslov integral operators with linear phase function, and it remains to apply the asymptotic formula and note that the Lefschetz number in this special case does not actually depend on h.

For the case in which the canonical transformation (9) may have manifolds of fixed points, the following theorem was proved in [7].

Theorem 1.4. *Suppose that the set $\operatorname{fix}(g)$ of fixed points of the canonical transformation g is a finite disjoint union of smooth compact connected manifolds G_k without boundary and that for all k the kernel of the operator $1-g_*$ at each point of G_k coincides with the tangent space to G_k (the nondegeneracy condition). Then the Lefschetz number of the diagram (11) has the asymptotics*

$$\mathcal{L} = \frac{1}{(2\pi h)^l}\left[\sum_k \exp\left\{\frac{i}{h}S_k\right\} \int_{G_k} \Big(\operatorname{Trace}\varphi_1(\alpha) - \operatorname{Trace}\varphi_2(\alpha)\Big)\, dm_k(\alpha) + O(h)\right],$$

where the sum is over the manifolds G_k of maximal dimension, $l = \max \dim G_k/2$, S_k is the value of the generating function on G_k (it is necessarily constant), and the dm_k are some nondegenerate volume forms on the respective G_k.

The symplectic geometry of the sets G_k can be rather complicated, and so it is difficult to find a closed-form expression for the measures dm_k in the general case. This was done in the following two extreme cases in [5]:

1. $\dim G_k$ is even and $\operatorname{rank} \omega^2|_{G_k} = \dim G_k$; that is, G_k is a symplectic submanifold of T^*M (here ω^2 is the standard symplectic form on T^*M).
2. $\dim G_k = \dim M$ and G_k is a Lagrangian submanifold of T^*M.

Let $\alpha \in G_k$. First, consider case 1. The symplectic plane $L = T_\alpha G_k \subset T_\alpha T^*M$ of dimension $l = \dim G_k$ in $T_\alpha T^*M$ and the skew-orthogonal plane L° give the decomposition $T_\alpha T^*M = L \oplus L^\circ$, and both L and L° are g_*-invariant. (Indeed, $1 - g_*$ vanishes on L, and so $g_* L = L$; the invariance of L° follows from the fact that g_* preserves the form ω^2.) Moreover, it is easily seen that

$$\ker\left((1 - g_*)\big|_{L^\circ}\right) = \{0\}$$

by virtue of the nondegeneracy condition.

Theorem 1.5. *Under the above conditions,*

$$dm_k(\alpha) = \frac{\left(\omega^2\big|_L\right)^{\wedge \frac{l}{2}}}{l! \sqrt{\det(1 - g_*)\big|_{L^\circ}}}$$

where the sign of the square root depends on the choices in the construction of the canonical operator.

Now consider case 2, where $L = T_\alpha G_k$ is Lagrangian.

Lemma 1.6. *There is a natural isomorphism*

$$T_\alpha T^*M/L \simeq L^*,$$

where L^* is the dual space. Furthermore,

$$\operatorname{Im}(1 - g_*) = L;$$

the mapping $1 - g_*$ factors through L and induces an isomorphism

$$1 - g_* \colon L^* \longrightarrow L.$$

Thus on L we have the well-defined nondegenerate symmetric bilinear form

$$B(x, y) = \langle (1 - g_*)^{-1} x, y \rangle,$$

where the angle brackets stand for the pairing between L and L^*. Let us reduce it to principal axes in some basis in a given fiber of L:

$$B(x, x) = \sum_{j=1}^{n} \varepsilon_j x_j^2,$$

where $\varepsilon_j = \pm 1$ and the x_j are the coordinates with respect to this basis.

Theorem 1.7. *In case 2, the form dm_k is given by*

$$dm_k(\alpha) = \sqrt{\prod \varepsilon_j} dx_1 \wedge \cdots \wedge dx_n,$$

where the choice of the branch of the square root is determined by the corresponding choices in the construction of the canonical operator.

2. The Atiyah–Bott–Lefschetz theorem for geometric endomorphisms on manifolds with conical singularities

2.1. Statement of the problem

In this section, we study the Lefschetz number for geometric endomorphisms on manifolds with conical singularities. We consider only endomorphisms of short (two-term) complexes, having in mind that the case of general finite complexes can be analyzed according to the scheme suggested in [16]. General information concerning manifolds with singularities and pseudodifferential operators on such manifolds can be found in [17–19] (see also references therein), and we assume that the reader is acquainted with the definitions.

Let M be a compact manifold with N conical singular points, which will be denoted by $\alpha_1, \ldots, \alpha_N$. The bases of the corresponding cones will be denoted by $\Omega_1, \ldots, \Omega_N$, respectively. Next, let

$$\widehat{D} = H^{s,\gamma}(E_1) \to H^{s-m,\gamma}(E_2), \quad s \in \mathbb{R},$$

be an elliptic differential operator of order m in weighted Sobolev spaces of sections of finite-dimensional vector bundles E_1 and E_2 over M. (Here $\gamma = (\gamma_1, \ldots, \gamma_N)$ is a given vector of weight exponents.) Recall that this means that the principal symbol of $\widehat{D}$ is invertible everywhere outside the zero section of the *compressed cotangent bundle* T^*M (e.g., see [25]) and the conormal symbols $\widehat{D}_0(\alpha_j, p)$ of $\widehat{D}$ at the conical points α_j are invertible on the weight lines $\mathfrak{L}_j = \{\operatorname{Im} p = \gamma_j\}, \quad j = 1, \ldots, N.$

Consider a commutative diagram of the form

$$\begin{array}{ccccccc} 0 & \longrightarrow & H^{s,\gamma}(E_1) & \xrightarrow{\widehat{D}} & H^{s-m,\gamma}(E_2) & \longrightarrow & 0 \\ & & \Big\downarrow{\scriptstyle \widehat{T}_1} & & \Big\downarrow{\scriptstyle \widehat{T}_2} & & \\ 0 & \longrightarrow & H^{s,\gamma}(E_1) & \xrightarrow{\widehat{D}} & H^{s-m,\gamma}(E_2) & \longrightarrow & 0, \end{array} \tag{13}$$

i.e., an endomorphism of the complex specified by $\widehat{D}$. (In contrast with the smooth case, here we are forced to use weighted Sobolev spaces rather than C^∞, since the kernel and cokernel of $\widehat{D}$ in general depend on the weight exponents.)

Here $\widehat{T}_j$ is a geometric endomorphism of the form

$$\left(\widehat{T}_j u\right)(x) = A_j(x)\, u(g(x)), \quad x \in M, \qquad j = 1, 2, \tag{14}$$

where

$$g : M \to M$$

is a smooth self-mapping of M and the $A_j : g^*E_j \longrightarrow E_j$ are bundle homomorphisms over M. Furthermore, we assume that g is a *diffeomorphism* in the category of manifolds with singularities; that is, the inverse mapping g^{-1} exists and is also a smooth self-mapping of M. Apparently, just as in usual Lefschetz theory, this restriction can be discarded, but this leads to some technical complications, which we wish to avoid here.

By definition, g has the properties $g(\{\alpha_1, \ldots, \alpha_N\}) \subset \{\alpha_1, \ldots, \alpha_N\}$ and $g(\operatorname{Int} M) \subset \operatorname{Int} M$ (i.e., preserves the sets of conical points and interior points) and has a special structure near the conical points. Essentially, this structure follows from the fact that g is a smooth mapping of the *manifold* $M^\wedge$ *with boundary* obtained by the blow-up of M at the conical points. Namely, let $g(\alpha) = \alpha'$, where $\alpha, \alpha' \in \{\alpha_1, \ldots, \alpha_N\}$ are conical points. Then in conical coordinate neighborhoods of the points α and α' the mapping g is given by the formulas

$$r' = rB\,(r, \omega)\,, \quad \omega' = C\,(r, \omega)\,, \tag{15}$$

where $\omega \in \Omega$ and $\omega' \in \Omega'$ are points on the bases of the corresponding model cones, r and r' are radial variables, and the functions $B(r, \omega) \geq 0$ and $C(r, \omega)$ are smooth up to $r = 0$; moreover, $B(r, \omega) > 0$ for $r > 0$. Since g is a diffeomorphism, it follows that $B(r, \omega)$ is positive also for $r = 0$. Furthermore, the mapping

$$g_\alpha \equiv C(0, \cdot) : \Omega \longrightarrow \Omega',$$

which will be referred to as the *boundary mapping* of g at α, is also a diffeomorphism.

The main task of this section is the computation of the Lefschetz number $\mathcal{L}$ of the endomorphism (13) under some additional assumptions.

First (in Subsection 2.2) we show that the Lefschetz number of the diagram (13) is completely determined by what happens in a neighborhood of fixed points of g. Then (in Subsec. 2.3) we compute the contributions of the fixed points under the additional assumption that they are nondegenerate. The contribution of interior fixed points has the standard form (the same as in the smooth theory), and the contribution of fixed singular points is expressed via the conormal symbols of the operators occurring in the diagram (13). Finally (in Subsection 2.4), under an additional tempered growth condition imposed on some characteristics of the conormal symbol, we simplify the expression for the contribution of a fixed singular point by reducing it to the trace of the sum of residues of some meromorphic operator family in a half-plane.

2.2. Localization and the contributions of fixed points

Here we show that the Lefschetz number $\mathcal{L}$ can be represented as the sum of contributions corresponding to connected components of the set $\operatorname{fix}(g)$ of fixed points of g. Although these contributions are defined here as integrals over some neighborhoods of these components with integrands ambiguous to a certain extent, the values of the integrals are independent of the freedom in the choice of the integrand as well as of the structure of the operators $\widehat{D}$, $\widehat{T}_1$, and $\widehat{T}_2$ outside an arbitrarily small neighborhood of the set $\operatorname{fix}(g)$. Formulas explicitly depending on the values at fixed points alone will be given later.

Our starting point is the trace formula (e.g., see [14])

$$\mathcal{L} = \operatorname{Trace} \widehat{T}_1(1 - \widehat{R}\widehat{D}) - \operatorname{Trace} \widehat{T}_2(1 - \widehat{D}\widehat{R}). \tag{16}$$

Here $\widehat{R}$ is an arbitrary almost inverse of $\widehat{D}$ modulo trace class operators. (In other words, the operators $1 - \widehat{R}\widehat{D}$ and $1 - \widehat{D}\widehat{R}$ are trace class.)

In turns out that for a special choice of $\widehat{R}$ the right-hand side of (16) splits into a sum of integrals over arbitrarily small neighborhoods of the components of $\operatorname{fix}(g)$. Let us equip M with a Riemannian metric $d\rho^2$ that has the product structure

$$d\rho^2 = dr^2 + r^2 d\omega^2$$

near the conical points in some conical coordinates. (Here $d\omega^2$ is some Riemannian metric on the base Ω of the cone.) Next, we introduce a measure $d\mu(x)$ that is smooth outside the conical points and has the form

$$d\mu(x) = \frac{dr}{r} \wedge d\omega$$

in the same coordinates near the conical points, where $d\omega$ is a smooth measure on Ω. Using this measure, one can treat kernels of pseudodifferential operators as (generalized) functions (or sections of appropriate bundles) on $M \times M$. Let

$$\operatorname{fix}(g) = K_1 \cup \ldots \cup K_k,$$

where the K_j, $j = 1, \ldots, k$, are disjoint compact sets.

Next, let $V_1, \ldots, V_k$ be sufficiently small neighborhoods of these sets. The function $\rho(x, g(x))$ is positive and continuous on the compact set $M \setminus (V_1 \cup \ldots \cup V_k)$ and hence has a nonzero minimum 2ε. Since $\widehat{D}$ is an elliptic differential operator on M, it follows that there exists an almost inverse $\widehat{R}$ of $\widehat{D}$ modulo trace class operators such that the kernel $R(x, y)$ of $\widehat{R}$ has the property

$$R(x, y) = 0 \quad \text{for } \rho(x, y) > \varepsilon. \tag{17}$$

(An operator $\widehat{R}$ with this property is said to be ε-narrow.) Then the kernels $K_1(x, y)$ and $K_2(x, y)$ of the operators $1 - \widehat{R}\widehat{D}$ and $1 - \widehat{D}\widehat{R}$ have the same property, since $\widehat{D}$ is a differential operator and does not enlarge supports. In terms of these kernels, formula (16) becomes

$$\mathcal{L} = \int_M \Big(\operatorname{Trace} A_1(x) K_1(g(x), x) - \operatorname{Trace} A_2(x) K_2(g(x), x) \Big)\, d\mu(x), \tag{18}$$

where Trace in the integrand is the matrix trace. With regard to (17), we can rewrite this expression in the form

$$\mathcal{L} = \sum_{j=1}^{k} \int_{V_j} (\operatorname{Trace} A_1(x) K_1(g(x), x) - \operatorname{Trace} A_2(x) K_2(g(x), x))\, d\mu(x), \tag{19}$$

since the integrand is identically zero outside the union of V_j.

It turns out that not only the sum as a whole but also each separate term is independent of the choice of an ε-narrow almost inverse operator (at least if the neighborhoods V_j are sufficiently small). This readily follows from the *locality* of the construction of an almost inverse operator: if there are two ε-narrow almost inverse operators, then one can construct a new ε-narrow almost inverse operator coinciding with the first operator in some V_{j_0} and with the second operator in all V_j with $j \neq j_0$. It follows that one can separately analyze each of the integrals specifying the contributions of the components of the set of fixed points and choose an ε-narrow almost inverse operator as is convenient near the component whose contribution is to be analyzed.

The expression (19) is not very convenient for the subsequent analysis, and we shall obtain a different, more convenient expression. One can assume that the operator $\widehat{R}$ is sufficiently narrow and the neighborhoods V_j are sufficiently small, so that not only V_j but also the sets

$$W_j = V_j \cup g(V_j)$$

are disjoint. Under these conditions, consider the jth term

$$\mathcal{L}_j = \int_{V_j} \Big(\operatorname{Trace} A_1(x) K_1\big(g(x), x\big) - \operatorname{Trace} A_2(x) K_2\big(g(x), x\big) \Big)\, d\mu(x) \qquad (20)$$

in (19). It will be referred to as the *contribution of the component* K_j *of the set of fixed points* to the Lefschetz number. To compute (20), we can modify the operator $\widehat{R}$ arbitrarily, provided that the integrand in (20) remains unchanged. Let f_j be a continuous function on M that is smooth in $\operatorname{Int} M$ and constant near each singular point and satisfies $f_j\big|_{V_j} \equiv 1$. Next, let ψ_j have similar properties and satisfy $\psi_j\big|_{g(V_j)} \equiv 1$. We replace the operator $\widehat{R}$ by $\psi_j \widehat{R} f_j$ and the identity operator by f_j (that is, assume that K_1 and K_2 are the kernels of the operators $f_j - \psi_j \widehat{R} f_j \widehat{D}$ and $f_j - \widehat{D} \psi_j \widehat{R} f_j$, respectively.) Then the integral (20) remains unchanged. Moreover, if $\operatorname{supp} f_j$ and $\operatorname{supp} \psi_j$ do not meet V_k and $g(V_k)$ for $k \neq j$, then one can extend the integration in (20) to the entire manifold M. (The integrand vanishes identically on $M \setminus V_j$.)

Formally L_j can be rewritten in the form

$$\mathcal{L}_j = \operatorname{Trace} \widehat{T}_1 (f_j - \psi_j \widehat{R} f_j \widehat{D}) - \operatorname{Trace} \widehat{T}_2 (f_j - \widehat{D} \psi_j \widehat{R} f_j). \qquad (21)$$

Although the operators $\widehat{T}_1 (f_j - \psi_j \widehat{R} f_j \widehat{D})$ and $\widehat{T}_2 (f_j - \widehat{D} \psi_j \widehat{R} f_j)$ are not trace class in general and do not have traces, the integrals of their kernels over the diagonal are well defined, and we use notation (21) for the corresponding integral.

2.3. Contributions of nondegenerate fixed points

Let $g : M \longrightarrow M$ be a diffeomorphism of a manifold with conical singularities, and let $x \in \operatorname{fix}(g)$ be a fixed point of g.

Definition 2.1. We say that x is a *nondegenerate fixed point* if one of the following conditions is satisfied:

1. x is an interior point, and $\det(1 - g_*(x)) \neq 0$.
2. x is a conical point, and in formulas (15), describing g in a neighborhood of x, either $B(0, \omega) > 1$ for all $\omega \in \Omega$ (a *repulsive* conical fixed point), or $B(0, \omega) < 1$ for all $\omega \in \Omega$ (an *attractive* conical fixed point).

Consider the geometric endomorphism (13), where the operators $\widehat{T}_{1,2}$ have the form (14) and $\widehat{D}$ is an elliptic differential operator on a manifold M with conical singularities.

Theorem 2.2. *Suppose that all fixed points of g are nondegenerate. Then the Lefschetz number of the diagram (13) has the form*

$$\begin{aligned}\mathcal{L} &= \sum_{x^*} \mathcal{L}_{\mathrm{int}}(x^*) + \sum_{\alpha^*} \mathcal{L}_{\mathrm{sing}}(\alpha^*) \\ &\equiv \sum_{x^*} \frac{\operatorname{Trace} A_1(x^*) - \operatorname{Trace} A_2(x^*)}{|\det(1 - g_*(x^*)|} \\ &\quad + \frac{1}{2\pi i} \sum_{\alpha^*} \operatorname{Trace} \int_{\operatorname{Im} p = \gamma(\alpha^*)} \widehat{T}_{10}(\alpha^*, p) \widehat{D}_0^{-1}(\alpha^*, p) \frac{\partial \widehat{D}_0}{\partial p}(\alpha^*, p)\, dp.\end{aligned} \tag{22}$$

Here the first term is the sum of contributions $\mathcal{L}_{\mathrm{int}}(x^*)$ of interior fixed points x^*, and the second term is the sum of contributions $\mathcal{L}_{\mathrm{sing}}(\alpha^*)$ of conical fixed points α^*. The integration in each term of the second sum is over the weight line determined by the weight exponent $\gamma(\alpha^*)$ of the corresponding conical point and the integral is understood as an oscillatory integral (see below) and gives a trace class operator in $L^2(\Omega)$, where Ω is the base of the cone at α^*, so that the trace is well defined.

Proof. To simplify the notation, we write, say, $\widehat{D}_0(p)$ instead of $\widehat{D}_0(\alpha^*, p)$ etc. and g_0 instead of g_{α^*}. Furthermore, without loss of generality we assume that all weight exponents are zero, and so the weight line is the real axis.

Prior to the proof, let us explain the meaning of the integral representing the contribution of a fixed conical point. To this end, we rewrite the integrand in (22) by using the following expression for the conormal symbol of $\widehat{T}_1$:

$$\widehat{T}_{10}(p) = A_{10}(\omega) e^{ib(0,\omega)p} g_0^*,$$

where $A_{10}(\omega) = A_1(x)|_{r=0}$ is the restriction of the bundle homomorphism A_1 to the singular point α^* and $b(r, \omega) = -\ln B(r, \omega)$ (see (15)). Now we have

$$\widehat{T}_{10}(p) \widehat{D}_0^{-1}(p) \frac{\partial \widehat{D}_0}{\partial p}(p) = A_{j0}(\omega) e^{ib(0,\omega)p} g_0^* \widehat{H}(p),$$

where

$$\widehat{H}(p) = \widehat{D}_0^{-1}(p) \frac{\partial \widehat{D}_0}{\partial p}(p).$$

Note the following two facts: (a) The family $\widehat{H}(p)$ is defined everywhere on the real line (by the ellipticity condition, there are no poles of $\widehat{D}_0^{-1}(p)$ there) and is a pseudodifferential operator of order -1 with parameter p on the real line; (b) since the conical fixed point is nonsingular, it follows that the function $b(0,\omega)$ is nonzero for all $\omega\in\Omega$. Using property (b), we write

$$\widehat{T}_{10}(p)=\left(\frac{1}{ib(0,\omega)}\right)^{l}\frac{\partial^{l}\widehat{T}_{10}}{\partial p^{l}}(p)$$

and perform the standard regularization of an oscillatory integral by l-fold integration by parts:

$$\frac{1}{2\pi i}\int_{-\infty}^{\infty}\widehat{T}_{10}(p)\widehat{H}(p)\,dp=\frac{1}{2\pi i}\int_{-\infty}^{\infty}\left(\frac{i}{b(0,\omega)}\right)^{l}\widehat{T}_{10}(p)\frac{\partial^{l}\widehat{H}(p)}{\partial p^{l}}\,dp. \qquad (23)$$

For sufficiently large l, the pseudodifferential operator $\partial^{l}\widehat{H}(p)/\partial p^{l}$ with parameter p of order $-1-l$ is trace class, and the integral on the right-hand side in (23) converges in the usual operator norm as well as in the trace norm (in every $H^{s}(\Omega)$), the result being independent of l (and consistent for various s). This provides the desired regularization. Simultaneously, we have shown that this integral specifies a trace class operator.

Let us compute the contributions of fixed points. As was mentioned in Subsec. 2.2, they can be computed separately as the corresponding trace integrals. In particular, for the interior fixed points the computation is the same as in the smooth case (see Subsec. 1.3), and it remains to compute the contribution of conical fixed points.

Let α^{*} be a conical fixed point. We use the conical coordinates (r,ω) in a neighborhood of α^{*}. Let us introduce a function $f(r)$ supported in a neighborhood of zero and equal to 1 for small r and a function $\psi(r)$ such that $\psi(r)f(r)=f(r)$, $\psi(r)f\big(g(r)\big)=f\big(g(r)\big)$, and $\psi(r)$ is also concentrated in a neighborhood of zero. One can choose these functions in a way such that

$$\psi(\lambda^{-1}r)f\big(g(\lambda^{-1}r)\big)=f\big(g(\lambda^{-1}r)\big)$$

for all $\lambda<1$. Then it follows from the results of Subsec. 2.2 that for all sufficiently small λ the contribution of α^{*} is given by the expression

$$\begin{aligned}\mathcal{L}_{\text{sing}}=\operatorname{Trace}\widehat{T}_{1}\big[U_{\lambda}f(r)U_{\lambda}^{-1}-U_{\lambda}\psi(r)U_{\lambda}^{-1}\widehat{R}U_{\lambda}f(r)U_{\lambda}^{-1}\widehat{D}\big]&\\ -\operatorname{Trace}\widehat{T}_{2}\big[U_{\lambda}f(r)U_{\lambda}^{-1}-\widehat{D}U_{\lambda}\psi(r)U_{\lambda}^{-1}\widehat{R}U_{\lambda}f(r)U_{\lambda}^{-1}\big],& \qquad (24)\end{aligned}$$

where, as before, Trace is the trace integral of the kernel of an operator and U_{λ} is the dilation operator

$$U_{\lambda}\varphi(\omega,r)=\varphi(\omega,r/\lambda).$$

In particular, the right-hand side of (24) is independent of λ. Hence we can pass to the limit as $\lambda\to 0$ in this expression, thus obtaining

$$\mathcal{L}_{\text{sing}}=\operatorname{Trace}\widehat{T}_{10}\big[f(r)-\psi(r)\widehat{R}_{0}f(r)\widehat{D}_{0}\big]-\operatorname{Trace}\widehat{T}_{20}\big[f(r)-\widehat{D}_{0}\psi(r)\widehat{R}_{0}f(r)\big]. \qquad (25)$$

Remark 2.3. We have used the identity

$$\operatorname{Trace}U_{\lambda}^{-1}WU_{\lambda}=\operatorname{Trace}W.$$

Here the operators $\widehat{T}_{j0}$, $\widehat{D}_0$, and $\widehat{R}_0$ are related to the corresponding conormal symbols by the Mellin transform:

$$\widehat{D}_0 = \mathcal{M}^{-1}\widehat{D}_0(p)\mathcal{M} \equiv \widehat{D}_0\left(ir\frac{\partial}{\partial r}\right)$$

etc. Note that $\widehat{D}_0\widehat{R}_0 = 1$ and $\widehat{R}_0\widehat{D}_0 = 1$, and so we can simplify the last expression by commuting the operator $\widehat{D}_0$ with the functions $f(r)$ and $\psi(r)$ in the first and second term, respectively. Since $f\psi = f$, we see that only terms containing commutators remain in the expression:

$$\mathcal{L}_{\text{sing}} = \operatorname{Trace}\widehat{T}_{10}\psi(r)\widehat{R}_0[\widehat{D}_0, f(r)] - \operatorname{Trace}\widehat{T}_{20}[\widehat{D}_0, \psi(r)]\widehat{R}_0 f(r).$$

Since $\widehat{D}_0$ is a differential operator and does not enlarge supports, it follows from the construction of f and ψ that the integrand in the second integral vanishes identically, while in the first integral the factor ψ is equal to unity on the support of the integrand and hence can be omitted. Consequently, we obtain

$$\mathcal{L}_{\text{sing}} = \operatorname{Trace}\widehat{T}_{10}\widehat{R}_0[\widehat{D}_0, f(r)] \tag{26}$$

A straightforward computation shows that the commutator $[\widehat{D}_0, f(r)]$ can be represented in the form

$$[\widehat{D}_0, f(r)] = i\frac{\partial\widehat{D}_0}{\partial p}\left(ir\frac{\partial}{\partial r}\right)r\frac{\partial f}{\partial r}(r) + \sum_{l=2}^{m}\widehat{B}_l\left(ir\frac{\partial}{\partial r}\right)\left[\left(r\frac{\partial}{\partial r}\right)^l f(r)\right],$$

where the $\widehat{B}_l(p)$ are some conormal symbols. By substituting this into (26), we obtain

$$\begin{aligned}\mathcal{L}_{\text{sing}} = i\operatorname{Trace}\widehat{T}_{10}\widehat{R}_0\frac{\partial\widehat{D}_0}{\partial p}\left(ir\frac{\partial}{\partial r}\right)r\frac{\partial f}{\partial r}(r)\\ + \sum_{l=2}^{m}\operatorname{Trace}\widehat{T}_{10}\widehat{R}_0\widehat{B}_l\left(ir\frac{\partial}{\partial r}\right)\left[\left(ir\frac{\partial}{\partial r}\right)^l f(r)\right].\end{aligned} \tag{27}$$

Let us compute the first term. We pass to the cylindrical coordinates ($r = e^{-t}$). Let $Z(\omega, \omega', t - t')$ be the kernel of the operator

$$\widehat{D}_0^{-1}\left(-i\frac{\partial}{\partial t}\right)\frac{\partial\widehat{D}_0}{\partial p}\left(-i\frac{\partial}{\partial t}\right).$$

Then

$$\begin{aligned}\operatorname{Trace}\widehat{T}_{10}\widehat{R}_0\frac{\partial\widehat{D}_0}{\partial p}\left(-i\frac{\partial}{\partial t}\right)\frac{\partial}{\partial t}f(e^{-t})\\ = \int\limits_{\Omega\times\mathbb{R}} A_0(\omega)Z(\omega, g_0(\omega), b(0,\omega))\frac{\partial}{\partial t}f(e^{-t})\,d\omega\,dt.\end{aligned} \tag{28}$$

The integral with respect to t is equal to -1 by the definition of $f(r)$, and we obtain

$$\operatorname{Trace}\widehat{T}_{10}\widehat{R}_0\frac{\partial\widehat{D}_0}{\partial p}\left(-i\frac{\partial}{\partial t}\right)\frac{\partial}{\partial t}f(e^{-t})=\int\limits_{\Omega}A_0(\omega)Z(\omega,g_0(\omega),b(0,\omega))\,d\omega. \tag{29}$$

Since $b(0,\omega)\neq 0$, it follows that $Z(\omega,g_0(\omega),b(0,\omega))$ is infinitely differentiable with respect to ω. (This is the kernel of a pseudodifferential operator away from the diagonal.) We multiply and divide the integrand by $b(0,\omega)^l$, where l is sufficiently large, and use the properties of the Fourier transform to obtain

$$\begin{aligned}&i\operatorname{Trace}\widehat{T}_{10}\widehat{R}_0\frac{\partial\widehat{D}_0}{\partial p}\left(-i\frac{\partial}{\partial t}\right)\frac{\partial}{\partial t}f(e^{-t})\\&\quad=\frac{1}{2\pi i}\operatorname{Trace}\int_{-\infty}^{\infty}\left(\frac{i}{b(0,\omega)}\right)^l\widehat{T}_{10}(p)\left(\frac{\partial}{\partial p}\right)^l\left(\widehat{D}_0^{-1}(p)\frac{\partial\widehat{D}_0(p)}{\partial p}\right)dp,\end{aligned} \tag{30}$$

that is, exactly the trace of the desired regularization (23). It remains to note that all other terms in (27) are zero, since, computing them in a similar way, we can use the fact that

$$\int_0^{\infty}\left(\frac{\partial}{\partial t}\right)^l f(e^{-t})\,dt=0$$

for $l\geq 2$ by our assumptions.

The proof of the theorem is complete. □

2.4. The contribution of conical fixed points: Further computation

Under certain conditions on the conormal symbol of $\widehat{D}$, the expression for the contribution of conical fixed points can be simplified dramatically.

Since $\widehat{D}$ is a *differential* operator, we see that its conormal symbol at each singular point is just a polynomial in p (a polynomial operator pencil on the corresponding manifold Ω_j). By the formal ellipticity, it is Agranovich–Vishik elliptic with parameter p [20] in some double sector of nonzero opening angle containing the real axis and is finite-meromorphically invertible in the entire complex plane. All but finitely many poles of the inverse family lie in the complement of the above-mentioned sector.

It follows that each strip $\{|\operatorname{Im}p|<R\}$ contains only finitely many poles of the family $\widehat{D}_0^{-1}(\alpha_j,p)$ and the principal part of the Laurent series at each pole is of finite rank. We impose some technical condition on the behavior of the number of these poles and some of their characteristics as $R\to\infty$.

Definition 2.4. We say that a conormal symbol $\widehat{B}(p)$ of order m is of *power type* if the following conditions are satisfied:

1. The orders of the poles of $\widehat{B}^{-1}(p)$ are bounded by some constant.
2. The number of the poles of $\widehat{B}^{-1}(p)$ in the strip $\{|\operatorname{Im}p|<R\}$ grows as $R\to\infty$ no faster than some power of R.

3. The ranks and the operator norms in $L^2(\Omega_j)$ of the coefficients of principal parts of the Laurent series of $\widehat{B}^{-1}(p)$ around the poles in the strip $\{|\operatorname{Im} p| < R\}$ are bounded above by some power of R as $R \to \infty$.
4. There exists a sequence of circles centered at the origin with radius tending to infinity such that $\widehat{B}^{-1}(p)$ has no poles on these circles and can be estimated for every s in the operator norm $H^s(\Omega_j) \to H^{s+m}(\Omega_j)$ by some power of the radius (which may depend on s).

Remark 2.5. The above conditions are rather natural. In some cases (e.g., for the Beltrami–Laplace operator associated with a conical metric on M) the power type property of the conormal symbol can be established with the use of results concerning the spectral asymptotics for self-adjoint elliptic pseudodifferential operators on closed manifolds (e.g, see [21–24]).

Theorem 2.6. *Suppose that the assumptions of Theorem 2.2 are satisfied and the conormal symbol $\widehat{D}_0(\alpha^*, p)$ of the operator $\widehat{D}$ at some fixed conical point α^* is of power type. Then the contribution of this conical point can be represented as the sum of traces of finite rank operators given by the residues of the integrand:*

$$\mathcal{L}_{\text{sing}}(\alpha^*) = \sum_{\operatorname{Im} p_j > \gamma(\alpha^*)} \operatorname{Trace} \operatorname*{Res}_{p=p_j} \left\{ \widehat{T}_{10}(\alpha^*, p) \widehat{D}_0^{-1}(\alpha^*, p) \frac{\partial \widehat{D}_0}{\partial p}(\alpha^*, p) \right\} \tag{31}$$

if α^* is an attractive point, and

$$\mathcal{L}_{\text{sing}}(\alpha^*) = - \sum_{\operatorname{Im} p_j < \gamma(\alpha^*)} \operatorname{Trace} \operatorname*{Res}_{p=p_j} \left\{ \widehat{T}_{10}(\alpha^*, p) \widehat{D}_0^{-1}(\alpha^*, p) \frac{\partial \widehat{D}_0}{\partial p}(\alpha^*, p) \right\} \tag{32}$$

if α^* is a repulsive point. Here the p_j are the poles of the family $\widehat{D}_0^{-1}(\alpha^*, p)$ and $\gamma(\alpha^*)$ is the weight exponent at the point α^*.

Proof. Without loss of generality, we can assume that $\gamma(\alpha^*) = 0$. Let $\widehat{D}_0(p)$ be the conormal symbol of $\widehat{D}$ at α^*. Suppose that $\widehat{D}_0(p)$ is of power type. Let S_k be a family of circles of radii $R_k \to \infty$ on which $\widehat{D}_0^{-1}(p)$ satisfies a power-law estimate of the norm according to item 4 of Definition 2.4. Since $\widehat{D}_0(p)$ is a polynomial of p, one can readily prove by induction that all derivatives $\partial^l \widehat{H}(p)/\partial p^l$ satisfy power-law estimates of the norm on these circles in appropriate pairs of spaces. Since the norm of the operator $\partial^l \widehat{H}(p)/\partial p^l$ decays polynomially as $p \to \infty$ in the above-mentioned double sector of nonzero angle containing the real axis, we see that the integration contour in the integral on the right-hand side of (23) can be closed in the half-plane where the exponential $e^{ib(0,\omega)p}$ decays. More precisely, consider the sequence of contours consisting of the segments $[-R_k, R_k]$ of the real axis and the corresponding half-circles S_k in the upper or lower half-plane. The integral over the half-circle tends to zero as $k \to \infty$ by virtue of the above considerations, and in the limit we find that the integral over the real axis is equal to the limit of the sums of residues in the half-disks bounded by these half-circles and the real axis. It follows from items 1–3 of Definition 2.4 that this limit is just equal to the sum of residues in the corresponding half-plane. (This sum converges absolutely.) These residues coincide with the residues of the integrand on the left-hand side, since formal integration by parts does not affect the residues at the poles. (This is a trivial consequence of the Cauchy integral formula applied to the integral over a small circle centered at a pole.) It remains to prove that one can transpose summation and trace computation. (Recall that so far we have established the convergence of the series in the operator rather than the trace norm.) This is however trivial. The terms of the series are finite rank operators, and for such operators the trace norm does not exceed the operator norm times the rank. It follows from items 1–3 that the total rank and the total operator norm of the coefficients of the principal parts of the Laurent series at the poles of the family $\widehat{H}(p)$ in the strip $|\operatorname{Im} p| < R$ grow at most polynomially in R, and the presence of the exponential factor $e^{ib(0,\omega)p}$ provides the absolute convergence of the series in the trace norm as well. □

2.5. Example

Now let us give a simple example illustrating the contribution of conical fixed points to the Lefschetz number. Suppose that M is a two-dimensional manifold with a conical singular point α, so that the base of the cone is diffeomorphic to a circle and in a neighborhood U of α one can introduce coordinates (r, ω), where $r \in [0, 1)$ is the distance from the conical point and $\omega \in S^1$ is a coordinate on the base of the cone. Consider a second-order elliptic differential operator $\widehat{D}$ on M and a geometric endomorphism $\widehat{T}$ of the form

$$\widehat{T}u(x) = u(g(x)), \quad x \in M,$$

where $g : M \to M$ is a given mapping such that $\widehat{D}\widehat{T} = \widehat{T}\widehat{D}$. Let us compute the contribution of the conical fixed point to the Lefschetz number of this geometric endomorphism assuming that the following conditions are satisfied in the neighborhood U:

1. The operator $\widehat{D}$ has the form
$$\widehat{D} = \left(r\frac{\partial}{\partial r}\right)^2 + \frac{\partial^2}{\partial \omega^2}.$$
2. The mapping g is given by the formula
$$g(r, \omega) = (\lambda r, \omega),$$
where $\lambda \neq 1$ is a positive number.

Clearly, the conical fixed point α is attractive for $\lambda < 1$ and repulsive for $\lambda > 1$. The case $\lambda = 1$ is degenerate and thereby excluded. The conormal symbols of $\widehat{D}$ and $\widehat{T}$ are equal to

$$\widehat{D}_0(p) = -p^2 + \frac{\partial^2}{\partial \omega^2}, \quad \widehat{T}_0(p) = \lambda^{-ip}.$$

The poles of $\widehat{D}_0^{-1}(p)$ are $p = ik$, $k \in \mathbb{Z}$, and the ranges of the corresponding residues are two-dimensional for $k \neq 0$ and one-dimensional for $k = 0$ (they coincide with the eigenspaces of the operator $d^2/d\omega^2$ on the circle). One can readily see that $\widehat{D}_0(p)$ is a family of power type and hence Theorem 2.6 applies. For example, for $\lambda < 1$ the contribution of the conical fixed point has the form

$$\mathcal{L}_{\text{sing}} = \sum_{k>\gamma} \operatorname{Trace} \operatorname*{Res}_{p=ik} \frac{2\lambda^{-ip} p}{p^2 - \frac{\partial^2}{\partial\omega^2}} = \sum_{k>\gamma} a_k \lambda^k,$$

where $a_k = 2$ for $k \neq 0$ and $a_k = 1$ for $k = 0$. The formula for $\lambda > 1$ is similar except that the sum is over $k < \gamma$ and is taken with the opposite sign.

3. The Atiyah–Bott–Lefschetz theorem for endomorphisms given by quantized canonical transformations

As was mentioned in the introduction, the Lefschetz formula necessarily becomes asymptotic for endomorphisms given by quantized canonical transformations. In other words, both the complex and the endomorphism depend on a real parameter $h \in (0, 1]$, and the Atiyah–Bott–Lefschetz theorem gives the asymptotic expansion of the Lefschetz number as $h \to 0$. Hence, prior to stating and proving the theorem mentioned in the section title, we briefly describe semiclassical pseudodifferential operators and quantized canonical transformations.

3.1. Semiclassical pseudodifferential operators

First, we introduce *symbol classes.*

Let $x = (x^1, \dots, x^n) \in \mathbb{R}^n_x$ and $\xi = (\xi_1, \dots, \xi_n) \in \mathbb{R}^n_\xi$. By

$$S^m = S^m(\mathbb{R}^n_x \times \mathbb{R}^n_\xi \times [0, 1]_h)$$

we denote the space of smooth functions $H(x, \xi, h)$ satisfying the estimates

$$\left| \frac{\partial^{|\alpha|+|\beta|+l} H}{\partial x^\alpha \partial \xi^\beta \partial h^l} \right| \leq C_{\alpha\beta l} (1 + |\xi|)^{m-|\beta|}, \quad |\alpha| + \beta + l = 0, 1, 2, \dots . \tag{33}$$

If these estimates hold, then the operator[1]

$$\widehat{H} = H\left(\overset{2}{x}, -ih\overset{1}{\frac{\partial}{\partial x}}, h \right) \tag{34}$$

is well defined and bounded in the spaces

$$\widehat{H} : H^s_h(\mathbb{R}^n) \to H^{s-m}_h(\mathbb{R}^n)$$

[1]We use the Feynman ordering of noncommuting operators; see [10, 13].

for every $s \in \mathbb{R}$. Here the norm in the Sobolev space $H_h^s(\mathbb{R}^n)$ with parameter h is defined by the formula

$$||u||_{s,h}^2 = \int \left| \left(1 - h^2 \frac{\partial^2}{\partial x^2} \right)^{s/2} u \right|^2 dx. \tag{35}$$

Now let

$$x = (r, \omega_1, \dots, \omega_{n-1}) \in \mathbb{R}_+^n = \{(r,\omega) | r \geq 0\}, \quad \xi = (p, q_1, \dots, q_{n-1}) \in \mathbb{C}_p \times \mathbb{R}_q^{n-1}.$$

By

$$S_\varepsilon^m = S_\varepsilon^m(\mathbb{R}_x^n \times \mathbb{C}_p \times \mathbb{R}_q^{n-1} \times [0,1)_h)$$

we denote the space of functions $H(x, \xi, h)$ satisfying the following conditions:

1. $H(x, \xi, h)$ is defined in the strip $\{|\operatorname{Im} p| < \varepsilon\}$ in $\mathbb{R}_x^n \times \mathbb{C}_p \times \mathbb{R}_q^{n-1} \times [0,1]_h$ and is smooth with respect to all variables and analytic in p in this strip.
2. $H(x, \xi, h) = 0$ for $r > R$, where R is sufficiently large.
3. The estimates (33) hold in the strip.

Under these assumptions, the operator

$$\widehat{H} = H\left(\overset{2}{r}, \overset{2}{\omega}, ihr\overset{1}{\frac{\partial}{\partial r}}, -ih\overset{1}{\frac{\partial}{\partial \omega}} \right) \equiv H\left(\overset{2}{e^{-t}}, \overset{2}{\omega}, -ih\overset{1}{\frac{\partial}{\partial t}}, -ih\overset{1}{\frac{\partial}{\partial \omega}} \right) \tag{36}$$

is well defined and bounded in the spaces

$$\widehat{H} : H_h^{s,\gamma}(\mathbb{R}_+ \times \mathbb{R}^{n-1}) \to H_h^{s-m,\gamma}(\mathbb{R}_+ \times \mathbb{R}^{n-1})$$

for any $s \in \mathbb{R}$ and $\gamma \in \mathbb{R}$ and for sufficiently small h. (It suffices to take $h < \varepsilon/|\gamma|$, so that the weight line $\mathfrak{L}_{h\gamma}$ will lie in the strip $\{|\operatorname{Im} p| < \varepsilon\}$.)

Here the norm in the weighted Sobolev spaces $H_h^{s,\gamma}(\mathbb{R}_+ \times \mathbb{R}^{n-1})$ is given by the expression

$$||u||_{s,\gamma,h}^2 = \int_0^\infty \int \left| \left(1 - h\frac{\partial^2}{\partial \omega^2} - \left(hr\frac{\partial}{\partial r} \right)^2 \right)^{s/2} (r^\gamma u) \right|^2 d\omega \frac{dr}{r}. \tag{37}$$

Now we define a semiclassical pseudodifferential operator of order m on a manifold M with conical singularities as an operator of the form (34) (respectively, (36)) in local coordinates on the smooth part of M (respectively, near conical singular operators) modulo integral operators $\widehat{Q}$ with smooth kernel such that

$$\widehat{Q} : H_h^{s,\gamma}(M) \to H_h^{s-N,\gamma}(M)$$

is compact for any s, γ, and N and has the norm $O(h^{N_1})$ for every N_1.

(The semiclassical Sobolev spaces $H_h^{s,\gamma}(M)$ are defined in a standard way: the norm is obtained with the help of a partition of unity from local expressions of the form (37) near conical points and (35) in the smooth part of M.)

In a usual way, one introduces the notion of the *conormal symbol* of a semiclassical pseudodifferential operator $\widehat{H}$ on M at a conical point $\alpha \in \{\alpha_1, \dots, \alpha_N\}$. The conormal symbol is an analytic family $\widehat{H}_0(p)$ of operators depending on the parameter p, $|\operatorname{Im} p| < \varepsilon$, and acting in function spaces on the base Ω of the corresponding cone.

3.2. Quantization of canonical transformations

We shall consider canonical transformations of the compressed cotangent space T^*M. An invariant definition of this space can be found in many papers (e.g., see [25]). It is a manifold with boundary

$$\partial T^*M = \bigsqcup_j (T^*\Omega_j \times \mathbb{R}),$$

equipped with a natural symplectic form ω^2 having a singularity on the boundary. Let us write out the expression of this form in special coordinates near the boundary (over a neighborhood U_j of a singular point α_j). We have the isomorphism

$$T^*U_j \simeq T^*\Omega_j \times [0,1) \times \mathbb{R}. \tag{38}$$

If in this expansion we denote the canonical coordinates on $T^*\Omega_j$ by (ω, q), the coordinate on $[0,1)$ by r, and the coordinate on $\mathbb{R}$ by p, then the equation of the boundary is $r = 0$ and the form ω^2 is given by the formula

$$\omega^2 = -\frac{dp \wedge dr}{r} + dq \wedge d\omega. \tag{39}$$

Canonical transformations are smooth mappings

$$g : T^*M \to T^*M$$

of manifolds with boundary such that

$$g^*\omega^2 = \omega^2$$

Under the assumption that the bases of all cones at singular points are connected, it is obvious that g is a diffeomorphism of the component of ∂T^*M over a conical point α onto the component of ∂T^*M over some (possibly, the same) conical point α_1. We write $\alpha_1 = g(\alpha)$.

Outside the singular points, the structure of canonical transformations is standard. Their structure near the conical points (i.e., near ∂T^*M) was described in [26]. It is convenient to describe this structure in the *cylindrical coordinates* (where the variable r is replaced by a new variable t according to the formula $r = e^{-t}$). Let (t, ω, p, q) and (τ, ψ, ξ, η) be cylindrical coordinates near α and $g(\alpha)$ on the first and second copies of T^*M, respectively. Then g can be represented as a mapping

$$g : (t, \omega, p, q) \to (\tau, \psi, \xi, \eta),$$

whose properties are described in the following theorem.

Theorem 3.1 (see [26]). *The following assertions hold.*

1. *The mapping g near a conical point α can be represented in the form*

$$\begin{aligned} \tau &= t + \chi(e^{-t}, \omega, p, q), & \psi &= \psi(e^{-t}, \omega, p, q), \\ \xi &= p + c + \widetilde{\xi}(e^{-t}, \omega, p, q), & \eta &= \eta(e^{-t}, \omega, p, q), \end{aligned} \tag{40}$$

where χ, ψ, $\tilde{\xi}$, *and* η *are smooth functions,* $\tilde{\xi}(0,\omega,p,q) = 0$, *and* c *is a constant, which will be referred to as the conormal shift of the mapping* g *at the point* α.

2. *The formulas*

$$\psi = \psi(0,\omega,p,q), \qquad \eta = \eta(0,\omega,p,q)$$

specify a family of canonical transformations

$$g(p) : T^*\Omega_\alpha \to T^*\Omega_{g(\alpha)},$$

where Ω_α *and* $\Omega_{g(\alpha)}$ *are the bases of the cones at the corresponding conical points. This family will be referred to as the conormal family of* g *at* α.

The following theorem describes some special coordinates that always exist on the graph of a canonical transformation.

Theorem 3.2. *Let*

$$g(\infty,\omega_0,p_0,q_0) = (\infty,\psi_0,\xi_0,\eta_0).$$

Then there exists a subset $I \subset \{1,\dots,n-1\}$ *such that the functions* $(\tau,\psi_I,\eta_{\bar{I}},p,q)$, *where* $\bar{I} = \{1,\dots,n-1\}\backslash I$, *for a system of local coordinates on the graph of* g *in a neighborhood of the point*

$$(\infty,\omega_0,p_0,q_0;\infty,\psi_0,\xi_0,\eta_0).$$

Moreover, in the corresponding neighborhood of the point $(\infty,\omega_0,p_0,q_0)$ *the transformation is defined by a generating function of the form*

$$S_I(\tau,\psi_I,\eta_{\bar{I}},p,q) = (p+c)\tau + S_{1I}(e^{-\tau},\psi_I,\eta_{\bar{I}},p,q)$$

via the usual formulas

$$t = \frac{\partial S_I}{\partial p},\ \xi = \frac{\partial S_I}{\partial \tau},\ \omega = \frac{\partial S_I}{\partial q},\ \eta_I = \frac{\partial S_I}{\partial \psi_I},\ \psi_{\bar{I}} = -\frac{\partial S_I}{\partial \eta_{\bar{I}}}.$$

Now let

$$g : T^*M \to T^*M$$

be a canonical transformation and a a smooth function on T^*M. Under some additional assumptions, we define an operator $T(g,a)$ (a quantized canonical transformation acting in the Sobolev spaces $H_h^{s,\gamma}(M)$).

Assumption 3.3. The transformation g is asymptotically first-order homogeneous with respect to the multiplicative action of the group $\mathbb{R}_+$ of positive numbers in the fibers of T_0^*M. The conormal shift of g is zero.

Assumption 3.4. The generating functions of g near the conical points are analytic in p in the strip $|\operatorname{Im} p| < \varepsilon$ for some $\varepsilon > 0$.

The graph

$$L_g \subset T^*M \times T^*M \tag{41}$$

of a canonical transformation g is a Lagrangian manifold with respect to the difference of symplectic forms on the first and second copies of T^*M. This Lagrangian manifold is equipped with the standard measure (volume form), namely, the nth exterior power of the symplectic form lifted from one of the copies of T^*M to L_g with the help of the standard projection.

Assumption 3.5. The manifold L_g satisfies the quantization conditions [9, 11].

Under these conditions, one can define the operator $T(g,a)$ (the quantized canonical transformation in Sobolev spaces $H_h^{s,\gamma}(M)$) for amplitudes a satisfying the following conditions.

Assumption 3.6. The function a belongs to the class $S_\varepsilon^m(T^*M)$ defined as the space of smooth functions on T^*M whose coordinate representatives belong to S^m for charts outside conical points and S_ε^m for conical charts.

Let K_g be the Maslov canonical operator on L_g [9, 11]. We define $\widehat{T}(g,a)$ as the integral operator with Schwartz kernel $[K_g(\pi_1^* a)](x,y)$ on the product $M \times M$, where $\pi_1 : L_g \to T^*M$ is the natural projection onto the first factor on the right-hand side in (41). (We assume that M is equipped with a smooth measure dx such that in the cylindrical coordinates near conical points one has $dx = dt \wedge d\omega$, where $d\omega$ is a smooth measure on the base Ω of the corresponding cone. Then Schwartz kernels can be treated as functions.)

If the support of a entirely lies in a cylindrical canonical chart with coordinates $(\tau, \psi_I, \eta_{\bar{I}}, p, q)$, then the operator $\widehat{T}(g,a)$ can be represented modulo compact operators and modulo $O(h)$ in the form

$$[\widehat{T}(g,a)u](\tau,\psi) = \left(\frac{i}{2\pi h}\right)^{n+|\bar{I}|/2} \iiint \exp\left\{\frac{i}{h}[S_I(\tau,\psi_I,\eta_{\bar{I}},p,q) + \psi_{\bar{I}}\eta_{\bar{I}}]\right\}$$
$$\times (\pi_1^* a)\left(\frac{D(\xi,\eta_I,\psi_{\bar{I}})}{D(p,q)}\right)^{1/2} (e^{-\tau},\psi_I,\eta_{\bar{I}},p,q)\tilde{u}(p,q)\,dp\,dq\,d\eta_{\bar{I}}, \quad (42)$$

where $\tilde{u}(p,q)$ is the semiclassical Fourier–Laplace transform of $u(t,\omega)$, $\pi_1^*(a)$ is expressed in the local coordinates of the canonical chart, the integral with respect to p is taken over the weight line $\mathfrak{L}_{h\gamma}$, and the argument of the Jacobian is taken in accordance with the construction of the canonical operator.

The following theorem was proved in [27].

Theorem 3.7. *Under the above assumptions, the operator $\widehat{T}(g,a)$ is continuous in the spaces*

$$\widehat{T}(g,a) : H_h^{s,\gamma}(M) \to H_h^{s-m,\gamma}(M)$$

for any $s \in \mathbb{R}$ and $\gamma \in \mathbb{R}$ provided that $h < \varepsilon/|\gamma|$.

For the operator $\widehat{T} = \widehat{T}(g,a)$, we defined the *conormal symbol* $\widehat{T}_0(p)$. To this end, we represent $\widehat{T}$ in a neighborhood of the conical point in the form

$$\widehat{T} = \widehat{T}\left(\overset{2}{r}, i\overset{1}{r\frac{\partial}{\partial r}}\right),$$

where the operator-valued symbol $\widehat{T}(r,p)$ of the operator $\widehat{T}$ is a family of operators acting in function spaces on the base Ω of the corresponding cone. Then

$$\widehat{T}_0(p) = \widehat{T}(0,p).$$

Let us write out the conormal symbol of the operator (42). Since the conormal shift is zero ($c = 0$), we have

$$S_I(\tau, \psi_I, \eta_{\bar{I}}, p, q) = p\tau + S_{1I}(e^{-\tau}, \psi_I, \eta_{\bar{I}}, p, q),$$

and the operator (42) can be rewritten in cylindrical coordinates in the form

$$[T(g,a)u](r,\psi) = \left(\frac{i}{2\pi h}\right)^{n-1+|\bar{I}|/2} \iint \exp\left\{\frac{i}{h}\left[S_{1I}\left(\overset{2}{r}, \psi_I, \eta_{\bar{I}}, i\overset{1}{r}\frac{\partial}{\partial r}, q\right) + \psi_I\eta_{\bar{I}}\right]\right\}$$
$$\times (\pi_1^* a)\frac{D(\xi,\eta_I,\psi_{\bar{I}})}{D(p,q)}\left(\overset{2}{r}, \psi_I, \eta_{\bar{I}}, i\overset{1}{r}\frac{\partial}{\partial r}, q\right)\check{u}(r,q)\,dq\,d\eta_I, \quad (43)$$

where $\check{u}(r,q)$ is the semiclassical Fourier transform of u with respect to ω. It follows that the conormal symbol $\widehat{T}_0(p)$ is given by the formula

$$[\widehat{T}_0(p)v](\psi) = \left(\frac{i}{2\pi h}\right)^{n-1+|\bar{I}|/2} \iint \exp\left\{\frac{i}{h}\left[S_{1I}\left(0, \psi_1, \eta_{\bar{I}}, p, q\right) + \psi_{\bar{I}}\eta_{\bar{I}}\right]\right\}$$
$$\times (\pi_1^* a)\left(\frac{D(\xi,\eta_I,\psi_{\bar{I}})}{D(p,q)}\right)^{1/2}(0, \psi_I, \eta_{\bar{I}}, p, q)\check{v}(q)\,dq\,d\eta_I, \quad (44)$$

or

$$\widehat{T}_0(p) = \widehat{T}(g(p), a(p)), \quad (45)$$

where $g(p)$ is the conormal family of g and $a(p)$ is the restriction of a to the boundary $r = 0$.

3.3. Main result

Let M be a compact manifold with conical singularities $\{\alpha_1, \dots, \alpha_N\}$, and let

$$\widehat{D} : C^\infty(M, E_1) \to C^\infty(M, E_2)$$

be a formally elliptic semiclassical pseudodifferential operator of order m on M. (Recall that the formal ellipticity means that the principal symbol $\sigma(\widehat{D})$ is invertible outside the zero section of T^*M.) Then the conormal symbol $\widehat{D}_0(p)$ is elliptic with parameter in the sense of Agranovich–Vishik [20] in some two-sided sector containing the real axis and is invertible sufficiently far from the origin in this sector. Outside the sector, for each $h \in (0,1]$ the operator $\widehat{D}_0^{-1}(p)$ has countably many poles with finite-dimensional principal parts of Laurent series.

Thus for each $h \in (0,1]$ in any interval $\{|\gamma| < \varepsilon\}$ there is at most finitely many γ such that the operator $\widehat{D}$ is not elliptic in the scale $\{H_h^{s,\gamma}(M)\}$.

Let $Z(\gamma)$ be the set of values of the parameter $h \in (0,1]$ for which $\widehat{D}_0^{-1}(p)$ has no poles on the weight line $\mathfrak{L}_{h\gamma}$ (or, equivalently, the operator $\widehat{D}$ is elliptic in the scale $H_h^{s,\gamma}(M)$). We say that γ is *admissible* if 0 is a limit point of $Z(\gamma)$. The following assertion is obvious.

Proposition 3.8. *The set of inadmissible γ is at most countable.*

In what follows, we deal only with admissible γ.

Now let g be a canonical transformation satisfying Assumptions 3.3–3.5, and let $a_i \in S^0(T^*M)$, $i = 1, 2$, be amplitudes satisfying Assumption 3.6. We set $\widehat{T}_i = \widehat{T}(g, a_i)$. Suppose that the diagram

$$\begin{array}{ccccccc} 0 & \longrightarrow & C^\infty(E) & \xrightarrow{\widehat{D}} & C^\infty(F) & \longrightarrow & 0 \\ & & \widehat{T}_1 \big\downarrow & & \big\downarrow \widehat{T}_2 & & \\ 0 & \longrightarrow & C^\infty(E) & \xrightarrow{\widehat{D}} & C^\infty(F) & \longrightarrow & 0 \end{array}$$

commutes. Then the diagram

$$\begin{array}{ccccccc} 0 & \longrightarrow & H_h^{s,\gamma}(E) & \xrightarrow{\widehat{D}} & H_h^{s-m,\gamma}(F) & \longrightarrow & 0 \\ & & \widehat{T}_1 \big\downarrow & & \big\downarrow \widehat{T}_2 & & \\ 0 & \longrightarrow & H_h^{s,\gamma}(E) & \xrightarrow{\widehat{D}} & H_h^{s-m,\gamma}(F) & \longrightarrow & 0 \end{array} \tag{46}$$

also commutes for all s, h, and $\gamma < \varepsilon/h$. For $h \in Z(\gamma)$, we have the well-defined Lefschetz number

$$\mathcal{L}(h) = \operatorname{Trace} \widehat{T}_1|_{\operatorname{Ker} \widehat{D}} - \operatorname{Trace} \widehat{T}_2|_{\operatorname{Coker} \widehat{D}}.$$

We shall obtain the asymptotics of $\mathcal{L}(h)$ as $h \to 0$, $h \in Z(\gamma)$, under some additional assumptions about g and $\widehat{D}_0(p)$.

Assumption 3.9. The transformation g is *nondegenerate* in the following sense.

1. For each interior fixed point $z = g(z) \in T^*M$, one has

$$\det(1 - g_*(z)) \neq 0.$$

2. The following condition is satisfied for each conical fixed point α: either

$$\chi(0, \omega, p, q) > 0$$

for all $p, \omega, q \in \mathbb{R} \times T^*\Omega$ (an *attractive* point), or

$$\chi(0, \omega, p, q) < 0$$

for all $p, \omega, q \in \mathbb{R} \times T^*\Omega$ (a *repulsive* point). Here $\chi(e^{-t}, \omega, p, q)$ is the function determining the t-component of g according to (40).

Assumption 3.10. At each conical point α, the conormal symbol $\widehat{D}_0(p)$ satisfies the following conditions:

(a) For every $\varepsilon > 0$, the number $N(h)$ of the poles (counting multiplicities) of the family $\widehat{D}_0^{-1}(p)$ in the strip $\{|\operatorname{Im} p| < \varepsilon\}$ satisfies the estimate

$$N(h) \leq C(\varepsilon) h^{-N_0}$$

for some $N_0 \in \mathbb{R}$.

(b) For every compact subset $K \subset \mathbb{C}$, the family $\widehat{D}_0^{-1}(p)$ satisfies the estimate

$$\|\widehat{D}_0^{-1}(p)\|_{H^s(\Omega) \to H^s(\Omega)} \leq C \operatorname{dist}(p, \operatorname{spec}(\widehat{D}_0))^{-N_1}, \quad p \in K,$$

where $\operatorname{dist}(p, \operatorname{spec}(\widehat{D}_0))$ is the distance from the point p to the spectrum $\operatorname{spec}(\widehat{D}_0)$ of the family $\widehat{D}_0^{-1}(p)$.

(c) The coefficients of principal parts of the Laurent series of the operator $\widehat{D}_0^{-1}(p)$ at the poles p_j are uniformly bounded in the operator norm in L^2 by Ch^{-N} for some N, where C is a constant independent of h and j.

Under these assumptions, the following theorem holds.

Theorem 3.11. *The Lefschetz number $\mathcal{L}(h)$ of the diagram* (46) *has the following asymptotics for a given admissible γ as $h \to 0$, $h \in Z(\gamma)$:*

$$\mathcal{L}(h) = \mathcal{L}_{\text{int}} + \sum_{\alpha_k} \mathcal{L}(\alpha_k) + O(h),$$

where $\mathcal{L}_{\text{int}}$ is the contribution of interior fixed points, given by the same formulas as in Theorem 1.3, *$\sum_{\alpha_k}$ extends over all conical fixed points α_k of g, and $\mathcal{L}(\alpha_k)$ is the contribution of α_k, which is given by the formula*

$$\mathcal{L}(\alpha_k) = \pm \sum_{\pm h\gamma_k < \pm \operatorname{Im} p_j < \varepsilon} \operatorname{Trace} \operatorname*{Res}_{p_j} \left\{ \widehat{T}_{10}(p) \widehat{D}_0^{-1}(p) \frac{\partial \widehat{D}_0(p)}{\partial p} \right\}. \tag{47}$$

(The upper sign corresponds to attractive points, and the lower sign to repulsive points.) Here $\widehat{T}_{10}(p)$ and $\widehat{D}_0(p)$ are the conormal symbols of the operators $\widehat{T}_1(p)$ and $\widehat{D}$, respectively, at the points α_k. The sum is taken over the poles of $\widehat{D}_0^{-1}(p)$ in the strip indicated in the subscript on the sum. The number $\varepsilon > 0$ is sufficiently small (and otherwise arbitrary).

Proof. To simplify the notation, we assume that M has a single conical point α (which is then necessarily a fixed point of g) and $\gamma = 0$. The proof consists of two parts.

1. For each given $h \in Z(\gamma)$, we construct a regularizer of $\widehat{D}$ depending on a parameter $\lambda \to \infty$ and obtain a preliminary formula for the Lefschetz number by letting $\lambda \to \infty$.
2. We study the asymptotics of the resulting expression as $h \to 0$, $h \in Z(\gamma)$.

1. We again use the trace formula

$$\mathcal{L}(h) = \operatorname{Trace}(\widehat{T}_1(1 - \widehat{R}\widehat{D})) - \operatorname{Trace}(\widehat{T}_2(1 - \widehat{D}\widehat{R})) \tag{48}$$

for the Lefschetz number. This time, we set

$$\widehat{R} = \psi_1 \widehat{R}_1 f_1 + \psi_2 \widehat{R}_2 f_2, \tag{49}$$

where $\widehat{R}_1$ is an arbitrary almost inverse of $\widehat{D}$ modulo trace class operators (it is a pseudodifferential operator on M for each $h \in Z(\gamma)$ but not a semiclassical pseudodifferential operator, since the dependence on h as $h \to 0$ need not be regular) and $\widehat{R}_2$ is a semiclassical pseudodifferential operator that is an almost inverse of $\widehat{D}$ modulo operators of large negative order (not necessarily compact). The second operator will be constructed in the form of a semiclassical pseudodifferential operator (see Lemma 3.14 below). For brevity, we refer to $\widehat{R}_2$ as an *interior regularizer*. In particular, it follows that the operators $\widehat{Q}' = 1 - \widehat{R}_2\widehat{D}$ and $\widehat{Q} = 1 - \widehat{D}\widehat{R}_2$ are semiclassical pseudodifferential operators. The functions f_1 and f_2 in (49) form a partition of unity such that $f_1 \equiv 1$ in a neighborhood of the conical point and $f_1 \equiv 0$ in a larger neighborhood; ψ_1 and ψ_2 are cutoff functions such that $\psi_i f_i = f_i$, $i = 1, 2$. Moreover, f_1, ψ_1, and ψ_2 depend only on the cylindrical variable t in a neighborhood of the conical point, and they also depend on the above-mentioned large parameter λ as follows:

$$f_1 = f_1(t - \lambda), \qquad \psi_1 = \psi_1(t - \lambda), \qquad \psi_2 = \psi_2(t - \lambda).$$

Lemma 3.12. *The functions f_1, f_2, ψ_1, and ψ_2 can be chosen to satisfy*

$$[\operatorname{supp} f_j \cup \pi g(\pi^{-1} \operatorname{supp} f_j)] \cap \operatorname{supp}(1 - \psi_j) = \emptyset, \quad j = 1, 2,$$

for all sufficiently large λ, where $\pi : T^*M \to M$ is the natural projection.

The proof follows from the structure of the canonical transformation in a neighborhood of the conical point (see Theorem 3.1).

Now we substitute the regularizer (49) into (48). After some computations, we obtain

$$\mathcal{L}(h) = \mathcal{L}_1 + \mathcal{L}_2 + \mathcal{L}_3,$$

where

$$\begin{aligned}
\mathcal{L}_1 &= \operatorname{Trace}(\widehat{T}_1(\widehat{R}_1 - \widehat{R}_2)[\widehat{D}, f_1]),\\
\mathcal{L}_2 &= \operatorname{Trace}(\widehat{T}_1\psi_2\widehat{Q}' f_2) - \operatorname{Trace}(\widehat{T}_2\psi_2\widehat{Q} f_2),\\
\mathcal{L}_3 &= \operatorname{Trace}(\widehat{T}_1\{(1-\psi_1)\widehat{R}_1[f_1, \widehat{D}] + (1-\psi_2)\widehat{R}_2[\widehat{D}, f_2]\})\\
&\quad + \operatorname{Trace}(\widehat{T}_2\{[\widehat{D}, \psi_1]\widehat{R}_1 f_1 + [\widehat{D}, \psi_2]\widehat{R}_2 f_2\})
\end{aligned}$$

(the argument of f_i and ψ_i in these formulas is $t - \lambda$).

Now let $\lambda \to \infty$. Computations similar to those in Section 2 show that

$$\lim_{\lambda\to\infty} \mathcal{L}_1 = \frac{1}{2\pi i}\int_{-\infty}^{\infty} \operatorname{Trace} \widehat{T}_{10}(p)(\widehat{R}_{10}(p) - \widehat{R}_{20}(p))\frac{\partial\, \widehat{D}_0(p)}{\partial p}\, dp. \tag{50}$$

(The subscript "0" indicates the conormal symbol; in particular, $\widehat{R}_{10}(p) = \widehat{D}_0^{-1}(p)$.)

To find the limit $\lim_{\lambda\to\infty} \mathcal{L}_2$, consider the compactly supported function

$$f(t) = f_1(t)f_2(t-\lambda).$$

Lemma 3.13. *One has*

$$\operatorname{Trace}(\widehat{T}_{10}\widehat{Q}'_0 f) = \operatorname{Trace}(\widehat{T}_{20}\widehat{Q}_0 f).$$

Using this lemma, we obtain

$$\lim_{\lambda\to\infty} \mathcal{L}_2 = \operatorname{Trace}(\widehat{T}_1\widehat{Q}' - \widehat{T}_{10}\widehat{Q}'_0 f_1(t)) - \operatorname{Trace}(\widehat{T}_2\widehat{Q} - \widehat{T}_{20}\widehat{Q}_0 f_1(t)).$$

Finally, the passage to the limit as $\lambda \to 0$ in the term $\mathcal{L}_3$ simply results in freezing the coefficients of $\widehat{T}_i$, $\widehat{D}$, and $\widehat{R}_i$ at the point $t = \infty$ $(r = 0)$. Thus we obtain

$$\begin{aligned}\lim_{\lambda\to\infty} \mathcal{L}_3 = \operatorname{Trace}(\widehat{T}_{10}\{(1-\psi_1)\widehat{R}_{10}[f_1, \widehat{D}_0] + (1-\psi_2)\widehat{R}_{20}[\widehat{D}_0, f_2]\}) \\ + \operatorname{Trace}(\widehat{T}_{20}\{[\widehat{D}_0, \psi_1]\widehat{R}_{10} f_1 + [\widehat{D}_0, \psi_2]\widehat{R}_{20} f_2\}),\end{aligned}$$

where $\widehat{D}_0 = \widehat{D}_0(-i\frac{\partial}{\partial t})$ etc. (Here the argument of ψ_i and f_i is t rather than $t-\lambda$.)

2. Now let us find the asymptotics of the Lefschetz number as $h \to 0$, $h \in Z(\gamma)$. We need to compute the asymptotics of the contributions

$$\mathcal{L}_{\text{int}} = \lim_{\lambda\to\infty} \mathcal{L}_2 \quad \text{and} \quad \mathcal{L}_{\text{cone}} = \lim_{\lambda\to\infty} \mathcal{L}_1.$$

Since $\widehat{R}_2$ is a semiclassical pseudodifferential operator, it follows that the asymptotics of $\mathcal{L}_{\text{int}}$ can be computed by the stationary phase method, which gives the standard expression for the contributions of interior stationary points (see Subsec. 1.3). To find the asymptotics of $\mathcal{L}_{\text{cone}}$, we take $\widehat{R}_2$ in a special form.

Lemma 3.14. *There exists a semiclassical pseudodifferential interior regularizer $\widehat{R}_2$ such that in a neighborhood of the conical point one has*

$$\widehat{R}_2 = \widehat{R}_2\left(t, -ih\frac{\partial}{\partial t}\right),$$

where the symbol $\widehat{R}_2(t,p)$ is holomorphic in the variable p in a sufficiently narrow strip $|\operatorname{Im} p| < \varepsilon$.

Now let us compute the integral (50) using the residue formula. Suppose that the conical point is attractive. It follows from Assumption 3.10 that for each $h \in Z(\gamma)$ there exists a $\rho(h) \in [\varepsilon/2, \varepsilon]$ such that the line $\operatorname{Im} p = \rho(h)$ does not contain poles of the family $\widehat{D}_0^{-1}(p)$ and the inequality $\widehat{D}_0^{-1}(p) \le C \cdot h^{-N_0 N_1}$ holds on this line for $|p| < R$. (For $|p| > R$, where R is sufficiently large, the decay of $\widehat{D}_0^{-1}(p)$ at infinity is guaranteed.)

Now let us consider the integral over the contour given by the union of two lines, $\operatorname{Im} p = 0$ and $\operatorname{Im} p = \rho(h)$, passed in opposite directions and use the Cauchy residue theorem. Then we obtain

$$\mathcal{L}_{\text{cone}} = \sum_{0<\operatorname{Im} p_j<\rho(h)} \operatorname{Trace} \operatorname*{Res}_{p_j} \left\{ \widehat{T}_{10}(p)\widehat{D}_0^{-1}(p)\frac{\partial \widehat{D}_0}{\partial p} \right\}$$
$$+ \frac{1}{2\pi i} \int_{\operatorname{Im} p=\rho(h)} \operatorname{Trace} \widehat{T}_{10}(p)(\widehat{R}_{10}(p) - \widehat{R}_{20}(p))\frac{\partial \widehat{D}_0(p)}{\partial p}\, dp.$$

(We have used the fact that $\widehat{R}_{20}(p)$ is holomorphic in the strip $0 \le \operatorname{Im} p \le \rho(h)$.)

It remains to estimate the integral over the line $\operatorname{Im} p = \rho(h)$. We have

$$\|\widehat{T}_{10}(p)\|_{L_2\to L_2} \le Ce^{-c_1/h}$$

for sufficiently small ε by virtue of the conditions imposed on the conical fixed point and the canonical transformation. (For small ε, the imaginary part of the generating function on the line $\operatorname{Im} p = \rho(h)$ is bounded below by const $\cdot\varepsilon$ with a positive constant.) Now

$$\widehat{R}_{10}(p) - \widehat{R}_{20}(p) = \widehat{D}_0^{-1}\widehat{Q}_{20},$$

and consequently,

$$|||\, \widehat{R}_{10}(p) - \widehat{R}_{20}(p)||| \le \|\widehat{D}_0^{-1}(p)\|_{L_2\to L_2}|||\, \widehat{Q}_{20}(p)|||,$$

where $|||\cdot|||$ is the trace norm of operators in L_2. Since

$$\|\widehat{D}_0^{-1}(p)\|_{L_2\to L_2} \le ch^{-N_1 N}$$

and

$$|||\, \widehat{Q}_{20}(p)||| \le C(1+|p|)^{-N_3},$$

where N_3 is arbitrarily large, we find that

$$|||\widehat{T}_{10}(p)(\widehat{R}_{10}(p) - \widehat{R}_{20}(p))\frac{\partial \widehat{D}_0(p)}{\partial p}||| \le C(1+|p|)^{-N_3}e^{-c_1/h}h^{N_1 N}$$

(with some other constant C), and so the second integral is $O(h^\infty)$.

Finally, one can show that the contribution of the term

$$\mathcal{L}_{\text{rem}} = \lim_{\lambda\to\infty} \mathcal{L}_3$$

is $O(h^\infty)$ by virtue of our assumptions on the conormal singularity, the canonical transformation, and the supports of f_i and ψ_i. The proof of the theorem is complete. □

Acknowledgments

A vast majority of the results reviewed here were obtained jointly with Professor V. E. Shatalov (Moscow) and Professor B.-W. Schulze of Potsdam University, Germany. We are grateful to Professor G. Rozenblum of Chalmers University, Göteborg, Sweden for numerous helpful discussions.

References

[1] M.F. Atiyah and R. Bott, *A Lefschetz Fixed Point Formula for Elliptic Complexes. I.* Ann. of Math. **86** (1967), 374–407.

[2] S. Lefschetz, *L'analysis situs et la géométrie algébrique.* Gauthier–Villars, Paris, 1924.

[3] V. Nazaikinskii, B.-W. Schulze, B. Sternin, and V. Shatalov, *The Atiyah–Bott–Lefschetz Fixed Point Theorem for Manifolds with Conical Singularities.* Ann. Global Anal. Geom. **17**:5 (1999), 409–439.

[4] V. Nazaikinskii, B.-W. Schulze, and B. Sternin, *A Semiclassical Quantization on Manifolds with Singularities and the Lefschetz Formula for Elliptic Operators.* Preprint No. 98/19, Univ. Potsdam, Institut für Mathematik, Potsdam, 1998.

[5] V. Nazaikinskii, *Semiclassical Lefschetz Formulas on Smooth and Singular Manifolds.* Russ. J. Math. Phys. **6**:2 (1999), 202–213.

[6] B. Sternin and V. Shatalov, *Atiyah–Bott–Lefschetz Fixed Point Theorem in Symplectic Geometry.* Dokl. Akad. Nauk **348**:2 (1996), 165–168.

[7] B. Sternin and V. Shatalov, *The Fixed Point Lefschetz Theorem for Quantized Canonical Transformations.* Funktsional. Anal. i Prilozhen. **32**:4 (1998), 35–48.

[8] B. Sternin and V. Shatalov, *Quantization of Symplectic Transformations and the Lefschetz Fixed Point Theorem.* Preprint No. 92, Max-Planck-Institut für Mathematik, Bonn, 1994.

[9] V.P. Maslov, *Perturbation Theory and Asymptotic Methods.* Moscow State University, Moscow, 1965.

[10] V.P. Maslov, *Operator Methods.* Nauka, Moscow, 1973.

[11] A. Mishchenko, V. Shatalov, and B. Sternin, *Lagrangian Manifolds and the Maslov Operator.* Springer-Verlag, Berlin–Heidelberg, 1990.

[12] V. Nazaikinskii, B. Sternin, and V. Shatalov, *Contact Geometry and Linear Differential Equations.* Walter de Gruyter, Berlin–New York, 1992.

[13] V. Nazaikinskii, B. Sternin, and V. Shatalov, *Methods of Noncommutative Analysis. Theory and Applications.* Walter de Gruyter, Berlin–New York, 1995.

[14] B.V. Fedosov, *Trace Formula for Schrödinger Operator.* Russ. J. Math. Phys. **1**:4 (1993), 447–463.

[15] B. Sternin and V. Shatalov, *Ten Lectures on Quantization.* Preprint No. 22, Université de Nice–Sophia Antipolis, Nice, 1994.

[16] B.-W. Schulze, *Elliptic Complexes on Manifolds with Conical Singularities.* In: Seminar Analysis of the Karl Weierstraß Institute 1986/87, number 106 in Teubner Texte zur Mathematik, Teubner, Leipzig, 1988. Pages 170–223.

[17] B.-W. Schulze, *Pseudodifferential Operators on Manifolds with Singularities.* North-Holland, Amsterdam, 1991.

[18] B.A. Plamenevskii, *Algebras of Pseudodifferantial Operators.* Kluwer, Dordrecht, 1989.

[19] B.-W. Schulze, B. Sternin, and V. Shatalov, *Differential Equations on Singular Manifolds. Semiclassical Theory and Operator Algebras.* Wiley–VCH, Berlin–New York, 1998.

[20] M. Agranovich and M. Vishik, *Elliptic Problems with Parameter and Parabolic Problems of General Type.* Uspekhi Mat. Nauk **19**:3 (1964), 53–161.

[21] M.A. Shubin, *Pseudodifferential Operators and Spectral Theory.* Springer-Verlag, Berlin–Heidelberg, 1985.

[22] M.S. Agranovich, B.Z. Katsenelenbaum, A.N. Sivov, and N.N. Voitovich, *Generalized Method of Eigenoscillations in Diffraction Theory.* Wiley–VCH, Berlin, 1999.

[23] A. Martinez, *Microlocal Exponential Estimates and Applications to Tunneling.* In: Microlocal Analysis and Spectral Theory, Kluwer, 1997. Pages 349–376.

[24] L. Hörmander, *The Spectral Function of an Elliptic Operator.* Acta Math. **121** (1968), 193–218.

[25] R. Melrose, *Transformation of Boundary Problems.* Acta Math. **147** (1981), 149–236.

[26] V. Nazaikinskii, B.-W. Schulze, B. Sternin, and V. Shatalov, *Quantization of Symplectic Transformations on Manifolds with Conical Singularities.* Preprint No. 97/23, Univ. Potsdam, Institut für Mathematik, Potsdam, 1997.

[27] V. Nazaikinskii, B.-W. Schulze, B. Sternin, and V. Shatalov, *A Lefschetz Fixed Point Theorem on Manifolds with Conical Singularities.* Preprint No. 97/20, Univ. Potsdam, Institut für Mathematik, Potsdam, 1997.

Vladimir Nazaikinskii
Institute for Problems in Mechanics
Russian Academy of Sciences
pr. Vernadskogo 101-1
119526 Moscow, Russia
e-mail: nazaikinskii@yandex.ru

Boris Sternin
Independent University of Moscow
Bol'shoi Vlas'evskii per. 11
119002 Moscow, Russia
e-mail: sternin@mail.ru

C^*-algebras and Elliptic Theory
Trends in Mathematics, 187–202

Residues and Index for Bisingular Operators

Fabio Nicola and Luigi Rodino

Abstract. We consider an algebra of pseudo-differential operators on the product of two manifolds which contains, in particular, the tensor products of usual pseudo-differential operators. For that algebra we discuss the existence of trace functionals like Wodzicki's residue and we prove a homological index formula for the elliptic elements.

Mathematics Subject Classification (2000). Primary 58J42; Secondary 35S05.

Keywords. Pseudo-differential operators, trace functionals, Wodzicki residue, index formula.

1. Introduction

In [14] (1975) the second author of the present paper considered a class of pseudo-differential operators, called bisingular operators, defined on the product of two compact manifolds $X_1 \times X_2$, with symbols satisfying in local product-type coordinates

$$|\partial_{\xi_1}^{\alpha_1}\partial_{\xi_2}^{\alpha_2}\partial_{x_1}^{\beta_1}\partial_{x_2}^{\beta_2}a(x_1,x_2,\xi_1,\xi_2)| \leq C_{\alpha_1,\alpha_2,\beta_1,\beta_2}\langle\xi_1\rangle^{m_1-|\alpha_1|}\langle\xi_2\rangle^{m_2-|\alpha_2|}.$$

The standard rules of the symbolic calculus can be recaptured for

$$A = a(x_1,x_2,D_1,D_2),$$

by considering the couple of vector-valued symbols

$$\sigma_1^{m_1}(A) : (x_1,\xi_1) \mapsto a(x_1,x_2,\xi_1,D_2),$$
$$\sigma_2^{m_2}(A) : (x_2,\xi_2) \mapsto a(x_1,x_2,D_1,\xi_2).$$

Let us limit attention to symbols $a(x_1,x_2,\xi_1,\xi_2)$ with asymptotic expansion in bi-homogeneous terms and write $HL^{m_1,m_2}(X_1\times X_2)$ for the corresponding class of operators. In this case $\sigma_1^{m_1}, \sigma_2^{m_2}$ can be re-defined more precisely as functions of x_1,ξ_1, homogeneous of order m_1 with respect to ξ_1 (functions of x_2,ξ_2 homogeneous of order m_2 with respect to ξ_2) with values in the space of the classical pseudo-differential operators $HL^{m_2}(X_2)$ (respectively $HL^{m_1}(X_1)$). Fredholm property, in suitable Sobolev spaces, is then proved for A under the assumption that the corresponding vector-valued symbols are elliptic, i.e., $(\sigma_1^{m_1}(A)(v_1))^{-1} \in HL^{-m_2}(X_2)$ for every $v_1 \in T^*X_1 \setminus 0$, $(\sigma_2^{m_2}(A)(v_2))^{-1} \in HL^{-m_1}(X_1)$ for every $v_2 \in T^*X_2 \setminus 0$, see the next Section 2 for details. A natural question in [14] was the computation of the index of $A \in HL^{m_1,m_2}(X_1 \times X_2)$, cf. also Pilidi [12], Rodino [15]. Because of the vector-valued setting, this turned out to be outside the range of the applications of the result of Atiyah and Singer [1], and a general formula was not attained. During the last 30 years, the index for vector-valued symbols and operators was the subject of deep investigation, in connection with hypoelliptic operators, operators on manifolds with singularities, etc., let us address for example to Fedosov, Schulze and Tarkhanov [5] and the references there. When aiming to the computation of the index of the bisingular operators, a very useful tool is given by the formula of Melrose and Nistor [9] in terms of residues. In fact, from a formal point of view there is a strong similarity with the proceeding in Lauter and Moroianu [7, 8], Nicola [11]; namely, one can define a couple of residue functionals for vector-valued symbols and deduce a general index formula, see next Section 3.4.

Before giving details, we would like to present some examples of bisingular operators; they were motivations for the analysis in [14], and still deserve some interest. First, bisingular partial differential operators on the product of two manifolds are locally of the form (all the coefficients are C^∞ in our setting):

$$A = \sum_{\substack{|\beta_1| \le m_1 \\ |\beta_2| \le m_2}} c_{\beta_1,\beta_2}(x_1,x_2) D_1^{\beta_1} D_2^{\beta_2}.$$

In this case

$$\sigma_1^{m_1}(A) = \sum\nolimits_{\substack{|\beta_1| = m_1 \\ |\beta_2| \le m_2}} c_{\beta_1,\beta_2}(x_1,x_2) \xi_1^{\beta_1} D_2^{\beta_2} \tag{1.1}$$

$$\sigma_2^{m_2}(A) = \sum\nolimits_{\substack{|\beta_1| \le m_1 \\ |\beta_2| = m_2}} c_{\beta_1,\beta_2}(x_1,x_2) D_1^{\beta_1} \xi_2^{\beta_2}, \tag{1.2}$$

and a full bi-homogeneous expansion is given by the terms

$$\sigma^{j,k}(A) = \sum_{\substack{|\beta_1| = j \\ |\beta_2| = k}} c_{\beta_1,\beta_2}(x_1,x_2) \xi_1^{\beta_1} \xi_2^{\beta_2}.$$

The Fredholm property of $A \in HL^{m_1,m_2}(X_1 \times X_2)$ is given by the condition $\sigma^{m_1,m_2}(A)(v_1,v_2) \neq 0$ for every $v_1 \in T^*X_1 \setminus 0$, $v_2 \in T^*X_2 \setminus 0$ and the invertibility of the operator-valued maps defined by (1.1),(1.2); we have ind(A)=0.

A simple example of the non-trivial index is given by the so-called double Cauchy integral operators, namely taking $X_1 = X_2 = \mathbb{S}^1$ we consider $A \in HL^{0,0}(\mathbb{S}^1 \times \mathbb{S}^1)$ of the form (in the following we regard $\mathbb{S}^1$ as unit circle in the complex plane and we understand counter-clockwise integration in the Cauchy sense):

$$\begin{aligned}Af(z_1,z_2) &= a_0(z_1,z_2)f(z_1,z_2) + \frac{1}{\pi i}\int_{\mathbb{S}^1}\frac{a_1(z_1,z_2,\zeta_1)}{\zeta_1 - z_1}f(\zeta_1,z_2)\,d\zeta_1\\ &\quad + \frac{1}{\pi i}\int_{\mathbb{S}^1}\frac{a_2(z_1,z_2,\zeta_2)}{\zeta_2 - z_2}f(z_1,\zeta_2)\,d\zeta_2\\ &\quad - \frac{1}{\pi^2}\iint_{\mathbb{S}^1\times\mathbb{S}^1}\frac{a_{12}(z_1,z_2,\zeta_1,\zeta_2)}{(z_1-\zeta_1)(z_2-\zeta_2)}f(\zeta_1,\zeta_2)d\zeta_1\,d\zeta_2. \qquad (1.3)\end{aligned}$$

In this case $\sigma_1^0(A)$, $\sigma_2^0(A)$ take values in $HL^0(\mathbb{S}^1)$. In view of the 0-homogeneity, we may limit to define them on $\mathbb{S}^*\mathbb{S}^1$, and identifying $\mathbb{S}^*\mathbb{S}^1$ with two copies of $\mathbb{S}^1$ we have

$$\begin{aligned}\sigma_1^0(A)(z_1,\pm 1)g(z_2) &= (a_0(z_1,z_2) \pm a_1(z_1,z_2,z_1))g(z_2)\\ &\quad + \frac{1}{\pi i}\int_{\mathbb{S}^1}\frac{a_2(z_1,z_2,\zeta_2) \pm a_{12}(z_1,z_2,z_1,\zeta_2)}{\zeta_2 - z_2}g(\zeta_2)\,d\zeta_2, \qquad (1.4)\end{aligned}$$

$$\begin{aligned}\sigma_2^0(A)(z_2,\pm 1)g(z_1) &= (a_0(z_1,z_2) \pm a_2(z_1,z_2,z_2))g(z_1)\\ &\quad + \frac{1}{\pi i}\int_{\mathbb{S}^1}\frac{a_1(z_1,\zeta_1,z_2) \pm a_{12}(z_1,z_2,\zeta_1,z_2)}{\zeta_1 - z_1}g(\zeta_1)\,d\zeta_1. \qquad (1.5)\end{aligned}$$

The Fredholm property of A in (1.3) depends on the invertibility of these Cauchy integral operators on $\mathbb{S}^1$, see [12],[15]. As for the index, in Section 5 we shall survey the results of [12],[15] and give an application of our formula.

Finally, concerning the generic pseudo-differential case, a simple example of operator $A \in HL^{m_1,m_2}(X_1 \times X_2)$ is given by the tensor product $A = A_1 \otimes A_1$, where $A_1 \in HL^{m_1}(X_1)$, $A_2 \in HL^{m_2}(X_2)$. Note that $A_1 \otimes A_2$ is not any more classical, if one at least of the factors is not a differential operator. Operators acting on sections of bundles will remain outside the present paper, however we should like also to recall the definition of the vector-tensor product

$$A_1 \boxtimes A_2 = \begin{pmatrix} A_1 \otimes I & -I \otimes A_2^* \\ I \otimes A_2 & A_1^* \otimes I \end{pmatrix}.$$

In Atiyah, Singer [1], I, pp. 512–515 (see also Hörmander [6], Theorem 19.2.7) it was observed that, for elliptic factors A_1, A_2 of order $m_1 > 0$, $m_2 > 0$, the standard symbol of $A_1 \boxtimes A_2$ is a matrix 2×2 of continuous functions. Then $A_1 \boxtimes A_2$ is Fredholm, being approximated uniformly by classical elliptic pseudo-differential operators; moreover we have $\operatorname{ind}(A_1 \boxtimes A_2) = \operatorname{ind}(A_1)\operatorname{ind}(A_2)$, generalizing the property of the Euler constant $\chi(X_1 \times X_2) = \chi(X_1)\,\chi(X_2)$. When $m_1 \leq 0$, $m_2 \leq 0$, the standard symbol of $A_1 \boxtimes A_2$ is not continuous, and the arguments in [1],[6] fail. However, the vector-valued symbols $\sigma_1^{m_1}(A_1 \boxtimes A_2)$, $\sigma_2^{m_2}(A_1 \boxtimes A_2)$ are invertible, therefore $A_1 \boxtimes A_2$ is still Fredholm as 2×2 system of operators in $HL^{m_1,m_2}(X_1 \times X_2)$, and the product formula for the index keeps valid, see [14] for details.

We end this introduction by recalling, for a pseudo-differential operator $A \in HL^m(X)$, some basic facts concerning residues and traces, which we shall apply in

Section 3,4 to $A \in HL^{m_1,m_2}(X_1 \times X_2)$. We refer to Wodzicki [18, 19] and Fedosov, Golse, Leichtnam and Schrohe [4] for details.

Let therefore X be a *compact* manifold, $\dim X = n$. We denote by $HL^{\mathbb{Z}}(X) := \cup_{m\in\mathbb{Z}} HL^m(X)$ the algebra of all operators of integer order.
Fix $Q \in HL^1(X)$, positive, elliptic and invertible. Then, for $A \in HL^{\mathbb{Z}}(X)$ the zeta-function

$$z \mapsto \mathrm{Tr}(AQ^{-z})$$

is well defined and holomorphic for large real part of z and admits a meromorphic extension to the whole plane, with at most simple poles at real, integer points. Then, the so-called *Wodzicki residue* is defined by

$$\mathrm{Res}\, A := \mathrm{Res}_{z=0}\mathrm{Tr}(AQ^{-z}).$$

The most important feature of this functional is that it verifies the trace property on $HL^{\mathbb{Z}}(X)$

$$\mathrm{Res}(A_1A_2) = \mathrm{Res}(A_2A_1), \quad \forall A_1, A_2 \in HL^{\mathbb{Z}}(X);$$

namely, it vanishes on commutators.

Actually, it turns out that it can be written down explicitly in terms of the symbol of the operator. More precisely we have

$$\mathrm{Res}\, A = (2\pi)^{-n} \int_{S^*X} a_{-n}(x,\xi)\iota_{\mathcal{R}}\omega^n, \tag{1.6}$$

where $a_{-n}(x,\xi)$ is the term homogeneous of degree $-n$ in the asymptotic expansion of the symbol of A and $\iota_{\mathcal{R}}\omega^n$ indicates the contraction of the nth power of the symplectic 2-form ω in T^*X with the radial vector field $\mathcal{R}$ in the fibers. In particular, we see that it vanishes on operators of order less than $-n$ and therefore on the ideal $\mathcal{I}$ of the smoothing operators (sometimes one expresses this fact by saying the it is *local*). As a consequence, it induces a trace on the quotient algebra $HL^{\mathbb{Z}}(X)/\mathcal{I}$. Moreover, when X is connected and $\dim X > 1$ this is the unique trace *on the quotient algebra.* It is remarkable the fact that, although the term a_{-n} does not have an invariant meaning, the integral (1.6) is however well defined. Also, from (1.6) it follows that the residue functional does not depend on the choice of Q.

It the following we shall also make use of the functional

$$\overline{\mathrm{Tr}}_Q(A) := \lim_{z\to 0}\left(\mathrm{Tr}(AQ^{-z}) - \mathrm{Res}\, A/z\right), \tag{1.7}$$

which expresses the regularized value of the trace $\mathrm{Tr}(AQ^{-z})$ at $z = 0$. Unlike Wodzicki's residue this functional is *not* a trace and depends on Q.

We should recall the remarkable relation between the Wodzicki residue and the Dixmier trace Tr_ω, which is established by Connes' trace theorem (see Connes [3], Proposition 5, page 307); namely, they coincide (up to a multiplicative constant) on operators of order $-n$ (where the latter is defined). Hence Wodzicki's residue can be regarded as an extension of the Dixmier trace to all operators of integer order.

Finally, as an application of the Wodzicki residue, we report on the index formula of Melrose and Nistor [9], which has been the main inspiration for our index formula in Section 4 (cf. also Lauter and Moroianu [7, 8]):

Let A be an elliptic classical pseudo-differential operator on X and let B be a parametrix. Then

$$\operatorname{ind}(A) = \operatorname{Res}([\log Q, B]A)$$

where Q is as above and $[\log Q, B] := \frac{d}{d\tau}(Q^\tau B Q^{-\tau})|_{\tau=0}$.

2. The class of bisingular operators

In this section we fix the notation used in this paper and we recall from Rodino [14] the definitions and the basic properties of the class of pseudo-differential operators that we are interested in here.

Let Ω_1, Ω_2 be open subsets of $\mathbb{R}^{n_1}, \mathbb{R}^{n_2}$ respectively.

Definition 2.1. For $m_1, m_2 \in \mathbb{R}$, we denote by $S^{m_1,m_2}(\Omega_1 \times \Omega_2)$ the space of all functions $a \in C^\infty(\Omega_1 \times \Omega_2 \times \mathbb{R}^{n_1} \times \mathbb{R}^{n_2})$ such that

$$|\partial_{\xi_1}^{\alpha_1} \partial_{\xi_2}^{\alpha_2} \partial_{x_1}^{\beta_1} \partial_{x_2}^{\beta_2} a(x_1, x_2, \xi_1, \xi_2)| \leq C_{\alpha_1,\alpha_2,\beta_1,\beta_2,K_1,K_2} \langle \xi_1 \rangle^{m_1 - |\alpha_1|} \langle \xi_2 \rangle^{m_2 - |\alpha_2|} \quad (2.1)$$

for all $\alpha_1, \alpha_2, \beta_1, \beta_2 \in \mathbb{Z}_+^n$,, $x_i \in K_i$, $\xi_i \in \mathbb{R}^{n_i}$, $i = 1, 2$, for arbitrary $K_i \subset\subset \Omega_i$, and with constants $C_{\alpha_1,\alpha_2,\beta_1,\beta_2,K_1,K_2} > 0$. (As usual, $\langle \xi \rangle := (1 + |\xi|^2)^{1/2}$).

The calculus corresponding to the estimates (2.1) is not temperate and, in fact, quite different from the usual one for pseudo-differential operators in Hörmander's classes $S^m_{1,0}$ (see Hörmander [6], Chapter XVIII). Moreover, an operator with symbol in $S^{m_1,m_2}(\Omega_1 \times \Omega_2)$ is not pseudolocal in general. However, it has the expected continuity property on suitably defined Sobolev spaces, see below. We refer directly to the paper [14] for the full symbolic calculus and the invariance properties with respect to changes of variables. Instead, we here fix the attention on the so-called *classical* symbols, namely symbols with a double asymptotic expansion in homogeneous terms. Since they were not introduced in [14] (in view of the application given there, the notion of principal symbol was sufficient), we will detail their definition now. Let us note a formal similarity with the classical operators with exit symbols in Schulze [16], Section 1.4.3.

Consider, for $i = 1, 2$, the so-called *radial compactification maps*

$$RC_i : \mathbb{R}^{n_i} \longrightarrow \mathbb{S}_+^{n_i} := \{\xi = (\xi', \xi_{n_i+1}) \in \mathbb{R}^{n_i+1} : \ |\xi| = 1, \xi_{n_i+1} \geq 0\},$$
$$RC_i(\xi) = (\xi/\langle \xi \rangle, 1/\langle \xi \rangle),$$

which define a diffeomorphism of $\mathbb{R}^{n_i}$ into the interior of the closed upper half-sphere $\mathbb{S}_+^{n_i} \subset R^{n_i+1}$. The maps

$$\widetilde{RC}_i : T^*\Omega_i = \Omega_i \times \mathbb{R}^{n_i} \longrightarrow \mathbb{S}_+^*\Omega_i := \Omega_i \times \mathbb{S}_+^{n_i}$$
$$\widetilde{RC}_i := \mathrm{Id} \times RC_i$$

are then induced.

We now look at the manifold with corners $\mathbb{S}_+^{n_1} \times \mathbb{S}_+^{n_2}$. Let ρ_1, ρ_2 be respective boundary defining functions for the two boundary hypersurfaces $\mathbb{S}^{n_1-1} \times \mathbb{S}_+^{n_2}$ and $\mathbb{S}_+^{n_1} \times \mathbb{S}^{n_2-1}$, defined in the following way: we take ρ_1 satisfying $(RC_1 \times \mathrm{Id})^*\rho_1)(\xi_1, \omega_2) = |\xi_1|^{-1}$, for $|\xi_1| \geq 1$, $\omega_2 \in \mathbb{S}_+^{n_2}$, and similarly for ρ_2. By means of the projection $\pi : \mathbb{S}_+^*\Omega_1 \times \mathbb{S}_+^*\Omega_2 \to \mathbb{S}_+^{n_1} \times \mathbb{S}_+^{n_2}$ we then define the functions $\tilde{\rho}_i = \pi^*\rho_i$, $i = 1, 2$, which are boundary defining functions for the two boundary hypersurfaces of the manifold with corners $\mathbb{S}_+^*\Omega_1 \times \mathbb{S}_+^*\Omega_2$.

Definition 2.2. A symbol $a \in S^{m_1,m_2}(\Omega_1 \times \Omega_2)$ is called a classical symbol if

$$a \in \left(\widetilde{RC}_1 \times \widetilde{RC}_2\right)^* \left(\tilde{\rho}_1^{-m_1} \tilde{\rho}_2^{-m_2} C^\infty(\mathbb{S}_+^*\Omega_1 \times \mathbb{S}_+^*\Omega_2)\right).$$

We denote by $HS^{m_1,m_2}(\Omega_1 \times \Omega_2)$ the space of these symbols and by $HL^{m_1,m_2}(\Omega_1 \times \Omega_2)$ the one of the corresponding pseudo-differential operators.

(We make clear that the space $C^\infty(\mathbb{S}_+^*\Omega_1 \times \mathbb{S}_+^*\Omega_2)$ of smooth functions on the manifold with corners $\mathbb{S}_+^*\Omega_1 \times \mathbb{S}_+^*\Omega_2$ is defined as the set of functions which admit smooth extensions to $(\Omega_1 \times \mathbb{S}^{n_1}) \times (\Omega_2 \times \mathbb{S}^{n_2})$.)

We observe, in particular, that symbols in $HS^{-\infty,-\infty}(\Omega_1 \times \Omega_2) = C^\infty(\Omega_1 \times \Omega_2; \mathcal{S}(\mathbb{R}^{n_1+n_2}))$ correspond, via the map $\widetilde{RC}_1 \times \widetilde{RC}_2$, to functions on $\mathbb{S}_+^*\Omega_1 \times \mathbb{S}_+^*\Omega_2$ which are smooth up to the boundary and vanish to infinite order there. Similar remarks apply to symbols in $HS^{-\infty,m_2}(\Omega_1 \times \Omega_2)$ and $HS^{m_1,-\infty}(\Omega_1 \times \Omega_2)$.

Consider now $m_1, m_2 \in \mathbb{Z}$. Given an operator $A = \mathrm{Op}(a) \in HL^{m_1,m_2}(\Omega_1 \times \Omega_2)$ there are well defined symbol maps (the first two are operator-valued)

$$\sigma_1^j(A) : T^*\Omega_1 \setminus 0 \to HL^{m_2}(\Omega_2),$$
$$\sigma_2^k(A) : T^*\Omega_2 \setminus 0 \to HL^{m_1}(\Omega_1),$$
$$\sigma^{j,k}(A) : (T^*\Omega_1 \setminus 0) \times (T^*\Omega_2 \setminus 0) \to \mathbb{C},$$

homogeneous respectively of degree j in ξ_1, k in ξ_2, (j,k) in ξ_1, ξ_2 separately, $j, k \in \mathbb{Z}$. Here we denoted by $HL^m(\Omega)$ the space of classical (or polyhomogeneous) pseudo-differential operators on an open subset $\Omega \subset \mathbb{R}^n$.

The construction of such maps goes as follows. As regards $\sigma_1^j(A)$, there exists $\tilde{a} \in \tilde{\rho}_1^{-m_1} \tilde{\rho}_2^{-m_2} C^\infty\left(\mathbb{S}_+^*\Omega_1 \times \mathbb{S}_+^*\Omega_2\right)\right)$ such that $a = \left(\widetilde{RC}_1 \times \widetilde{RC}_2\right)^* \tilde{a}$. We now perform a formal Taylor expansion at $\tilde{\rho}_1 = 0$ and write $\tilde{a} = \sum_{j \leq m_1} \tilde{a}_j \tilde{\rho}_1^{-j}$, where $\tilde{a}_j$ are defined on $\partial(\mathbb{S}_+^*\Omega_1) \times \mathbb{S}_+^*\Omega_2 = (\Omega_1 \times \mathbb{S}^{n_1-1}) \times \mathbb{S}_+^*\Omega_2$. We then consider the function $\left(\mathrm{Id} \times \widetilde{RC}_2\right)^* \tilde{a}_j : (\Omega_1 \times \mathbb{S}^{n_1-1}) \times T^*\Omega_2 \to \mathbb{C}$, and we extend it to a homogeneous function of degree j with respect to $\xi_1 \in \mathbb{R}^{n_1} \setminus \{0\}$. Finally we define

$$\sigma_1^j(A)(x_1, \xi_1) := \left(\left(\mathrm{Id} \times \widetilde{RC}_2\right)^* \tilde{a}_j\right)(x_1, \xi_1; x_2, D_{x_2}) \in HL^{m_2}(\Omega_2),$$

$(x_1, \xi_1) \in T^*\Omega_1 \setminus 0$. By reversing the role of the couples of variables (x_1, ξ_1) and (x_2, ξ_2) we obtain the maps $\sigma_2^k(A)(x_2, \xi_2)$, $(x_2, \xi_2) \in T^*\Omega_2 \setminus 0$. Finally the construction of $\sigma^{j,k}(A)$ is obtained similarly by taking a double Taylor expansion of $\tilde{a}$ at $\tilde{\rho}_1 = \tilde{\rho}_2 = 0$.

Remark 2.3. The following "compatibility relations" are satisfied:

$$\sigma^k\left(\sigma_1^j(A)(x_{,1},\xi_1)\right)(x_2,\xi_2) = \sigma^j\left(\sigma_2^k(A)(x_2,\xi_2)\right)(x_1,\xi_1),$$

for every $(x_1,\xi_1) \in T^*\Omega_1 \setminus 0$, $(x_2,\xi_2) \in T^*\Omega_1 \setminus 0$. Here $\sigma^j(A)$, for a classical operator $A \in HL^m(\Omega)$, $m \in \mathbb{Z}$, denotes the homogeneous term of degree j in the asymptotic expansion of its symbol.

In the usual way, these classes of pseudo-differential operators can be transferred on the product $X_1 \times X_2$ of two manifolds X_1, X_2 by means of local coordinate charts. We denote by $HL^{m_1,m_2}(X_1 \times X_2)$ the space of the *classical* pseudo-differential operators of order (m_1,m_2) on $X_1 \times X_2$. From now on X_1 and X_2 will be *compact* manifolds, with $\dim X_1 = n_1$, $\dim X_2 = n_2$.
One also introduces the scale of Sobolev spaces

$$H^{s_1,s_2}(X_1 \times X_2) := H^{s_1}(X_1)\hat{\otimes} H^{s_2}(X_2).$$

We have $H^{s_1,s_2}(X_1 \times X_2) \subset H^{s_1',s_2'}(X_1 \times X_2)$ if $s_1 \geq s_1'$, $s_2 \geq s_2'$, and the inclusion is compact if $s_1 > s_1'$, $s_2 > s_2'$. Moreover, any operator $A \in HL^{m_1,m_2}(X_1 \times X_2)$ acts $H^{s_1,s_2}(X_1 \times X_2) \to H^{s_1-m_1,s_2-m_2}(X_1 \times X_2)$ continuously.
As one expects, for $A \in HL^{m_1,m_2}(X_2 \times X_2)$ the *principal symbols*

$$\sigma_1^{m_1}(A) : T^*X_1 \setminus 0 \longrightarrow HL^{m_2}(X_2),$$
$$\sigma_2^{m_2}(A) : T^*X_2 \setminus 0 \longrightarrow HL^{m_1}(X_1),$$
$$\sigma^{m_1,m_2}(A) : (T^*X_1 \setminus 0) \times (T^*X_2 \setminus 0) \longrightarrow \mathbb{C},$$

are invariantly defined as smooth functions, homogeneous respectively of degree m_1 in ξ_1, m_2 in ξ_2, (j,k) in ξ_1, ξ_2 separately. Moreover they are multiplicative, in the sense that $\sigma_1^{m_1+m_1'}(AB) = \sigma_1^{m_1}(A)\sigma_1^{m_1'}(B)$ if $A \in HL^{m_1,m_2}(X_1 \times X_2)$, $B \in HL^{m_1',m_2'}(X_1 \times X_2)$, and so on.

So far, we considered operators acting on scalar-valued functions, but what we said holds without any change for operators acting on sections of *trivial* bundles. Instead, in dealing with general bundles, the definition of the principal symbols would be more sophisticated (see [14]). Hence, for simplicity, we will limit ourselves in the following to considering operators acting on scalar-valued functions.

Here is the notion of ellipticity for operators in $HL^{m_1,m_2}(X_1 \times X_2)$.

Definition 2.4. We say that an operator $A \in HL^{m_1,m_2}(X_1 \times X_2)$ is elliptic if $\sigma^{m_1,m_2}(A)(v_1,v_2) \neq 0$ for every $v_1 \in T^*X_1 \setminus 0$, $v_2 \in T^*X_2 \setminus 0$, and if the operators $\sigma_1^{m_1}(A)(v_1) \in HL^{m_2}(X_2)$ and $\sigma_2^{m_2}(A)(v_2) \in HL^{m_1}(X_1)$ (which are therefore elliptic) are invertible for every $v_1 \in T^*X_1 \setminus 0$, $v_2 \in T^*X_2 \setminus 0$, with inverses in $HL^{-m_2}(X_2)$, $HL^{-m_1}(X_1)$ respectively.

As a consequence of the symbolic calculus developed in [14], we have a parametrix and Fredholm properties for elliptic elements in $HL^{m_1,m_2}(X_1 \times X_2)$.

Theorem 2.5. *Let $A \in HL^{m_1,m_2}(X_1 \times X_2)$ be elliptic. There exists*

$$B \in HL^{-m_1,-m_2}(X_1 \times X_2)$$

such that

$$\begin{cases} AB = \mathrm{Id} + K_1 \\ BA = \mathrm{Id} + K_2, \end{cases}$$

where Id *is the identity on* $C^\infty(X_1 \times X_2)$*, and* K_1 *and* K_2 *are compact operators on* $H^{s_1,s_2}(X_1 \times X_2)$ *for every* $s_1, s_2 \in \mathbb{R}$ *(more precisely,* $K_1, K_2 \in HL^{-1,-1}(X_1 \times X_2)$*). Then* A *as a map from* $H^{s_1,s_2}(X_1 \times X_2)$ *to* $H^{s_1-m_1,s_2-m_2}(X_1 \times X_2)$ *is a Fredholm operator.*

As usual, by means of a formal Neumann series one can construct a parametrix B for which the operators K_1, K_2 in Theorem 2.5 are in $HL^{-p,-p}(X_1 \times X_2)$, with p arbitrarily large.

It follows from Theorem 2.5 that for an elliptic operator in $HL^{m_1,m_2}(X_1 \times X_2)$ we can therefore consider its *index*, namely the integer number

$$\mathrm{ind}(A) := \dim \operatorname{Ker} A - \dim \operatorname{Coker} A \in \mathbb{Z}.$$

It turns out that it only depends on the homotopy class of the joint symbol $(\sigma_1^{m_1}, \sigma_2^{m_2})$ in the space of elliptic symbols. In the sequel we will give a formula for the index of an elliptic operator in terms of residue type functionals we are going to construct in the next section.

3. Residue traces

We begin with a sufficient condition for an operator in $HL^{m_1,m_2}(X_1 \times X_2)$ to be a trace class operator on $L^2(X_1 \times X_2)$.

Proposition 3.1. *Lat* $A \in HL^{m_1,m_2}(X_1 \times X_2)$*, with* $m_1 < -n_1$*,* $m_2 < -n_2$*. Then* A *is trace class on* $L^2(X_1 \times X_2)$ *and*

$$\mathrm{Tr} A = \int_{X_1 \times X_2} K|_\Delta, \tag{3.1}$$

where the density $K|_\Delta$ *is the restriction to the diagonal* $\Delta \subset (X_1 \times X_2) \times (X_1 \times X_2)$ *of the kernel* K *of* A*.*

Proof. The proof goes exactly as the classical one, see for example Shubin [17], Proposition 27.2.

As an alternative one sees at once that, if $m_1 < -n_1$, $m_2 < -n_2$, any symbol in $S^{m_1,m_2}(\Omega_1 \times \Omega_2)$ with compact support with respect to (x_1, x_2) is integrable, together with all its derivatives. Hence, by means of a partition of unity, the first part of the statement follows, e.g., from Robert [13], Thm. (II-53).
As regards formula (3.1), certainly it holds for operators in $HL^{-\infty,-\infty}(X_1 \times X_2)$ and then extends by continuity to operators in $HL^{m_1,m_2}(X_1 \times X_2)$ for every $m_1 < -n_1$, $m_2 < -n_2$ (whose kernels are continuous densities). □

We would like to extend the functional (3.1) further. Following an idea of Melrose and Nistor [9] (applied there to the b-calculus) we will construct the desired extension by means of a suitably defined "double zeta-function" associated with any operator $A \in HL^{m_1,m_2}(X_1 \times X_2)$ of integer order.

Precisely, let $Q_1 \in HL^1(X_1)$, $Q_2 \in HL^1(X_2)$ be classical, positive, elliptic and invertible operators.

Theorem 3.2. *Let $A = A(z,\tau)$, $(z,\tau) \in \mathbb{C}^2$, be a holomorphic family of operators in $HL^{m_1,m_2}(X_1 \times X_2)$. The double zeta-function*

$$(z,\tau) \mapsto \mathrm{Tr}(A(z,\tau)Q_1^{-z} \otimes Q_2^{-\tau})$$

is holomorphic for $\Re z > m_1 + n_1$, $\Re\tau > m_2 + n_2$, and extends to a meromorphic function with at most simple poles at $z = n_1 + m_1 - j$, $\tau = m_2 + n_2 - k$, $j, k \in \mathbb{Z}_+$.

Proof. The theorem follows from the calculus in [14] and Seeley's results on the kernels of complex powers of pseudo-differential operators, see, e.g., Shubin [17], Chapter II. □

Hence, for any given

$$A \in HL^{\mathbb{Z},\mathbb{Z}}(X_1 \times X_2) := \cup_{m_1 \in \mathbb{Z}} \cup_{m_2 \in \mathbb{Z}} HL^{m_1,m_2}(X_1 \times X_2),$$

in a neighborhood of $0 \in \mathbb{C}^2$ we can write

$$z\tau\mathrm{Tr}(AQ_1^{-z} \otimes Q_2^{-\tau}) = \mathrm{Tr}_{1,2}(A) + \tau\widehat{\mathrm{Tr}}_1(A) + z\widehat{\mathrm{Tr}}_2(A) + \tau^2 V + \tau z V' + z^2 V'', \quad (3.2)$$

with V, V', V'' holomorphic, *defining in this way the functionals* $\mathrm{Tr}_{1,2}(A)$, $\widehat{\mathrm{Tr}}_1(A)$, *and* $\widehat{\mathrm{Tr}}_2(A)$.

For $i = 1, 2$, let $\iota_{\mathcal{R}_i}\omega_i^{n_i}$ be the contraction of the n_ith power of the symplectic form ω_i on T^*X_i with the radial vector field $\mathcal{R}_i$ in the fibers.

Theorem 3.3. *The functionals in* (3.2) *have the following explicit expressions:*

$$\mathrm{Tr}_{1,2}(A) = (2\pi)^{-n_1-n_2} \int\limits_{\mathbb{S}^*X_1 \times \mathbb{S}^*X_2} \sigma^{-n_1,-n_2}(A)\, \iota_{\mathcal{R}_1}\omega_1^{n_1} \iota_{\mathcal{R}_2}\omega_2^{n_2}, \quad (3.3)$$

$$\widehat{\mathrm{Tr}}_1(A) = (2\pi)^{-n_1} \int_{\mathbb{S}^*X_1} \overline{\mathrm{Tr}}_{Q_2}\, \sigma_1^{-n_1}(A)\, \iota_{\mathcal{R}_1}\omega_1^{n_1}, \quad (3.4)$$

$$\widehat{\mathrm{Tr}}_2(A) = (2\pi)^{-n_2} \int_{\mathbb{S}^*X_2} \overline{\mathrm{Tr}}_{Q_1}\, \sigma_2^{-n_2}(A)\, \iota_{\mathcal{R}_2}\omega_2^{n_2}, \quad (3.5)$$

where the functional $\overline{\mathrm{Tr}}_Q$ is defined in (1.7).

Proof. We denote by $Q_{1,z}(x_1, y_1)$, $Q_{2,\tau}(x_2, y_2)$, $A(x_1, y_1, x_2, y_2)$ the kernels of the operators Q_1^{-z}, $Q_2^{-\tau}$, A respectively. They are here regarded as distributions, after trivializing the density bundles on X_1 and X_2 by Riemannian volume densities dV_1 and dV_2.

We notify the reader that, to avoid weighting down this proof, the arguments carried out below will be, in most cases, quite formal.

We have

$$\mathrm{Tr}_{1,2}(A) = \lim_{\tau\to 0}\left(\tau \lim_{z\to 0} z\mathrm{Tr}(AQ_1^{-z} \otimes Q_2^{-\tau})\right). \quad (3.6)$$

On the other hand,

$$\mathrm{Tr}(AQ_1^{-z} \otimes Q_2^{-\tau}) = \int K_z(x_2, y_2) Q_{2,\tau}(y_2, x_2) dV_2(x_2) dV_2(y_2), \tag{3.7}$$

with

$$\begin{aligned} K_z(x_2, y_2) &:= \int A(x_1, y_1, x_2, y_2) Q_{1,z}(y_1, x_1) dV_1(x_1) dV_1(y_1) \\ &= \frac{(2\pi)^{-n_1}}{z} \int_{\mathbb{S}^* X_1} \sigma_1^{-n_1}(x_1, \xi_1; x_2, y_2) \iota_{\mathcal{R}_1} \omega_1^{n_1} + F(z, x_2, y_2), \end{aligned} \tag{3.8}$$

where $\sigma_1^{-n_1}(x_1, \xi_1; x_2, y_2)$ denotes the kernel of the operator $\sigma_1^{-n_1}(x_1, \xi_1)$ and F is holomorphic in a neighborhood of $0 \in \mathbb{C}$. The last equality follows from the classical Wodzicki's results [18, 19] (see also Fedosov, Golse, Leichtnam and Schrohe [4]). Hence,

$$\begin{aligned} &\lim_{z \to 0} z \mathrm{Tr}(AQ_1^{-z} \otimes Q_2^{-\tau}) = \\ &= (2\pi)^{-n_1} \int_{\mathbb{S}^* X_1} \sigma_1^{-n_1}(x_1, \xi_1; x_2, y_2) Q_{2,\tau}(y_2, x_2) \iota_{\mathcal{R}_1} \omega_1^{n_1} dV_2(x_2) dV_2(y_2) \\ &= (2\pi)^{-n_1} \int_{\mathbb{S}^* X_1} (\mathrm{Res}\, \sigma_1^{-n_1}(x_1, \xi_1)/\tau + \overline{\mathrm{Tr}}_{Q_2} \sigma_1^{-n_1}(x_1, \xi_1)) \iota_{\mathcal{R}_1} \omega_1^{n_1} + G(\tau) \end{aligned} \tag{3.9}$$

where $G(\tau)$ is holomorphic in a neighborhood of $0 \in \mathbb{C}$. By using (3.6) and (3.9) we deduce

$$\mathrm{Tr}_{1,2}(A) = (2\pi)^{-n_1} \int_{\mathbb{S}^* X_1} \mathrm{Res}\, \sigma_1^{-n_1}(x_1, \xi_1) \iota_{\mathcal{R}_1} \omega_1^{n_1}, \tag{3.10}$$

namely (3.3).

The functional $\widehat{\mathrm{Tr}}_1(A)$ can be obtained as

$$\widehat{\mathrm{Tr}}_1(A) = \lim_{\tau \to 0} \Big(\lim_{z \to 0} z \mathrm{Tr}(AQ_1^{-z} \otimes Q_2^{-\tau}) - \mathrm{Tr}_{1,2}(A)/\tau \Big).$$

Therefore, applying (3.9) and (3.3) gives at once (3.4). Similarly one deduces (3.5). □

In particular, we see that the functional $\mathrm{Tr}_{1,2}$ does not depend on the choice of the operators Q_1 and Q_2.

Remark 3.4. The restrictions Tr_1, Tr_2 of $\widehat{\mathrm{Tr}}_1$ and $\widehat{\mathrm{Tr}}_2$ to the subalgebras

$$HL^{\mathbb{Z}, -n_2-1}(X_1 \times X_2) := \cup_{m_1 \in \mathbb{Z}} HL^{m_1, -n_2-1}(X_1 \times X_2)$$

and

$$HL^{-n_1-1, \mathbb{Z}}(X_1 \times X_2) := \cup_{m_2 \in \mathbb{Z}} HL^{-n_1-1, m_2}(X_1 \times X_2)$$

respectively are given by

$$\mathrm{Tr}_1(A) = (2\pi)^{-n_1} \int_{\mathbb{S}^* X_1} \mathrm{Tr}\, \sigma_1^{-n_1}(A)\, \iota_{\mathcal{R}_1} \omega_1^{n_1}, \quad A \in HL^{\mathbb{Z}, -n_2-1}(X_1 \times X_2), \tag{3.11}$$

$$\mathrm{Tr}_2(A) = (2\pi)^{-n_2} \int_{\mathbb{S}^* X_2} \mathrm{Tr}\, \sigma_2^{-n_2}(A)\, \iota_{\mathcal{R}_2} \omega_2^{n_2}, \quad A \in HL^{-n_1-1, \mathbb{Z}}(X_1 \times X_2). \tag{3.12}$$

In particular, they do not depend on the choice of Q_1 and Q_2.

Theorem 3.5. *The functionals* $\mathrm{Tr}_{1,2}$, Tr_1 *and* Tr_2 *define traces on the algebras* $HL^{\mathbb{Z},\mathbb{Z}}(X_1 \times X_2)$,

$$HL^{\mathbb{Z}, -\infty}(X_1 \times X_2) := \cup_{m_1 \in \mathbb{Z}} \cap_{m_2 \in \mathbb{Z}} HL^{m_1, m_2}(X_1 \times X_2)$$

and

$$HL^{-\infty, \mathbb{Z}}(X_1 \times X_2) := \cup_{m_2 \in \mathbb{Z}} \cap_{m_1 \in \mathbb{Z}} HL^{m_1, m_2}(X_1 \times X_2)$$

respectively. Since they vanish on the ideal of smoothing operators, they also descend to the quotient algebras.

Proof. The trace property for these functionals follows from their definition (3.2) in terms of the usual trace of trace class operators together with Theorem 3.2 and the analytic continuation principle (in fact the arguments in Lemma 6 of [9] can be easily adapted to our situation). □

4. Index formula

We can now apply the residue functionals constructed in the previous section to prove a homological index formula in the spirit of Melrose and Nistor [9], Lauter and Moroianu [7, 8].

Let Q_1 and Q_2 be as in the previous section. We observe that there exist well-defined exterior derivatives on the algebra $HL^{\mathbb{Z},\mathbb{Z}}(X_1 \times X_2)$ of operators of integer order, given by

$$HL^{\mathbb{Z},\mathbb{Z}}(X_1 \times X_2) \ni A \mapsto [\log(Q_1 \otimes \mathrm{Id}), A] := \frac{d}{d\tau}\Big((Q_1^{\tau} \otimes \mathrm{Id}) A (Q_1^{-\tau} \otimes \mathrm{Id}) - A \Big)\Big|_{\tau=0} \tag{4.1}$$

and

$$HL^{\mathbb{Z},\mathbb{Z}}(X_1 \times X_2) \ni A \mapsto [\log(\mathrm{Id} \otimes Q_2), A] := \frac{d}{d\tau}\Big((\mathrm{Id} \otimes Q_2^{\tau}) A (\mathrm{Id} \otimes Q_2^{-\tau}) - A \Big)\Big|_{\tau=0}. \tag{4.2}$$

The derivatives with respect to τ in (4.1) and (4.2) are well defined as derivatives of holomorphic families of fixed order. Indeed, we see that for $A \in HL^{m_1,m_2}(X_1 \times X_2)$, $m_1, m_2 \in \mathbb{Z}$, and every $\tau \in \mathbb{C}$ it turns out

$$\sigma_1^{m_1}\Big((Q_1^{\tau} \otimes \mathrm{Id}) A (Q_1^{-\tau} \otimes \mathrm{Id}) - A \Big)(v_1) = 0, \quad \forall v_1 \in T^* X_1 \setminus 0$$

and

$$\sigma_2^{m_2}\Big((\mathrm{Id} \otimes Q_2^{\tau}) A (\mathrm{Id} \otimes Q_2^{-\tau}) - A \Big)(v_2) = 0, \quad \forall v_2 \in T^* X_2 \setminus 0.$$

Hence, for every τ the operators $(\mathrm{Id}\otimes Q_2^{\tau})A(\mathrm{Id}\otimes Q_2^{-\tau})-A$ and $(\mathrm{Id}\otimes Q_2^{\tau})A(\mathrm{Id}\otimes Q_2^{-\tau})-A$ have order (m_1-1,m_2) and (m_1,m_2-1) respectively.
It also follows that

$$[\log(\mathrm{Id}\otimes Q_2),A]\in HL^{m_1-1,m_2}(X_1\times X_2)$$

and

$$[\log(\mathrm{Id}\otimes Q_2),A]\in HL^{m_1,m_2-1}(X_1\times X_2).$$

The starting point is then the following result.

Lemma 4.1. *Let $A\in HL^{m_1,m_2}(X_1\times X_2)$ be elliptic and let $B\in HL^{-m_1,-m_2}$ be a parametrix which inverts A up to trace class remainders. We have*

$$\mathrm{ind}(A)=\mathrm{Tr}[A,B].$$

Proof. It is a consequence of a classical result of Functional Analysis, see, e.g., Hörmander [6], Proposition 19.1.14. □

Theorem 4.2. *Let $A\in HL^{m_1,m_2}(X_1\times X_2)$ be elliptic and let $B\in HL^{-m_1,-m_2}$ be a parametrix which inverts A up to trace class remainders. Then we have*

$$\mathrm{ind}(A)=\widehat{\mathrm{Tr}}_1\big(A[\log(Q_1\otimes\mathrm{Id}),B]\big)+\widehat{\mathrm{Tr}}_2\big([\log(\mathrm{Id}\otimes Q_2),B]A\big). \tag{4.3}$$

Proof. We have

$$\begin{aligned}\mathrm{ind}(A)=\mathrm{Tr}[A,B]&=\mathrm{Tr}((AB-BA)Q_1^{-z}\otimes Q_2^{-\tau})|_{z=0,\tau=0}\\&=\mathrm{Tr}(A(B-(Q_1^{-z}\otimes\mathrm{Id})B(Q_1^{z}\otimes\mathrm{Id}))Q_1^{-z}\otimes Q_2^{-\tau}\\&\qquad+((\mathrm{Id}\otimes Q_2^{\tau})B(\mathrm{Id}\otimes Q_2^{-\tau})-B)AQ_1^{-z}\otimes Q_2^{-\tau})|_{z=0,\tau=0}\\&=\mathrm{Tr}(A(z[\log(Q_1\otimes\mathrm{Id}),B]+z^2F(z))Q_1^{-z}\otimes Q_2^{-\tau}\\&\qquad+(\tau[\log(\mathrm{Id}\otimes Q_2),B]+\tau^2G(\tau))AQ_1^{-z}\otimes Q_2^{-\tau})|_{z=0,\tau=0}\end{aligned} \tag{4.4}$$

where F and G are holomorphic families, of fixed order. We should make clear that we are computing the value at $z=0,\tau=0$ of the *meromorphic extension* of the above trace (the equalities above are valid in the domain of holomorphy). The index formula (4.3) then follows from the definition of the functionals $\widehat{\mathrm{Tr}}_1$ and $\widehat{\mathrm{Tr}}_2$. □

5. An example

Lat us return to the double Cauchy integral operator A in (1.3). First, we re-write A in a new form, suited for the computation of the index. Define, for $r=1,2$ the Plemelj's projections:

$$P^r_{z_1}f(z_1,z_2)=\tfrac{1}{2}f(z_1,z_2)+\tfrac{(-1)^{r+1}}{2\pi i}\int_{\mathbb{S}^1}\tfrac{f(\zeta_1,z_2)}{\zeta_1-z_1}d\zeta_1,$$
$$P^r_{z_2}f(z_1,z_2)=\tfrac{1}{2}f(z_1,z_2)+\tfrac{(-1)^{r+1}}{2\pi i}\int_{\mathbb{S}^1}\tfrac{f(z_1,\zeta_2)}{\zeta_2-z_2}d\zeta_2.$$

Under the invertibility assumption for σ_1^0, σ_2^0 in (1.4), (1.5), we have

$$A = A_0(\mathrm{Id} + K_{11}P^1_{z_1})(\mathrm{Id} + K_{12}P^2_{z_1})(\mathrm{Id} + K_{21}P^1_{z_2})(\mathrm{Id} + K_{22}P^2_{z_2}) + K, \tag{5.1}$$

where

$$A_0 = \sum_{s,t=1}^{2} u_{st}(z_1, z_2)P^s_{z_1}P^t_{z_2} \tag{5.2}$$

and for $j = 1, 2$

$$K_{1j}f(z_1, z_2) = \int_{\mathbb{S}^1} k_{1j}(z_1, z_2, \zeta_2)f(z_1, \zeta_2)\,d\zeta_2, \tag{5.3}$$
$$K_{2j}f(z_1, z_2) = \int_{\mathbb{S}^1} k_{2j}(z_1, z_2, \zeta_1)f(\zeta_1, z_2)\,d\zeta_1. \tag{5.4}$$

The operator K is regularizing on $\mathbb{S}^1 \times \mathbb{S}^1$. All the functions u_{st}, k_{1j}, k_{2j} are C^∞, and can be computed in terms of a_0, a_1, a_2, a_{12} in (1.3), see [12] and [15], Remark 9.2. We have also from [12],[15] that all the factors in the right-hand side of (5.1) are elliptic in $HL^{0,0}(\mathbb{S}^1 \times \mathbb{S}^1)$, hence Fredholm. Therefore the computation of $\mathrm{ind}(A)$ is reduced to the computation of the index of each factor.
Concerning $\mathrm{ind}(A_0)$ we address the reader to Proposition 8.4 in [15]; in short, the ellipticity of A_0 implies $u_{st}(z_1, z_2) \neq 0$ and we may define the homotopy numbers

$$o^{(h)}_{st} = (2\pi)^{-1}[\arg u_{st}(z_1, z_2)]_{\mathbb{S}^1_{z_h}}, \qquad s, t, h = 1, 2.$$

We have $o^{(1)}_{21} = o^{(1)}_{11}$, $o^{(1)}_{22} = o^{(1)}_{12}$, $o^{(2)}_{12} = o^{(2)}_{11}$, $o^{(2)}_{22} = o^{(2)}_{21}$, as it follows again from the ellipticity assumption, and

$$\mathrm{ind}(A_0) = \left(o^{(1)}_{11} - o^{(1)}_{22}\right)\left(o^{(2)}_{12} - o^{(2)}_{21}\right).$$

Concerning the index of the other factors, a sketch of the computation was given in [12]. We shall use our formula (4.3) for obtain a more explicit result. We limit attention to

$$A = \mathrm{Id} + K_{11}P^1_{z_1} \in HL^{0,-\infty}(\mathbb{S}^1 \times \mathbb{S}^1),$$

the arguments for the other 3 factors being similar. To connect with the preceding sections, we use the local coordinate θ_1, $z_1 = e^{i\theta_1}$, and write ξ_1 for the corresponding dual variable. We have

$$\sigma_1^0(A)(\theta_1, \xi_1) = \begin{cases} \mathrm{Id} + a(\theta_1), & \xi_1 > 0 \\ 0, & \xi_1 < 0, \end{cases}$$

where

$$a(\theta_1)g(z_2) = \int_{\mathbb{S}^1} k_{11}(e^{i\theta_1}, z_2, \zeta_2)g(\zeta_2)\,d\zeta_2$$

with $k_{11}(z_1, z_2, \zeta_2)$ as in (5.3).
We therefore have $a(\theta_1) \in HL^{-\infty}(\mathbb{S}^1)$, with $\mathrm{Id} + a(\theta_1)$ invertible for every $\theta_1 \in \mathbb{S}^1$, in view of the ellipticity of A. Moreover $\sigma_2^0(A) = \mathrm{Id}$. If $B \in HL^{0,0}(\mathbb{S}^1 \times \mathbb{S}^1)$ is a parametrix, we have $\sigma_1^0(B) = \sigma_1^0(A)^{-1}$ and $\sigma_2^0(B) = \mathrm{Id}$, see Section 2.

Let $Q_1 \in HL^1(\mathbb{S}^1)$ be a classical, positive and invertible operator, with principal symbol given by $|\xi_1|$, and similarly for Q_2. Then

$$\begin{aligned}
&\sigma_1^{-1}([\log(Q_1 \otimes \mathrm{Id}), B])(\theta_1, \xi_1)\\
&\quad = \begin{cases} \frac{1}{i\xi_1}\frac{d(\mathrm{Id}+a(\theta_1))^{-1}}{d\theta_1}, & \xi_1 > 0\\ 0, & \xi_1 < 0\end{cases}\\
&\quad = \begin{cases} -\frac{1}{i\xi_1}(\mathrm{Id}+a(\theta_1))^{-1}\frac{da(\theta_1)}{d\theta_1}(\mathrm{Id}+a(\theta_1))^{-1}, & \xi_1 > 0\\ 0, & \xi_1 < 0\end{cases}.
\end{aligned} \tag{5.5}$$

Moreover, we have

$$\sigma_1^{-1}(A[\log(Q_1 \otimes \mathrm{Id}), B]) = \sigma_1^0(A)\sigma_1^{-1}([\log(Q_1 \otimes \mathrm{Id}), B]). \tag{5.6}$$

On the other hand it turns out

$$\sigma_2^{-1}([\log(\mathrm{Id} \otimes Q_2), B]A) = \underbrace{\sigma_2^{-1}([\log(\mathrm{Id} \otimes Q_2), B])}_{=0}\,\underbrace{\sigma_2^0(A)}_{=\mathrm{Id}} = 0. \tag{5.7}$$

It follows from (5.7) and (3.11) that

$$\begin{aligned}
\mathrm{ind}(A) &= \widehat{\mathrm{Tr}}_1\big(A[\log(Q_1 \otimes \mathrm{Id}), B]\big)\\
&= (2\pi)^{-1}\int_{\mathbb{S}^*\mathbb{S}^1} \mathrm{Tr}\,\sigma_1^{-1}\big(A[\log(Q_1 \otimes \mathrm{Id}), B]\big)(\theta_1, \xi_1)\,\iota_{\xi_1\frac{\partial}{\partial\xi_1}} d\xi_1 \wedge d\theta_1.
\end{aligned}$$

By virtue of (5.6) and (5.5) we obtain

$$\mathrm{ind}(A) = -\frac{1}{2\pi i}\int_0^{2\pi} \mathrm{Tr}\,\frac{da(\theta_1)}{d\theta_1}(\mathrm{Id}+a(\theta_1))^{-1}d\theta_1 = -\frac{1}{2\pi i}\int_{\mathbb{S}^1} \mathrm{Tr}\,\frac{da}{dz_1}(\mathrm{Id}+a)^{-1}dz_1. \tag{5.8}$$

To give an example of non-trivial index, fix $J > 0$ and consider

$$Af(z_1, z_2) = f(z_1, z_2) + \frac{1}{2\pi i}\int_{\mathbb{S}^1} (z_1^{-J} - 1)\,\zeta_2^{-1}P^1_{z_1}f(z_1, \zeta_2)\,d\zeta_2,$$

so that

$$a\,g(z_2) = \frac{1}{2\pi i}\int_{\mathbb{S}^1} (z_1^{-J} - 1)\,\zeta_2^{-1}g(\zeta_2)\,d\zeta_2.$$

The preceding formula gives then $\mathrm{ind}(A) = J$. The reader will check directly that $\mathrm{Ker}\,A$ is spanned by the linearly independent functions $f(z_1, z_2) = z_1^j - z_1^{j-J}$, $j = 1, \dots, J$, and $\dim \mathrm{Coker}\,A = 0$.

Remark 5.1. After completing the present paper, R. Melrose called our attention on R. Melrose, F. Rochon [10], where an index theorem in K-theory is given in the context of the algebras of pseudo-differential operators of fibred-cusp type, generalizing the K-theory of Atiyah and Singer in the boundaryless case. The results there intersect our results, with some difference of language. In particular, in

Appendix A of [10] the authors consider operators which locally, for $x = (x_1, x_2) \in \Omega = \Omega_1 \times \Omega_2$, $\xi = (\xi_1, \xi_2) \in \mathbb{R}^{n_1+n_2}$, satisfy the estimates

$$|\partial_{\xi_1}^{\alpha_1}\partial_{\xi_2}^{\alpha_2}\partial_{x_1}^{\beta_1}\partial_{x_2}^{\beta_2}a(x_1, x_2, \xi_1, \xi_2)| \leq C_{\alpha_1,\alpha_2,\beta_1,\beta_2,K_1,K_2}\langle\xi_1\rangle^{m_1-|\alpha_1|}\langle\xi\rangle^{m_2-|\alpha_2|}. \quad (5.9)$$

This differs from our (2.1) because of the last term in the right-hand side, in correspondence to the ξ_2 variables, involving all the dual variables. By some work, it seems possible to connect our operator A from (2.1) with operators from (5.9), key point being to split A into terms with classical symbols, symbols of type (5.9) and similar, by interchanging the role of ξ_1 and ξ_2. We shall not develop further this in the present paper.

We can be more precise, however, in the case when $m_1 = -\infty$, i.e., estimates are satisfied for any m_1; then (2.1) and (5.9) coincide. For such operators we may therefore appeal directly to the results of [10] in terms of K-theory, and this gives as a particular case our winding-number formula (5.8).

References

[1] M.F. Atiyah, I.M. Singer, *The index of elliptic operators, I, II and III*, Ann. of Math., **87** (1968), 484–604.

[2] P. Boggiatto, F. Nicola, *Non-commutative residues for anisotropic pseudo-differential operators in* $\mathbb{R}^n$, J. Funct. Anal., **203** (2003), 305–320.

[3] A. Connes, *Noncommutative Geometry*, Academic Press, New York, 1994.

[4] B.V. Fedosov, F. Golse, E. Leichtnam, E. Schrohe, *The noncommutative residue for manifolds with boundary*, J. Funct. Anal., **142** (1996), 1–31.

[5] B.V. Fedosov, B.-W. Schulze, N. Tarkhanov, *On the index of elliptic operators on a wedge*, J. Funct. Anal., , bf 157 (1998), 164–209.

[6] L. Hörmander, *The Analysis of Linear Partial Differential Operators*, Vol.III, Springer-Verlag, 1985.

[7] R. Lauter, S. Moroianu, *Homology of pseudo-differential operators on manifolds with fibered boundaries*, J. Reine Angew. Math., **547** (2002), 207–234.

[8] R. Lauter, S. Moroianu, *The index of cusp operators on manifolds with corners*, Ann. Glob. Anal. Geom., **21** (2002), 31–49.

[9] R. Melrose, V. Nistor, *Homology of pseudo-differential operators* I, *Manifolds with boundary*, Preprint, MIT 1996.

[10] R. Melrose, F. Rochon, *Index in K-theory for families of fibred cusp operators*, Preprint 2005, available at www-math.mit.edu/~rbm.

[11] F. Nicola, *Trace functionals for a class of pseudo-differential operators in* $\mathbb{R}^n$, Math. Phys. Anal. Geom., **6** (2003), 89–105.

[12] V.S. Pilidi, *The index of the bisingular operators*, Funkcional Anal. i Prilozen, **7**(4) (1973), 93-94; Funct. Anal. Appl., **7** (1973), 337–338.

[13] D. Robert, Autour de l'approximation semi-classique, Birkhäuser, Boston, MA, 1987.

[14] L. Rodino, *A class of pseudo-differential operators on the product of two manifolds and applications*, Ann. Scuola Norm. Sup. Pisa, ser. IV, **2** (1975), 287–302.

[15] L. Rodino, *Polysingular integral operators*, Ann. Mat. Pura Appl., **124** (1980), 59–106.

[16] B.-W. Schulze, *Boundary value problems and singular pseudo-differential operators*, J. Wiley, Chichester, 1998.

[17] M.A. Shubin, *Pseudo-differential operators and spectral theory*, Springer-Verlag, Berlin, 1987.

[18] M. Wodzicki, *Spectral Asymmetry and Noncommutative Residue*, Thesis, Stekhlov Institute of Mathematics, Moscow, 1984.

[19] M. Wodzicki, *Local invariants and spectral asymmetry*, Invent. Math., **75** (1984), 143–178.

Fabio Nicola
Dipartimento di Matematica
Politecnico di Torino,
Corso Duca degli Abruzzi, 24
I-10129 Torino, Italy
e-mail: `fabio.nicola@polito.it`

Luigi Rodino
Dipartimento di Matematica
Università di Torino,
Via Carlo Alberto 10,
I-10123 Torino, Italy
e-mail: `luigi.rodino@unito.it`

C^*-algebras and Elliptic Theory
Trends in Mathematics, 203–212

On the Hopf-type Cyclic Cohomology with Coefficients

I.M. Nikonov and G.I. Sharygin

Abstract. In this note we discuss the Hopf-type cyclic cohomology with coefficients, introduced in the paper [1]: we calculate it in a couple of interesting examples and propose a general construction of coupling between algebraic and coalgebraic version of such cohomology, taking values in the usual cyclic cohomology of an algebra.

Mathematics Subject Classification (2000). Primary 16E40; Secondary 19D55.

Keywords. Cyclic homology, Hopf algebras, Weil complex.

1. Motivations and definitions

The notion of the Hopf-type cyclic cohomology was first introduced in the papers [4, 5] where it was motivated by the purposes of the index theory. It turned out that the constructed cohomology is related in an intrinsic way to much of the structure of a Hopf algebra $\mathcal{H}$, multiplication, comultiplication and antipode. Another peculiarity of this construction is that it deals with some additional data, modular pair in involution (δ, σ) (see below), and therefore this cohomology was denoted as $HC^*_{(\delta,\sigma)}(\mathcal{H})$. It turned out that for any algebra $\mathcal{A}$ over $\mathcal{H}$ and an "equivariant trace" t on $\mathcal{A}$ there exists a homomorphism t_* from $HC^*_{(\delta,\sigma)}(\mathcal{H})$ to the usual cyclic cohomology of $\mathcal{A}$.

This construction was many times generalized later. So in [6] there was given a construction of a kind of dual theory, i.e., of Hopf-type cyclic homology of a Hopf algebra. In [3] a construction extending the homomorphisms t_* to higher traces was proposed and in the series of papers [7, 8, 9] a number of generalizations of the original constructions to a wider class of objects was given.

The first author was supported by the grant RFFI 05-01-00923 and the Presidential grant NSh-1988.2003.1
The second author was partly supported by the grant RFFI 04-01-00702.

The most general of the existing constructions of the Hopf-type (co)homology (we assume that the Hopf algebra $\mathcal{H}$ has invertible antipode S) was given in the paper [1]. This construction uses as the coefficients the stable anti-Yetter-Drinfeld modules/comodules over $\mathcal{H}$. Recall that a right-left $\mathcal{H}$-module and comodule (right module and left comodule) $\mathcal{M}$ is called anti-Yetter-Drinfeld if

$$(mh)^{(-1)} \otimes (mh)^{(0)} = S(h^{(3)})m^{(-1)}h^{(1)} \otimes m^{(0)}h^{(2)}, \tag{1}$$

for all $h \in \mathcal{H},\ m \in \mathcal{M}$, and it is stable if (for all $m \in \mathcal{M}$)

$$m^{(0)}m^{(-1)} = m. \tag{2}$$

There are similar definitions for the modules with left action and left coaction, left action and right coaction and right action and coaction. We shall often abbreviate the title "stable anti-Yetter-Drinfeld" to SAYD. For instance, let σ be a group-like element in $\mathcal{H}$ and $\delta : \mathcal{H} \to \mathbb{C}$ a character. With these data one can in an evident way define a (right-left) module comodule $\mathbb{C}_{(\delta,\,\sigma)}$. It turns out that $\mathbb{C}_{(\delta,\,\sigma)}$ is SAYD iff $(\delta,\,\sigma)$ is a modular pair in involution (one can take this for the definition of the latter).

Given a SAYD $\mathcal{M}$ over $\mathcal{H}$, one can define for arbitrary $\mathcal{H}$-module algebra $\mathcal{A}$ or a $\mathcal{H}$-module coalgebra $\mathcal{C}$ their Hopf-type cohomology with coefficients in $\mathcal{M}$, $HC^*_{\mathcal{H}}(\mathcal{A},\,\mathcal{M})$ and $HC^*_{\mathcal{H}}(\mathcal{C},\,\mathcal{M})$.

Let us first give the definition for the case of coalgebras. Suppose we are given a SAYD module $\mathcal{M}$ and a coalgebra $\mathcal{C}$, with a left $\mathcal{H}$-action $\mathcal{H}\otimes\mathcal{C} \to \mathcal{C}$, correlating with the coalgebraic structure on $\mathcal{C}$, i.e., $(hc)^{(1)} \otimes (hc)^{(2)} = h^{(1)}c^{(1)} \otimes h^{(2)}c^{(2)}$ for all $h \in \mathcal{H},\ c \in \mathcal{C}$. We assume that $\mathcal{M}$ is a right $\mathcal{H}$-module and left comodule, but similar constructions exist for all other sorts of SAYD modules.

Now, $HC^*_{\mathcal{H}}(\mathcal{C},\,\mathcal{M})$ is defined by the following construction. First one considers paracocyclic module $C^*(\mathcal{C},\,\mathcal{M})$:

$$C^n(\mathcal{C},\,\mathcal{M}) = \mathcal{M} \otimes C^{\otimes n+1}, \tag{3}$$

and the cyclic operations are defined by the formulas

$$\delta_i(m \otimes c_0 \otimes \cdots \otimes c_n) = \begin{cases} m \otimes c_0 \otimes \cdots \otimes c_i^{(1)} \otimes c_i^{(2)} \otimes \cdots \otimes c_n, & 0 \le i \le n, \\ m^{(0)} \otimes c_0^{(1)} \otimes c_1 \otimes \cdots \otimes c_n \otimes m^{(-1)}c_0^{(2)}, & i = n+1, \end{cases} \tag{4}$$

$$\sigma_i(m\otimes c_0 \otimes \cdots \otimes c_n) = m \otimes c_0 \otimes \cdots \otimes \epsilon(c_i) \otimes \cdots \otimes c_n, \tag{5}$$

$$\tau_n(m\otimes c_0 \otimes \cdots \otimes c_n) = m^{(0)} \otimes c_1 \otimes \cdots \otimes c_n \otimes m^{(-1)}c_0. \tag{6}$$

Recall that "para(co)cyclic" means that all the usual (co)cyclic relations are satisfied, probably except for $\tau_n^{n+1} = 1$. $\mathcal{C}$ being a $\mathcal{H}$-module, one can extend the action of the Hopf algebra to the tensor power of $\mathcal{C}$ diagonally and consider the factor-module $C^*_{\mathcal{H}}(\mathcal{C},\,\mathcal{M})$, $C^n_{\mathcal{H}}(\mathcal{C},\mathcal{M}) = \mathcal{M} \otimes_{\mathcal{H}} (\mathcal{C}^{\otimes n+1})$. Now it is easy to show, that $C^*_{\mathcal{H}}(\mathcal{C},\,\mathcal{M})$ is preserved by the action of the (para)cyclic operations introduced above iff $\mathcal{M}$ is anti-Yetter-Drinfeld. And if $\mathcal{M}$ is also stable then $C^*_{\mathcal{H}}(\mathcal{C}, \mathcal{M})$ with the operations restricted on it from $C^*(\mathcal{C}, \mathcal{M})$ is cyclic.

By definition Hopf-type cyclic (respectively, periodic) (co)homology of $\mathcal{C}$ with coefficients in $\mathcal{M}$, denoted $HC^*_{\mathcal{H}}(\mathcal{C}, \mathcal{M})$ (resp. $HP^*_{\mathcal{H}}(\mathcal{C}, \mathcal{M})$) is the cyclic (resp. periodic) (co)homology of the cocyclic module $C^*_{\mathcal{H}}(\mathcal{C}, \mathcal{M})$. This means that one introduces the mixed complex with differentials b and B associated in a usual way with the (co)cyclic module $C^*_{\mathcal{H}}(\mathcal{C}, \mathcal{M})$ (see, e.g., [10]) and takes the cohomology of the corresponding total complex (resp., periodic super-complex).

Somewhat dually one can define for arbitrary $\mathcal{H}$-module algebra $\mathcal{A}$ (i.e an algebra on which $\mathcal{H}$ acts in such a way that $h(ab) = h^{(1)}(a)h^{(2)}(b)$), first, a para-cocyclic module $C^*(\mathcal{A}, \mathcal{M})$, $C^n(\mathcal{A}, \mathcal{M}) = \mathrm{Hom}(\mathcal{M} \otimes \mathcal{A}^{\otimes n+1}, \mathbb{C})$. The cyclic operations on $C^*(\mathcal{A}, \mathcal{M})$ are defined by the formulas

$$(\delta_i f)(m \otimes a_0 \otimes \cdots \otimes a_n) = f(m \otimes a_0 \otimes \cdots \otimes a_i a_{i+1} \otimes \cdots \otimes a_n), 0 \le i < n, \quad (7)$$

$$(\delta_n f)(m \otimes a_0 \otimes \cdots \otimes a_n) = f(m^{(0)} \otimes (S^{-1}(m^{(-1)})a_n)a_0 \otimes \cdots \otimes a_{n-1}), \quad (8)$$

$$(\sigma_i f)(m \otimes a_0 \otimes \cdots \otimes a_n) = f(m \otimes a_0 \otimes \cdots \otimes a_i \otimes 1 \otimes \cdots \otimes a_n),\ 0 \le i \le n, \quad (9)$$

$$(\tau_n f)(m \otimes a_0 \otimes \cdots \otimes a_n) = f(m \otimes S^{-1}(m^{(-1)})a_n \otimes a_0 \otimes \cdots \otimes a_{n-1}). \quad (10)$$

Now one can assume that $\mathcal{H}$ acts on $\mathbb{C}$ via the counit. It also acts on the tensor product of $\mathcal{A}$ and $\mathcal{M}$ diagonally. This allows one to define the module $C^*_{\mathcal{H}}(\mathcal{A}, \mathcal{M})$ by the formula $C^n_{\mathcal{H}}(\mathcal{A}, \mathcal{M}) = \mathrm{Hom}_{\mathcal{H}}(M \otimes A^{\otimes n+1}, \mathbb{C})$. Once again, it is possible to show that the operations (7)–(10) can be restricted to $C^n_{\mathcal{H}}(\mathcal{A}, \mathcal{M})$ iff the module $\mathcal{M}$ is anti-Yetter-Drinfeld and that $C^n_{\mathcal{H}}(\mathcal{A}, \mathcal{M})$ equipped with these operations is cyclic if $\mathcal{M}$ is stable. As before, one defines the cyclic (resp. periodic) cohomology of $\mathcal{A}$, $HC^*_{\mathcal{H}}(\mathcal{A}, \mathcal{M})$ ($HP^*_{\mathcal{H}}(\mathcal{A}, \mathcal{M})$) as the cyclic (periodic) cohomology of the corresponding mixed complex.

2. Examples

Let us take in the construction above $\mathcal{C} = \mathcal{H}$. In this case one can write down the cocyclic module $C^*_{\mathcal{H}}(\mathcal{C}, \mathcal{M}) = C^*_{\mathcal{H}}(\mathcal{H}, \mathcal{M})$ in an explicit way. To this end one considers the following isomorphism:

$$C^n_{\mathcal{H}}(\mathcal{H}, \mathcal{M}) = \mathcal{M} \otimes_{\mathcal{H}} \mathcal{H}^{\otimes n+1} \overset{\Phi}{\cong} \mathcal{M} \otimes \mathcal{H}^{\otimes n}, \quad (11)$$

$$m \otimes_{\mathcal{H}} (h_0 \otimes \cdots \otimes h_n) \overset{\Phi}{\mapsto} m h_0^{(1)} \otimes S(h_0^{(2)})(h_1 \otimes \cdots \otimes h_n). \quad (12)$$

In this formula we assume that $\mathcal{H}$ acts "diagonally" on its own tensor powers. The inverse of Φ is given by the equation

$$m \otimes g_1 \otimes \cdots \otimes g_n \overset{\Phi^{-1}}{\mapsto} m \otimes_{\mathcal{H}} (1 \otimes g_1 \otimes \cdots \otimes g_n). \quad (13)$$

One checks that under this identification the cocyclic maps on $\mathcal{M} \otimes_{\mathcal{H}} (\mathcal{H}^{\otimes n+1})$ take the following form on $\mathcal{M} \otimes \mathcal{H}^{\otimes n}$:

$$\delta_i(m \otimes h_1 \otimes \cdots \otimes h_n) = \begin{cases} m \otimes 1 \otimes h_1 \otimes \cdots \otimes h_n, & i = 0, \\ m \otimes h_1 \otimes \cdots \otimes h_i^{(1)} \otimes h_i^{(2)} \otimes \cdots \otimes h_n, & 1 \leq i \leq n, \\ m^{(0)} \otimes h_1 \otimes \cdots \otimes h_n \otimes m^{(-1)}, & i = n+1, \end{cases} \tag{14}$$

$$\sigma_i(m \otimes h_1 \otimes \ldots \otimes h_n) = m \otimes h_1 \otimes \cdots \otimes \epsilon(h_{i+1}) \otimes \cdots \otimes h_n, \; 0 \leq i \leq n-1, \tag{15}$$

$$\tau_n(m \otimes h_1 \otimes \ldots \otimes h_n) = m^{(0)} h_1^{(1)} \otimes S(h_1^{(2)})(h_2 \otimes \cdots \otimes h_n \otimes m^{(-1)}). \tag{16}$$

The following statement is a complete analogue of the Lemma 5.1 from [3]:

Lemma 1. *Take $\mathcal{C} = \mathcal{H}$ in the definition of $HC^*_{\mathcal{H}}(\mathcal{C}, \mathcal{M})$. Then for an arbitrary SAYD $\mathcal{M}$ the Hopf-type Hochschild cohomology $HH^*_{\mathcal{H}}(\mathcal{H}, \mathcal{M})$ (defined as the cohomology of the underlying cocyclic module with respect to the differential $b = \sum(-1)^i \delta_i$) is equal to the cotorsion groups of the left $\mathcal{H}$-comodule $\mathcal{M}$, $Cotor^*_{\mathcal{H}}(\mathbb{C}, \mathcal{M})$.*

Proof. As in the cited paper, the proof is obtained by mere inspection of definitions. Actually, from the formulas (14) it follows that the Hochschild complex in the considered case is isomorphic to the cobar-resolution $F(\mathcal{M}, \mathcal{H}, \mathbb{C})$ of $\mathcal{H}$-comodule $\mathcal{M}$. □

Let now the Hopf algebra $\mathcal{H}$ be commutative. In this case every left comodule $\mathcal{M}$ over $\mathcal{H}$ can be given the structure of (right-left) SAYD module over the same Hopf algebra simply by putting $m \cdot h = \epsilon(h)m$ (where ϵ is the counit of $\mathcal{H}$). Indeed the stability is evident, and we check the anti-Yetter-Drinfeld property: $(m \cdot h)^{(-1)} \otimes (m \cdot h)^{(0)} = \epsilon(h)(m^{(-1)} \otimes m^{(0)})$, and $S(h^{(3)})m^{(-1)}h^{(1)} \otimes m^{(0)} \cdot h^{(2)} = m^{(-1)}h^{(1)}S(h^{(3)}) \otimes \epsilon(h^{(2)})m^{(0)} = m^{(-1)}h^{(1)}S(h^{(2)}) \otimes m^{(0)} = \epsilon(h)m^{(-1)} \otimes m^{(0)}$.

Similarly, if $\mathcal{H}$ is cocommutative then every (right) $\mathcal{H}$-module $\mathcal{M}$ can in a natural way be regarded as a SAYD module with the left coaction, given by $m \mapsto 1 \otimes m$: $S(h^{(1)})\,1\,h^{(3)} \otimes mh^{(2)} = S(h^{(1)})h^{(2)} \otimes mh^{(3)} = 1 \otimes m\epsilon(h^{(1)})h^{(2)} = 1 \otimes mh$. Observe that in this case the Hopf-type Hochschild cohomology of $\mathcal{H}$ with coefficients in $\mathcal{M}$ is related with the $\mathcal{H}$-cotorsion of $\mathbb{C}$ (where $\mathbb{C}$ is given the structure of $\mathcal{H}$-comodule in a trivial way):

$$HH^*_{\mathcal{H}}(\mathcal{H}, \mathcal{M}) = Cotor^*_{\mathcal{H}}(\mathbb{C}, \mathcal{M}) = Cotor^*_{\mathcal{H}}(\mathbb{C}, \mathbb{C} \otimes \mathcal{M}) = Cotor^*_{\mathcal{H}}(\mathbb{C}, \mathbb{C}) \otimes \mathcal{M}. \tag{17}$$

Let us give few examples.

Example. Let $\mathcal{H}$ be equal to $\mathbb{C}[\Gamma]$ the group algebra of a discrete group Γ. Then $\mathcal{H}$ is cocommutative and we can apply the observation from the paragraph above. The $\mathbb{C}[\Gamma]$-modules are the same as representations of the group Γ and we have for a representation V: $HH^0_{\mathbb{C}[\Gamma]}(\mathbb{C}[\Gamma], V) = Cotor^0_{\mathbb{C}[\Gamma]}(\mathbb{C}, \mathbb{C}) \otimes V = V$, and is equal to

0 if $n \neq 0$. Now from the Connes' exact sequence it follows that

$$HC^{2n}_{\mathbb{C}[\Gamma]}(\mathbb{C}[\Gamma],\,V) = V \qquad \text{and} \qquad HC^{2n+1}_{\mathbb{C}[\Gamma]}(\mathbb{C}[\Gamma],\,V) = 0$$

and similarly, for the periodic cohomology

$$HP^{0}_{\mathbb{C}[\Gamma]}(\mathbb{C}[\Gamma],\,V) = V \qquad \text{and} \qquad HP^{1}_{\mathbb{C}[\Gamma]}(\mathbb{C}[\Gamma],\,V) = 0.$$

Example. Let now $\mathcal{C} = \mathcal{H}$ be equal to the universal enveloping algebra $U(\mathfrak{g})$ of a Lie algebra $\mathfrak{g}$. As in the previous example, this Hopf algebra is cocommutative and we can use the the observation, preceding the Example 1. As one knows, the $U(\mathfrak{g})$-modules are in 1-1 correspondence with the representations of the Lie algebra $\mathfrak{g}$.

Proposition 2. For any representation V of the Lie algebra $\mathfrak{g}$:

$$HP^{*}_{U(g)}(U(\mathfrak{g}),\,V) = \bigoplus_{i \equiv * \mathrm{mod} 2} H_i(\mathfrak{g},\,V)$$

where on the right stands the Lie algebra homology of $\mathfrak{g}$ with coefficients in V.

Proof. The proof is obtained by a slight modification of the reasoning in [3]. First, we observe that $HH^{*}_{U(\mathfrak{g})}(U(\mathfrak{g}),\,V) = V \otimes \Lambda^{*}(\mathfrak{g})$ and the isomorphism is given by the formula $\theta : V \otimes \Lambda^{*}\mathfrak{g} \to V \otimes U(\mathfrak{g})^{\otimes n}$:

$$\theta(v \otimes X_1 \wedge X_2 \wedge \cdots \wedge X_n) = \sum_{\sigma \in \Sigma_n} (-1)^{\varepsilon(\sigma)} v \otimes X_{\sigma(1)} \otimes X_{\sigma(2)} \otimes \cdots \otimes X_{\sigma(n)}. \quad (18)$$

In this formula we regard $\mathfrak{g}$ as a subspace in $U(\mathfrak{g})$. Due to the equation (17), the general case follows from the case of trivial module $\mathbb{C}$ over $\mathfrak{g}$. And this is proven in [3].

Now the proof of the proposition follows from the fact that θ intertwines the B differential in the mixed complex, constructed from $C^{*}_{U(\mathfrak{g})}(U(\mathfrak{g}),\,V)$ with the Chevalley differential in the Lie algebra complex $V \otimes \Lambda^{*}(\mathfrak{g})$. □

Example. **Hopf-cyclic cohomology for Hopf fibration.** Let $\mathcal{H} = \mathcal{O}(SU(2))$ and $\mathcal{M} = \mathcal{O}(S^2)$ be algebras of polynomial functions on algebraical varieties $SU(2)$ and S^2. The group product in $SU(2)$ defines a Hopf algebra structure on $\mathcal{H}$ and the action of $SU(2)$ on the sphere $S^2 = SU(2)/U(1)$ induces a left coaction of Hopf algebra $\mathcal{H}$ on $\mathcal{M}$. On the other hand, there is an embedding of S^2 into $SU(2)$ whose image is the subspace of traceless unitary matrices. It defines a homomorphism $\pi : \mathcal{H} \to \mathcal{M}$ and turns $\mathcal{M}$ into a right $\mathcal{H}$-module. One can check easily that the module $\mathcal{M}$ with the action and coaction of $\mathcal{H}$ given above is an SAYD-module. Since $\mathcal{H}$ itself is a $\mathcal{H}$-module coalgebra, one can consider the cyclic cohomology of this coalgebra with coefficients in the module $\mathcal{M}$. The calculation of the cohomology groups $HC^{*}_{\mathcal{H}}(\mathcal{H},\,\mathcal{M})$ will be our main goal now.

Firstly, take the Hopf algebra $\mathcal{H}$ as the coefficient module with (coleft-right) SAYD-module structure given by the following formulas:

$$m \cdot h = mh \qquad \text{for each } m \in \mathcal{H},\ h \in \mathcal{H}$$

$$\Delta_{\mathcal{H}}(m) = S(m^{(3)})m^{(1)} \otimes m^{(2)} \qquad \text{for each } m \in \mathcal{H}.$$

In this case for any $\mathcal{H}$-module coalgebra $\mathcal{C}$ the cyclic module $C^n(\mathcal{C}, \mathcal{H})$ coincides with the usual cyclic module for coalgebra $\mathcal{C}$ due to the natural isomorphism $\mathcal{H} \otimes_{\mathcal{H}} \mathcal{C}^{\otimes n} = \mathcal{C}^{\otimes n}$, which is easily seen to commute with the cyclic structures. So we have the following proposition.

Proposition 3. *For any $\mathcal{H}$-module coalgebra $\mathcal{C}$ one has $HC^*_{\mathcal{H}}(\mathcal{C}, \mathcal{H}) = HC^*(\mathcal{C})$.*

Hence, in the case $\mathcal{C} = \mathcal{H} = \mathcal{O}(SU(2))$ the equality $HC^*_{\mathcal{H}}(\mathcal{H}, \mathcal{H}) = HC^*(\mathcal{O}(SU(2)))$ is true. In order to calculate the latter cohomology group we note that by Peter-Weil Theorem the coalgebra $\mathcal{O}(SU(2))$ splits into a direct sum of matrix coalgebras and each of them is linearly generated by matrix elements of some irreducible representation of the group $SU(2)$. Since the cyclic cohomology is Morita-invariant and commutes with direct sums, one has

$$HC^n(\mathcal{H}) = R(SU(2)) \otimes_{\mathbb{Z}} \mathbb{C}, \quad n \text{ even},$$
$$HC^n(\mathcal{H}) = 0, \quad n \text{ odd},$$

where $R(SU(2))$ is the representation ring of the group $SU(2)$. As it is widely known, $R(SU(2))$ is a free commutative group and has basis τ_n, $n \geq 0$, that consists of the irreducible representations. The representation τ_n has dimension $n+1$.

Consider now the coefficient module $\mathcal{M} = \mathcal{O}(S^2)$. Due to homomorphism theorem one has the equality $\mathcal{M} = \mathcal{H}/\ker\pi$ where $\ker\pi = x_0\mathcal{H}$ and $x_0 \in \mathcal{H}$ is the trace function on $SU(2)$. Then

$$C^*_{\mathcal{H}}(\mathcal{H}, \mathcal{M}) = C^*_{\mathcal{H}}(\mathcal{H}, \mathcal{H})/C^*_{\mathcal{H}}(\mathcal{H}, x_0\mathcal{H}) = C^*(\mathcal{H})/x_0C^*(\mathcal{H}),$$

where the multiplication of x_0 in $C^n(\mathcal{H}) = \mathcal{H}^{\otimes n}$ is constructed by means of the diagonal. Note that this multiplication is compatible with the cyclic module structure on $C^*(\mathcal{H})$ and that the element x_0 is not zero divisor in $\mathcal{H}$. Hence, $HC^*_{\mathcal{H}}(\mathcal{H}, x_0\mathcal{H}) = HC^*_{\mathcal{H}}(\mathcal{H}, \mathcal{H})$ and we have the following cohomology exact sequence:

$$\cdots \to HC^n(\mathcal{H}) \xrightarrow{x_0^*} HC^n(\mathcal{H}) \to HC^n_{\mathcal{H}}(\mathcal{H}, \mathcal{M}) \to \ldots.$$

The isomorphism $HC^{2k}(\mathcal{H}) = R(SU(2)) \otimes \mathbb{C}$ turns the element x_0 into τ_1 and the tensor product of representations is given by the formula:

$$\tau_1 \cdot \tau_n = \tau_{n+1} + \tau_{n-1}.$$

Thus, the map x_0^* is injective and

$$HC^n_{\mathcal{H}}(\mathcal{H}, \mathcal{M}) = \mathbb{C}, \quad n \text{ even},$$
$$HC^n_{\mathcal{H}}(\mathcal{H}, \mathcal{M}) = 0, \quad n \text{ odd}.$$

3. The construction of higher pairing

In this section we give a very brief outline of the construction, giving the analog of pairing between the higher equivariant traces on $\mathcal{H}$ and the Hopf-type cohomology proposed in [3] and generalizing the pairing of the zero-order (Hopf-type) cohomology of a $\mathcal{H}$-module algebra with the cohomology of a $\mathcal{H}$-module coalgebra of the paper [1]. We don't give proofs here and don't address the natural question, whether the pairing that we introduce coincides with that, constructed in the paper [11]. These questions we postpone for a paper to follow ([14]).

We begin with recalling the construction of the non-commutative Weil algebra associated with a coalgebra $\mathcal{C}$. This algebra was first introduced (for $\mathcal{C}$ being a Hopf algebra) in [3] (and independently by the second author in [13]). It is the free (non-commutative) graded unital algebra $W(\mathcal{C})$ generated by degree-1 elements i_h, $h \in \mathcal{C}$, and degree-2 elements w_h, $h \in \mathcal{C}$ (both symbols are linear in h). The differential $\partial = \partial_0 + d$ in $W(\mathcal{C})$ is given by formulae

$$di_h = w_h, \qquad \partial_0 i_h = -i_{h^{(1)}} i_{h^{(2)}}, \tag{19}$$

$$dw_h = 0, \qquad \partial_0 w_h = w_{h^{(1)}} i_{h^{(2)}} - i_{h^{(1)}} w_{h^{(2)}}. \tag{20}$$

One easily proves that $W(\mathcal{C})$ is acyclic. Moreover, even the "commuted" complex associated to this algebra $W(\mathcal{C})_\sharp = W(\mathcal{C})/[W(\mathcal{C}), W(\mathcal{C})]$ is acyclic. In addition, this algebra has the following universal property: for any differential graded algebra Ω and any *linear map* $\rho : \mathcal{C} \to \Omega^1$, there exists a unique *homomorphism of differential graded algebras* $\tilde{\rho} : W(\mathcal{C}) \to \Omega$, such that the restriction of $\tilde{\rho}$ to the degree 1 part of $W(\mathcal{C})$ is given by $\tilde{\rho}(i_h) = \rho(h)$.

There is a natural ideal $I(\mathcal{C})$ inside $W(\mathcal{C})$, namely the ideal, generated by the elements w_h. One denotes by $W_n(\mathcal{C})$ the factor-algebra $W(\mathcal{C})/I^{n+1}(\mathcal{C})$ and by $W_n(\mathcal{C})_\sharp$ the "commuted" complex, associated to it. The cohomology of $W_n(\mathbb{C})_\sharp$ were first computed by Quillen in [12]. In [3] there were calculated the cohomology of the coinvariant space of $W_n(\mathcal{H})_\sharp$ (for a Hopf algebra $\mathcal{H}$) with respect to the diagonal action of $\mathcal{H}$ and a character δ of $\mathcal{H}$. It turned out that (see [3], Theorem 7.3) this cohomology coincides (up to a change of dimensions) with the Connes-Moscovici cohomology $HC^*_{(\delta,\,1)}(\mathcal{H})$ (if $(\delta, 1)$ is a modular pair in involution). The following construction is a direct generalization of this result.

Let $\mathcal{M}$ be a (right-left) SAYD over a Hopf algebra $\mathcal{H}$ and $\mathcal{C}$ – a $\mathcal{H}$-module coalgebra. Consider the complex $W_n(\mathcal{C}, \mathcal{M}) = \mathcal{M} \otimes_{\mathcal{H}} W_n(\mathcal{C})$. Now one can consider the following operators on $W_n(\mathcal{C}, \mathcal{M})$, analogous to those defined by Crainic in [3, Section 8]:

$$\partial_0^{\mathcal{M}} = 1 \otimes \partial_0, \quad d^{\mathcal{M}} = 1 \otimes d, \quad \partial^{\mathcal{M}} = \partial_0^{\mathcal{M}} + d^{\mathcal{M}},$$

$$t^{\mathcal{M}}(m \otimes ax) = (-1)^{|a||x|}(m^{(0)} \otimes x(m^{(-1)} \cdot a)),$$

$$b_t^{\mathcal{M}}(m \otimes ax) = t^{\mathcal{M}}(m \otimes \partial_0(a)x),$$

$$\varphi_1^{\mathcal{M}}(m \otimes i_c x) = 0, \quad \varphi_1^{\mathcal{M}}(m \otimes w_c x) = \frac{1}{n(x)+1} t^{\mathcal{M}}(m \otimes w_c x),$$

$$\varphi_0^{\mathcal{M}}(m \otimes x) = \frac{1}{n(x)} \sum_{i=1}^{|x|} \lambda_i(x)(t^{\mathcal{M}})^i(m \otimes x),$$

where $m \in \mathcal{M}$, $a = i_c$ or w_c, $c \in \mathcal{C}$, $x \in W_n(\mathcal{C})$ and for $x = a_1 \dots a_p$

$$\lambda_i(x) = \#\{j \le i \,|\, a_j \text{ is of type } w_c\}, \quad n(x) = \lambda_p(x).$$

These operators satisfy the relation of [3, Lemma 8.2].

One makes $W_n(\mathcal{C}, \mathcal{M})$ a $W(\mathcal{C})$-bimodule by the following rule: the right multiplication by elements of $W(\mathcal{C})$ is given by the action of $W(\mathcal{C})$ on the second tensor, and the left action is defined by the formula

$$\alpha \cdot (m \otimes \beta) = m^{(0)} \otimes m^{(-1)}(\alpha)\beta, \tag{21}$$

where $\alpha \in W(\mathcal{C})$, $\beta \in W_n(\mathcal{C})$ and $\mathcal{H}$ acts on $W(\mathcal{C})$ and $W_n(\mathcal{C})$ diagonally. This formula is well-defined because $\mathcal{M}$ is a SAYD-module. Denote by $W_n(\mathcal{C}, \mathcal{M})_\sharp$ the factor complex $W_n(\mathcal{C}, \mathcal{M})/[W(\mathcal{C}), W_n(\mathcal{C}, \mathcal{M})]$ and by $HC^*_{\mathcal{H}}(\mathcal{C}, \mathcal{M}; n)$ its cohomology. One can prove the following theorem, analogous to the Theorem 7.3 in [3]:

Theorem 4.

$$HC^*_{\mathcal{H}}(\mathcal{C}, \mathcal{M}; n) = HC^{*-2n}_{\mathcal{H}}(\mathcal{C}, \mathcal{M}). \tag{22}$$

The proof literally coincides with the reasoning in [3, Section 8] after substituting $W_n(\mathcal{C}, \mathcal{M})$ instead of $W_n(\mathcal{C})$ and operators $\partial_0^{\mathcal{M}}$, $d^{\mathcal{M}}, \dots$ defined above instead of Crainic's operators ∂_0, $d, \dots$.

Let now $\mathcal{A}$ be a $\mathcal{C}$-module algebra (i.e., $\mathcal{C}$ acts on $\mathcal{A}$ so that for all $a, b \in \mathcal{A}$, $c \in \mathcal{C}$, $c(ab) = c^{(1)}(a)c^{(2)}(b)$). Let $\mathcal{H}$ act on $\mathcal{A}$ so that this action respects the action of $\mathcal{C}$ (i.e., for all $h \in \mathcal{H}$, $c \in \mathcal{C}$, $a \in \mathcal{A}$, $h(c(a)) = (h(c))(a)$). In this case it is possible to consider both $HC^*_{\mathcal{H}}(\mathcal{C}, \mathcal{M})$ and $HC^*_{\mathcal{H}}(\mathcal{A}, \mathcal{M})$. We are going to define the pairing

$$HC^*_{\mathcal{H}}(\mathcal{C}, \mathcal{M}) \otimes HC^*_{\mathcal{H}}(\mathcal{A}, \mathcal{M}) \to HC^*(\mathcal{A}),$$

where on the right-hand side stands the usual cyclic cohomology of the algebra $\mathcal{A}$.

To this end we shall need a suitable description of cycles in $C^*_{\mathcal{H}}(\mathcal{A}, \mathcal{M})$. So we introduce the notion of higher $\mathcal{H}$-equivariant $\mathcal{M}$-twisted traces on $\mathcal{A}$ ($\mathcal{M}$-traces for short). Let

$$0 \to \mathcal{I} \to \mathcal{R} \to \mathcal{A} \to 0 \tag{23}$$

be an exact sequence of $\mathcal{H}$-algebras over the coalgebra $\mathcal{C}$, splitting as $\mathcal{C}$- and $\mathcal{H}$-module sequence. To define the even $\mathcal{M}$-traces one shall consider the $\mathcal{R}$-module $\mathcal{M} \otimes_{\mathcal{H}} (\mathcal{R}/\mathcal{I}^{n+1})$ where the right action of $\mathcal{R}$ is defined in an evident way and the left action – by a formula similar to (21) (this is well defined because $\mathcal{M}$ is a SAYD-module). By definition degree $2n$ $\mathcal{M}$-traces on $\mathcal{A}$ are the functionals on $\mathcal{M} \otimes_{\mathcal{H}} (\mathcal{R}/\mathcal{I}^{n+1})$ for some choice of the extension (23) vanishing on the subspace generated by the commutators $[\mathcal{M} \otimes_{\mathcal{H}} (\mathcal{R}/\mathcal{I}^{n}), \mathcal{R}]$. We shall denote the space $\mathcal{M} \otimes_{\mathcal{H}} (\mathcal{R}/\mathcal{I}^{n+1})/[\mathcal{M} \otimes_{\mathcal{H}} (\mathcal{R}/\mathcal{I}^{n+1}), \mathcal{R}]$ by $\mathcal{R}_{n,\mathcal{M},\sharp}$.

Similarly, the degree $2n-1$ $\mathcal{M}$-traces on $\mathcal{A}$ are the linear functionals on the $\mathcal{M} \otimes_{\mathcal{H}} \mathcal{I}^n$ vanishing on the subspace $[\mathcal{M} \otimes_{\mathcal{H}} \mathcal{I}^n, \mathcal{I}]$.

The following proposition is similar to the description of cyclic cocycles given in [2].

Proposition 5. *There is an epimorphism from the space of higher $\mathcal{M}$-traces of degree k on the space of cocycles in $C^*_{\mathcal{H}}(\mathcal{A}, \mathcal{M})$. The cohomology classes determined by two cocycles are cohomologous iff the corresponding traces are homotopic* *(here the notion of homotopy is defined by mimicking the corresponding definition in* [15]*).*

Now we can define the pairing. We shall do it only for the even $\mathcal{M}$-traces (odd traces are treated similarly). Let $\tau : \mathcal{M} \otimes_{\mathcal{H}} (\mathcal{R}/\mathcal{I}^{n+1}) \to \mathbb{C}$ be such a trace and $[\tau]$ the cohomology class that it defines. First, we choose a $\mathcal{C}$-linear $\mathcal{H}$-equivariant splitting $\rho : \mathcal{A} \to \mathcal{R}$ in the exact sequence (23). This splitting can be regarded as a map $\bar{\rho} : \mathcal{C} \to \mathrm{Hom}(B(\mathcal{A}), \mathcal{R})$ where $B(\mathcal{A})$ is the bar-resolution of $\mathcal{A}$. The space $\mathrm{Hom}(B(\mathcal{A}), \mathcal{R})$ can be given a structure of the differential graded algebra (differential and grading are induced from $B(\mathcal{A})$ and the algebra structure is determined by the fact that bar-resolution of any algebra always bears the structure of graded colagebra). By the universal property of $W(\mathcal{C})$ this map can be extended to the map $\tilde{\bar{\rho}} : W(\mathcal{C}) \to \mathrm{Hom}(B(\mathcal{A}), \mathcal{R})$. Tensoring this map with $\mathcal{M}$, one obtains a map

$$\tilde{\bar{\rho}}_{\mathcal{M}} : \mathcal{M} \otimes_{\mathcal{H}} W(\mathcal{C}) \to \mathrm{Hom}(B(\mathcal{A}), \mathcal{M} \otimes_{\mathcal{H}} \mathcal{R}).$$

One easily checks that the n+1 power of $I(\mathcal{C})$ is mapped by $\tilde{\bar{\rho}}_{\mathcal{M}}$ to $\mathrm{Hom}(B(\mathcal{A}), \mathcal{M} \otimes_{\mathcal{H}} \mathcal{I}^{n+1})$. Let $B(\mathcal{A})^{\natural}$ denote the cocenter of coalgebra $B(\mathcal{A})$ (i.e., the space of elements in coalgebra, on which the comultiplication is commutative). Then it is easy to see that the commutators $[W(\mathcal{C}), W_n(\mathcal{C}, \mathcal{M})]$ are mapped to $\mathrm{Hom}(B(\mathcal{A})^{\natural}, [\mathcal{M} \otimes_{\mathcal{H}} (\mathcal{R}/\mathcal{I}^{n+1}), \mathcal{R}])$. Thus, we obtain a well-defined map of chain complexes from $W_n(\mathcal{C}, \mathcal{M})_{\sharp}$ to $\mathrm{Hom}(B(\mathcal{A})^{\natural}, \mathcal{R}_{n,\mathcal{M},\sharp})$ which we denote by $\rho_{\sharp,\mathcal{M}}$. Let ω be a degree m cycle in $W_n(\mathcal{C}, \mathcal{M})_{\sharp}$, $[\omega]$ – the cohomology class that it defines. We should associate to ω and τ a degree m cyclic cocycle $\omega \cdot \tau$ of the algebra $\mathcal{A}$. By definition we put

$$(\omega \cdot \rho)(a_0, \dots, a_{m-1}) = \sum_{i=0}^{m-1} (-1)^i \tau(\rho_{\sharp,\mathcal{M}}(\omega)(a_i, \dots, a_{m-1}, a_0, \dots, a_{i-1})). \quad (24)$$

Proposition 6. *The formula* (24) *determines a well-defined map* $HC^*_{\mathcal{H}}(\mathcal{C}, \mathcal{M}) \otimes HC^*_{\mathcal{H}}(\mathcal{A}, \mathcal{M}) \to HC^*(\mathcal{A})$ *(i.e., it does not depend on the choice of the splitting* ρ*, on the choice of* τ *in the homotopy equivalence class and on the choice of* ω *in the class* $[\omega]$*).*

The proof has no difference with the proof of [3, Theorem 7.5].

References

[1] P.M. Hajac, M. Khalkali, B. Rangipour, M. Sommerhäuser, *Hopf-cyclic homology and cohomology with coefficients.* C.R. Math. Acad. Sci. Paris **338** (2004), 667–672.

[2] J. Cuntz, D. Quillen, *Cyclic homology and nonsingularity.* J. Amer. Math. Soc. **8** (1995), 373–442.

[3] M. Crainic, *Cyclic cohomology of Hopf algebras.* J. Pure Appl. Algebra **166** (2002), 29–66.

[4] A. Connes, H. Moscovici, *Hopf algebras, cyclic cohomology and the transverse index theorem.* Commun. Math. Phys. **198** (1998), 199–246.

[5] A. Connes, H. Moscovici, *Cyclic cohomology and Hopf algebras.* Lett. Math. Phys. **52** (2000), 97–108.

[6] R. Taileffer, *Cyclic homology of Hopf algebras.* K-Theory **24** (2001), 69–85.

[7] M. Akbarpour, M. Khalkhali, *Equivariant cyclic cohomology of Hopf module algebras.* J. reine angew. Math. **559** (2003), 137–152.

[8] M. Khalkhali, B. Rangipour, *Invariant cyclic homology.* K-Theory **28** (2003), 183–205.

[9] M. Khalkhali, B. Rangipour, *A new cyclic module for Hopf algebras.* K-theory **27** (2002), 111–131.

[10] J.-L. Loday, *Cyclic homology.* 2-nd Edition, Springer-Verlag, 1998.

[11] M. Khalkhali, B. Rangipour, *Cup Products in Hopf-Cyclic Cohomology.* C.R. Math. Acad. Sci. Paris **340** (2005), 9–14.

[12] D. Quillen, *Chern-Simons forms and cyclic cohomology.* The interface of Mathematics and particle Physics, (Oxford, 1988), 117–134.

[13] G. Sharygin, *A new construction of characteristic classes for noncommutative algebraic principal bundles.* Banach Center Publ. **61** (2003), 219–230.

[14] I. Nikonov, G. Sharygin, *Pairings in Hopf-type cyclic cohomology with coefficients.* (in preparation)

[15] D. Quillen, *Algebra cochains and cyclic cohomology.* Publ. Math. I.H.E.S. **68** (1989), 139–174.

I.M. Nikonov
Dept. of Differential Geometry
Moscow State University
Leninskie Gory
Moscow, 119992, Russia
e-mail: nikonov@mech.math.msu.su

G.I. Sharygin
ITEP
ul. B. Cheremushkinskaja, 25
Moscow, 117218, Russia
e-mail: sharygin@itep.ru

C^*-algebras and Elliptic Theory
Trends in Mathematics, 213–245

The Thom Isomorphism in Gauge-equivariant K-theory

Victor Nistor and Evgenij Troitsky

Abstract. In a previous paper [14], we have introduced the gauge-equivariant K-theory group $K^0_{\mathcal{G}}(X)$ of a bundle $\pi_X : X \to B$ endowed with a continuous action of a bundle of compact Lie groups $p : \mathcal{G} \to B$. These groups are the natural range for the analytic index of a family of gauge-invariant elliptic operators (*i.e.*, a family of elliptic operators invariant with respect to the action of a bundle of compact groups). In this paper, we continue our study of gauge-equivariant K-theory. In particular, we introduce and study products, which helps us establish the Thom isomorphism in gauge-equivariant K-theory. Then we construct push-forward maps and define the topological index of a gauge-invariant family.

Mathematics Subject Classification (2000). Primary 19Kxx; Secondary 46L80, 58Jxx.

Keywords. Elliptic operator, group action, family of operators, index formula, twisted K-theory, Thom isomorphism.

1. Introduction

In this paper we establish a Thom isomorphism theorem for gauge equivariant K-theory. Let $p : \mathcal{G} \to B$ be a bundle of *compact* groups. Recall that this means that each fiber $\mathcal{G}_b := p^{-1}(b)$ is a compact group and that, locally, $\mathcal{G}$ is of the form $U \times G$, where $U \subset B$ open and G a fixed compact group. Let X and B be locally compact spaces and $\pi_X : X \to B$ be a continuous map. In the present paper, as in [14], this map will be supposed to be a locally trivial bundle. A part of present results can be extended to the case of a general map. This, as well as the proof of a general index theorem, will be the subject of [15].

E.T. was partially supported by RFFI Grant 05-01-00923, Grant for the support of leading scientific schools and Grant "Universities of Russia" YP.04.02.530. The present joint research was started under the hospitality of MPIM (Bonn).
V.N. was partially supported by the NSF grant DMS 0200808 (Operator Algebras).

Assume that $\mathcal{G}$ acts on X. This action will be always fiber-preserving. Then we can associate to the action of $\mathcal{G}$ on X *$\mathcal{G}$-equivariant K-theory groups* $K^i_{\mathcal{G}}(X)$ as in [14]. We shall review and slightly generalize this definition in Section 2.

For X compact, the group $K^0_{\mathcal{G}}(X)$ is defined as the Grothendieck group of $\mathcal{G}$-equivariant vector bundles on X. If X is not compact, we define the groups $K^0_{\mathcal{G}}(X)$ using fiberwise one-point compactifications. We shall call these groups simply *gauge-equivariant K-theory groups* of X when we do not want to specify $\mathcal{G}$. The reason for introducing the gauge-equivariant K-theory groups is that they are the natural range for the index of a gauge-invariant families of elliptic operators. In turn, the motivation for studying gauge-invariant families and their index is due to their connection to spectral theory and boundary value problems on non-compact manifolds. Some possible connections with Ramond-Ramond fields in String Theory were mentioned in [8, 14]. See also [1, 9, 12, 13].

In this paper, we continue our study of gauge-equivariant K-theory. We begin by providing two alternative definitions of the relative $K_{\mathcal{G}}$-groups, both based on complexes of vector bundles. (In this paper, all vector bundles are complex vector bundles, with the exception of the tangent bundles and where explicitly stated.) These alternative definitions, modeled on the classical case [2, 10], provide a convenient framework for the study of products, especially in the relative or non-compact cases. The products are especially useful for the proof of the Thom isomorphism in gauge-equivariant theory, which is one of the main results of this paper. Let $E \to X$ be a $\mathcal{G}$-equivariant complex vector bundle. Then the Thom isomorphism is a natural isomorphism

$$\tau_E : K^i_{\mathcal{G}}(X) \to K^i_{\mathcal{G}}(E). \tag{1}$$

(There is also a variant of this result for spin^c-vector bundles, but since we will not need it for the index theorem [15], we will not discuss it in this paper.) The Thom isomorphism allows us to define Gysin (or push-forward) maps in K-theory. As it is well known from the classical work of Atiyah and Singer [4], the Thom isomorphism and the Gysin maps are some of the main ingredients used for the definition and study of the topological index. In fact, we shall proceed along the lines of that paper to define the topological index for gauge-invariant families of elliptic operators. Some other approaches to Thom isomorphism in general settings of Noncommutative geometry were the subject of [6, 7, 11, 16, 12], and many other papers.

Gauge-equivariant K-theory behaves in many ways like the usual equivariant K-theory, but exhibits also some new phenomena. For example, the groups $K^0_{\mathcal{G}}(B)$ may turn out to be reduced to $K^0(B)$ when $\mathcal{G}$ has "a lot of twisting" [14, Proposition 3.6]. This is never the case in equivariant K-theory when the action of the group is trivial but the group itself is not trivial. In [14], we addressed this problem in two ways: first, we found conditions on the bundle of groups $p : \mathcal{G} \to B$ that guarantee that $K^0_{\mathcal{G}}(X)$ is not too small (this condition is called *finite holonomy* and is recalled below), and, second, we studied a substitute of $K^0_{\mathcal{G}}(X)$ which is never too small (this substitute is $K(C^*(\mathcal{G}))$, the K-theory of the C^*-algebra of the bundle of compact groups $\mathcal{G}$).

In this paper, we shall again need the finite holonomy condition, so let us review it now. To define the finite holonomy condition, we introduced the *representation covering* of $\mathcal{G}$, denoted $\widehat{\mathcal{G}} \to B$. As a space, $\widehat{\mathcal{G}}$ is the union of all the representation spaces $\widehat{\mathcal{G}}_b$ of the fibers $\mathcal{G}_b$ of the bundle of compact groups $\mathcal{G}$. One measure of the twisting of the bundle $\mathcal{G}$ is the holonomy associated to the covering $\widehat{\mathcal{G}} \to B$. We say that $\mathcal{G}$ has *representation theoretic finite holonomy* if $\widehat{\mathcal{G}}$ is a union of compact-open subsets. (An equivalent condition can be obtained in terms of the fundamental groups when B is path-connected, see Proposition 2.3 below.)

Let $C^*(\mathcal{G})$ be the enveloping C^*-algebra of the bundle of compact groups $\mathcal{G}$. We proved in [14, Theorem 5.2] that

$$K^j_{\mathcal{G}}(B) \cong K_j(C^*(\mathcal{G})), \tag{2}$$

provided that $\mathcal{G}$ has representation theoretic finite holonomy. This guarantees that $K^j_{\mathcal{G}}(B)$ is not too small. It also points out an alternative, algebraic definition of the groups $K^i_{\mathcal{G}}(X)$.

The structure of the paper is as follows. We start from the definition of gauge-equivariant K-theory and some basic results from [14], most of them related to the "finite holonomy condition," a condition on bundles of compact groups that we recall in Section 2. In Section 3 we describe an equivalent definition of gauge-equivariant K-theory in terms of complexes of vector bundles. This will turn out to be especially useful when studying the topological index. In Section 4 we establish the Thom isomorphism in gauge-equivariant K-theory, and, in Section 5, we define and study the Gysin maps. The properties of the Gysin maps allow us to define in Section 6 the topological index and establish its main properties.

2. Preliminaries

We now recall the definition of gauge-equivariant K-theory and some basic results from [14]. An important part of our discussion is occupied by the discussion of the finite holonomy condition for a bundle of compact groups $p : \mathcal{G} \to B$.

> *All vector bundles in this paper are assumed to be* **complex** *vector bundles, unless otherwise mentioned and excluding the tangent bundles to the various manifolds appearing below.*

2.1. Bundles of compact groups and finite holonomy conditions

We begin by introducing bundles of compact and locally compact groups. Then we study finite holonomy conditions for bundles of compact groups.

Definition 2.1. Let B be a locally compact space and let G be a locally compact group. We shall denote by $\mathrm{Aut}(G)$ the group of automorphisms of G. A *bundle of locally compact groups $\mathcal{G}$ with typical fiber G over B* is, by definition, a fiber bundle $\mathcal{G} \to B$ with typical fiber G and structural group $\mathrm{Aut}(G)$.

We fix the above notation. Namely, *from now on and throughout this paper, unless explicitly otherwise mentioned, B will be a compact space and $\mathcal{G} \to B$ will be a bundle of compact groups with typical fiber G.*

We need now to introduce the representation theoretic holonomy of a bundle of Lie group with compact fibers $p : \mathcal{G} \to B$. Let $\mathrm{Aut}(G)$ be the group of automorphisms of G. By definition, there exists then a principal $\mathrm{Aut}(G)$-bundle $\mathcal{P} \to B$ such that

$$\mathcal{G} \cong \mathcal{P} \times_{\mathrm{Aut}(G)} G := (\mathcal{P} \times G)/\,\mathrm{Aut}(G).$$

We shall fix $\mathcal{P}$ in what follows.

Let $\widehat{\mathcal{G}}$ be the (disjoint) union of the sets $\widehat{\mathcal{G}}_b$ of equivalence classes of irreducible representations of the groups $\mathcal{G}_b$. Using the natural action of $\mathrm{Aut}(G)$ on $\widehat{G}$, we can naturally identify $\widehat{\mathcal{G}}$ with $\mathcal{P} \times_{\mathrm{Aut}(G)} \widehat{G}$ as fiber bundles over B.

Let $\mathrm{Aut}_0(G)$ be the connected component of the identity in $\mathrm{Aut}(G)$. The group $\mathrm{Aut}_0(G)$ will act trivially on the set $\widehat{G}$, because the later is discrete. Let

$$H_R := \mathrm{Aut}(G)/\,\mathrm{Aut}_0(G), \quad \mathcal{P}_0 := \mathcal{P}/\,\mathrm{Aut}_0(G), \quad \text{and} \quad \widehat{\mathcal{G}} \simeq \mathcal{P}_0 \times_{H_R} \widehat{G}.$$

Above, $\widehat{\mathcal{G}}$ is defined because $\mathcal{P}_0$ is an H_R-principal bundle. The space $\widehat{\mathcal{G}}$ will be called the *representation space of* $\mathcal{G}$ and the covering $\widehat{\mathcal{G}} \to B$ will be called *the representation covering associated to* $\mathcal{G}$.

Assume now that B is a path-connected, locally simply-connected space and fix a point $b_0 \in B$. We shall denote, as usual, by $\pi_1(B, b_0)$ the fundamental group of B. Then the bundle $\mathcal{P}_0$ is classified by a morphism

$$\rho : \pi_1(B, b_0) \to H_R := \mathrm{Aut}(G)/\,\mathrm{Aut}_0(G), \tag{3}$$

which will be called *the holonomy of the representation covering of* $\mathcal{G}$.

For our further reasoning, we shall sometimes need the following finite holonomy condition.

Definition 2.2. We say that $\mathcal{G}$ has *representation theoretic finite holonomy* if every $\sigma \in \widehat{\mathcal{G}}$ is contained in a compact-open subset of $\widehat{\mathcal{G}}$.

In the cases we are interested in, the above condition can be reformulated as follows [14].

Proposition 2.3. *Assume that B is path-connected and locally simply-connected. Then $\mathcal{G}$ has* representation theoretic finite holonomy *if, and only if* $\pi_1(B, b_0)\sigma \subset \widehat{G}$ *is a finite set for any irreducible representation σ of G.*

From now on we shall assume that $\mathcal{G}$ has representation theoretic finite holonomy.

2.2. Gauge-equivariant K-theory

Let us now define the gauge equivariant K-theory groups of a "$\mathcal{G}$-fiber bundle" $\pi_Y : Y \to B$. All our definitions are well known if B is reduced to a point (cf. [2, 10]). First we need to fix the notation.

If $f_i : Y_i \to B$, $i = 1, 2$, are two maps, we shall denote by

$$Y_1 \times_B Y_2 := \{(y_1, y_2) \in Y_1 \times Y_2,\ f_1(y_1) = f_2(y_2)\} \tag{4}$$

their fibered product. Let $p : \mathcal{G} \to B$ be a bundle of locally compact groups and let $\pi_Y : Y \to B$ be a continuous map. We shall say that $\mathcal{G}$ *acts on* Y if each group $\mathcal{G}_b$ acts continuously on $Y_b := \pi^{-1}(b)$ and the induced map μ

$$\mathcal{G} \times_B Y := \{(g, y) \in \mathcal{G} \times Y,\ p(g) = \pi_Y(y)\} \ni (g, y) \longrightarrow \mu(g, y) := gy \in Y$$

is continuous. If $\mathcal{G}$ acts on Y, we shall say that Y is *a $\mathcal{G}$-space.* If, in addition to that, $Y \to B$ is also locally trivial, we shall say that Y is *a $\mathcal{G}$-fiber bundle*, or, simply, a *$\mathcal{G}$-bundle.* This definition is a particular case of the definition of the action of a groupoid on a space.

Let $\pi_Y : Y \to B$ be a $\mathcal{G}$-space, with $\mathcal{G}$ a bundle of compact groups over B. Recall that a vector bundle $\tilde{\pi}_E : E \to Y$ is *a $\mathcal{G}$-equivariant vector bundle* (or simply *a $\mathcal{G}$-equivariant vector bundle*) if

$$\pi_E := \pi_Y \circ \tilde{\pi}_E : E \to B$$

is a $\mathcal{G}$-space, the projection

$$\tilde{\pi}_E : E_b := \pi_E^{-1}(b) \to Y_b := \pi_Y^{-1}(b)$$

is $\mathcal{G}_b := p^{-1}(b)$ equivariant, and the induced action $E_y \to E_{gy}$ of $g \in \mathcal{G}$, between the corresponding fibers of $E \to Y$, is linear for any $y \in Y_b$, $g \in \mathcal{G}_b$, and $b \in B$.

To define gauge-equivariant K-theory, we first recall some preliminary definitions from [14]. Let $\tilde{\pi}_E : E \to Y$ be a $\mathcal{G}$-equivariant vector bundle and let $\tilde{\pi}_{E'} : E' \to Y'$ be a $\mathcal{G}'$-equivariant vector bundle, for two bundles of compact groups $\mathcal{G} \to B$ and $\mathcal{G}' \to B'$. We shall say that $(\gamma, \phi, \eta, \psi) : (\mathcal{G}', E', Y', B') \to (\mathcal{G}, E, Y, B)$ is a *γ-equivariant morphism of vector bundles* if the following five conditions are satisfied:

(i) $\gamma : \mathcal{G}' \to \mathcal{G}$, $\phi : E' \to E$, $\eta : Y' \to Y$, and $\psi : B \to B'$,
(ii) all the resulting diagrams are commutative,
(iii) $\phi(ge) = \gamma(g)\phi(e)$ for all $e \in E'_b$ and all $g \in \mathcal{G}'_b$,
(iv) γ is a group morphism in each fiber, and
(v) ϕ is a vector bundle morphism.

We shall say that $\phi : E \to E'$ *is a γ-equivariant morphism of vector bundles* if, by definition, it is part of a morphism $(\gamma, \phi, \eta, \psi) : (\mathcal{G}', E', Y', B') \to (\mathcal{G}, E, Y, B)$. Note that η and ψ are determined by γ and ϕ.

As usual, if $\psi : B' \to B$ is a continuous [respectively, smooth] map, we define the *inverse image* $(\psi^*(\mathcal{G}), \psi^*(E), \psi^*(Y), B')$ of a $\mathcal{G}$-equivariant vector bundle $E \to Y$ by $\psi^*(\mathcal{G}) = \mathcal{G} \times_B B'$, $\psi^*(E) = E \times_B B'$, and $\psi^*(Y) = Y \times_B B'$. If $B' \subset B$ and ψ is the embedding, this construction gives the restriction of a $\mathcal{G}$-equivariant vector bundle $E \to Y$ to a closed, invariant subset $B' \subset B$ of the base of $\mathcal{G}$, yielding a $\mathcal{G}_{B'}$-equivariant vector bundle. Usually $\mathcal{G}$ will be fixed, however.

Let $p : \mathcal{G} \to B$ be a bundle of compact groups and $\pi_Y : Y \to B$ be a $\mathcal{G}$-space. The set of isomorphism classes of $\mathcal{G}$-equivariant vector bundles $\tilde{\pi}_E : E \to Y$ will be denoted by $\mathcal{E}_{\mathcal{G}}(Y)$. On this set we introduce a monoid operation, denoted "+," using the direct sum of vector bundles. This defines a monoid structure on the set $\mathcal{E}_{\mathcal{G}}(Y)$ as in the case when B consists of a point.

Definition 2.4. Let $\mathcal{G} \to B$ be a bundle of compact groups acting on the $\mathcal{G}$-space $Y \to B$. Assume Y to be compact. The *$\mathcal{G}$-equivariant K-theory group* $K^0_{\mathcal{G}}(Y)$ is defined as the group completion of the monoid $\mathcal{E}_{\mathcal{G}}(Y)$.

When working with gauge-equivariant K-theory, we shall use the following terminology and notation. If $E \to Y$ is a $\mathcal{G}$-equivariant vector bundle on Y, we shall denote by $[E]$ its class in $K^0_{\mathcal{G}}(Y)$. Thus $K^0_{\mathcal{G}}(Y)$ consists of differences $[E] - [E^1]$. The groups $K^0_{\mathcal{G}}(Y)$ will also be called *gauge equivariant* K-theory groups, when we do not need to specify $\mathcal{G}$. If B is reduced to a point, then $\mathcal{G}$ is group, and the groups $K^0_{\mathcal{G}}(Y)$ reduce to the usual equivariant K-groups.

We have the following simple observations on gauge-equivariant K-theory. First, the familiar functoriality properties of the usual equivariant K-theory groups extend to the gauge equivariant K-theory groups. For example, assume that the bundle of compact groups $\mathcal{G} \to B$ acts on a fiber bundle $Y \to B$ and that, similarly, $\mathcal{G}' \to B'$ acts on a fiber bundle $Y' \to B'$. Let $\gamma : \mathcal{G} \to \mathcal{G}'$ be a morphism of bundles of compact groups and $f : Y \to Y'$ be a γ-equivariant map. Then we obtain a natural group morphism

$$(\gamma, f)^* : K^0_{\mathcal{G}'}(Y') \to K^0_{\mathcal{G}}(Y). \tag{5}$$

If γ is the identity morphism, we shall denote $(\gamma, f)^* = f^*$.

A $\mathcal{G}$-equivariant vector bundle $E \to Y$ on a $\mathcal{G}$-space $Y \to B$, Y compact, is called *trivial* if, by definition, there exists a $\mathcal{G}$-equivariant vector bundle $E' \to B$ such that E is isomorphic to the pull-back of E' to Y. Thus $E \simeq Y \times_B E'$. If $\mathcal{G} \to B$ has representation theoretic finite holonomy and Y is a compact $\mathcal{G}$-bundle, then every $\mathcal{G}$-equivariant vector bundle over Y can be embedded into a trivial $\mathcal{G}$-equivariant vector bundle. This embedding will necessarily be as a direct summand.

If $\mathcal{G} \to B$ does not have finite holonomy, it is possible to provide examples if $\mathcal{G}$-equivariant vector bundles that do not embed into trivial $\mathcal{G}$-equivariant vector bundles [14]. Also, a related example from [14] shows that the groups $K^0_{\mathcal{G}}(Y)$ can be fairly small if the holonomy of $\mathcal{G}$ is "large."

A further observation is that it follows from the definition that the tensor product of vector bundles defines a natural ring structure on $K^0_{\mathcal{G}}(Y)$. We shall denote the product of two elements a and b in this ring by $a \otimes b$ or, simply, ab, when there is no danger of confusion. In particular the groups $K^i_{\mathcal{G}}(X)$ for $\pi_X : X \to B$ are equipped with a natural structure of $K^0_{\mathcal{G}}(B)$-module obtained using the pull-back of vector bundles on B, namely, $ab := \pi_X^*(a) \otimes b \in K^0_{\mathcal{G}}(X)$ for $a \in K^0_{\mathcal{G}}(B)$ and $b \in K^0_{\mathcal{G}}(X)$.

The definition of the gauge-equivariant groups extends to non-compact $\mathcal{G}$-spaces Y as in the case of equivariant K-theory. Let Y be a $\mathcal{G}$-bundle. We shall denote then by $Y^+ := Y \cup B$ the compact space obtained from Y by the one-point compactification of each fiber (recall that B is compact). The need to consider the space Y^+ is the main reason for considering also non longitudinally smooth fibers bundles on B. Then

$$K^0_{\mathcal{G}}(Y) := \ker\big(K^0_{\mathcal{G}}(Y^+) \to K^0_{\mathcal{G}}(B)\big).$$

Also as in the classical case, we let

$$K^n_{\mathcal{G}}(Y, Y') := K^0_{\mathcal{G}}((Y \setminus Y') \times \mathbb{R}^n)$$

for a $\mathcal{G}$-subbundle $Y' \subset Y$. Then [14] we have the following periodicity result

Theorem 2.5. *We have natural isomorphisms*

$$K^n_{\mathcal{G}}(Y, Y') \cong K^{n-2}_{\mathcal{G}}(Y, Y').$$

Gauge-equivariant K-theory is functorial with respect to open embeddings. Indeed, let $U \subset X$ be an open, $\mathcal{G}$-equivariant subbundle. Then the results of [14, Section 3] provide us with a natural map morphism

$$i_* : K^n_{\mathcal{G}}(U) \to K^n_{\mathcal{G}}(X). \tag{6}$$

In fact, i_* is nothing but the composition $K^n_{\mathcal{G}}(U) \cong K^n_{\mathcal{G}}(X, X \setminus U) \to K^n_{\mathcal{G}}(X)$.

2.3. Additional results

We now prove some more results on gauge-equivariant K-theory.

Let $\mathcal{G} \to B$ and $\mathcal{H} \to B$ be two bundles of compact groups over B. Recall that an $\mathcal{H}$-bundle $\pi_X : X \to B$ is called *free* if the action of each group $\mathcal{H}_b$ on the fiber X_b is free (*i.e.*, $hx = x$, $x \in X_b$, implies that h is the identity of $\mathcal{G}_b$.) We shall need the following result, which is an extension of a result in [10, page 69]. For simplicity, we shall write $\mathcal{G} \times \mathcal{H}$ instead of $\mathcal{G} \times_B \mathcal{H}$.

Theorem 2.6. *Suppose $\pi_X : X \to B$ is a $\mathcal{G} \times \mathcal{H}$-bundle that is free as an $\mathcal{H}$-bundle. Let $\pi : X \to X/\mathcal{H}$ be the (fiberwise) quotient map. For any $\mathcal{G}$-equivariant vector bundle $\tilde{\pi}_E : E \to X/\mathcal{H}$, we define the* induced vector bundle

$$\pi^*(E) := \big\{(x, \varepsilon) \in X \times E,\ \pi(x) = \tilde{\pi}_E(\varepsilon)\big\} \to X,$$

with the action of $\mathcal{G} \times \mathcal{H}$ given by $(g, h) \cdot (x, \varepsilon) := ((g, h)x, g\varepsilon)$. Then π^* gives rise to a natural isomorphism $K^0_{\mathcal{G}}(X/\mathcal{H}) \to K^0_{\mathcal{G} \times \mathcal{H}}(X)$.

Proof. Let $\pi^* : K^0_{\mathcal{G}}(X/\mathcal{H}) \to K^0_{\mathcal{G} \times \mathcal{H}}(X)$ be the induction map, as above. We will construct a map $r : K^0_{\mathcal{G} \times \mathcal{H}}(X) \to K^0_{\mathcal{G}}(X/\mathcal{H})$ satisfying $\pi^* \circ r = \mathrm{Id}$ and $r \circ \pi^* = \mathrm{Id}$. Let $\pi_F : F \to X$ be a $\mathcal{G} \times \mathcal{H}$-vector bundle. Since the action of $\mathcal{H}$ on X is free, the induced map $\overline{\pi}_F : F/\mathcal{H} \to X/\mathcal{H}$ of quotient spaces is a (locally trivial) $\mathcal{G}$-bundle. Clearly, this construction is invariant under homotopy, and hence we can define $r[F] := [F/\mathcal{H}]$.

Let us check now that r is indeed an inverse of π^*. Denote by $F \ni f \to \mathcal{H}f \in F/\mathcal{H}$ the quotient map. Let $F \to X$ be a $\mathcal{G} \times \mathcal{H}$-vector bundle. To begin with, the total space of $\pi^* \circ r(F)$ is

$$\big\{(x, \mathcal{H}f) \in X \times (F/\mathcal{H}),\ \pi^*(x) = \overline{\pi}_F(\mathcal{H}f)\big\},$$

by definition. Then the map $F \ni f \to (\pi_F(f), \mathcal{H}f) \in \pi^* \circ r(F)$ is an isomorphism. Hence, $\pi^* r = \mathrm{Id}$.

Next, consider a $\mathcal{G}$-vector bundle $\pi_E : E \to X/\mathcal{H}$. The total space of $r \circ \pi^*(E)$ is then

$$\{(\mathcal{H}y, \varepsilon) \in (X/\mathcal{H}) \times E,\ \mathcal{H}y = \pi_E(\varepsilon))\},$$

because $\mathcal{H}$ acts only on the first component of $\pi^*(E)$. Then

$$E \to r \circ \pi^*(E), \qquad \varepsilon \mapsto (\pi(\varepsilon), \varepsilon)$$

is an isomorphism. Hence, $r \circ \pi^* = \mathrm{Id}$. □

See also [14, Theorem 3.5].

Corollary 2.7. *Let P be a $\mathcal{G} \times \mathcal{H}$-bundle that is free as an $\mathcal{H}$-bundle. Also, let W be an $\mathcal{H}$-bundle, then there is a natural isomorphism*

$$K_{\mathcal{G}\times\mathcal{H}}(P \times_B W) \cong K_{\mathcal{G}}(P \times_{\mathcal{H}} W).$$

Proof. Take $X := P \times_B W$ in the previous theorem. Then X is a free $\mathcal{H}$-bundle, because P is, and $X/\mathcal{H} =: P \times_{\mathcal{H}} W$. □

In the following section, we shall also need the following quotient construction associated to a trivialization of a vector bundle over a subset. Namely, if $Y \subset X$ is a $\mathcal{G}$-invariant, closed subbundle, then we shall denote by $X/_BY$ the fiberwise quotient space over B, that is the quotient of X with respect to the equivalence relation $\sim$, $x \sim y$ if, and only if, $x, y \in Y_b$, for some $b \in B$.

If E is a $\mathcal{G}$-equivariant vector bundle over a $\mathcal{G}$-bundle X, together with a $\mathcal{G}$-equivariant trivialization over a $\mathcal{G}$-subbundle $Y \subset X$, then we can generalize the quotient (or collapsing) construction of [2, §1.4] to obtain a vector bundle over $X/_BY$, where by $X/_BY$ we denote the fiberwise quotient bundle over B, as above.

Lemma 2.8. *Suppose that X is a $\mathcal{G}$-bundle and that $Y \subset X$ is a closed, $\mathcal{G}$-invariant subbundle. Let $E \to X$ be a $\mathcal{G}$-vector bundle and $\alpha : E|_Y \cong Y \times_B \mathcal{V}$ be a $\mathcal{G}$-equivariant trivialization, where $\mathcal{V} \to B$ is a $\mathcal{G}$-equivariant vector bundle. Then we can naturally associate to (E, α) a naturally defined vector bundle $E/\alpha \to X/_BY$ that depends only on homotopy class of α.*

Proof. Let $p : Y \times_B \mathcal{V} \to \mathcal{V}$ be the natural projection. Introduce the following equivalence relation on E:

$$e \sim e' \Leftrightarrow e, e' \in E|_Y \text{ and } p\alpha(e) = p\alpha(e').$$

Let then E/α be equal to $E/\sim$. This is locally trivial vector bundle over $X/_BY$. Indeed, it is necessary to verify this only in a neighborhood of $Y/_BY \cong B$. Let U be a $\mathcal{G}$ invariant open subset of X such that α can be extended to an isomorphism $\widetilde{\alpha} : E|_U \cong U \times_B \mathcal{V}$. We obtain an isomorphism

$$\alpha' : (E|_U)/\alpha \cong (U/_BY) \times_B \mathcal{V}, \qquad \alpha'(e) = \widetilde{\alpha}(e).$$

Moreover, $(U/_BY) \times_B \mathcal{V}$ is a locally trivial $\mathcal{G}$-equivariant vector bundle.

Suppose that α_0 and α_1 are homotopic trivializations of $E|_Y$, that is, trivializations such that there exists a trivialization $\beta : E \times I|_{Y\times I} \cong Y \times I \times_B \mathcal{V}$, $\beta|_{E\times\{0\}} = \alpha_0$ and $\beta|_{E\times\{1\}} = \alpha_1$. Let

$$f : X/_BY \times I \to (X \times I)/_{(B\times I)}(Y \times I).$$

Then the bundle $f^*((E \times I)/\beta)$ over $X/_BY \times I$ satisfies

$$f^*((E \times I)/\beta)|_{(X/_BY)\times\{i\}} = E/\alpha_i, \quad i = 0, 1.$$

Hence, $E/\alpha_0 \cong E/\alpha_1$. □

3. K-theory and complexes

For the purpose of defining the Thom isomorphism, it is convenient to work with an equivalent definition of gauge-equivariant K-theory in terms of complexes of vector bundles. This will turn out to be especially useful when studying the topological index.

The statements and proofs of this section, except maybe Lemma 3.4, follow the classical ones [2, 10], so our presentation will be brief.

3.1. The $L^n_{\mathcal{G}}$-groups

We begin by adapting some well-known concepts and constructions to our settings.

Let $X \to B$ be a locally compact, paracompact $\mathcal{G}$-bundle. A *finite complex of $\mathcal{G}$-equivariant vector bundles* over X is a complex

$$(E^*; d) = \Big(\dots \xrightarrow{d_{i-1}} E^i \xrightarrow{d_i} E^{i+1} \xrightarrow{d_{i+1}} \dots\Big), \quad i \in \mathbb{Z},$$

of $\mathcal{G}$-equivariant vector bundles over X with only finitely many E^i's different from zero. Explicitly, E^i are $\mathcal{G}$-equivariant vector bundles, d_i's are $\mathcal{G}$-equivariant morphisms, $d_{i+1}d_i = 0$ for every i, and $E^i = 0$ for $|i|$ large enough. We shall also use the notation $(E^*; d) = \big(E^0, \dots, E^n,\ d_i : E^i|_Y \to E^{i+1}|_Y\big)$, if $E^i = 0$ for $i < 0$ and for $i > n$.

As usual, a *morphism of complexes* $f : (E^*; d) \to (F^*; \delta)$ is a sequence of morphisms $f_i : E^i \to F^i$ such that $f_{i+1}d_i = \delta_{i+1}f_i$, for all i. These constructions yield the category of finite complexes of $\mathcal{G}$-equivariant vector bundles. Isomorphism in this category will be denoted by $(E^*; d) \cong (F^*; \delta)$.

In what follows, we shall consider a pair (X, Y) of $\mathcal{G}$-bundles with X is a compact $\mathcal{G}$-bundle, unless explicitly otherwise mentioned.

Definition 3.1. Let X be a compact $\mathcal{G}$-bundle and Y be a closed $\mathcal{G}$-invariant subbundle. Denote by $C^n_{\mathcal{G}}(X, Y)$ the set of (isomorphism classes of) sequences

$$(E^*; d) = \big(E^0, E^1, \dots, E^n;\ d_k : E^k|_Y \to E^{k+1}|_Y\big)$$

of $\mathcal{G}$-equivariant vector bundles over X such that $(E^k|_Y; d)$ is exact if we let $E^j = 0$ for $j < 0$ or $j > n$.

We endow $C^n_{\mathcal{G}}(X,Y)$ with the semigroup structure given by the direct sums of complexes. An element in $C^n_{\mathcal{G}}(X,Y)$ is called *elementary* if it is isomorphic to a complex of the form

$$\ldots \to 0 \to E \xrightarrow{\text{Id}} E \to 0 \to \ldots,$$

Two complexes $(E^*;d),(F^*,\delta) \in C^n_{\mathcal{G}}(X,Y)$ are called *equivalent* if, and only if, there exist elementary complexes $Q^1,\ldots,Q^k,P^1,\ldots,P^m \in C^n_{\mathcal{G}}(X,Y)$ such that

$$E \oplus Q_1 \oplus \cdots \oplus Q_k \cong F \oplus P_1 \oplus \cdots \oplus P_m.$$

We write $E \simeq F$ in this case. The semigroup of equivalence classes of sequences in $C^n_{\mathcal{G}}(X,Y)$ will be denoted by $L^n_{\mathcal{G}}(X,Y)$.

We obtain, from definition, natural injective semigroup homomorphisms

$$C^n_{\mathcal{G}}(X,Y) \to C^{n+1}_{\mathcal{G}}(X,Y) \quad \text{and} \quad C_{\mathcal{G}}(X,Y) := \bigcup_n C^n_{\mathcal{G}}(X,Y).$$

The equivalence relation $\sim$ commutes with embeddings, so the above morphisms induce morphisms $L^n_{\mathcal{G}}(X,Y) \to L^{n+1}_{\mathcal{G}}(X,Y)$. Let $L^\infty_{\mathcal{G}}(X,Y) := \varinjlim L^n_{\mathcal{G}}(X,Y)$.

Lemma 3.2. *Let $E \to X$ and $F \to X$ be $\mathcal{G}$-vector bundles. Let $\alpha : E|_Y \to F|_Y$ and $\beta : E \to F$ be surjective morphisms of $\mathcal{G}$-equivariant vector bundles. Also, assume that α and $\beta|_Y$ are homotopic in the set of surjective $\mathcal{G}$-equivariant vector bundle morphisms. Then there exists a surjective morphism of $\mathcal{G}$-equivariant vector bundles $\widetilde{\alpha} : E \to F$ such that $\widetilde{\alpha}|_Y = \alpha$. The same result remains true if we replace "surjective" with "injective" or with "isomorphism" everywhere.*

Proof. Let $Z := (Y \times [0,1]) \cup (X \times \{0\})$ and $\pi : Z \to X$ be the projection. Let $\pi^*(E) \to Z$ and $\pi^*(F) \to Z$ be the pull-backs of E and F. The homotopy in the statement of the Lemma defines a surjective morphism $a : \pi^*(E) \to \pi^*(F)$ such that $a|_{Y\times\{1\}} = \alpha$ and $a|_{X\times\{0\}} = \beta$. By [14, Lemma 3.12], the morphism a can be extended to a surjective morphism over $(U \times [0,1]) \cup (X \times \{0\})$, where U is an open $\mathcal{G}$-neighborhood of Y. (In fact, in that Lemma we considered only the case of an isomorphism, but the case of a surjective morphism is proved in the same way.) Let $\phi : X \to [0,1]$ be a continuous function such that $\phi(Y) = 1$ and $\phi(X \setminus U) = 0$. By averaging, we can assume ϕ to be $\mathcal{G}$-equivariant. Then define $\widetilde{\alpha}(x) = a(x,\phi(x))$, for all $x \in X$. □

Remark 3.3. Suppose that X is a compact $\mathcal{G}$-space and $Y = \emptyset$. Then we have a natural isomorphism $\chi_1 : L^1_{\mathcal{G}}(X,\emptyset) \to K^0_{\mathcal{G}}(X)$ taking the class of $(E^0,E^1;0)$ to the element $[E^0] - [E^1]$.

We shall need the following lemma.

Lemma 3.4. *Let $p : \mathcal{G} \to B$ be a bundle of compact groups and $\pi_X : X \to B$ be a compact $\mathcal{G}$-bundle. Assume that π_X has a cross-section, which we shall use to identify B with a subset of X. Then the sequence*

$$0 \to L^1_{\mathcal{G}}(X,B) \to L^1_{\mathcal{G}}(X) \to L^1_{\mathcal{G}}(B)$$

is exact.

Proof. Suppose that $E = (E^1, E^0; \phi)$ defines an element of $L^1_{\mathcal{G}}(X)$ such that its image in $L^1_{\mathcal{G}}(B)$ is zero. Then the definition of $E \sim 0$ in $L^1_{\mathcal{G}}(B)$ shows that the restrictions of E^1 and E^0 to B are isomorphic over B. Hence, the above sequence is exact at $L^1_{\mathcal{G}}(X)$.

Suppose now that $(E^1, E^0; \phi)$ represents a class in $L^1_{\mathcal{G}}(X, B)$ such that its image in $L^1_{\mathcal{G}}(X)$ is zero. This means (keeping in mind Remark 3.3) that there exists a $\mathcal{G}$-equivariant vector bundle $\widetilde{P}$ and an isomorphism $\widetilde{\psi} : E^1 \oplus \widetilde{P} \cong E^0 \oplus \widetilde{P}$. Let us define $P := \widetilde{P} \oplus \pi_X^*(E^0|_B) \oplus \pi_X^*(\widetilde{P}|_B)$, where $\pi_X : X \to B$ is the canonical projection, as in the statement of the Lemma. Also, define $\psi = \widetilde{\psi} \oplus \mathrm{Id} : E^1 \oplus P \to E^0 \oplus P$, which is also an isomorphism.

We thus obtain that $T := \psi(\phi \oplus \mathrm{Id})^{-1}$ is an automorphism of $(E^0 \oplus P)|_B$, which has the form $\begin{pmatrix} \beta & 0 \\ 0 & \mathrm{Id} \end{pmatrix}$ with respect to the decomposition

$$(E^0 \oplus P)|_B = (E^0 \oplus \widetilde{P})|_B \oplus (E^0 \oplus \widetilde{P})|_B.$$

The automorphism $T := \psi(\phi \oplus \mathrm{Id})^{-1}$ is homotopic to the automorphism T_1 defined by the matrix

$$\begin{pmatrix} \mathrm{Id} & 0 \\ 0 & \beta \end{pmatrix}.$$

Since T_1 extends to an automorphism of $E^0 \oplus P$ over X, namely $\begin{pmatrix} \mathrm{Id} & 0 \\ 0 & \pi_X^*(\beta) \end{pmatrix}$, Lemma 3.2 gives that the automorphism $\psi(\phi \oplus \mathrm{Id})^{-1}$ also can be extended to X, such that over B we have the following commutative diagram:

$$\begin{array}{ccc} (E^1 \oplus P)|_B & \xrightarrow{\phi \oplus \mathrm{Id}} & (E^0 \oplus P)|_B \\ \Big\downarrow{\scriptstyle \psi|_B} & & \Big\downarrow{\scriptstyle \alpha|_B = \begin{pmatrix} \beta & 0 \\ 0 & \mathrm{Id} \end{pmatrix}} \\ (E^0 \oplus P)|_B & \xrightarrow{\mathrm{Id}} & (E^0 \oplus P)|_B. \end{array}$$

Hence, $(E^1, E^0, \phi) \oplus (P, P, \mathrm{Id}) \cong (E^0 \oplus P, E^0 \oplus P, \mathrm{Id})$ and so is zero in $L^1_{\mathcal{G}}(X, B)$. □

3.2. Euler characteristics

We now generalize the above construction to other groups $L^n_{\mathcal{G}}$, thus proving the existence and uniqueness of Euler characteristics.

Definition 3.5. Let X be a compact $\mathcal{G}$-space and $Y \subset X$ be a $\mathcal{G}$-invariant subset. An *Euler characteristic* χ_n is a natural transformation of functors $\chi_n : L^n_{\mathcal{G}}(X, Y) \to K^0_{\mathcal{G}}(X, Y)$, such that for $Y = \emptyset$ it takes the form

$$\chi_n(E) = \sum_{i=0}^{n} (-1)^i [E^i],$$

for any sequence $E = (E^*; d) \in L^n_{\mathcal{G}}(X, Y)$.

Lemma 3.6. *There exists a unique natural transformation of functors* (i.e., *an Euler characteristic*)

$$\chi_1 : L^1_{\mathcal{G}}(X,Y) \to K^0_{\mathcal{G}}(X,Y),$$

which, for $Y = \emptyset$, *has the form indicated in* 3.3.

Proof. To prove the uniqueness, suppose that χ_1 and χ_1' are two Euler characteristics on $L^1_{\mathcal{G}}$. Then $\chi_1'\chi_1^{-1}$ is a natural transformation of $K^0_{\mathcal{G}}$ that is equal to the identity on each $K^0_{\mathcal{G}}(X)$. Let us consider the long exact sequence

$$\cdots \to K^{n-1}_{\mathcal{G}}(Y,Y') \to K^{n-1}_{\mathcal{G}}(Y) \to K^{n-1}_{\mathcal{G}_1}(Y') \\ \to K^n_{\mathcal{G}}(Y,Y') \to K^n_{\mathcal{G}}(Y) \to K^n_{\mathcal{G}}(Y') \to \dots \quad (7)$$

associated to a pair (Y,Y) of $\mathcal{G}$-bundles (see [14], Equation (10) for a proof of the exactness of this sequence). The map $K^0_{\mathcal{G}}(X,B) \to K^0_{\mathcal{G}}(X)$ from this exact sequence is induced by $(X,\emptyset) \to (X,B)$, and hence, in particular, it is natural. Assume that $\pi_X : X \to B$ has a cross-section. Then the exact sequence (7) for $(Y,Y') = (X,B)$ yields a natural exact sequence $0 \to K^0_{\mathcal{G}}(X,B) \to K^0_{\mathcal{G}}(X)$. This in conjunction with Lemma 3.4 shows that $\chi_1'\chi_1^{-1}$ is the identity on $K^0_{\mathcal{G}}(X,B)$. (Recall that we agreed to denote by $X/_BY$ the fiberwise quotient space over B, that is the quotient of X with respect to the equivalence relation $\sim$, $x \sim y$ if, and only if, $x,y \in Y_b$, for some $b \in B$.) Finally, since the map $(X,Y) \to (X/_BY,B)$ induces an isomorphism of $K^0_{\mathcal{G}}$-groups [14, Theorem 3.19], $\chi_1'\chi_1^{-1}$ is the identity on $K^0_{\mathcal{G}}(X,Y)$ for all pairs (X,Y).

To prove the existence of the Euler characteristic χ_1, let $(E^1,E^0,\alpha) := (\alpha : E^1 \to E^0)$ represent an element of $L^1_{\mathcal{G}}(X,Y)$. Suppose that X_0 and X_1 are two copies of X and $Z := X_0 \cup_Y X_1 \to B$ is the $\mathcal{G}$-bundle obtained by identifying the two copies of $Y \subset X_i$, $i = 0,1$. The identification of $E^1|_Y$ and $E^0|_Y$ with the help of α gives rise to an element $[F^0] - [F^1] \in K^0_{\mathcal{G}}(Z)$ defined as follows. By adding some bundle to both E^i's, we can assume that E^1 is trivial (that is, it is isomorphic to the pull-back of a vector bundle on B). Then E^1 extends to a trivial $\mathcal{G}$-vector bundle $\widetilde{E}^1 \to Z$. We define $F^0 := E^0 \cup_\alpha E^1$ and $F^1 := \widetilde{E}^1$.

The exact sequence (7) and the natural $\mathcal{G}$-retractions $\pi_i : Z \to X_i$, give natural direct sum decompositions

$$K^0_{\mathcal{G}}(Z) = K^0_{\mathcal{G}}(Z,X_i) \oplus K^0_{\mathcal{G}}(X_i), \qquad i = 0,1. \quad (8)$$

The natural map $(X_0,Y) \to (Z,X_1)$ induces an isomorphism

$$k : K^0_{\mathcal{G}}(Z,X_1) \to K^0_{\mathcal{G}}(X_0,Y).$$

Let us define then $\chi_1(E^0,E^1,\alpha)$ to be equal to the image under k of the $K^0_{\mathcal{G}}(Z,X_1)$-component of $(E^0,E^1;\alpha)$ (with respect to (8)). It follows from its definition that this map is natural, respects direct sums, and is independent with respect to the addition of elementary elements. Our proof is completed by observing that $\chi_1(E^1,E^0;\alpha) = [E^0] - [E^1]$ when $Y = \emptyset$. □

We shall also need the following continuity property of the functor $L^1_{\mathcal{G}}$. Recall that we have agreed to denote by $X/_BY$ the fiberwise quotient bundle over B.

Lemma 3.7. *The natural homomorphism*

$$\Pi^* : L^1_{\mathcal{G}}(X/_BY, Y/_BY) = L^1_{\mathcal{G}}(X/_BY, B) \to L^1_{\mathcal{G}}(X, Y)$$

is an isomorphism for all pairs (X, Y) of compact $\mathcal{G}$-bundles.

Proof. Lemmata 3.6 and 3.4 give the following commutative diagram

$$\begin{array}{ccc} L^1_{\mathcal{G}}(X/_BY, B) & \xrightarrow{\Pi^*} & L^1_{\mathcal{G}}(X, Y) \\ \cong \big\downarrow \chi_1 & & \big\downarrow \chi_1 \\ K^0_{\mathcal{G}}(X/_BY, B) & \xrightarrow[\cong]{\Pi^*} & K^0_{\mathcal{G}}(X, Y). \end{array}$$

From this we obtain injectivity.

To prove surjectivity, suppose that E^1 and E^0 are $\mathcal{G}$-equivariant vector bundles over X and $\alpha : E^1|_Y \to E^0|_Y$ is an isomorphism of the restrictions. Let $P \to X$ be a $\mathcal{G}$-bundle such that there is an isomorphism $\beta : E^1 \oplus P \cong F$, where F is a trivial bundle (*i.e.*, isomorphic to a pull back from B). Then $(E^1, E^0, \alpha) \sim (F, E^0 \oplus P, \gamma)$, where $\gamma = (\alpha \oplus \mathrm{Id})\beta^{-1}$. The last object is the image of $(F, (E^0 \oplus P)/\gamma, \gamma/\gamma)$ (see Lemma 2.8). □

We obtain the following corollaries.

Corollary 3.8. *The Euler characteristic $\chi_1 : L^1_{\mathcal{G}}(X, Y) \to K^0_{\mathcal{G}}(X, Y)$ is an isomorphism and hence it defines an equivalence of functors.*

Proof. This follows from Lemmas 3.7 and 3.6. □

Lemma 3.9. *The class of (E^1, E^0, α) in $L^1_{\mathcal{G}}(X, Y)$ depends only on the homotopy class of the isomorphism α.*

Proof. Let $Z = X \times [0, 1]$, $W = Y \times [0, 1]$. Denote by $p : Z \to X$ the natural projection and assume that α_t is a homotopy, where $\alpha_0 = \alpha$. Then α_t gives rise an isomorphism $\beta : p^*(E^1)|_W \cong p^*(E^0)|_W$, and hence to an element $(p^*(E^1), p^*(E^0), \beta)$ of $L^1_{\mathcal{G}}(Z, W)$. If

$$i_t : (X, Y) \to (X \times \{t\}, Y \times \{t\}) \subset (Z, W), \quad t \in [0, 1],$$

are the standard inclusions, then $(E^1, E^0, \alpha_t) = i_t^*(p^*(E^1), p^*(E^0), \beta)$. Consider the commutative diagram

$$\begin{array}{ccccc} L^1_{\mathcal{G}}(X, Y) & \xleftarrow{i_0^*} & L^1_{\mathcal{G}}(Z, W) & \xrightarrow{i_1^*} & L^1_{\mathcal{G}}(X, Y) \\ \big\downarrow \chi_1 & & \big\downarrow \chi_1 & & \big\downarrow \chi_1 \\ K^0_{\mathcal{G}}(X, Y) & \xleftarrow{i_0^*} & K^0_{\mathcal{G}}(Z, W) & \xrightarrow{i_1^*} & K^0_{\mathcal{G}}(X, Y). \end{array}$$

The vertical morphisms and and the morphisms of the bottom line of the above diagram are isomorphisms. Hence, the arrows of the top line are isomorphisms too. The composition $i_0^*(i_1^*)^{-1}$ is identity for the bottom line, hence it is the identity for the top line too. □

The following theorem reduces the study of the functors $L^n_{\mathcal{G}}$, $n > 1$, to the study of $L^1_{\mathcal{G}}$.

Theorem 3.10. *The natural map $j_n : L^n_{\mathcal{G}}(X,Y) \to L^{n+1}_{\mathcal{G}}(X,Y)$ is an isomorphism.*

Proof. Let $E = (E^0, E^1, \dots, E^{n+1}; d_k)$, $d_k : E^k|_Y \to E^{k+1}|_Y$ represent an element of the semigroup $L^{n+1}_{\mathcal{G}}(X,Y)$. To prove the surjectivity of j_n, let us first notice that E is equivalent to the complex

$$(E^0, \dots, E^{n-2}, E^{n-1} \oplus E^{n+1}, E^n \oplus E^{n+1}, E^{n+1}; \\ d_0, \dots, d_{n-2} \oplus 0, d_{n-1} \oplus \mathrm{Id}, d_n \oplus 0).$$

The maps $d_n \oplus 0 : (E^n \oplus E^{n+1})|_Y \to E^{n+1}|_Y$ and $0 \oplus \mathrm{Id} : (E^n \oplus E^{n+1})|_Y \to E^{n+1}|_Y$ are homotopic within the set of surjective, $\mathcal{G}$-equivariant vector bundle morphisms $(E^n \oplus E^{n+1})|_Y \to E^{n+1}|_Y$. Hence, by Lemma 3.2, $d_n \oplus 0$ can be extended to a surjective morphism $b : E^n \oplus E^{n+1} \to E^{n+1}$ of $\mathcal{G}$-equivariant vector bundles (over the whole of X). So, the bundle $E^n \oplus E^{n+1}$ is isomorphic to $\ker(b) \oplus E^{n+1}$. Hence, the E is equivalent to

$$(E^0, \dots, E^{n-2}, E^{n-1} \oplus E^{n+1}, \ker(b), 0; d_0, \dots, d_{n-2} \oplus 0, d_{n-1}, 0).$$

This proves the surjectivity of j_n.

To prove the injectivity of j_n, it is enough to define, for any n, a left inverse $q_n : L^n_{\mathcal{G}}(X,Y) \to L^1_{\mathcal{G}}(X,Y)$ to $s_n := j_{n-1} \circ \dots \circ j_1$. Suppose that $(E^*; d)$ represents an element of semigroup $L^n_{\mathcal{G}}(X,Y)$. Choose $\mathcal{G}$-invariant Hermitian metrics on E^i and let $d_i^* : E^{i+1}|_Y \to E^i|_Y$ be the adjoint of d_i. Let

$$F^0 := \bigoplus_i E^{2i}, \quad F^1 := \bigoplus_i E^{2i+1}, \quad b : F^0|_Y \to F^1|_Y, \quad b = \sum_i (d_{2i} + d^*_{2i+1}).$$

A standard verification shows that b is an isomorphism. Since all invariant metrics are homotopic to each other, Lemma 3.9 shows that $(E; d) \to (F, b)$ defines a morphism $q_n : L^n_{\mathcal{G}}(X,Y) \to L^1_{\mathcal{G}}(X,Y)$. This is the desired left inverse for s_n. □

Let us observe that the proof of the above theorem and Lemma 3.9 give the following corollary.

Corollary 3.11. *The class of $E = (E^i, d_i)$ in $L^n_{\mathcal{G}}(X,Y)$ does not change if we deform the differentials d_i continuously.*

We are now ready to prove the following basic result.

Theorem 3.12. *For each n there exists a unique Euler characteristic*

$$\chi_n : L^n_{\mathcal{G}}(X,Y) \cong K^0_{\mathcal{G}}(X,Y).$$

In particular, $L^\infty_{\mathcal{G}}(X,Y) \cong K^0_{\mathcal{G}}(X,Y)$ *and* $L^n_{\mathcal{G}}(X,Y)$ *has a natural group structure for any closed,* $\mathcal{G}$*-invariant subbundle* $Y \subset X$.

Proof. The statement is obtained from the lemmas we have proved above as follows. First of all, Theorem 3.10 allows us to define

$$\chi_n := \chi_1 \circ j_1^{-1} \circ \dots j_{n-1}^{-1} : L^n_{\mathcal{G}}(X,Y) \to K^0_{\mathcal{G}}(X,Y).$$

Lemma 3.7 shows that χ_n is an isomorphism. The uniqueness of χ_n is proved in the same way as the uniqueness of χ_1 (Lemma 3.6). □

3.3. Globally defined complexes

Theorem 3.12 provides us with an alternative definition of the groups $K^0_{\mathcal{G}}(X,Y)$. We now derive yet another definition of these groups that is closer to what is needed in applications and is based on differentials defined on X, not just on Y.

Let $(E;d)$ be a complex of $\mathcal{G}$-equivariant vector bundles over a $\mathcal{G}$-space X. A point $x \in X$ will be called a *point of acyclicity of* $(E;d)$ if the restriction of $(E;d)$ to x, *i.e.*, the sequence of linear spaces

$$(E;d)_x = \Big(\dots \xrightarrow{(d_i)_x} E^i_x \xrightarrow{(d_{i+1})_x} E^{i+1}_x \xrightarrow{(d_{i+2})_x} \dots\Big),$$

is exact. The *support* $\operatorname{supp}(E;d)$ of the finite complex $(E;d)$ is the complement in X of the set of its points of acyclicity. This definition and the following lemma hold also for X non-compact.

Lemma 3.13. *The support* $\operatorname{supp}(E;d)$ *is a closed* $\mathcal{G}$*-invariant subspace of* X.

Proof. The fact that $\operatorname{supp}(E;d)$ is closed is classical (see [2, 10] for example). The invariance should be checked up over one fiber of X at $b \in B$. But this is once again a well-known fact of equivariant K-theory (see e.g. [10]). □

Lemma 3.14. *Let* $E^n, \dots, E^0$ *be* $\mathcal{G}$*-equivariant vector bundles over* X *and let* Y *be a closed,* $\mathcal{G}$*-invariant subbundle of* X*. Suppose there are given morphisms* $d_i : E^i|_Y \to E^{i-1}|_Y$ *such that* $(E^i|_Y, d_i)$ *is an exact complex. Then the morphisms* d_i *can be extended to morphisms defined over* X *such that we still have a complex of* $\mathcal{G}$*-equivariant vector bundles.*

Proof. We will show that we can extend each d_i to a morphism $r_i : E^i \to E^{i-1}$ such that $r_{i-1} \circ r_i = 0$. Let us find a $\mathcal{G}$-invariant open neighborhood U of Y in X such that for any i there exists an extension s_i of d_i to U with (E,s) still an exact sequence. The desired r_i will be defined then as $r_i = \rho s_i$, where $\rho : X \to [0,1]$ is a continuous function such that $\rho = 1$ on Y and $\operatorname{supp} \rho \subset U$.

Let us construct U by induction over i. Assume that for the closure $\overline{U}_i$ of some open $\mathcal{G}$-neighborhood of Y in X we can extend d_j to s_j, $j = 1, \dots, i$, such that, on $\overline{U}_i$, the sequence

$$E^i \xrightarrow{s_i} E^{i-1} \xrightarrow{s_{i-1}} \dots \to E^0 \to 0$$

is exact. Suppose $K_i := \ker(s_i|\overline{U}_i)$. Then d_{i+1} determines a cross section of the bundle $\operatorname{Hom}_{\mathcal{G}}(E^i, K_i)|_Y$. This section can be extended to an open $\mathcal{G}$-neighborhood

V of Y in $\overline{U}_i$. We hence obtain an extension $s_{i+1} : E^{i+1} \to K_i$ of $d_{i+1} : E^{i+1} \to K_i$ over V. Since $d_{i+1}|_Y$ is surjective (with range K_i), the morphism s_{i+1} will be surjective $\overline{U}_{i+1}$ for some open $U_{i+1} \subset U_i$. □

The above lemma suggests the following definition.

Definition 3.15. Let X be a compact $\mathcal{G}$-bundle and $Y \subset X$ be a $\mathcal{G}$-invariant subbundle. We define $E^n_{\mathcal{G}}(X, Y)$ to be the semigroup of homotopy classes of complexes of $\mathcal{G}$-equivariant vector bundles of length n over X such that their restrictions to Y are acyclic (*i.e.*, exact).

We shall say that two complexes are *homotopic* if they are isomorphic to the restrictions to $X \times \{0\}$ and $X \times \{1\}$ of a complex defined over $X \times I$ and acyclic over $Y \times I$.

Remark 3.16. By Corollary 3.11, the restriction of morphisms induces a morphism $\Phi^n : E^n_{\mathcal{G}}(X, Y) \to L^n_{\mathcal{G}}(X, Y)$.

Theorem 3.17. *Let X be a compact $\mathcal{G}$-bundle and $Y \subset X$ be a $\mathcal{G}$-invariant subbundle. Then the natural transformation Φ_n, defined in the above remark, is an isomorphism.*

Proof. The surjectivity of Φ_n follows from 3.14. The injectivity of Φ_n can be proved in the same way as [2, Lemma 2.6.13], keeping in mind Lemma 3.14.

More precisely, we need to demonstrate that differentials of any complex over $\mathcal{G}$-subbundle $(X \times \{0\}) \cup (X \times \{1\}) \cup (Y \times I)$ of $X \times I$, which is acyclic over $Y \times I$, can be extended to a complex over the entire $X \times I$. The desired construction has the following three stages. First, let V be a $\mathcal{G}$-invariant neighborhood of Y such that the restriction of our complex is still acyclic on $(V \times \{0\}) \cup (V \times \{1\}) \cup (Y \times I)$ as well as on its closure $(\bar{V} \times \{0\}) \cup (\bar{V} \times \{1\}) \cup (Y \times I)$. By Lemma 3.14, one can extend the differentials d_i to $\mathcal{G}$-equivariant morphisms r_i over $\bar{V} \times I$ that still define a complex. Second, let ρ_1, ρ_2 be a $\mathcal{G}$-invariant partition of unity subordinated to the covering $\{V \times I, (X \setminus Y) \times I\}$ of $X \times Y$. Let us extend original differentials to

$$(X \times [0, 1/4]) \cup (X \times [3/4, 1]) \cup (V \times I)$$

by taking $d_i(x, t) := d_i(x, 0)$ for $t \leq \frac{1}{4} \cdot \rho_2(x)$, $x \in X \setminus Y$. We proceed similarly near $t = 1$. Also,

$$d_i(x, t) := r_i\left(x, \frac{(t - \frac{1}{4} \cdot \rho_2(x))}{(1 - \frac{1}{2} \cdot \rho_2(x))}\right), \quad x \in V.$$

Finally, by multiplying the differential d_i with a function $\tau : X \times I \to I$ that is equal to 1 on the original subset of definition of the differential and is equal to 0 outside

$$(X \times [0, 1/4]) \cup (X \times [3/4, 1]) \cup (V \times I),$$

we obtain the desired extension. □

3.4. The non-compact case

In the case of a locally compact, paracompact $\mathcal{G}$-bundle X, we change the definitions of $L^n_{\mathcal{G}}$ and $E^n_{\mathcal{G}}$ as follows. In the definition of $L^n_{\mathcal{G}}$, the morphisms d_i have to be defined and to form an exact sequence off the interior of some compact $\mathcal{G}$-invariant subset C of $X \smallsetminus Y$ (the complement of Y in X). In the definition of $E^n_{\mathcal{G}}$, the complexes have to be exact outside some compact $\mathcal{G}$-invariant subset of $X \smallsetminus Y$. In other words, $L^n_{\mathcal{G}}(X, Y) = L^n_{\mathcal{G}}(X^+, Y^+)$.

Since the proof of Lemma 3.14 is still valid, we have the analogue of Theorem 3.17: there is a natural isomorphism

$$L^n_{\mathcal{G}}(X, Y) \cong E^n_{\mathcal{G}}(X, Y).$$

The proof of the other statement can be extended also to the non-compact case. The only difference is that we have to replace Y with $X \setminus U$, where U is an open, $\mathcal{G}$-invariant subset with compact closure. Then, when we study two element sequences $E = (E^i, d_i)$, we have to take the unions of the corresponding open sets. Of course, these sets are not bundles, unlike Y, but for our argument using extensions this is not a problem. This ultimately gives

$$K^0_{\mathcal{G}}(X, Y) \cong L^n_{\mathcal{G}}(X, Y) \cong E^n_{\mathcal{G}}(X, Y), \quad n \geq 1. \tag{9}$$

As we shall see below, the liberty of using these equivalent definitions of $K^0_{\mathcal{G}}(X, Y)$ is quite convenient in applications, especially when studying products.

4. The Thom isomorphism

In this section, we establish the Thom isomorphism in gauge-equivariant K-theory. We begin with a discussion of products and of the Thom morphism.

4.1. Products

Let $\pi_X : X \to B$ be a $\mathcal{G}$-space, $\tilde{\pi}_F : F \to X$ be a complex $\mathcal{G}$-vector bundle over X, and $s : X \to F$ be a $\mathcal{G}$-invariant section. We shall denote by $\Lambda^i F$ the ith exterior power of F, which is again a complex $\mathcal{G}$-equivariant vector bundle over X. As in the proof of the Thom homomorphism for ordinary vector bundles, we define the complex $\Lambda(F, s)$ of $\mathcal{G}$-equivariant vector bundles over X by

$$\Lambda(F, s) := (0 \to \Lambda^0 F \xrightarrow{\alpha^0} \Lambda^1 F \xrightarrow{\alpha^1} \ldots \xrightarrow{\alpha^{n-1}} \Lambda^n F \to 0), \tag{10}$$

where $\alpha^k(v_x) = s(x) \wedge v_x$ for $v_x \in \Lambda^k F^x$ and $n = \dim F$. It is immediate to check that $\alpha^{j+1}(x)\alpha^j(x) = 0$, and hence that $(\Lambda(F, s), \alpha)$ is indeed a complex.

The Künneth formula shows that the complex $\Lambda(F, s)$ is acyclic for $s(x) \neq 0$, and hence $\operatorname{supp}(\Lambda(F, s)) := \{x \in X | s(x) = 0\}$. If this set is compact, then the results of Section 3 will associate to the complex $\Lambda(F, s)$ of Equation (10) an element

$$[\Lambda(F, s)] \in K^0_{\mathcal{G}}(X). \tag{11}$$

Let X be a $\mathcal{G}$-bundle and $\pi_F : F \to X$ be a $\mathcal{G}$-equivariant vector bundle over X. The point of the above construction is that $\pi_F^*(F)$, the lift of F back to itself, has a canonical section whose support is X. Let us recall how this is defined. Let $\pi_{FF} : \pi_F^*(F) \to F$ be the $\mathcal{G}$-vector bundle over F with total space

$$\pi_F^*(F) := \{(f_1, f_2) \in F \times F,\ \pi_F(f_1) = \pi_F(f_2)\}$$

and $\pi_{FF}(f_1, f_2) = f_1$. The vector bundle $\pi_{FF} : \pi_F^*(F) \to F$ has the canonical section

$$s_F : F \to \pi_F^* F, \qquad s_F(f) = (f, f).$$

The support of s_F is equal to X. Hence, if X is a compact space, using again the results of Section 3, especially 3.17, we obtain an element

$$\lambda_F := [\Lambda(\pi_F^*(F), s_F)] \in K^0_{\mathcal{G}}(F). \tag{12}$$

Recall that the tensor product of vector bundles defines a natural product $ab = a \otimes b \in K^0_{\mathcal{G}}(X)$ for any $a \in K^0_{\mathcal{G}}(B)$ and any $b \in K^0_{\mathcal{G}}(X)$, where $\pi_X : X \to B$ is a compact $\mathcal{G}$-space, as above.

Recall that all our vector bundles are assumed to be complex vector bundles, except for the ones coming from geometry (tangent bundles, their exterior powers) and where explicitly mentioned. Due to the importance that F be complex in the following definition, we shall occasionally repeat this assumption.

Definition 4.1. Let $\pi_F : F \to X$ be a (complex) $\mathcal{G}$-equivariant vector bundle. Assume the $\mathcal{G}$-bundle $X \to B$ is compact and let $\lambda_F \in K^0_{\mathcal{G}}(F)$ be the class defined in Equation (12), then the mapping

$$\phi^F : K^0_{\mathcal{G}}(X) \to K^0_{\mathcal{G}}(F), \qquad \phi^F(a) = \pi_F^*(a) \otimes \lambda_F,$$

is called the *Thom morphism.*

As we shall see below, the definition of the Thom homomorphism extends to the case when X is not compact, although the Thom element itself is not defined if X is not compact.

The definition of the Thom homomorphism immediately gives the following proposition. We shall use the notation of Proposition 4.1.

Proposition 4.2. *The Thom morphism $\phi^F : K^0_{\mathcal{G}}(X) \to K^0_{\mathcal{G}}(F)$ is a morphism of $K^0_{\mathcal{G}}(B)$-modules.*

Let $\iota : X \hookrightarrow F$ be the zero section embedding of X into F. Then ι induces homomorphisms

$$\iota^* : K^0_{\mathcal{G}}(F) \to K^0_{\mathcal{G}}(X) \quad \text{and} \quad \iota^* \circ \phi^F : K^0_{\mathcal{G}}(X) \to K^0_{\mathcal{G}}(X).$$

It follows from the definition that $\iota^* \phi^F(a) = a \cdot \sum_{i=0}^n (-1)^i \Lambda^i F$.

4.2. The non-compact case

We now consider the case when X is locally compact, but not necessarily compact. The complex $\Lambda(\pi_F^*(F), s_F)$ has a non-compact support, and hence it does not define an element of $K_{\mathcal{G}}^0(F)$. However, if $a = [(E, \alpha)] \in K_{\mathcal{G}}^0(X)$ is represented by the complex (E, α) of vector bundles with compact support (Section 3), then we can still consider the tensor product complex

$$\big(\pi_F^*(\mathcal{E}),\, \pi_F^*(\alpha)\big) \otimes \Lambda(\pi_F^* F, s_F).$$

From the Künneth formula for the homology of a tensor product we obtain that the support of a tensor product complex is the intersection of the supports of the two complexes. In particular, we obtain

$$\begin{aligned}\operatorname{supp}\{(\pi_F^* E, \pi_F^*\alpha) \otimes \Lambda(\pi_F^* F, s_F)\} \subset \operatorname{supp}(\pi_F^* E, \pi_F^*\alpha) \cap \operatorname{supp}\Lambda(\pi_F^* F, s_F) \subset \\ \subset \operatorname{supp}(\pi_F^* E, \pi_F^*\alpha) \cap X = \operatorname{supp}(E, \alpha). \quad (13)\end{aligned}$$

Thus, the complex $(\pi_F^*\mathcal{E}, \pi_E^*\alpha) \otimes \Lambda(\pi_F^* F, s_F)$ has compact support and hence defines an element in $K_{\mathcal{G}}^0(F)$.

Proposition 4.3. *The homomorphism of $K_{\mathcal{G}}^0(B)$-modules*

$$\phi^F : K_{\mathcal{G}}^0(X) \to K_{\mathcal{G}}^0(F), \quad \phi^F(a) = [(\pi_F^*\mathcal{E}, \pi_F^*\alpha) \otimes \Lambda(\pi_F^* F, s_F)], \tag{14}$$

defined in Equation (13) *extends the Thom morphism to the case of not necessarily compact X. The Thom morphism ϕ^F satisfies*

$$i^*\phi^F(a) = a \cdot \sum_{i=0}^{n} (-1)^i \Lambda^i F \tag{15}$$

in the non-compact case as well.

Let $F \to X$ be a $\mathcal{G}$-equivariant vector bundle and $F^1 = F \times \mathbb{R}$, regarded as a vector bundle over $X \times \mathbb{R}$. The periodicity isomorphisms in gauge-equivariant K-theory groups [14, Theorem 3.18]

$$K_{\mathcal{G}}^{i \pm 1}(X \times \mathbb{R}, Y \times \mathbb{R}) \simeq K_{\mathcal{G}}^i(X, Y)$$

can be composed with ϕ^{F^1}, the Thom morphism for F^1, giving a morphism

$$\phi^F : K_{\mathcal{G}}^i(X) \to K_{\mathcal{G}}^i(F)\,, \quad i = 0, 1. \tag{16}$$

This morphism is the Thom morphism for K^1.

Let $p_X : X \to B$ and $p_Y : Y \to B$ be two *compact* $\mathcal{G}$-fiber bundles. Let $\pi_E : E \to X$ and $\pi_F : F \to Y$ be two complex $\mathcal{G}$-equivariant vector bundles. Denote by $p_1 : X \times_B Y \to X$ and by $p_2 : X \times_B Y \to Y$ the projections onto the two factors and define $E \boxtimes F := p_1^* E \otimes p_2^* F$. The $\mathcal{G}$-equivariant vector bundle $E \boxtimes F$ will be called the *external tensor product of E and F over B*. It is a vector bundle over $X \times_B Y$. Then the formula

$$K_{\mathcal{G}}^0(X) \otimes K_{\mathcal{G}}^0(Y) \ni [E] \otimes [F] \to [E] \boxtimes [F] := [E \boxtimes F] \in K_{\mathcal{G}}^0(X \times_B Y)$$

defines a product $K_{\mathcal{G}}^0(X) \otimes K_{\mathcal{G}}^0(Y) \to K_{\mathcal{G}}^0(X \times_B Y)$.

In particular, consider two complex $\mathcal{G}$-equivariant vector bundles $\pi_E : E \to X$ and $\pi_F : F \to X$. Then $E \oplus F = E \times_X F$ and we obtain a product

$$K^i_{\mathcal{G}}(E) \otimes K^j_{\mathcal{G}}(F) \ni [E] \otimes [F] \to [E] \boxtimes [F] := [E \boxtimes F] \to K^{i+j}_{\mathcal{G}}(E \oplus F).$$

Using also periodicity, we obtain the product

$$\boxtimes : K^i_{\mathcal{G}}(E) \otimes K^j_{\mathcal{G}}(F) \to K^{i+j}_{\mathcal{G}}(E \oplus F). \tag{17}$$

This product is again seen to be given by the tensor product of the (lifted) complexes (when representing K-theory classes by complexes) as in the classical case.

The external product $\boxtimes$ behaves well with respect to the "Thom construction" in the following sense. Let F^1 and F^2 be two complex bundles over X, and s_1, s_2 two corresponding sections of these bundles. Then

$$\Lambda(F^1 \oplus F^2, s_1 \otimes 1 + 1 \otimes s_2) = \Lambda(F^1, s_1) \boxtimes \Lambda(F^2, s_2). \tag{18}$$

In particular, if X is compact, we obtain

$$\lambda_E \boxtimes \lambda_F = \lambda_{E \oplus F}. \tag{19}$$

We shall write $s_1 + s_2 = s_1 \otimes 1 + 1 \otimes s_2$, for simplicity.

The following theorem states that the Thom class is multiplicative with respect to direct sums of vector bundles (see also [5]).

Theorem 4.4. *Let $E, F \to X$ be two $\mathcal{G}$-equivariant vector bundles, and regard $E \oplus F \to E$ as the $\mathcal{G}$-equivariant vector bundle $\pi_E^*(F)$ over E. Then $\phi^{\pi_E^*(F)} \circ \phi^E = \phi^{E \oplus F}$.*

The above theorem amounts to the commutativity of the diagram

$$\begin{array}{ccc} K^*_{\mathcal{G}}(X) & \xrightarrow{\ \phi^E\ } & K^*_{\mathcal{G}}(E) \\ & {\scriptstyle \phi^{E\oplus F}} \searrow \quad \swarrow {\scriptstyle \phi^{\pi_E^*(F)}} & \\ & K^*_{\mathcal{G}}(E \oplus F) & \end{array} \tag{20}$$

Proof. Let $F^1 := \pi_E^*(F) = E \oplus F$, regarded as a vector bundle over E. Consider the projections

$$\pi_E : E \to X, \quad \pi_F : F \to X, \quad \pi_{E \oplus F} : E \oplus F \to X, \quad t = \pi_{F^1} : E \oplus F \to E.$$

Let $x \in K^0_{\mathcal{G}}(X)$. Then $\phi^E(x) = \pi_E^*(x) \otimes \Lambda(\pi_E^*(E), s_E)$. Now we use that $t^* \pi_E^*(x) = \pi^*_{E \oplus F}(x)$ and $t^* \Lambda(\pi_E^*(E), s_E) = \Lambda(\pi^*_{E \oplus F}(E), s_E \circ t)$. Since $s_E \circ t + s_{F^1} = s_{E \oplus F}$, Equations (18) and (19) then give

$$\Lambda(\pi^*_{E \oplus F}(E \oplus F), s_{E \oplus F}) = \Lambda(\pi^*_{E \oplus F}(E), s_E \circ t) \otimes \Lambda(\pi^*_{E \oplus F}(F), s_{F^1}).$$

Putting together the above calculations we obtain

$$\begin{aligned}\phi^{F^1}\phi^E(x) &= t^*(\phi^E(x)) \otimes \Lambda(t^*(F^1), s_{F^1}) \\ &= t^*\pi_E^*(x) \otimes t^*(\Lambda(\pi_E^*(E), s_E)) \otimes \Lambda(t^*(F^1), s_{F^1}) \\ &= \pi_{E\oplus F}^*(x) \otimes \Lambda(\pi_{E\oplus F}^* E, s_E \circ t) \otimes \Lambda(\pi_{E\oplus F}^* F, s_{F^1}) \\ &= \pi_{E\oplus F}^*(x) \otimes \Lambda(\pi_{E\oplus F}^*(E \oplus F), s_{E\oplus F}) = \phi_1^{E\oplus F}(x).\end{aligned}$$

The proof is now complete. □

We are now ready to formulate and prove the Thom isomorphism in the setting of gauge-equivariant vector bundles. Recall that the Thom morphism was introduced in Definition 4.1.

Theorem 4.5. *Let $X \to B$ be a $\mathcal{G}$-bundle and $F \to X$ a complex $\mathcal{G}$-equivariant vector bundle, then $\phi^F : K^i_{\mathcal{G}}(X) \to K^i_{\mathcal{G}}(F)$ is an isomorphism.*

Proof. Assume first that F is a trivial bundle, that is, that $F = X \times_B \mathcal{V}$, where $\mathcal{V} \to B$ is a complex, finite-dimensional $\mathcal{G}$-equivariant vector bundle. We continue to assume that B is compact.

Let us denote by $\underline{\mathbb{C}} := B \times \mathbb{C}$ the 1-dimensional $\mathcal{G}$-bundle with the trivial action of $\mathcal{G}$ on $\mathbb{C}$. Also, let us denote by $P(\mathcal{V} \oplus \underline{\mathbb{C}})$ the projective space associated to $\mathcal{V} \oplus \underline{\mathbb{C}}$. As a topological space, $P(\mathcal{V} \oplus \underline{\mathbb{C}})$ identifies with the fiberwise one-point compactification of $\mathcal{V}$. The embeddings $\mathcal{V} \subset P(\mathcal{V}\oplus\underline{\mathbb{C}})$ and $\mathcal{V}\times_B X \subset P(\mathcal{V}\oplus\underline{\mathbb{C}})\times_B X$ then gives rise to the following natural morphism (Equation (6))

$$j : K^0_{\mathcal{G}}(\mathcal{V}) \to K^0_{\mathcal{G}}(P(\mathcal{V} \oplus \underline{\mathbb{C}})), \qquad j : K^0_{\mathcal{G}}(\mathcal{V} \times_B X) \to K^0_{\mathcal{G}}(P(\mathcal{V} \oplus \underline{\mathbb{C}}) \times_B X).$$

Let X be compact and let $x \in K^0_{\mathcal{G}}(P(\mathcal{V} \oplus \underline{\mathbb{C}}) \times_B X)$ be arbitrary. The fibers of the projectivization $P(\mathcal{V} \oplus \underline{\mathbb{C}})$ are complex manifolds, so we can consider the analytical index of the correspondent family of Dolbeault operators over $P(V \oplus 1)$ with coefficients in x (cf. [3, page 123]). This index is an element of $K^0_{\mathcal{G}}(X)$ by the results of [14]. Taking the composition with j (cf. [3, page 122-123]), we get a family of mappings $\alpha_X : K^0_{\mathcal{G}}(\mathcal{V} \times_B X) \to K^0_{\mathcal{G}}(X)$, having the following properties:

(i) α_X is functorial with respect to $\mathcal{G}$-equivariant morphisms;
(ii) α_X is a morphism of $K^0_{\mathcal{G}}(X)$-modules;
(iii) $\alpha_B(\lambda_V) = 1 \in K^0_{\mathcal{G}}(B)$.

Let $X^+ := X \cup B$ be the fiberwise one-point compactification of X. The commutative diagram

$$\begin{array}{ccccc} 0 \to K^0_{\mathcal{G}}(\mathcal{V} \times_B X) & \longrightarrow & K^0_{\mathcal{G}}(\mathcal{V} \times_B X^+) & \longrightarrow & K^0_{\mathcal{G}}(\mathcal{V}) \\ & & \downarrow \alpha_{X^+} & & \downarrow \alpha_B \\ 0 \to K^0_{\mathcal{G}}(X) & \longrightarrow & K^0_{\mathcal{G}}(X^+) & \longrightarrow & K^0_{\mathcal{G}}(B) \end{array}$$

with exact lines allows us to define α_X for X non-compact as the restriction and corestriction of α_{X^+}.

Let $x \in K^0_{\mathcal{G}}(X)$, then by (ii)

$$\alpha_X(\lambda_V x) = \alpha_X(\lambda_{\mathcal{V}})x = x, \qquad \alpha\phi = \mathrm{Id}. \tag{21}$$

Let $q := \pi_F : F = \mathcal{V} \times_B X \to X$, $p : X \times_B \mathcal{V} \to X$, $q_1 : \mathcal{V} \times_B X \times_B \mathcal{V} \to \mathcal{V}$ (the projection onto the first entry), and $q_2 : X \times_B \mathcal{V} \to B$. Let us denote by $\widetilde{y} \in K^0_{\mathcal{G}}(X \times_B \mathcal{V})$ the element obtained from y under the mapping $X \times_B \mathcal{V} \to \mathcal{V} \times_B X$, $(x, v) \mapsto (-v, x)$ (such that $\mathcal{V} \times_B X \times_B \mathcal{V} \to \mathcal{V} \times_B X \times_B \mathcal{V}$, $(u, x, v) \mapsto (-v, x, u)$ is homotopic to the identity).

Let $y \in K^0_{\mathcal{G}}(\mathcal{V} \times_B X)$, then once again by (i), (ii) and then by (iii)

$$\begin{aligned}
\phi(\alpha_X(y)) = \pi_E^*\alpha_X(y) \otimes q^*\lambda_{\mathcal{V}} &= \alpha_{X \times_B \mathcal{V}}(p_1^* y) \otimes q^*\lambda_{\mathcal{V}} = \alpha_{X \times_B \mathcal{V}}(p_1^* y \otimes q^*\lambda_V) \\
= \alpha_{X \times_B \mathcal{V}}(y \boxtimes \lambda_V) &= \alpha_{X \times_B \mathcal{V}}(\lambda_V \boxtimes \widetilde{y}) = \alpha_{X \times_B \mathcal{V}}(q_1^*\lambda_V \otimes \widetilde{y}) \\
= \alpha_{X \times_B \mathcal{V}}(q_1^*\lambda_V) \otimes \widetilde{y} &= q_2^*\alpha_B(\lambda_V) \otimes \widetilde{y} = q_2^*(1) \otimes \widetilde{y} = \widetilde{y} \in K^0_{\mathcal{G}}(X \times_B \mathcal{V}),
\end{aligned} \tag{22}$$

We obtain that $\phi \circ \alpha_X$ is an isomorphism. Since $\alpha_X \circ \phi = \mathrm{Id}$, α_X is the two-sides inverse of ϕ, and hence the automorphism $\phi \circ \alpha_X$ is the identity.

The proof for a general (complex) $\mathcal{G}$-equivariant vector bundle $F \to X$ can be done as in [3, p. 124]. However, we found it more convenient to use the following argument. Embed first F into a trivial bundle $E = \mathcal{V} \times_B X$. Let ϕ^1 and ϕ^2 be the Thom maps associated to the bundles $E \to F$ and $E \to X$. Then by Theorem 4.4 the diagram

$$\begin{array}{ccc}
K^0_{\mathcal{G}}(F) & \xrightarrow{\ \phi^1\ } & K^0_{\mathcal{G}}(E) \\
 & {\scriptstyle \phi_F}\nwarrow \quad \nearrow{\scriptstyle \phi^2} & \\
 & K^0_{\mathcal{G}}(X) &
\end{array}$$

is commutative, while ϕ^2 is an isomorphism by the first part of the proof. Therefore ϕ_F is injective. The same argument show that ϕ^1 is injective. But ϕ^1 must also be surjective, because ϕ^2 is an isomorphism. Thus, ϕ^1 is an isomorphism, and hence ϕ_F is an isomorphism too. □

5. Gysin maps

We now discuss a few constructions related to the Thom isomorphism, which will be necessary for the definition of the topological index. The most important one is the Gysin map. For several of the constructions below, the setting of $\mathcal{G}$-spaces and even $\mathcal{G}$-bundles is too general, and we shall have to consider *longitudinally smooth* $\mathcal{G}$-fiber bundles $\pi_X : X \to B$. The main reason why we need longitudinally smooth bundles to define the Gysin map is the same as in the definition of the Gysin map for embeddings of smooth manifolds. We shall denote by $T_{\mathrm{vert}}X$ the *vertical* tangent bundle to the fibers of $X \to B$. All tangent bundles below will be *vertical* tangent bundles.

Let X and Y be longitudinally smooth $\mathcal{G}$-fiber bundles, $i : X \to Y$ be an equivariant fiberwise embedding, and $p_T : T_{\mathrm{vert}}X \to X$ be the vertical tangent bundle to X. Assume Y is equipped with a $\mathcal{G}$-invariant Riemannian metric and let $p_N : N_{\text{vert}} \to X$ be the fiberwise normal bundle to the image of i

Let us choose a function $\varepsilon : X \to (0, \infty)$ such that the map of N_{vert} to itself

$$n \mapsto \varepsilon \frac{n}{1 + |n|}$$

is $\mathcal{G}$-equivariant and defines a $\mathcal{G}$-diffeomorphism $\Phi : N_{\text{vert}} \to W$ onto an open tubular neighborhood $W \supset X$ in Y.

Let $(N \oplus N)_{\text{vert}} := N_{\text{vert}} \oplus N_{\text{vert}}$. The embedding $i : X \to Y$ can be written as a composition of two fiberwise embeddings $i_1 : X \to W$ and $i_2 : W \to Y$. Passing to differentials we obtain

$$T_{\text{vert}}X \xrightarrow{di_1} T_{\text{vert}}W \xrightarrow{di_2} T_{\text{vert}}Y \quad \text{and} \quad d\Phi : T_{\text{vert}}N \to T_{\text{vert}}W,$$

where we use the simplified notation $T_{\text{vert}}N = T_{\text{vert}}N_{\text{vert}}$.

Lemma 5.1. (cf. [10, page 112]) *The manifold $T_{\text{vert}}N$ can be identified with $p_T^*(N \oplus N)_{\text{vert}}$ with the help of a $\mathcal{G}$-equivariant diffeomorphism ψ that makes the following diagram commutative*

$$\begin{array}{ccc} p_T^*(N \oplus N)_{\text{vert}} & \xleftarrow{\quad\psi\quad} & T_{\text{vert}}N \\ \downarrow & & \downarrow \\ T_{\text{vert}}X & & N_{\text{vert}} \\ & {\searrow}^{p_T} \quad {}^{p_N}{\swarrow} & \\ & X. & \end{array}$$

Proof. The vertical tangent bundle $T_{\text{vert}}N \to N_{\text{vert}}$ and the vector bundle

$$p_N^*(T_{\text{vert}}X) \oplus p_N^*(N_{\text{vert}}) \to N_{\text{vert}}$$

are isomorphic as $\mathcal{G}$-equivariant vector bundles over N_{vert}.

Indeed, a point of the total space $T_{\text{vert}}N$ is a pair of the form $(n_1, t + n_2)$, where both vectors are from the fiber over the point $x \in X$. Similarly, we represent elements $p_T^*(N \oplus N)_{\text{vert}}$ as pairs of the form $(t, n_1 + n_2)$. Let us define ψ by the equality $\psi(n_1, t + n_2) = (t, n_1 + n_2)$. □

With the help of the relation $i \cdot (n_1, n_2) = (-n_2, n_1)$, we can equip

$$p_T^*(N \oplus N)_{\text{vert}} = p_T^*(N_{\text{vert}}) \oplus p_T^*(N_{\text{vert}}) \tag{23}$$

with a structure of a complex manifold. Then we can consider the Thom homomorphism

$$\phi : K^0_{\mathcal{G}}(T_{\text{vert}}X) \to K^0_{\mathcal{G}}(p_T^*(N \oplus N)_{\text{vert}}).$$

Since $T_{\text{vert}}W$ is an open $\mathcal{G}$-stable subset of $T_{\text{vert}}Y$ and $di_2 : T_{\text{vert}}W \to T_{\text{vert}}Y$ is a fiberwise embedding, by Equation (6), we obtain the homomorphism $(di_2)_* : K^0_{\mathcal{G}}(T_{\text{vert}}W) \to K^0_{\mathcal{G}}(T_{\text{vert}}Y)$.

Definition 5.2. Let $i : X \to Y$ be an equivariant embedding of $\mathcal{G}$-bundles. The *Gysin homomorphism* is the mapping

$$i_! : K^0_{\mathcal{G}}(T_{\mathrm{vert}}X) \to K^0_{\mathcal{G}}(T_{\mathrm{vert}}Y), \qquad i_! = (di_2)_* \circ (d\Phi^{-1})^* \circ \psi^* \circ \phi.$$

In other words, the Gysin map is obtained by passage to K-groups in the upper line of the diagram

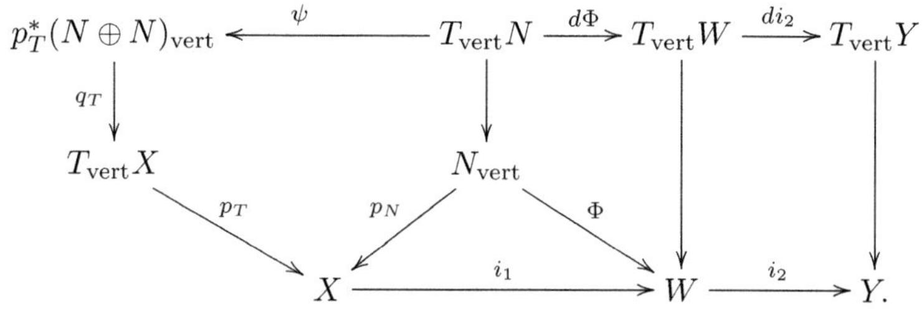

Another choice of metric and neighborhood W induces the homotopic map and (by the item 3 of Theorem 5.3 below) the same homomorphism.

Theorem 5.3 (Properties of Gysin homomorphism). *Let $i : X \to Y$ be a $\mathcal{G}$-embedding. Then*

(i) *$i_!$ is a homomorphism of $K^0_{\mathcal{G}}(B)$-modules.*
(ii) *Let $i : X \to Y$ and $j : Y \to Z$ be two fiberwise $\mathcal{G}$-embeddings, then $(j \circ i)_! = j_! \circ i_!$.*
(iii) *Let fiberwise embeddings $i_1 : X \to Y$ and $i_2 : X \to Y$ be $\mathcal{G}$-homotopic in the class of embeddings. Then $(i_1)_! = (i_2)_!$.*
(iv) *Let $i_! : X \to Y$ be a fiberwise $\mathcal{G}$-diffeomorphism, then $i_! = (di^{-1})^*$.*
(v) *A fiberwise embedding $i : X \to Y$ can be represented as a composition of embeddings X in N_{vert} (as the zero section $s_0 : x \to N$) and $N_{\mathrm{vert}} \to Y$ by $i_2 \circ \Phi : N_{\mathrm{vert}} \to Y$. Then $i_! = (i_2 \circ \Phi)_!(s_0)_!$.*
(vi) *Consider the complex bundle $p_T^*(N_{\mathrm{vert}} \otimes \mathbb{C})$ over $T_{\mathrm{vert}}X$. Let us form the complex $\Lambda(p_T^*(N_{\mathrm{vert}} \otimes \mathbb{C}); 0)$:*

$$0 \to \Lambda^0(p_T^*(N_{\mathrm{vert}} \otimes \mathbb{C})) \xrightarrow{0} \dots \xrightarrow{0} \Lambda^k(p_T^*(N_{\mathrm{vert}} \otimes \mathbb{C})) \to 0$$

with noncompact support. If $a \in K^0_{\mathcal{G}}(T_{\mathrm{vert}}X)$, then the complex

$$a \otimes \Lambda(p_T^*(N_{\mathrm{vert}} \otimes \mathbb{C}), 0)$$

has compact support and defines an element of $K^0_{\mathcal{G}}(T_{\mathrm{vert}}X)$. Then

$$(di)^* i_!(a) = a \cdot \Lambda(p_T^*(N_{\mathrm{vert}} \otimes \mathbb{C}); 0),$$

where di is the differential of the embedding i.
(vii) *$i_!(x(di)^*y) = i_!(x) \cdot y$, where $x \in K^0_{\mathcal{G}}(T_{\mathrm{vert}}X)$ and $y \in K^0_{\mathcal{G}}(T_{\mathrm{vert}}Y)$.*

Proof. (i) This follows from the definition of $i_!$.

(ii) To simplify the argument, let us identify the tubular neighborhood with the normal bundle. Then $(j \circ i)_!$ is the composition

$$K^0_{\mathcal{G}}(T_{\text{vert}}X) \xrightarrow{\phi} K^0_{\mathcal{G}}(T_{\text{vert}}N \oplus T_{\text{vert}}N''_{\text{vert}}) \to K^0_{\mathcal{G}}(T_{\text{vert}}Z),$$

where N'_{vert} is the fiberwise normal bundle of Y in Z, $N''_{\text{vert}} = N'_{\text{vert}}|_X$, and for the sum of tangent bundles to the vertical normal bundles $T_{\text{vert}}N \oplus T_{\text{vert}}N''_{\text{vert}}$ is considered in the same way as a complex bundle over $T_{\text{vert}}X$, as in Equation (23). On the other hand, $j_! \circ i_!$ represents the composition

$$K^0_{\mathcal{G}}(T_{\text{vert}}X) \xrightarrow{\phi} K^0_{\mathcal{G}}(T_{\text{vert}}N) \to K^0_{\mathcal{G}}(T_{\text{vert}}Y) \xrightarrow{\phi} K^0_{\mathcal{G}}(T_{\text{vert}}N'_{\text{vert}}) \to K^0_{\mathcal{G}}(T_{\text{vert}}Z).$$

By the properties of ϕ, the following diagram is commutative

$$\begin{array}{ccccc}
K^0_{\mathcal{G}}(T_{\text{vert}}X) & \xrightarrow{\quad\phi\quad} & K^0_{\mathcal{G}}(T_{\text{vert}}N) & \longrightarrow & K^0_{\mathcal{G}}(T_{\text{vert}}Y) \\
& \searrow^{\phi} & \downarrow{\scriptstyle\phi} & & \downarrow{\scriptstyle\phi} \\
& & K^0_{\mathcal{G}}(T_{\text{vert}}N \oplus T_{\text{vert}}N'_{\text{vert}}) & \xrightarrow{\phi} & K^0_{\mathcal{G}}(T_{\text{vert}}N'_{\text{vert}}) \\
& & \downarrow & & \downarrow \\
& & K^0_{\mathcal{G}}(T_{\text{vert}}Z) & =\!=\!=\!= & K^0_{\mathcal{G}}(T_{\text{vert}}Z).
\end{array}$$

This completes the proof of (ii).

(iii) The morphism q_T depends only on the homotopy class of the embeddings used to define it. The assertion thus follows from the homotopy invariance of K-theory.

(iv) In this case $N = X$, $W = Y$, $\Phi = i$, $i_2 = \text{Id}_Y$, and the formula is obvious.

(v) This follows from (ii).

(vi) By definition,

$$(di)^* \circ i_! = (di_1)^* \circ (di_2)^* \circ (di_2)_* \circ (d\phi^{-1})^* \circ \psi^* \circ \phi^* = (\psi \circ d\Phi^{-1} \circ di_1)^* \circ \phi,$$

where $i_1 : X \to W$, $i_2 : W \to Y$. Let $(n_1, t + n_2) \in T_{\text{vert}}N = p_N^*(T_{\text{vert}}X) \oplus p_N^*(N_{\text{vert}})$, where n_1 is the shift under the exponential mapping and $t + n_2$ is a vertical tangent vector to W. If $d\Phi(n_1, t + n_2)$ is in $T_{\text{vert}}X$, then $n_1 = n_2 = 0$. Hence,

$$d\Phi^{-1} di_1(t) = (0, t + 0), \quad \psi \circ d\Phi^{-1} \circ di_1(t) = (t, 0 + 0).$$

Therefore, $\psi \circ d\Phi^{-1} \circ di_1 : T_{\text{vert}}X \to p_T^*(N_{\text{vert}} \oplus N_{\text{vert}})$ is the embedding of the zero section. Since $\phi(a) = a \cdot \Lambda(q_T^* p_T^*(N_{\text{vert}} \otimes \mathbb{C}), s_{p_T^*(N_{\text{vert}} \otimes \mathbb{C})})$, it follows that $(di)^* \circ i_!(a) = a \cdot \Lambda(p_T^*(N_{\text{vert}} \otimes \mathbb{C}), 0)$.

(vii) The mapping $di_1 \circ q_T \circ \psi \circ d\Phi^{-1} : T_{\mathrm{vert}}W \to T_{\mathrm{vert}}W$ is homotopic to the identical mapping. Hence,

$$\begin{aligned}
& i_!(x \cdot (di)^* y) = (di_2)_* (d\Phi^{-1})^* \psi^* \phi(x \cdot (di)^* y) \\
& = (di_2)_* (d\Phi^{-1})^* \psi^* [(q_T^*(x) \lambda_{p_T^*(N_{\mathrm{vert}} \otimes \mathbb{C})})(q_T^*(di)^* y)] \\
& = (di_2)_* [(d\Phi^{-1})^* \psi^* (q_T^*(x) \lambda_{p_T^*(N_{\mathrm{vert}} \otimes \mathbb{C})}) \underbrace{(d\Phi^{-1})^* \psi^* q_T^* (di_1)^*}_{\mathrm{Id}} (di_2)^* y] \\
& = [(di_2)_* (d\Phi^{-1})^* \psi^* (q_T^*(x) \lambda_{p_T^*(N_{\mathrm{vert}} \otimes \mathbb{C})})] \, [(di_2)_* (di_2)^* y] = i_!(x) \cdot y.
\end{aligned}$$

The proof is now complete. □

We shall need also the following properties of the Gysin map. If $X = B$, the trivial longitudinally smooth $\mathcal{G}$-bundle, we shall identify $T_{\mathrm{vert}}X = B$ and $T_{\mathrm{vert}}\mathcal{V} = \mathcal{V} \otimes \mathbb{C}$ for a real bundle $\mathcal{V} \to B$.

Theorem 5.4. *Suppose that $\mathcal{V} \to B$ is a $\mathcal{G}$-equivariant* real *vector bundle and that $X = B$. Then the mapping*

$$i_! : K^0_{\mathcal{G}}(B) = K^0_{\mathcal{G}}(TX_{\mathrm{vert}}) \to K^0_{\mathcal{G}}(T_{\mathrm{vert}}\mathcal{V}) = K^0_{\mathcal{G}}(\mathcal{V} \otimes \mathbb{C})$$

coincides with the Thom homomorphism $\phi^{\mathcal{V} \otimes \mathbb{C}}$.

Proof. The assertion follows from the definition of $i_!$. More precisely, let $X = B \hookrightarrow \mathcal{V}$, $N = \mathcal{V}$ be the zero section embedding. In the definition of the Thom homomorphism, W can be chosen to be equal to the bundle D_1 of interiors of the balls of radius 1 in $\mathcal{V}$ with respect to an invariant metric. In this case, the diagram from the definition of the Gysin homomorphism 5.2 takes the following form

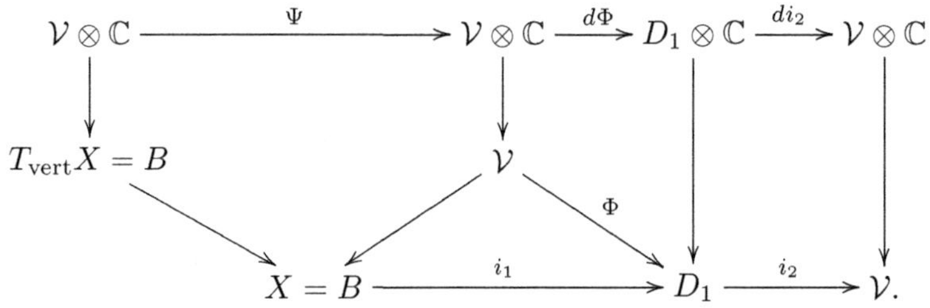

In our case $\Psi = \mathrm{Id}$ and $di_2 \circ d\Phi$ is homotopic to Id, since this map has the form $v \otimes z \mapsto (v \otimes z)/(1 + |v \otimes z|)$. Hence, $i_! = \phi$. □

Theorem 5.5. *Suppose that $\mathcal{V}'$ and $\mathcal{V}''$ are $\mathcal{G}$-equivariant $\mathbb{R}$-vector bundles over B and that $i : X \to \mathcal{V}'$ is an embedding. Let $k : X \to \mathcal{V}' \oplus \mathcal{V}''$, $k(x) = i(x) + 0$. Let ϕ be the Thom homomorphism of the complex bundle*

$$T_{\mathrm{vert}}(\mathcal{V}' \oplus \mathcal{V}'') = \mathcal{V}' \otimes \mathbb{C} \oplus \mathcal{V}'' \otimes \mathbb{C} \longrightarrow T_{\mathrm{vert}}\mathcal{V}' = \mathcal{V}' \otimes \mathbb{C}.$$

Then the following diagram is commutative

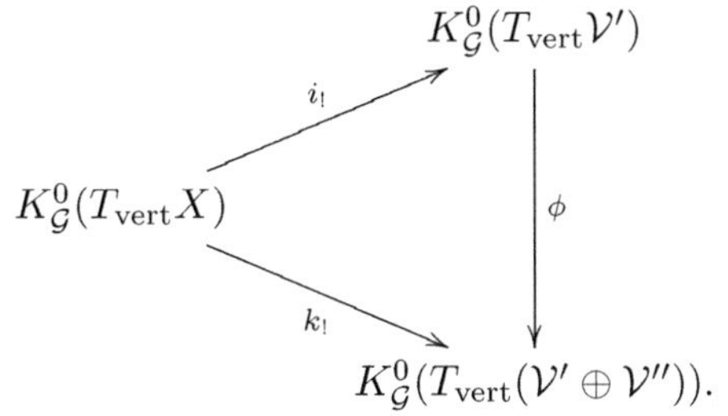

Proof. To prove the statement, let us consider the embedding $i : X \to \mathcal{V}'$ and denote, as before, by N_{vert} the fiberwise normal bundle and by W the tubular neighborhood appearing in the definition of the Thom homomorphism associated to i. Then $N_{\mathrm{vert}} \oplus \mathcal{V}''$ is a fiberwise normal bundle for the embedding k with the tubular neighborhood $W \oplus D(\mathcal{V}'')$, where $D(\mathcal{V}'')$ is a ball bundle. If $a \in K^0_{\mathcal{G}}(T_{\mathrm{vert}}X)$, then

$$k_!(a) = (di_2 \oplus 1)_* \circ (d\Phi^{-1} \oplus 1)^* \circ (\psi \oplus 1)^* \circ \phi^{N \oplus N \oplus \mathcal{V}'' \oplus \mathcal{V}''}(a).$$

We have $\phi^{N \oplus N \oplus \mathcal{V}'' \oplus \mathcal{V}''} = \phi^{N \oplus N} \circ \phi^{\mathcal{V}'' \oplus \mathcal{V}''}$, by Theorem 4.4. Since $a = a \cdot \underline{\mathbb{C}}$, where $\underline{\mathbb{C}}$ is the trivial line bundle, we obtain

$$\begin{aligned} k_!(a) &= (di_2)_* \circ (d\Phi^{-1})^* \circ \Psi^* \circ \phi^{N_{\mathrm{vert}} \oplus N_{\mathrm{vert}}}(a) \circ \phi^{\mathcal{V}'' \oplus \mathcal{V}''}(\underline{\mathbb{C}}) \\ &= i_!(a) \cdot \lambda_{T(\mathcal{V}' \oplus \mathcal{V}'')_{\mathrm{vert}}} = \phi(i_!(a)). \end{aligned} \tag{24}$$

The proof is now complete. □

6. The topological index

We begin with a "fibered Mostow-Palais theorem" that will be useful in defining the index.

Theorem 6.1. *Let $\pi_X : X \to B$ be a compact $\mathcal{G}$-fiber bundle. Then there exists a real $\mathcal{G}$-equivariant vector bundle $\mathcal{V} \to B$ and a fiberwise smooth $\mathcal{G}$-embedding $X \to \mathcal{V}$. After averaging one can assume that the action of $\mathcal{G}$ on $\mathcal{V}$ is orthogonal.*

Proof. Fix $b \in B$ and let U be an equivariant trivialization neighborhood of b for both X and $\mathcal{G}$. By the Mostow-Palais theorem, there exists a representation of $\mathcal{G}_b$ on a finite dimensional vector space V_b and a smooth $\mathcal{G}_b$-equivariant embedding $i_b : X_b \to V_b$. This defines an embedding

$$\psi : \pi_X^{-1}(U_b) \simeq U_b \times X_b \to U_b \times V_b, \tag{25}$$

which is $\mathcal{G}$-equivariant in an obvious sense.

We can cover B with finitely many open sets U_{b_j}, as above, corresponding to the points b_j, $j = 1, \ldots, N$. Denote by V_j the corresponding representations and by ψ_j the corresponding embeddings, as in Equation (25). Let $W := \oplus V_j$. Also,

Let ϕ_j be a partition of unity subordinated to the covering by $U_j = U_{b_j}$. We define then

$$\Psi := \oplus(\phi_j \circ \pi_X)\psi_j : X \to B \times W,$$

which is a $\mathcal{G}$-equivariant embedding of X into the trivial $\mathcal{G}$-equivariant vector bundle $\mathcal{V} := B \times W$, as desired. □

Let us now turn to the definition of the topological index. Let $X \to B$ be a compact, longitudinally smooth $\mathcal{G}$-bundle. From Theorem 6.1 it follows that there exists an $\mathcal{G}$-equivariant *real* vector bundle $\mathcal{V} \to B$ and a fiberwise smooth $\mathcal{G}$-equivariant embedding $i : X \to \mathcal{V}$. We can assume that $\mathcal{V}$ is endowed with an orthogonal metric and that $\mathcal{G}$ preserves this metric. Thus, the Gysin homomorphism

$$i_! : K^0_{\mathcal{G}}(T_{\text{vert}}X) \to K^0_{\mathcal{G}}(T_{\text{vert}}\mathcal{V}) = K^0_{\mathcal{G}}(\mathcal{V} \otimes \mathbb{C})$$

is defined (see Section 4). Since $T_{\text{vert}}\mathcal{V} = \mathcal{V} \otimes \mathbb{C}$ is a complex vector bundle, we have the following Thom isomorphism (see Section 4):

$$\phi : K^0_{\mathcal{G}}(B) \xrightarrow{\sim} K^0_{\mathcal{G}}(T_{\text{vert}}\mathcal{V}).$$

Definition 6.2. The *topological index* is by definition the morphism:

$$\text{t-ind}^X_{\mathcal{G}} : K^0_{\mathcal{G}}(T_{\text{vert}}X) \to K^0_{\mathcal{G}}(B), \qquad \text{t-ind}^X_{\mathcal{G}} := \phi^{-1} \circ i_!.$$

The topological index satisfies the following properties.

Theorem 6.3. *Let $X \to B$ be a longitudinally smooth bundle and*

$$\text{t-ind}^X_{\mathcal{G}} : K^0_{\mathcal{G}}(T_{\text{vert}}X) \to K^0_{\mathcal{G}}(B)$$

be its associated topological index. Then

(i) $\text{t-ind}^X_{\mathcal{G}}$ *does not depend on the choice of the $\mathcal{G}$-equivariant vector bundle $\mathcal{V}$ and on the embedding $i : X \to \mathcal{V}$.*
(ii) $\text{t-ind}^X_{\mathcal{G}}$ *is a $K^0_{\mathcal{G}}(B)$-homomorphism.*
(iii) *If $X = B$, then the map*

$$\text{t-ind}^X_{\mathcal{G}} : K^0_{\mathcal{G}}(B) = K^0_{\mathcal{G}}(T_{\text{vert}}X) \to K^0_{\mathcal{G}}(B)$$

coincides with $\mathrm{Id}_{K^0_{\mathcal{G}}(B)}$.
(iv) *Suppose X and Y are compact longitudinally smooth $\mathcal{G}$-bundles, $i : X \to Y$ is a fiberwise $\mathcal{G}$-embedding. Then the diagram*

$$\begin{array}{ccc} K^0_{\mathcal{G}}(T_{\text{vert}}X) & \xrightarrow{i_!} & K^0_{\mathcal{G}}(T_{\text{vert}}Y) \\ & \searrow^{\text{t-ind}^X_{\mathcal{G}}} \quad \swarrow^{\text{t-ind}^Y_{\mathcal{G}}} & \\ & K^0_{\mathcal{G}}(B). & \end{array}$$

commutes.

Proof. To prove (i), let us consider two embeddings

$$i_1 : X \to \mathcal{V}',$$
$$i_2 : X \to \mathcal{V}''$$

into $\mathcal{G}$-equivariant vector bundles. Denote by $j = i_1 + i_2$ the induced embedding $j : X \to \mathcal{V}' \oplus \mathcal{V}''$. It is sufficient to show that i_1 and j define the same topological index. Let us define a homotopy of $\mathcal{G}$-embeddings by the formula

$$j_s(x) = i_1(x) + s \cdot i_2(x) \ : \ X \to \mathcal{V}' \oplus \mathcal{V}'', \quad 0 \le s \le 1.$$

Then, by Theorems 5.3(iii) and 5.5, the indices for j and j_0 coincide. Let us show now that $j_0 = i_1 + 0$ and i_1 define the same topological indexes. For this purpose consider the diagram

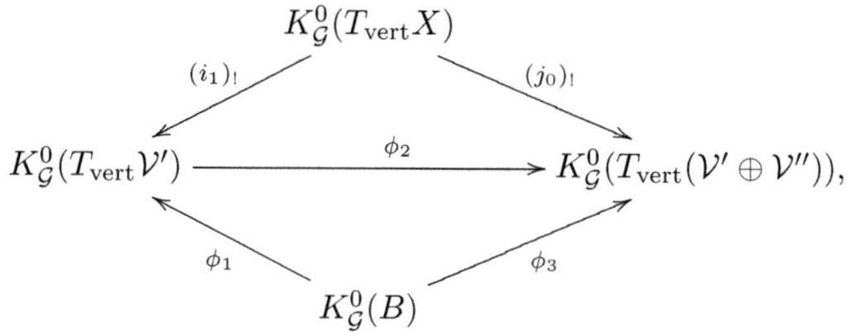

where ϕ_i are the corresponding Thom homomorphisms. The upper triangle is commutative by Theorem 5.4.2, and the lower is commutative by Theorem 4.4. Hence $\phi_1^{-1} \circ (i_1)_! = \phi_3^{-1} \circ (j_0)_1$ as desired.

(ii) follows from 4.2 and 5.3(i). Property (iii) follows from the definition of the index and from 5.4.

To prove (iv), let us consider the diagram

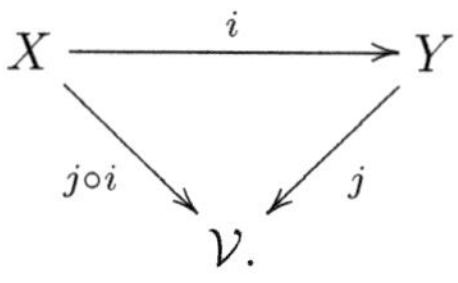

We now use 5.3(ii). This gives the commutative diagram

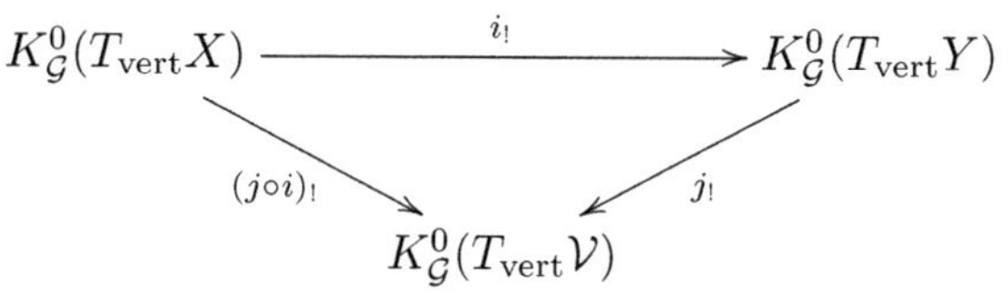

or

$$
\begin{array}{ccccc}
K^0_{\mathcal{G}}(T_{\mathrm{vert}}X) & & \xrightarrow{\quad i_! \quad} & & K^0_{\mathcal{G}}(T_{\mathrm{vert}}Y) \\
& \searrow^{(j\circ i)_!} & & \swarrow^{j_!} & \\
& & K^0_{\mathcal{G}}(T_{\mathrm{vert}}\mathcal{V}) & & \\
& {}_{\text{t-ind}^X_{\mathcal{G}}}\searrow & \uparrow \phi & \swarrow_{\text{t-ind}^Y_{\mathcal{G}}} & \\
& & K^0_{\mathcal{G}}(B). & &
\end{array}
$$

This completes the proof. $\square$

We now investigate the behavior of the topological index with respect to fiber products of bundles of compact groups.

Theorem 6.4. *Let $\pi : P \to X$ be a principal right $\mathcal{H}$-bundle with a left action of $\mathcal{G}$ commuting with $\mathcal{H}$. Suppose F is a longitudinally smooth $(\mathcal{G}\times\mathcal{H})$-bundle. Let us denote by Y the space $P \times_{\mathcal{H}} F$. Let $j : X' \to X$ and $k : F' \to F$ be fiberwise $\mathcal{G}$- and $(\mathcal{G}\times\mathcal{H})$-embeddings, respectively. Let $\pi' : P' \to X'$ be the principal $\mathcal{H}$-bundle induced by j on X'. Assume that $Y' := P' \times_{\mathcal{H}} F'$. The embeddings j and k induce $\mathcal{G}$-embedding $j * k : Y' \to Y$. Then the diagram*

$$
\begin{array}{ccc}
K^0_{\mathcal{G}}(T_{\mathrm{vert}}X) \otimes_{K^0_{\mathcal{G}}(B)} K^0_{\mathcal{G}\times\mathcal{H}}(T_{\mathrm{vert}}F) & \xrightarrow{\ \gamma\ } & K^0_{\mathcal{G}}(T_{\mathrm{vert}}Y) \\
\uparrow {\scriptstyle j_!\otimes k_!} & & \uparrow {\scriptstyle (j*k)_!} \\
K^0_{\mathcal{G}}(T_{\mathrm{vert}}X') \otimes_{K^0_{\mathcal{G}}(B)} K^0_{\mathcal{G}\times\mathcal{H}}(T_{\mathrm{vert}}F') & \xrightarrow{\ \gamma\ } & K^0_{\mathcal{G}}(T_{\mathrm{vert}}Y')
\end{array}
$$

is commutative.

Let us remark that in the statement of this theorem there is no compactness assumption on $X, X', F,$ and F', since there is no compactness assumption in the definition of the Gysin homomorphism. This is unlike in the definition of the topological index where we start with a compact $\mathcal{G}$-bundle $X \to B$.

Proof. Let us use the definition of γ:

$$
\begin{array}{ccccc}
K^0_{\mathcal{G}}(T_{\mathrm{vert}}X) \otimes K^0_{\mathcal{G}\times\mathcal{H}}(T_{\mathrm{vert}}F) & \longrightarrow & K^0_{\mathcal{G}}(T_{\mathrm{vert}}X) \otimes K^0_{\mathcal{G}\times\mathcal{H}}(P \times T_{\mathrm{vert}}F) & \xrightarrow{\cong} & \\
{\scriptstyle j_!\otimes k_!}\uparrow & \boxed{1} & \uparrow{\scriptstyle \varepsilon} & \boxed{2} & \\
K^0_{\mathcal{G}}(T_{\mathrm{vert}}X') \otimes K^0_{\mathcal{G}\times\mathcal{H}}(T_{\mathrm{vert}}F') & \longrightarrow & K^0_{\mathcal{G}}(T_{\mathrm{vert}}X') \otimes K^0_{\mathcal{G}\times\mathcal{H}}(P' \times T_{\mathrm{vert}}F') & \xrightarrow{\cong} &
\end{array}
$$

$$
\begin{array}{ccc}
\cong K^0_{\mathcal{G}}(T_{\mathrm{vert}}X) \otimes K^0_{\mathcal{G}}(P \times_{\mathcal{H}} T_{\mathrm{vert}}F) & \to & K^0_{\mathcal{G}}(\pi_1^* T_{\mathrm{vert}}X) \otimes K^0_{\mathcal{G}}(P \times_{\mathcal{H}} T_{\mathrm{vert}}F) \to \\
{\scriptstyle \beta}\uparrow & \boxed{3} & \uparrow{\scriptstyle \alpha} \\
\cong K^0_{\mathcal{G}}(TX') \otimes K^0_{\mathcal{G}}(P' \times_{\mathcal{H}} T_{\mathrm{vert}}F') & \to & K^0_{\mathcal{G}}((\pi_1')^* T_{\mathrm{vert}}X') \otimes K^0_{\mathcal{G}}(P' \times_{\mathcal{H}} T_{\mathrm{vert}}F') \to
\end{array}
$$

$$\begin{array}{c} \to K^0_{\mathcal{G}}((\pi_1^* T_{\mathrm{vert}} X) \times (P \times_{\mathcal{H}} T_{\mathrm{vert}} F)) \to \\ \boxed{4} \\ \to K^0_{\mathcal{G}}((\pi_1')^* T_{\mathrm{vert}} X' \times (P' \times_{\mathcal{H}} T_{\mathrm{vert}} F')) \to \end{array} \tag{26}$$

$$\begin{array}{ccc} \to K^0_{\mathcal{G}}(\pi_1^* T_{\mathrm{vert}} X \oplus (P \times_{\mathcal{H}} T_{\mathrm{vert}} F)) & = & K^0_{\mathcal{G}}(T_{\mathrm{vert}} Y) \\ \boxed{4} & & \uparrow (j * k)_! \\ \to K^0_{\mathcal{G}}((\pi_1')^* T_{\mathrm{vert}} X' \oplus (P' \times_{\mathcal{H}} T_{\mathrm{vert}} F')) & = & K^0_{\mathcal{G}}(T Y'_{\mathrm{vert}}), \end{array}$$

where projections $\pi_1 : Y = P \times_{\mathcal{H}} F \to X$ and $\pi_1' : Y' = P \times_{\mathcal{H}} F' \to X'$ are defined as above. Here we use the isomorphism $K^0_{\mathcal{G}\times\mathcal{H}}(P \times W) \cong K^0_{\mathcal{G}}(P \times_{\mathcal{H}} W)$ for a free $\mathcal{H}$-bundle P (see Theorem 2.6). Let us remind the diagram, which was used for the definition of the Gysin homomorphism of an embedding $j : X' \to X$:

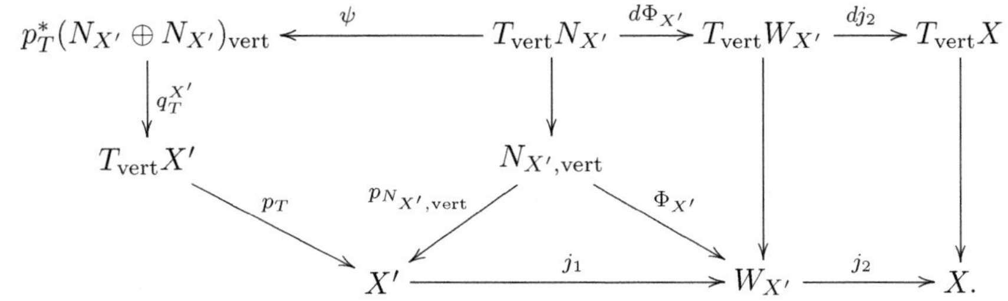

From the similar diagrams for $k_!$ and $(j * k)_!$ and the explicit form of these maps, it follows that the square $\boxed{4}$ in (26) is commutative if, and only if, α has the following form:

$$\begin{aligned} \alpha(\sigma \otimes \rho) = (\pi_1^*)\Big\{(dj_2)_* \, (d\Phi_{X'}^{-1})^* \, \psi_{X'}^*\Big\} \circ \phi^S(\sigma) \otimes \\ \otimes (\pi^* j_2 \times_{\mathcal{H}} dk_2)_* \left((\pi^* \Phi_{X'} \times_{\mathcal{H}} d\Phi_{F'})^{-1} \right)^* (1 \times_{\mathcal{H}} \psi_{F'})^* \, \phi^R(\rho), \end{aligned}$$

where S and T are bundles of the form

$$S : \begin{array}{c} \pi_1^* \left((p_T^{X'})^* \{N_{X'} \oplus N_{X'}\} \right) \\ \downarrow (\pi_1')^* q_T^{X'} \\ (\pi_1')^* \, (T_{\mathrm{vert}} X'), \end{array} \qquad R : \begin{array}{c} \pi^* N_{X'} \times_{\mathcal{H}} (p_T^{F'})^* \, (N_{F'} \oplus N_{F'}) \\ \downarrow (\pi')^*(p_{N_{X'}}) \times_{\mathcal{H}} q_T^{F'} \\ \pi^* \, X' \times_{\mathcal{H}} T_{\mathrm{vert}} F' = P' \times_{\mathcal{H}} T_{\mathrm{vert}} F'. \end{array}$$

Hence the square $\boxed{3}$ in (26) is commutative if, and only if, the homomorphism β has the form

$$\beta(\tau \otimes \rho) = j_!(\tau) \otimes \ \otimes (\pi^* j_2 \times_{\mathcal{H}} dk_2)_* \Big((\pi^* \Phi_{X'} \times_{\mathcal{H}} d\Phi_{F'})^{-1} \Big)^* \, (1 \times_{\mathcal{H}} \psi_{F'})^* \, \phi^R(\rho),$$

where $\tau \in K^0_{\mathcal{G}}(TX'), \rho \in K^0_{\mathcal{G}}(P' \times_{\mathcal{H}} TF')$. In turn, the square $\boxed{2}$ in (26) is commutative if, and only if, the homomorphism ε has the form

$$\varepsilon(\tau \otimes \delta) = j_!(\tau) \otimes \otimes (\pi^* j_2 \times_{\mathcal{H}} dk_2)_* \Big((\pi^* \Phi_{X'} \times_{\mathcal{H}} d\Phi_{F'})^{-1} \Big)^* \, (1 \times_{\mathcal{H}} \psi_{F'})^* \, \phi_{\mathbb{C}}^{\widetilde{R}}(\delta),$$

where $\tau \in K^0_{\mathcal{G}}(TX'), \delta \in K^0_{G\times\mathcal{H}}(P' \times TF')$, and $\widetilde{R}$ is the bundle

$$\widetilde{R}: \quad \begin{array}{c} \pi^* N_{X'} \times (p_T^{F'})^* (N_{F'} \oplus N_{F'}) \\ \downarrow (\pi')^*(p_N) \times q_T^{F'} \\ P' \times TF'. \end{array}$$

Suppose $\delta = [\underline{\mathbb{C}}]\widehat{\otimes}\omega$, where $[\underline{\mathbb{C}}] \in K^0_{\mathcal{G}\times\mathcal{H}}(P')$, $\underline{\mathbb{C}}$ is the one-dimensional trivial bundle and $\omega \in K^0_{\mathcal{G}\times\mathcal{H}}(TF')$. Then

$$\begin{aligned} \varepsilon(\tau \otimes \delta) &= j_!(\tau) \otimes \left\{\pi^*(j_2)_*(\Phi_{X'}^{-1})^*[\underline{\mathbb{C}}]\widehat{\otimes}k_!(\omega)\right\} = \\ &= j_!(\tau) \otimes \left\{[\underline{\mathbb{C}}]\widehat{\otimes}k_!(\omega)\right\}. \end{aligned}$$

Since the map $K^0_{\mathcal{G}\times\mathcal{H}}(TF) \to K^0_{\mathcal{G}\times\mathcal{H}}(P \times TF)$ (as well as the lower line in (26)) has the form $\omega \mapsto [\underline{\mathbb{C}}]\widehat{\otimes}\omega$, we have proved the commutativity of $\boxed{1}$ in (26). □

From this theorem we obtain the following corollary.

Corollary 6.5. *Let M be a compact smooth H-manifold, let $\mathcal{H} = B \times H$, and let P be a principal longitudinally smooth $\mathcal{H}$-bundle over X carrying also an action of $\mathcal{G}$ commuting with the action of $\mathcal{H}$. Also, let $X \to B$ be a compact longitudinally smooth $\mathcal{G}$-bundle. Let $Y := P \times_H M \to X$ be associated longitudinally smooth $\mathcal{G}$-bundle. Taking $F = B \times M$, we define $T_M Y := T_F Y$. Then $T_M Y$ is a $\mathcal{G}$-invariant real subbundle of $T_{\mathrm{vert}} Y$ and $T_M Y = P \times_H TM$. Let $j : X' \to X$ be a fiberwise $\mathcal{G}$-equivariant embedding and let $k : M' \to M$ be an H-embedding. Denote by $\pi' : P' \to X'$ the principal $\mathcal{H}$-bundle induced by j on X' and assume that $Y' := P' \times_H M'$. The embeddings j and k induce $\mathcal{G}$-embedding $j * k : Y' \to Y$. Then the diagram*

$$\begin{array}{ccc} K^0_{\mathcal{G}}(T_{\mathrm{vert}}X) \otimes K^0_H(TM) & \xrightarrow{\gamma} & K^0_{\mathcal{G}}(T_{\mathrm{vert}}Y) \\ \uparrow{\scriptstyle j_!\otimes k_!} & & \uparrow{\scriptstyle (j*k)_!} \\ K^0_{\mathcal{G}}(T_{\mathrm{vert}}X') \otimes K^0_H(TM') & \xrightarrow{\gamma} & K^0_{\mathcal{G}}(T_{\mathrm{vert}}Y') \end{array}$$

is commutative.

References

[1] M. Atiyah and G. Segal, *Twisted K-theory.* E-print math.KT/0407054.

[2] M.F. Atiyah, *K-theory.* W.A. Benjamin, Inc., New York – Amsterdam, 1967.

[3] ———, *Bott periodicity and the index of elliptic operators.* Quart. J. Math. Oxford **19** (1968), 113–140.

[4] M.F. Atiyah and I. M. Singer, *The index of elliptic operators. I.* Ann. of Math. (2) **87** (1968), 484–530.

[5] C. Carvalho, *C^*-algebraic K-theory and index theory.* PhD dissertation, Oxford University, Department of Math., 2003, (to appear in K-Theory).

[6] A. Connes, *An analogue of the Thom isomorphism for crossed products of a C^*-algebra by an action of* $\mathbb{R}$. Adv. Math. **39** (1981), 31–55.

[7] Th. Fack and G. Skandalis, *Connes' analogue of the Thom isomorphism for the Kasparov groups.* Invent. Math. **64** (1981), 7–14.

[8] D.S. Freed, *Twisted K-theory and loop groups.* International Congress of Mathematicians (2002, Beijing), Higher Education Press, Beijing, 2002, pp. 419–430.

[9] D.S. Freed, M. J. Hopkins, and C. Teleman, *Twisted equivariant K-theory with complex coefficients.* E-print, math.AT/0206257.

[10] T. Friedrich, *Vorlesungen über K-Theorie.* Teubner, Leipzig, 1987.

[11] M. Hilsum and G. Skandalis, *Invariance par homotopie de la signature à coefficients dans un fibré presque plat.* J. Reine Angew. Math. **423** (1992), 73–99.

[12] V. Mathai, R.B. Melrose, and I.M. Singer, *The index of projective families of elliptic operators.* E-print, math.DG/0206002.

[13] V. Nistor, *An index theorem for gauge-invariant families: The case of solvable groups.* Acta Math. Hungar. **99** (2003), no. 1–2, 155–183.

[14] V. Nistor and E. Troitsky, *An index for gauge-invariant operators and the Dixmier-Douady invariant.* Trans. Am. Math. Soc. **356** (2004), no. 1, 185–218.

[15] V. Nistor and E. Troitsky, *An index theorem for gauge-invariant families: The case of compact groups*, (tentative title, work in progress).

[16] E. V. Troitsky, *"Twice" equivariant C^*-index theorem and the index theorem for families.* Acta. Appl. Math. **68** (2001), 39–70.

Victor Nistor
Department of Mathematics
Pennsylvania State University
e-mail: `nistor@math.psu.edu`
URL: `http://www.math.psu.edu/nistor`

Evgenij Troitsky
Dept. of Mech. and Math.
Moscow State University
119899 Moscow, Russia
e-mail: `troitsky@mech.math.msu.su`
URL: `http://mech.math.msu.su/~troitsky`

C^*-algebras and Elliptic Theory
Trends in Mathematics, 247–289

Pseudodifferential Subspaces and Their Applications in Elliptic Theory

Anton Savin and Boris Sternin

Abstract. The aim of this paper is to explain the notion of subspace defined by means of pseudodifferential projection and give its applications in elliptic theory. Such subspaces are indispensable in the theory of well-posed boundary value problems for an arbitrary elliptic operator, including the Dirac operator, which has no classical boundary value problems. Pseudodifferential subspaces can be used to compute the fractional part of the spectral Atiyah–Patodi–Singer eta invariant, when it defines a homotopy invariant (Gilkey's problem). Finally, we explain how pseudodifferential subspaces can be used to give an analytic realization of the topological K-group with finite coefficients in terms of elliptic operators. It turns out that all three applications are based on a theory of elliptic operators on closed manifolds acting in subspaces.

Mathematics Subject Classification (2000). Primary 58J20; Secondary 58J28, 58J32, 19K56.

Keywords. elliptic operator, boundary value problem, pseudodifferential subspace, dimension functional, η-invariant, index, mod$n$-index, parity condition.

Introduction

Pseudodifferential subspaces and boundary value problems. A subspace (of some space of functions) is said to be *pseudodifferential* if it is determined by a projection that is a pseudodifferential operator.

The notion of pseudodifferential projections (and subspaces) goes back to the work of Hardy, who defined the celebrated Hardy space as the range of a pseudodifferential projection in the space of square integrable functions on the circle; since then, pseudodifferential subspaces were used in the theory of Toeplitz operators (Gohberg–Krein [37]), in the proof of the finiteness theorem for classical boundary value problems (Calderón, Seeley [26, 67, 69]), in the construction of asymptotics of eigenvalues (Birman–Solomyak [14]) and in other places.

The work was partially supported by RFBR grants NN 05-01-00982, 03-02-16336, 06-01-00098 and presidential grant MK-1713.2005.1.

Subspaces also have significant applications in studies related to topology. Let us give some examples. In the paper of Wojciechowski [73] (an extended account of the results can be found in [18]) it was shown that the space of projections that differ from a fixed (say, pseudodifferential) projection by a compact operator is a classifying space for K-theory. This result is similar to the Atiyah–Jänich theorem [6], which gives a realization of the classifying space for K-theory in terms of Fredholm operators. Subspaces defined by pseudodifferential projections served as a prototype for the definition by Kasparov [42] and Brown–Douglas–Fillmore [25] of the odd analytic K-homology.

Pseudodifferential subspaces are also important in modern elliptic theory. Indeed, it is well known that by no means all elliptic operators on a manifold with boundary have well-defined (Fredholm) boundary value problems. In other words, for a general elliptic operator one cannot impose boundary conditions such that the so-called Shapiro–Lopatinskii condition is satisfied (e.g., see [41]). Unfortunately, the class of operators for which such conditions do not exist includes many important operators such as Cauchy–Riemann, Dirac, signature operators and others. The following question emerges naturally: is there a natural extension of elliptic theory which enables one to define Fredholm boundary value problems for geometric operators?

The answer to this question is contained in this paper. Let us explain it here in a few words. The simplest examples like the Cauchy–Riemann operator show that although these operators do not have Fredholm problems in Sobolev spaces, Fredholm problems do exist if the boundary values belong to some *subspaces* of the Sobolev spaces, e.g., the Hardy space (for the Cauchy–Riemann operator).

Important examples of well-defined boundary value problems were defined for Dirac operators in the series of papers by Atiyah, Patodi, and Singer [2, 3, 4] on spectral asymmetry. In particular, it was shown that for a suitable choice of the subspace the boundary value problem has the Fredholm property and its index was calculated. Actually, a closer look shows that pseudodifferential subspaces appear already in classical boundary value problems as the so-called Calderón–Seeley subspaces. In this case, the Shapiro–Lopatinskii (Atiyah–Bott) condition requires this subspace to be isomorphic to a Sobolev space. Of course, this is a very restrictive condition. Calderón–Seeley subspaces rarely satisfy it. On the other hand, in the framework of pseudodifferential subspaces this restrictive condition is absent and the existence of well-defined boundary value problems for arbitrary operators is a trivial statement: it suffices to take the Calderón–Seeley subspace as the space of boundary values of the problem.

Homotopy invariants of pseudodifferential subspaces and the Atiyah–Patodi–Singer η-invariant. The question of finding homotopy invariants of pseudodifferential subspaces is very important and interesting. It turns out that such invariants can be obtained from suitable index formulas for elliptic operators acting in subspaces. For sufficiently large classes of operators, the index is the sum of contributions of the principal symbol and of the subspaces. More precisely, there are index formulas

[64, 65]

$$\operatorname{ind}(D, \widehat{L}_1, \widehat{L}_2) = f(\sigma) + d(\widehat{L}_1) - d(\widehat{L}_2), \tag{0.1}$$

where the triple $D, \widehat{L}_1, \widehat{L}_2$ defines an elliptic operator in subspaces $\widehat{L}_1$ and $\widehat{L}_2$, σ is the principal symbol of D, d is the dimension functional defined on the class of pseudodifferential subspaces, and f is a functional on the set of principal symbols of elliptic operators. If the subspaces $\widehat{L}_{1,2}$ satisfy the so-called parity condition (see below), $d(\widehat{L})$ is equal to the η-invariant of a self-adjoint operator having $\widehat{L}$ as its positive spectral subspace.

Let us note that the functional d is not a homotopy invariant of the *principal symbol* of the projection defining the subspace. However, one can show that the fractional part of this functional is homotopy invariant. Thus we have the problem of computing this fractional part in topological terms.

The interest in this problem originates from the fact that, as we mentioned earlier, if the parity condition is satisfied, then the functional d coincides with the η-invariant of Atiyah, Patodi, and Singer. The computation of the fractional part of the η-invariant has important applications (see below).

It is well known that the η-invariant of an elliptic self-adjoint operator is only a spectral invariant. However, in some classes of operators its fractional part defines a homotopy invariant. If the η-invariant is a homotopy invariant, one obtains the problem of computing it in topological terms. The first computation of the η-invariant was made in [3, 4], where operators with coefficients in flat bundles were considered.

Another interesting class of operators with homotopy invariant η-invariants was found by P. Gilkey [35]. This class consists of differential operators with the parity of their order opposite to the parity of the dimension of the manifold (*Gilkey's parity condition*). The homotopy invariant fractional part of the η-invariant is computed in this case in terms of subspaces (see [60], [62]). Such computations have important applications in geometry. Here we confine ourselves to a brief survey of results directly related to the present paper. For other aspects of the η-invariant, we refer the reader to remarkable surveys [70], [15], [52], [58].

For example, in the theory of pin^c bordisms (the distinctive feature of this theory is that the bordism group has elements of all arbitrary large orders 2^n) there is a natural question: what numerical invariants can be used to detect torsion elements of high order? It turns out that the answer can be given in terms of the η-invariant. The point is that odd-dimensional pin^c-manifolds bear the natural Dirac operator [33]. The fractional part of the η-invariant of this operator gives a (fractional) genus of pin^c-manifolds. It is proved in [12] that the Stiefel–Whitney numbers and this fractional genus classify pin^c-manifolds up to bordism.

We would like to note that formulas like (0.1) hold for many more operators than those specified by the parity condition. However, the invariants appearing in such formulas may not coincide with the η-invariant. Their determination is a very interesting question. For instance, an analog of the index formula (0.1) for the Dirac operator (its positive spectral subspace does not satisfy Gilkey's condition) involves the Kreck–Stolz invariant [45] and the Eells–Kuiper invariant [29] (see section 3 of the present paper). Let us recall that the former distinguishes homotopy classes of positive scalar curvature metrics, while the latter deals with the 28 exotic 7-spheres of Milnor.

Pseudodifferential subspaces and mod n **index theory.** The K-group of the cotangent bundle T^*X of a closed manifold classifies elliptic operators on X up to stable homotopy. It follows, in particular, that any element of the K-group $K_c(T^*X)$ can be realized as an elliptic operator on X. Does the same statement hold for the K-groups with finite coefficients $\mathbb{Z}_n$? We show (see also [60]) that the answer is "yes": on a smooth manifold, the elements of the K-group with coefficients are realized by pseudodifferential operators in subspaces.

We would like to mention that a similar problem appeared earlier. For instance, in the theory of spectral boundary value problems or b-pseudodifferential operators, Freed and Melrose studied so-called "modulo n" index theory in [30, 31], where manifolds with $\mathbb{Z}_n$-singularities appear as a geometric model [55, 71, 50]. Topological aspects of such manifolds were also studied. Mironov [49] defined the product of $\mathbb{Z}_n$-manifolds, which is again a $\mathbb{Z}_n$-manifold (the classification problem for products in the smooth situation is studied, for example, in [22]). From this point of view, the following problem suggested by Buchstaber is of interest: when is the index (modulo n) of elliptic operators multiplicative under Mironov's product? We note that for the signature operator the answer is "yes" [50].

Outline of the paper. In the first section, we explain the theory of elliptic operators in subspaces of Sobolev space on a closed manifold without boundary. Here we give two important results. The first concerns the necessary and sufficient conditions for the decomposition of the index of an elliptic operator into the sum of homotopy invariant contributions of the principal symbol and the subspaces in which it acts. The second result is the index formula for elliptic operators in pseudodifferential subspaces. Let us note one important fact: the index of an elliptic operator in subspaces is not determined by the principal symbol of the operator, but also depends on the subspaces.[1] In the second section, we discuss the theory of boundary value problems for elliptic operators in subspaces. We show how pseudodifferential subspaces appear in classical boundary value problems (i.e., boundary value porblems satisfying the Shapiro–Lopatinskii condition), Atiyah–Patodi–Singer spectral boundary value problems [2], and boundary value problems for general elliptic operators [66]. We give index formulas for general operators and for specific operators. The third section explains the application of pseudodifferential subspaces to the problem of computation of the fractional part of the η-invariant for the case in which the latter has the homotopy invariance property. We give formulas for the η-invariant in terms of Poincaré duality in K-theory. Finally, in Section 4 we explain index theory "modulo n" of Freed and Melrose on $\mathbb{Z}_n$-manifolds and index theory "modulo n" on manifolds without boundary.

---

[1] In fact, even the *complete symbols* of the operator and the projections defining the subspaces are insufficient to determine the index.

1. Elliptic theory in subspaces on a closed manifold

In this section, we introduce elliptic operators acting in subspaces on a closed manifolds. We use this theory (which is very simple from the analytic point of view) to illustrate the main topological aspects of index theory in subspaces.

1.1. Statement of problems in subspaces

Subspaces and symbols. Let E be a complex vector bundle over a closed manifold M.

Definition 1. [13] A linear subspace $\widehat{L} \subset C^\infty(M,E)$ is said to be *pseudodifferential* if it can be represented as the range

$$\widehat{L} = \operatorname{Im} P$$

of a projection $P : C^\infty(M,E) \to C^\infty(M,E)$, $P^2 = P$, that is a classical (see [43]) pseudodifferential operator of order zero.

Just as pseudodifferential operators are distinguished in the set of all linear operators by the property that they have symbols (which is a function on the cotangent bundle of the manifold), pseudodifferential subspaces also have symbols.

Definition 2. *The symbol L of a pseudodifferential subspace $\widehat{L}$* is the vector bundle

$$L = \operatorname{Im} \sigma(P) \subset \pi^* E, \qquad L \in \operatorname{Vect}(S^*M)$$

over the cosphere bundle S^*M, defined as the range of the principal symbol of P. Here $\pi : S^*M \to M$ is the natural projection.

The symbol of a subspace does not depend on the choice of a projection.

Example 1. *The Hardy space $\widehat{\mathcal{H}} \subset C^\infty(\mathbb{S}^1)$* of boundary values of holomorphic functions in the unit disc $D^2 \subset \mathbb{C}$ is pseudodifferential. Indeed, the orthogonal projection P onto the Hardy space is a pseudodifferential operator of order zero with principal symbol equal to (e.g., see [56])

$$\sigma(P)(\varphi,\xi) = \begin{cases} 1, & \xi = 1, \\ 0, & \xi = -1, \end{cases} \qquad (\varphi,\xi) \in S^*\mathbb{S}^1 = \mathbb{S}^1_+ \sqcup \mathbb{S}^1_-. \tag{1.1}$$

This gives us

$$\mathcal{H}(\varphi,\xi) = \operatorname{Im} \sigma(P)(\varphi,\xi) = \begin{cases} \mathbb{C}, & \text{if } \xi = 1, \\ 0, & \text{if } \xi = -1. \end{cases}$$

To put this another way, the symbol is one-dimensional on the first component of the cosphere bundle and zero-dimensional on the second component. However, in higher dimensions the space of boundary values of holomorphic functions is no longer pseudodifferential. The projections defining such subspaces are called *Szegö projections.* The notion of symbol of such operators and subspaces requires more subtle techniques (see [23]) and goes beyond the scope of the present paper.

Example 2. The space of sections of a vector bundle E is defined by the identity projection, and the symbol coincides with the pullback of E to S^*M.

Subspaces and self-adjoint operators. There is a convenient way to construct subspaces starting from self-adjoint elliptic operators. If A is an elliptic self-adjoint operator of order ≥ 0 on M, then the *nonnegative spectral subspace* denoted by $\widehat{L}_+(A)$ is the subspace generated by eigenvectors of A with nonnegative eigenvalues.

For example, the nonnegative spectral subspace of $-id/d\varphi$ on the circle is the Hardy space. It turns out that in the general case the spectral subspace is pseudodifferential and its symbol can be identified easily.

Proposition 1. *The symbol of the spectral subspace is equal to*

$$L_+(A) = L_+(\sigma(A)), \tag{1.2}$$

where $L_+(\sigma(A)) \in \mathrm{Vect}(S^*M)$ *is the subbundle in* π^*E *generated by eigenvectors of* $\sigma(A)$ *with positive eigenvalues.*

Formula (1.2) can be obtained if we rewrite the projection $\Pi_+(A)$ defining $\widehat{L}_+(A)$ as

$$\Pi_+(A) = \frac{|A| + A}{2|A|}, \quad |A| = \sqrt{A^2}$$

(we assume that A is invertible). By a theorem of Seeley [68], the symbol of the absolute value of an operator is equal to the absolute value of the symbol. Hence $\Pi_+(A)$ is a pseudodifferential operator with symbol

$$\sigma(\Pi_+(A)) = \frac{|\sigma(A)| + \sigma(A)}{2|\sigma(A)|} = \Pi_+(\sigma(A)).$$

This implies (1.2). □

Example 3. The space of closed forms of degree k on a compact manifold M without boundary is pseudodifferential, since it is the spectral subspace of the elliptic self-adjoint operator $d\delta - \delta d$ of order two (δ is the adjoint of the exterior derivative d).

It follows from Proposition 1 that an arbitrary smooth subbundle $L \subset \pi^*E$ is the symbol of a pseudodifferential subspace [13]. To prove this, it suffices to define an elliptic self-adjoint operator A with $L_+(\sigma(A)) = L$. This is obviously possible.

The infinite Grassmannian and the relative index of subspaces. We point out that there are many subspaces with the same symbol. For example, we do not change the symbol if we add any finite-dimensional subspace to a given subspace.

It is useful to make a comparison with the usual operators. Here the space of operators with a given symbol is contractible. In the case of subspaces, the corresponding space has a nontrivial topology. In more detail, let us fix the symbol L of a subspace. Let Gr_L be the (infinite) *Grassmannian* of subspaces with symbol equal to L.

Theorem 1. (Wojciechowski [73]) *Suppose that* $0 < \dim L < \dim E$. *Then the Grassmannian* Gr_L *is a classifying space for* K*-theory; i.e., the set of homotopy classes of maps* $[X, \mathrm{Gr}_L]$ *is isomorphic to the group* $K(X)$ *for any compact space* X.

The classifying map can be given explicitly in terms of one very important invariant of subspaces [24]. *The relative index of a pair of subspaces* $\widehat{L}_{1,2}$ with the same principal symbol is the index of the following Fredholm operator:

$$\operatorname{ind}(\widehat{L}_1, \widehat{L}_2) \stackrel{def}{=} \operatorname{ind}(P_{\widehat{L}_2} : \widehat{L}_1 \to \widehat{L}_2) \in \mathbb{Z},$$

where $P_{\widehat{L}_2}$ is the orthogonal projection onto $\widehat{L}_2$. The relative index is sometimes referred to as the relative dimension of subspaces, since if one of the subspaces is inside another, then it coincides with the corresponding codimension.

Now we use the relative index to give an explicit formula for the isomorphism in the theorem of Wojciechowski: the map takes a family $\{\widehat{L}_x\}_{x\in X}$ of subspaces to the relative index $\operatorname{ind}(\widehat{L}_x, \widehat{L}) \in K(X)$ with some given subspace $\widehat{L}$. It follows from this theorem that the Grassmannian has countably many connected components and two subspaces are homotopic if and only if their relative index is zero.

Operators in subspaces.

Definition 3. [64] *A pseudodifferential operator of order* m *in subspaces* is a triple $(D, \widehat{L}_1, \widehat{L}_2)$, where

$$D : \widehat{L}_1 \longrightarrow \widehat{L}_2$$

is a linear operator acting between pseudodifferential subspaces. We assume that D is a restriction of a pseudodifferential operator of order m acting in the ambient spaces of sections.

For operators in subspaces, it is easy to prove most of the analytic facts of elliptic theory, such as the symbolic calculus, ellipticity, smoothness of solutions and so on.

Definition 4. *The symbol of an operator in subspaces* is the vector bundle homomorphism

$$\sigma(D) : L_1 \longrightarrow L_2. \tag{1.3}$$

The symbol is well defined, since the condition $D\widehat{L}_1 \subset \widehat{L}_2$ can be restated in terms of the projections defining $\widehat{L}_{1,2}$ as $P_2 D P_1 = D P_1$. If we consider the symbols of operators, the latter equality gives (1.3). Note finally that an arbitrary homomorphism (1.3) is the symbol of some operator in subspaces.

Elliptic operators.

Definition 5. An operator in subspaces is *elliptic* if its principal symbol (1.3) is an isomorphism.

Theorem 2. *An elliptic operator D of order m has the Fredholm property as an operator*

$$D : H^s(M, E_1) \supset \overline{\widehat{L}_1} \longrightarrow \overline{\widehat{L}_2} \subset H^{s-m}(M, E_2),$$

in the closures of the subspaces $\widehat{L}_{1,2} \subset C^\infty(M, E_{1,2})$ in the Sobolev norm.

To prove the theorem, it suffices to take as a regularizer an arbitrary operator from $\widehat{L}_2$ to $\widehat{L}_1$ with symbol $\sigma(D)^{-1} : L_2 \longrightarrow L_1$. The desired properties of the regularizer follow from the standard composition formulas. □

The index does not depend on the Sobolev smoothness exponent s and is denoted by $\operatorname{ind}(D, \widehat{L}_1, \widehat{L}_2)$.

As opposed to analytical aspects, the topological aspects of index theory in subspaces have new effects compared with the Atiyah–Singer theory. We will describe these effects in the next section.

1.2. Index decompositions and dimensions of infinite-dimensional subspaces

The most important property of the index of pseudodifferential operators is its homotopy invariance, i.e., constancy for continuous deformations of operators. For operators in subspaces, the index remains constant also for deformations of the subspaces.

Proposition 2. *For a continuous family of Fredholm operators*

$$D_t : \operatorname{Im} P_t \longrightarrow \operatorname{Im} Q_t, \qquad t \in [0,1], \ \operatorname{Im} P_t \in H_1, \operatorname{Im} Q_t \in H_2$$

in subspaces $\operatorname{Im} P_t, \operatorname{Im} Q_t$ of some fixed Hilbert spaces, the index remains constant. By continuity we mean the continuity of the family $D_t : H_1 \to H_2$ and continuity of the families P_t, Q_t.

The *proof* of this proposition can be obtained if we reduce our family to a family in fixed spaces. A reduction can be done by virtue of the following well-known fact: for a continuous family of projections, there exists a continuous family of invertible operators U_t realizing the equivalence of projections $P_t = U_t P_0 U_t^{-1}$ (e.g., see [17]). □

As soon as the index is homotopy invariant, we arrive at the index problem: the index has to be computed in topological terms. However, unlike the index of the usual elliptic operators in sections of vector bundles, the index of operators in subspaces is not determined by the principal symbol of the operator. For example, all finite-dimensional operators have zero principal symbol, but their index can be any number. A closer look at the problem shows that the index is determined if we fix the the principal symbol and the subspaces

$$\operatorname{ind}\left(D, \widehat{L}_1, \widehat{L}_2\right) = f\left(\sigma(D), \widehat{L}_1, \widehat{L}_2\right).$$

Index decomposition problem. Thus there are two sorts of contributions to the index: of the finite-dimensional data of the problem (the principal symbol) and infinite-dimensional (the subspaces). A natural question arises: is the index equal to the sum

$$\operatorname{ind}\left(D,\widehat{L}_1,\widehat{L}_2\right)=f_1\left(\sigma\left(D\right)\right)+f_2\left(\widehat{L}_1,\widehat{L}_2\right), \tag{1.4}$$

of these two contributions? If such a decomposition of the index is possible, then how to obtain the corresponding index formula? Since the index is a homotopy invariant, we will also require that both contributions are homotopy invariant.

Let us analyze the index decomposition (1.4). We first make an obvious remark. If the subspaces were of finite dimension, then the index would be equal to zero plus the difference of dimensions of the spaces. This observation enables us to give the following important reformulation of the index decomposition problem.

Dimension functionals.

Definition 6. A homotopy invariant functional d on the set of subspaces is a *dimension functional* if it has the following property: for two subspaces with equal symbols,

$$d(\widehat{L}_1)-d(\widehat{L}_2)=\operatorname{ind}(\widehat{L}_1,\widehat{L}_2).$$

Remark 1. Usually dimension functionals are defined in terms of trace functionals. Namely, if $T:A\longrightarrow\mathbb{C}$ is a trace functional (this means that T is linear and $T(ab)=T(ba)$) on an operator algebra A that extends the usual operator trace on finite-rank operators. Then a dimension functional of a subspace $\widehat{L}=\operatorname{Im}P$ defined as the range of projection $P\in A$ is defined as $d(\widehat{L}):=T(P)$. Such extensions of the operator trace were studied by Kontsevich-Vishik [44] for algebras of pseudodifferential operators on smooth manifolds. We would also like to refer the reader to [38, 39, 46, 48, 51, 54] for some of the studies of traces on more general operator algebras and applications.

Lemma 1. *There exists an index decomposition* (1.4) *for operators in subspaces* $D:\widehat{L}\to C^\infty(M,F)$ *if and only if there exists a dimension functional on the set of subspaces.*

For the proof, it suffices to show that the difference $\operatorname{ind}(D,\widehat{L})-d(\widehat{L})$ does not depend on the choice of the subspaces and is a homotopy invariant of the symbol. This is proved using the logarithmic property of the index in subspaces: if we take an elliptic operator and replace a subspace by a different subspace with the same principal symbol, then the index is changed by the relative index of subspaces. □

The assumption in the lemma that one of the subspaces is the space of vector bundle sections does not restrict generality, since an arbitrary operator $D:\widehat{L}_1\to\widehat{L}_2$ can be reduced to such a form by adding the identity operator in the orthogonal complement $\widehat{L}_2^\perp$.

Obstruction to index decomposition. As a rule, pseudodifferential subspaces are infinite-dimensional (in the usual sense). Hence it is no wonder that there is an obstruction to the existence of dimension functionals. It is most convenient to describe this obstruction using self-adjoint operators.

Suppose that the desired dimension functional exists. Consider a family of elliptic self-adjoint operators $A_t, t \in [0,1]$. Let us examine what happens with the corresponding family of spectral subspaces $\widehat{L}_+(A_t)$. This family may have discontinuities for smooth variations of the parameter: if some eigenvalue changes its sign, then the spectral subspace changes by a jump (a finite-dimensional subspace is either added to it if the sign changes from minus to plus, or subtracted in the opposite case.Thus the value of the dimension functional of spectral subspaces has to change by the algebraic number of eigenvalues of the family that cross zero during the homotopy:

$$d(\widehat{L}_+(A_1)) - d(\widehat{L}_+(A_0)) = \left\{ \begin{array}{c} \text{algebraic number of eigenvalues} \\ \text{crossing zero during the homotopy} \end{array} \right\}. \tag{1.5}$$

It turns out that there exist *periodic* homotopies of operators ($A_0 = A_1$) for which the number on the right-hand side in (1.5) is nonzero (simple examples can be found in [59]). Thus we arrive at a contradiction. This shows that a universal dimension functional does not exist.

In other words, to define a dimension functional, one cannot consider the entire Grassmannian; rather one has to search for a dimension functional on some smaller classes of subspaces. It is not hard to give a criterion for the existence of such decompositions. Before we formulate the corresponding result exactly, let us introduce one notion appearing in this criterion.

Spectral flow [4]**.** Let $A_t, t \in [0,1]$ be a continuous family of elliptic self-adjoint operators. Then the number on the right-hand side of (1.5) is called the *spectral flow* of the family and denoted by $\operatorname{sf}\{A_\tau\}_{\tau\in[0,1]}$.

Note that this definition makes sense only in the case of general position, when the graph of the spectrum of the family is transversal to the line $\lambda = 0$. A well-defined formula for the spectral flow can be obtained if we put the objects in general position (see [47], [57]). In our situation, this can be done explicitly: we take a small perturbation of the straight line $\lambda = 0$ that makes it a broken line, see Fig. 1, with alternating horizontal and vertical segments. We only assume that the horizontal segments do not meet the spectrum of the family.

Denote the coordinates of vertices of our broken line as $\{(\tau_i, \lambda_i)\}_{i=0,N}$. Let us use this broken line to compute[2] the spectral flow as the net number of eigenvalues passing the broken line from below. In terms of relative indices, the corresponding formula is the sum over vertices

$$\operatorname{sf}\{A_\tau\}_{\tau\in[0,1]} = \sum_{i=1}^{N-1} \operatorname{ind}(\operatorname{Im}\Pi_{\lambda_i}(A_{\tau_i}), \operatorname{Im}\Pi_{\lambda_{i-1}}(A_{\tau_i})), \tag{1.6}$$

[2]Or, speaking rigorously, define.

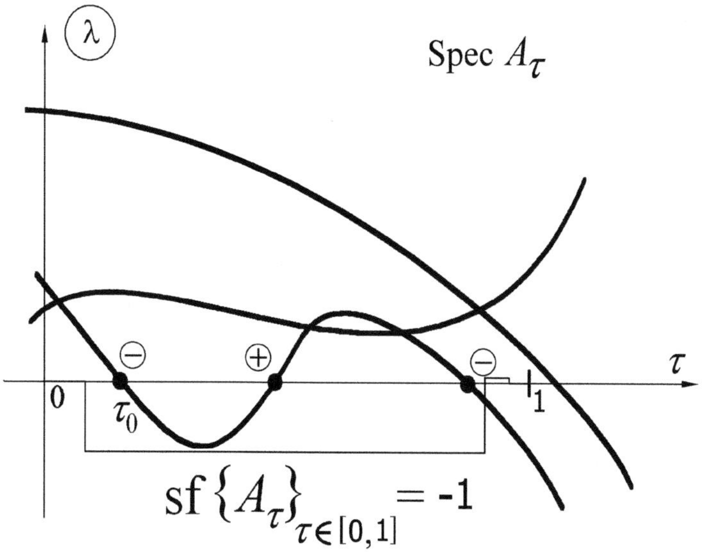

FIGURE 1. Spectral flow

where $\Pi_\lambda(\cdot)$ is the spectral projection of a self-adjoint operator corresponding to eigenvalues greater than or equal to λ. One can show that the spectral flow is well defined; i.e., this number does not depend on the choice of a broken line.

Using this formula as the definition of the spectral flow, it is not hard to obtain Eq. (1.5). Let us now state the necessary and sufficient conditions of the existence of index decompositions.

Criterion of index decompositions [59]**.** Let us fix a subspace Σ in the space of symbols of all pseudodifferential subspaces on M, and let Gr_Σ be the Grassmannian of all pseudodifferential subspaces with symbols in Σ.

Theorem 3. *There exists a dimension functional on the Grassmannian* Gr_Σ *if and only if for every periodic family* $\{A_\tau\}_{\tau\in\mathbb{S}^1}$ *of elliptic self-adjoint operators one has*

$$\mathrm{sf}\{A_\tau\}_{\tau\in\mathbb{S}^1} = 0$$

whenever the symbols of the spectral projections of the operators A_t *belong to* Σ *for all* t.

Sketch of proof. The *necessity* of the vanishing of the spectral flow follows from Eq. (1.5).

Sufficiency. For each connected component $\Sigma_\alpha \subset \Sigma$, let us choose one elliptic operator A_α with the symbol of the spectral subspace $L_+(A_\alpha)$ in Σ_α. We shall consider these operators as reference points; in particular, we define the dimension functional to be zero on them.

Let now A be an elliptic operator on M. Then its principal symbol is an element of some Σ_α. Hence there is a homotopy $\{A_t\}_{t\in[0,1]}$ between A and A_α. Now we can define the dimension functional to be equal to the spectral flow of the homotopy:

$$d\left(\widehat{L}_+(A)\right) \overset{\text{def}}{=} \mathrm{sf}\{A_t\}_{t\in[0,1]}. \qquad \square$$

The assumptions of this theorem can be verified effectively. Indeed, the spectral flow of a periodic family $\mathcal{A} = \{A_t\}_{t\in\mathbb{S}^1}$ of elliptic self-adjoint operators on a closed manifold M is computed by the Atiyah–Patodi–Singer formula [4]

$$\mathrm{sf}\,\{A_t\}_{t\in\mathbb{S}^1} = \left\langle \mathrm{ch} L_+(\mathcal{A}) \mathrm{Td}\,(T^*M\otimes\mathbb{C}),\left[S^*M\times\mathbb{S}^1\right]\right\rangle. \tag{1.7}$$

Here $\mathrm{ch}L_+(\mathcal{A})\in H^{ev}\left(S^*M\times\mathbb{S}^1\right)$ is the Chern character of the bundle

$$L_+(\mathcal{A})\in \mathrm{Vect}\left(S^*M\times\mathbb{S}^1\right)$$

defined by the principal symbol of the family, Td is the Todd class, while the pairing with the fundamental class is denoted by $\langle,\left[S^*M\times\mathbb{S}^1\right]\rangle$.

Thus as a corollary we obtain the following criterion for the existence of index decompositions.

Theorem 4. (on index decompositions) *There exists an index decomposition for elliptic operators in subspaces of the Grassmannian* Gr_Σ *if and only if for an arbitrary periodic family of elliptic self-adjoint operators whose spectral subspaces have symbols in* Σ *the spectral flow is zero.*

Let us consider examples in which this condition is satisfied.

Example 4. (Gilkey's parity condition) Let Σ be the set of symbols of spectral subspaces of elliptic self-adjoint *differential* operators. The spectral flow of periodic families of elliptic operators from this class will be zero if the so-called *parity condition* is satisfied [35]:

$$\mathrm{ord}A+\dim M\equiv 1 (\mathrm{mod}\ 2).$$

For example, for first-order operators the spectral flow of a periodic family A_t is equal to the index of the *differential* operator $\partial/\partial t + A_t$ on the odd-dimensional manifold $M\times\mathbb{S}^1$. It is well known that such indices are trivial (e.g., see [56]).

Actually, the "differentiality" of operators in the parity condition has a geometric origin. Namely, the principal symbol of a differential operator of even order is invariant under the involution $\alpha:(x,\xi)\mapsto(x,-\xi)$. Therefore, the symbol of the spectral subspace is also invariant

$$\alpha^*L=L,\quad L\in\mathrm{Vect}\,(S^*M). \tag{1.8}$$

Such symbols are called *even*. Similarly, the symbols of spectral subspaces of odd-order differential operators are called *odd*. Odd symbols satisfy the condition

$$\alpha^*L\oplus L=\pi^*E,$$

where E stands for the ambient bundle ($L\subset\pi^*E$). In other words, the fibers of an odd symbol L are complementary subspaces at antipodal points of the cosphere bundle. This explains why the natural analog of Gilkey's parity condition for pseudodifferential operators requires that the symbol is even in odd dimensions and odd otherwise.

Let us restrict ourselves to these classes Σ (further examples and explicit index formulas will appear later in the paper, see also [59]).

1.3. Example. Index under Gilkey's parity condition

In this section, we obtain index decompositions for operators in even and odd subspaces. We first consider the even case.

Dimension of even subspaces. Let $\widehat{\mathrm{Even}}\,(M)$ be the set of even pseudodifferential subspaces on a manifold M. The Grothendieck group of the semigroup of homotopy classes of even subspaces is denoted by $K\big(\widehat{\mathrm{Even}}\,(M)\big)$.

Proposition 3. [64] *On an odd-dimensional manifold, one has*

$$(\mathbb{Z}\oplus K(M))\otimes\mathbb{Z}\,[1/2]\simeq K\big(\widehat{\mathrm{Even}}\,(M)\big)\otimes\mathbb{Z}\,[1/2]\,. \tag{1.9}$$

Here $\mathbb{Z}\,[1/2]$ is the ring of dyadic numbers $k/2^n$, $k,n\in\mathbb{Z}$. The map takes each natural number k to a projection of rank k and each vector bundle $E\in\mathrm{Vect}(M)$ to a projection that defines E as a subbundle in a trivial bundle.

Corollary 1. *In odd dimensions, there exists a unique dimension functional (see Definition 6)*

$$d:\widehat{\mathrm{Even}}\,(M)\longrightarrow\mathbb{Z}\,[1/2]$$

that is additive and satisfies the normalization condition

$$d\,(C^\infty(M,E))=0. \tag{1.10}$$

The starting point of the proof is the exact sequence

$$0\longrightarrow\mathbb{Z}\longrightarrow K(\widehat{\mathrm{Even}}(M))\longrightarrow K(P^*M)\longrightarrow 0, \tag{1.11}$$

where $P^*X=S^*X/\mathbb{Z}_2$ is the projectivization of the cosphere bundle. The first map corresponds to the embedding of finite-dimensional subspaces in the even subspaces. The second is induced by the symbol map.

This sequence admits further simplification. Namely, the projection $P^*M\to M$ induces an isomorphism on K-groups modulo 2-torsion if $\dim M$ is odd [35]. Thus taking the tensor product of (1.11) by $\mathbb{Z}[1/2]$ (the product preserves the exactness!) we obtain the exact sequence

$$0\longrightarrow\mathbb{Z}\,[1/2]\longrightarrow K(\widehat{\mathrm{Even}}(M))\otimes\mathbb{Z}\,[1/2]\longrightarrow K(M)\otimes\mathbb{Z}\,[1/2]\longrightarrow 0.$$

The latter sequence is easy to split. The splitting map

$$K^0\,(M)\otimes\mathbb{Z}\,[1/2]\longrightarrow K\big(\widehat{\mathrm{Even}}\,(M)\big)\otimes\mathbb{Z}\,[1/2]\,.$$

takes each vector bundle to the projection onto the space of its sections. The splitting gives the desired isomorphism (1.9). □

Index formula. To obtain an index formula for operators in even subspaces, it is also necessary to define a homotopy invariant of the principal symbol of the operator.

It turns out that the principal symbol of an elliptic operator in even subspaces defines the usual elliptic symbol, i.e., the symbol of elliptic operator in vector bundle sections. Indeed, for a symbol (L_1 and L_2 are even)

$$\sigma(D) : L_1 \to L_2,$$

the composition $\alpha^*\left(\sigma^{-1}(D)\right)\sigma(D)$ takes L_1 to itself. Thus one defines the elliptic symbol

$$\alpha^*\left(\sigma^{-1}(D)\right)\sigma(D) \oplus 1 : \pi^* E \longrightarrow \pi^* E, \tag{1.12}$$

where we make use of the decomposition $\pi^* E = L_1 \oplus L_1^{\perp}$ into complementary bundles.

Theorem 5. [64] *The following index formula holds:*

$$\operatorname{ind}(D, \widehat{L}_1, \widehat{L}_2) = \frac{1}{2}\operatorname{ind}_t[\alpha^*\left(\sigma^{-1}(D)\right)\sigma(D) \oplus 1] + d(\widehat{L}_1) - d(\widehat{L}_2) \tag{1.13}$$

provided that the subspaces are even and the dimension of the manifold is an odd number. Here ind_t *is the topological index of Atiyah and Singer.*

Proof (sketch). 1) Let us take the contributions of the subspaces to the left-hand side of (1.13). Then we interpret the formula as an equality of two homotopy invariants of the principal symbol. Thus it is sufficient to verify the formula for one representative in each homotopy class of principal symbols. 2) The simplest representative can be obtained by Proposition 3. Namely, in geometric terms this proposition claims that the direct sum of 2^N copies of the symbol of the subspace is homotopic to the symbol lifted from the base. Such a homotopy can be lifted to a homotopy of operators in subspaces. 3) For an operator acting in spaces of vector bundle sections, both sides of (1.13) are computed by the Atiyah–Singer formula. They turn out to be equal. □

Remark 2. The contribution of the principal symbol to the index is of course computed by the Atiyah–Singer formula. However, there is a direct geometric realization of this contribution. Namely, consider the so-called *blow-up* ${}^{p}T^*M$ (e.g., see [19]) of the cotangent bundle T^*M along the zero section $M \subset T^*M$

$${}^{p}T^*M = \{(x, \gamma, \xi) \in P^*M \times T^*M | \, \xi \in \gamma\}.$$

In other words ${}^{p}T^*M$ is obtained from the cotangent bundle by two operations: we first delete a tubular neighborhood of the zero section and then identify antipodal points on the boundary (see Fig. 2).

The principal symbol of an operator in even subspaces defines a virtual vector bundle over the blow-up, and the contribution of the principal symbol to the index is expressed by the cohomological formula [64]

$$\operatorname{ind}_t\left[\alpha^*\left(\sigma^{-1}(D)\right)\sigma(D) \oplus 1\right] = \left\langle \operatorname{ch}\left[\sigma(D)\right] \operatorname{Td}\left(T^*M \otimes \mathbb{C}\right), \left[{}^{p}T^*M\right]\right\rangle.$$

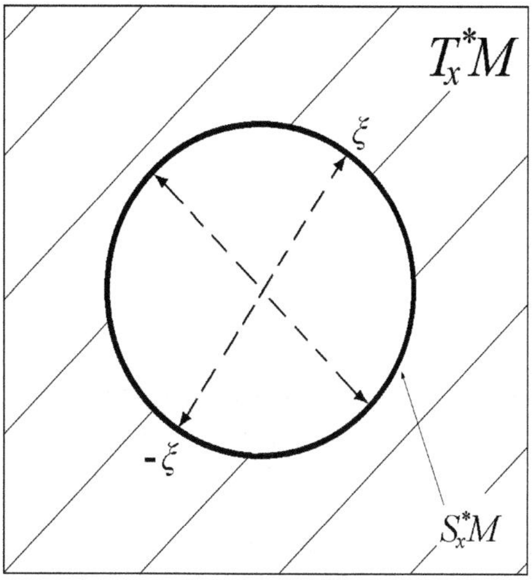

FIGURE 2. The blow-up ${}^pT^*M$ of the cotangent bundle

Thus at the cohomology level the only difference of this topological expression from the Atiyah–Singer formula is a different domain of integration.

Odd theory [65]. The main results of elliptic theory in even subspaces, like the dimension functional and the index formula, have analogs in elliptic theory in odd subspaces modulo some modifications: on an even-dimensional manifold, there exists a unique additive dimension functional of odd subspaces subject to the normalization

$$d(\widehat{L}) + d(\widehat{L}^\perp) = 0,$$

where $\widehat{L}^\perp$ is the complementary bundle. The index formula in odd subspaces is (cf. (1.13))

$$\operatorname{ind}(D, \widehat{L}_1, \widehat{L}_2) = \frac{1}{2}\operatorname{ind}_t[\alpha^*\sigma(D) \oplus \sigma(D)] + d(\widehat{L}_1) - d(\widehat{L}_2).$$

Note that the proofs in the odd case are technically more complicated, since the symbols of odd subspaces cannot be interpreted as vector bundles over the projective space. For example, one has the following interesting fact.

Proposition 4. *The dimension of an odd bundle $L \subset \pi^*E$ over a manifold M of dimension n satisfies*

$$\left.\begin{array}{l} n = 2k \\ n = 2k+1 \end{array}\right\} \Rightarrow \dim L \text{ is divisible by } 2^{k-1}. \tag{1.14}$$

The proof is based on the well-known property of odd functions [34]: an odd function on $\mathbb{S}^n$ defines a section of the Hopf bundle

$$\gamma = \mathbb{S}^n \times \mathbb{C}/\left\{(x,t) \sim (-x,-t)\right\},$$

while an invertible vector-valued function defines a trivialization $l\gamma \simeq \mathbb{C}^l$. On the other hand, the Hopf bundle gives the generator of

$$K\left(\mathbb{RP}^{2k}\right) \simeq K\left(\mathbb{RP}^{2k+1}\right) \simeq \mathbb{Z}_{2^k}$$

(the Adams theorem). Therefore, $2^{\lfloor n/2 \rfloor}$ divides $\dim E$, and we obtain the desired relation (1.14), since an odd vector bundle defined by the projection $p(\xi)$ gives us an invertible odd function

$$i\tau + (2p(\xi) - 1)|\xi|. \qquad \square$$

2. Boundary value problems and subspaces

2.1. Classical boundary value problems

Let D

$$D : C^\infty(M, E) \longrightarrow C^\infty(M, F)$$

be an elliptic differential operator of order $m \geq 1$ on a manifold M with boundary $X = \partial M$. Such operators are never Fredholm: the kernel is infinite-dimensional. To define a Fredholm operator, D should be equipped with boundary conditions.

Boundary conditions. It is convenient to define the boundary conditions using the boundary operator

$$j_X^{m-1} : C^\infty(M, E) \longrightarrow C^\infty(X, E^m|_X),$$

which is defined in terms of the trivialization $X \times [0, 1) \subset M$ of a neighborhood of the boundary with normal coordinate t. The operator

$$j_X^{m-1} u = \left(u|_X, -i\frac{\partial}{\partial t}u\Big|_X, \ldots, \left(-i\frac{\partial}{\partial t}\right)^{m-1} u\Big|_X \right)$$

takes each function u to the value at the boundary of its jet in the normal direction.

Definition 7. *A classical boundary value problem* (see [41]) for a differential operator D is a system of equations of the form

$$\begin{cases} Du = f, & u \in H^s(M, E), f \in H^{s-m}(M, F), \\ Bj_X^{m-1}u = g, & g \in H^\sigma(X, G), \end{cases} \tag{2.1}$$

where

$$B : \bigoplus_{k=0}^{m-1} H^{s-1/2-k}(X, E|_X) \longrightarrow H^\sigma(X, G) \tag{2.2}$$

is a pseudodifferential operator at the boundary; here we assume that the orders of the components $B_k : H^{s-1/2-k}(X, E|_X) \to H^\sigma(X, G)$ are $s - 1/2 - k - \sigma$.

If the smoothness exponent of the Sobolev space is sufficiently large, $s > m - 1/2$, then the operator (D, B) is well defined.

Ellipticity of boundary value problems and the Calderón subspace. The ellipticity condition for classical boundary value problems, known as the *Shapiro–Lopatinskii condition*, can easily be obtained with the use of the following result (see [67], [41]).

Theorem 6. (on the Calderón–Seeley subspace) *Let D be an elliptic differential operator on a manifold with boundary. Then the following assertions hold.*

1. *The cokernel of D is finite-dimensional.*
2. *The Calderón–Seeley space $j_X^{m-1}(\ker D)$ of jets at the boundary of the elements of the kernel is a pseudodifferential subspace. The boundary operator is Fredholm*

$$j_X^{m-1} : \ker D \longrightarrow j_X^{m-1} \ker D.$$

Denote the Calderón–Seeley subspace by $\widehat{L}_+(D)$. Its symbol $L_+(D)$ is a vector bundle over S^*X.

Definition 8. (*Shapiro–Lopatinskii condition*) A boundary value problem is *elliptic* if the restriction of the principal symbol of the boundary condition B to the Calderón subspace is an isomorphism

$$\sigma(B) : L_+(D) \longrightarrow \pi^*G. \tag{2.3}$$

In other words, the ellipticity of the boundary value problem is equivalent to the ellipticity of the boundary operator as an operator in subspaces. □

Theorem 7. *A boundary value problem (D, B) for an elliptic operator D has the Fredholm property if and only if it is elliptic.*

This finiteness theorem follows directly from the properties of the Calderón subspace and the finiteness theorem for operators in subspaces.

The symbol of the Calderón–Seeley subspace can be computed easily. Let $(x, \xi') \in S^*X$ be a point on the cosphere bundle of the boundary. Let

$$L_+(D)_{x,\xi'} \subset E_x^m,$$

be the subspace of Cauchy data of solutions $u(t)$ of the ordinary differential equation

$$\sigma(D)\left(x, 0, \xi', -i\frac{d}{dt}\right) u(t) = 0, \qquad (x, \xi') \in S^*X$$

with constant coefficients on the half-line $\{t \geq 0\}$, that are bounded as $t \to +\infty$. Globally, this family of subspaces defines the smooth vector bundle

$$L_+(D) \subset \pi^* \left. E^m \right|_X, \qquad \pi : S^*X \to X.$$

It can be proved [67] that this bundle is the symbol of the Calderón–Seeley subspace.

Example 5. For the Laplace operator, the bundle $L_+(\Delta)$ coincides with the image of the diagonal embedding $\mathbb{C} \subset \mathbb{C} \oplus \mathbb{C}$. For the Cauchy–Riemann operator $\partial/\partial\bar{z}$ in the unit disk, L_+ is not constant:

$$L_+\left(\frac{\partial}{\partial \bar{z}}\right)\Big|_{\mathbb{S}^1_+} \simeq \mathbb{C}, \qquad L_+\left(\frac{\partial}{\partial \bar{z}}\right)\Big|_{\mathbb{S}^1_-} = 0, \text{ where } S^*\mathbb{S}^1 = \mathbb{S}^1_+ \sqcup \mathbb{S}^1_-.$$

The Atiyah–Bott obstruction and index theorem for boundary value problems. The Shapiro–Lopatinskii condition (2.3) is a restrictive condition on the class of operators D, for which one can define elliptic boundary conditions. Indeed, if an elliptic boundary condition for D exists, then the bundle $L_+(D) \in \operatorname{Vect}(S^*X)$ is a pullback of some bundle on the base X. Such a pullback exists by no means for all operators (the simplest example for which the pullback does not exist is given by the Cauchy–Riemann operator).

The essence of this restriction was uncovered by Atiyah and Bott [7]. They showed that, up to a certain stabilization, the operators possessing elliptic boundary conditions are precisely those with symbols at the boundary homotopic to the symbols independent of the covariables. The situation can be represented by the following K-theory exact sequence:

$$\to K_c(T^*(M \setminus \partial M)) \longrightarrow K_c(T^*M) \xrightarrow{\partial} K_c(\partial T^*M) \to \cdots$$

Namely, elliptic symbols $\sigma(D)$ on M define elements of the group in the center (via the difference construction). On the other hand, the elements of the leftmost group correspond to symbols that are independent of the covariables in a neighborhood of the boundary. Thus the Atiyah–Bott result says that the existence of an elliptic boundary value problem is equivalent to the property that our element comes from $K_c(T^*(M \setminus \partial M))$, while the element obtained by the boundary map ∂ is the obstruction to the existence of elliptic boundary conditions (*the Atiyah–Bott obstruction*). Moreover, Atiyah and Bott showed that the choice of an elliptic boundary condition B explicitly determines some specific element in $K_c(T^*(M \setminus \partial M))$.

Let us note that there is a well-defined topological index map on $K_c(T^*(M \setminus \partial M))$, which, together with the Atiyah–Bott construction, gives an index formula for classical boundary value problems. The reader can find the proof of the index theorem for boundary value problems in [41].

Example 6. Consider the Euler operator

$$d + \delta : \Lambda^{\mathrm{ev}}(M) \longrightarrow \Lambda^{\mathrm{odd}}(M) \tag{2.4}$$

on a compact manifold with boundary. Here $\Lambda^{\mathrm{ev}\,/\,\mathrm{odd}}(M)$ are the spaces of even (odd) differential forms. As elliptic boundary conditions, we can take the absolute boundary conditions

$$j^*(*\omega) = g, \qquad j^* : \Lambda^{\mathrm{odd}}(M) \to \Lambda^{\mathrm{odd}}(X) \tag{2.5}$$

(where $*$ is the Hodge star operator). By Hodge theory on manifolds with boundary (e.g., see [36], [28]), the index of (2.4), (2.5) is equal to the Euler characteristic of M:

$$\operatorname{ind}(d+\delta, j^**) = \chi(M).$$

However, the classical theory has one very significant drawback. Among the classical operators considered in index theory, only the Euler operator admits classical elliptic boundary value problems. The Dirac, Hirzebruch and Todd operators do not admit elliptic boundary conditions: even at a point $x \in M$ the principal symbol of these operators is a rational generator of $K_c(T^*_x M) \simeq \mathbb{Z}$ (e.g., see [56]) and hence is by no means homotopic to a constant symbol.

2.2. Spectral problems of Atiyah, Patodi, and Singer and general boundary value problems in subspaces

We saw in the previous section that many elliptic operators (e.g., Dirac and signature operator) do not have elliptic boundary conditions, since the Atiyah–Bott obstruction for these operators does not vanish. Since these operators are very important in applications, there naturally emerges a question of defining a class of elliptic boundary value problems for *general* elliptic operators, in particular those with a nontrivial Atiyah–Bott obstruction. Such a class of boundary value problems is naturally constructed using the following reasoning.

Recall that the ellipticity condition for a boundary value problem (D, B) requires the isomorphism

$$L_+(D) \overset{\sigma(B)}{\longrightarrow} \pi^* G \tag{2.6}$$

defined by the principal symbol of the boundary condition B. Meanwhile, the obstruction explained by the asymmetry of (2.6): the a priori general bundle $L_+(D)$ over S^*X must be isomorphic to a bundle of a very special form, i.e., a bundle lifted from X. Hence it is clear that the obstruction will disappear if we manage to make a generalization of the notion of boundary conditions such that we could insert an arbitrary vector bundle on S^*X into the right-hand side of (2.6). The simplest realization of this idea is given by the so-called spectral boundary value problems.

Atiyah–Patodi–Singer spectral boundary value problems [2]. Let D be an elliptic differential operator of order one. We shall assume that near the boundary it has a decomposition

$$D|_{\partial M\times[0,1)} = \gamma\left(\frac{\partial}{\partial t} + A\right),$$

where A is an elliptic self-adjoint operator on $X = \partial M$. The *spectral boundary value problem for* D is the system of equations

$$\begin{cases} Du = f, & u \in H^s(M, E), f \in H^{s-1}(M, F), \\ \Pi_+(A)\, u|_X = g, & g \in \operatorname{Im}\Pi_+(A). \end{cases} \tag{2.7}$$

This boundary value problem has the Fredholm property. The reader can prove the coincidence of the bundles $L_+(D)$ and $\operatorname{Im}\sigma(\Pi_+(A))$. Hence (2.6) is the identity map in this case. The statement of spectral problems for differential operators of any order can be found in [53].

General boundary value problems [18, 66]. For an elliptic operator D, consider the boundary value problems

$$\begin{cases} Du = f, & u \in H^s(M,E), f \in H^{s-m}(M,F), \\ Bj_X^{m-1}u = g, & g \in \operatorname{Im} P \subset H^\sigma(X,G), \end{cases} \tag{2.8}$$

which differ from classical boundary value problems (2.1) in the space of boundary data $\operatorname{Im} P$, which is a subspace of the Sobolev space at the boundary and is determined by a pseudodifferential projection P of order zero.

Definition 9. Boundary value problem (2.8) is said to be *elliptic* if the principal symbol of the operator of boundary conditions defines a vector bundle isomorphism

$$\sigma(B) : L_+(D) \to \operatorname{Im} \sigma(P),$$

i.e., the restriction of B to the Calderón subspace is an elliptic operator in subspaces.

The following finiteness theorem holds.

Theorem 8. *Boundary value problem* (2.8) *defines a Fredholm operator if and only if it is elliptic.*

The proof can be obtained from the theorem on the Calderón–Seeley subspace. □

Order reduction of boundary value problems. It is possible to reduce orders of boundary value problems. For classical boundary value problems, one can reduce the boundary value problem using order reduction to a pseudodifferential operator which is a multiplication operator near the boundary and does not require boundary conditions (see [41] or [61] for the description of the reduction procedure; note that the index of such zero-order operators is computed by the Atiyah–Singer formula [9], cf. [27]). For boundary value problems in subspaces, the same method enables one to reduce an arbitrary boundary value problem to a spectral problem for a first-order operator [61]. In addition, the pseudodifferential subspace of the spectral problem can be chosen to coincide with subspace of boundary data of the original problem. For this reason, we consider only spectral problems in the rest of this section.

2.3. Index of boundary value problems in subspaces

We have seen that subspaces are useful if we study analytical properties of spectral problems. In this section, we show that subspaces are also important in the study of topological aspects of these problems: many index formulas for operators in subspaces on closed manifolds (see Section 1) have natural analogs for boundary value problems. To save space, we will give only the formulations of the results.

The index of spectral boundary value problems is not determined by the principal symbol of the operator D. To have a definite index, we have to fix the principal symbol and the spectral subspace. It is impossible to decompose the index as a sum of homotopy invariant contributions of the symbol and the subspace. A decomposition exists if and only if for the class of spectral subspaces at the boundary there exists a dimension functional. Let us give two examples when explicit index formulas can be obtained.

2.4. Examples. The index of operators with parity condition. The index of the signature operator

The index of spectral problems in even subspaces. Consider spectral boundary value problems $(D, \Pi_+(A))$ on an even-dimensional manifold M and suppose additionally that the spectral subspace $\operatorname{Im}\Pi_+(A)$ is even. Finally, we assume that the principal symbol of A is an even function of the covariables.

It turns out that in this case $\sigma(D)$ has a natural continuation to the double of M. Recall that the double

$$2M = M \bigcup\nolimits_{\partial M} M$$

is obtained by gluing two copies of M along the boundary.

To construct the desired continuation, we consider two copies of the manifold. We take the symbol $\sigma(D)$ on the first copy and $\alpha^*\sigma(D)$ on the second copy. Here $\alpha : S^*M \longrightarrow S^*M$ is the antipodal involution of M. Near the boundary, the symbols $\sigma(D)$ and $\alpha^*\sigma(D)$ are

$$i\tau + a(x,\xi) \quad \text{and} \quad -i\tau + a(x,\xi).$$

It is clear that they are mapped one into another as we glue neighborhoods of the boundary:

$$x \to x, \quad t \to -t.$$

Thus the two symbols define an elliptic symbol $\sigma(D) \cup \alpha^*\sigma(D)$ on the double of M. This symbol defines the difference element

$$[\sigma(D) \cup \alpha^*\sigma(D)] \in K_c(T^*2M).$$

in the K-group with compact supports of the cotangent bundle of the double. We define the *topological index* of D to be half the usual topological index of the element on the double

$$\operatorname{ind}_t D \overset{def}{=} \frac{1}{2}\operatorname{ind}_t[\sigma(D) \cup \alpha^*\sigma(D)].$$

Theorem 9. [64] *For spectral boundary value problems in even subspaces, one has*

$$\operatorname{ind}(D, \Pi_+(A)) = \operatorname{ind}_t D - d(\operatorname{Im}\Pi_+(A)).$$

The proof is by analogy with the proof in the case of closed manifolds: one uses homotopies to reduce the spectral problem to the simplest form. In this case, the simplest spectral problem is a classical boundary value problem; i.e., its spectral subspace is the space of sections of a vector bundle. □

Remark 3. A similar index formula is valid for operators in odd subspaces. In this case, one defines the operator $D \cup \alpha^* D^{-1}$ on the double with symbol equal to $\alpha^* \sigma(D)^{-1}$ on the second copy of the manifold.

The index of the signature operator [2]. On a $4k$-dimensional oriented manifold M, consider the signature operator

$$d + d^* : \Lambda^+(M) \longrightarrow \Lambda^-(M),$$

where the $\Lambda^\pm(M)$ are subspaces of forms invariant under the involution

$$\alpha : \Lambda^*(M) \longrightarrow \Lambda^*(M), \quad \alpha|_{\Lambda^p(M)} = (-1)^{\frac{p(p-1)}{2}+k} * .$$

On the boundary of M, we have $\Lambda^\pm(M)|_{\partial M} \simeq \Lambda^*(\partial M)$. If we take a product metric in a neighborhood of the boundary, then the signature operator is equal to

$$\frac{\partial}{\partial t} + A$$

modulo vector bundle isomorphisms (see [2]). The *tangential signature operator* A acts on the boundary

$$A : \Lambda^*(\partial M) \longrightarrow \Lambda^*(\partial M), \quad A\omega = (-1)^{k+p}(d * - \varepsilon * d)\,\omega,$$

where for a form $\omega \in \Lambda^{2p}(\partial M)$ of even degree we have $\varepsilon = 1$, while for $\omega \in \Lambda^{2p-1}(\partial M)$ – $\varepsilon = -1$. This operator is elliptic and self-adjoint.

The index of the spectral boundary value problem can be computed by de Rham–Hodge theory:

$$\operatorname{ind}(d + d^*, \Pi_+) = \operatorname{sign} M - \dim H^*(\partial M)/2,$$

where $\operatorname{sign} M$ is the signature of a manifold with boundary.

We will obtain the index decomposition for the Dirac operator later in Section 3.8, since it involves a new invariant – the η-invariant of Atiyah, Patodi, and Singer.

3. The spectral η-invariant of Atiyah, Patodi, and Singer

3.1. Definition of the η-invariant

Let A be an elliptic self-adjoint operator of a positive order on a closed manifold M. Let us define the *spectral η-function*

$$\eta(A, s) = \sum_{\lambda_j \in \operatorname{Spec} A, \lambda_j \neq 0} \operatorname{sgn}\lambda_i \, |\lambda_i|^{-s} \equiv \operatorname{Tr}\left(A\left(A^2\right)^{-s/2-1/2}\right).$$

It is analytic in the half-plane $\operatorname{Re} s > \dim M/\operatorname{ord} D$ (for these parameter values, the series is absolutely convergent).

Definition 10. [2] *The η-invariant* of the operator A is

$$\eta(A) = \frac{1}{2}\left(\eta(A, 0) + \dim\ker A\right) \in \mathbb{R}. \tag{3.1}$$

Remark 4. The spectral η-invariant can be understood as a kind of infinite-dimensional analog of the notion of signature of a quadratic form, since in finite dimensions a self-adjoint operator defines a quadratic form and the η-invariant of an invertible operator is equal to the signature modulo the factor 1/2.

Of course, for (3.1) to make sense, it is necessary to have the analytic continuation of the η-function to $s = 0$.

Theorem 10. [4],[32] *The η-function extends to a meromorphic function on the complex plane with possible poles at $s_j = \frac{\operatorname{ord} D - j}{\dim M}$, $j \in \mathbb{Z}_+$. At $s = 0$, the function is analytic.*

Let us note that the meromorphic continuation is a consequence of the expression of the η-function in terms of the ζ-function

$$\zeta(A, s) = \sum_{\lambda_j \in \operatorname{Spec} A} \lambda_i^{-s}$$

of positive operators:

$$\eta(A, s) = \frac{\zeta(A_+, s) - \zeta(A_-, s)}{2^s - 1}, \qquad \text{where } A_\pm = \frac{(3|A| \pm A)}{2}.$$

The meromorphic continuation for the ζ-function is well known (e.g., see [68]).

However, the analyticity of the η-function at the origin is more intricate. More precisely, the residue is equal to

$$\operatorname*{Res}_{s=0} \eta(A, s) = \frac{\zeta(A_+, 0) - \zeta(A_-, 0)}{\ln 2}. \tag{3.2}$$

The ζ-invariants in this formula can be expressed as integrals over M of some complicated expressions in the complete symbol of A. The integrand is in general nonzero! Nevertheless, Atiyah, Patodi, and Singer proved for odd-dimensional manifolds [4] and Gilkey [32] proved for even-dimensional manifolds that the residue is zero. Hence the η-function is holomorphic at the origin and the η-invariant is well defined.

Rather surprisingly, up to now there is no purely analytic proof of the analyticity of the η-function at the origin. The results cited earlier all rely on global topological methods. However, the triviality of the residue is proved by an explicit analytic computation for Dirac type operators in [16].

Example 7. On a circle of length 2π with coordinate φ, consider

$$A_t = -i\frac{d}{d\varphi} + t.$$

Here t is a real constant. Let us compute the η-invariant. The spectrum of A is the lattice $t + \mathbb{Z}$. Thus the η-invariant is a periodic function of t (with period 1). Assume that $0 < t < 1$. Gathering the eigenvalues in pairs, we obtain

$$\eta(A_t, s) = \sum_{n \geq 1} \left[(n+t)^{-s} - (n-t)^{-s} \right] + t^{-s}.$$

This series is absolutely convergent on the semiaxis $s > 0$, and the limit as $s \to +0$ is $-2t+1$ (we use the Taylor expansion for the expression in the brackets); hence

$$\eta(A_t) = \frac{\eta(A_t, 0) + \dim\ker A_t}{2} = \frac{1}{2} - \{t\},$$

where $\{\} \in [0,1)$ is the fractional part. Thus we see that for our smooth elliptic family A_t the family of η-invariants is only piecewise smooth. Moreover, the jumps (they are integral) happen as some eigenvalue of the operator changes its sign.

The behavior of the η-invariant under deformations of the operator. In the last example, we observed the piecewise smooth variation of the η-invariant for smooth variation of operators. It turns out that the η-invariant has similar properties in the general case. More precisely, the following result holds.

Proposition 5. [4] *Let $A_t, t \in [0,1]$, be a smooth family of elliptic self-adjoint operators. Then the function $\eta(A_t)$ is piecewise smooth. It decomposes as the sum*

$$\eta(A_{t'}) - \eta(A_0) = \mathrm{sf}\,(A_t)_{t\in[0,t']} + \int_0^{t'} \omega(t_0)\, dt_0, \tag{3.3}$$

of a locally constant function, the spectral flow of Section 1.2, *and the smooth function*

$$\omega(t_0) = \left.\frac{d}{dt}\zeta(B_{t,t_0})\right|_{t=t_0} \in C^\infty[0,1],$$

where we use the ζ-invariant of the auxiliary family $B_{t,t_0} = |A_{t_0}| + P_{\ker A_{t_0}} + (t-t_0)\dot{A}_{t_0}$. Here $P_{\ker A}$ is the projection onto the kernel of A.

Proof (sketch). If the family is invertible, then one can easily write out the derivatives of the η- and ζ-functions:

$$\frac{d}{dt}\zeta(B_t, s) = -s\mathrm{Tr}\left(\dot{B}_t\, B_t^{-s-1}\right), \qquad \frac{d}{dt}\eta(A_t, s) = -s\mathrm{Tr}\left(\dot{A}_t\, (A_t^2)^{-\frac{1}{2}(s+1)}\right).$$

It is clear now that (3.3) holds for $s = t = 0$.

If the family is not invertible, then the decomposition (3.3) can be obtained making use of broken lines from the definition of spectral flow (see Fig. 1). This technique reduces us to the case of invertible families. □

Remark 5. (Singer) These properties motivate an interesting interpretation of the η-invariant, which is similar to the interpretation of index as the invariant labelling the connected components of the space of Fredholm operators. Consider the space of self-adjoint Fredholm operators. Atiyah and Singer [10] proved that this space consists of three connected components. Two components correspond to semibounded operators and are contractible. However, the third component (containing operators with spectrum unbounded in both directions) has a nontrivial topology. Let us denote it by $\mathcal{F}_s$. This space is a classifying space for odd K-theory:

$$[X, \mathcal{F}_s] \simeq K^1(X).$$

In particular, its first cohomology is $H^1(\mathcal{F}_s) \simeq \mathbb{Z}$. The generator of this group is given by the spectral flow of periodic families

$$[\mathrm{sf}] \in H^1(\mathcal{F}_s),$$

more precisely, the value of the cocycle sf on a loop $(A_t)_{t\in\mathbb{S}^1}$ is equal to the spectral flow along the loop. It turns out that the η-invariant provides a de Rham representative of this cohomology class (at least on the subspace of pseudodifferential operators). More precisely, let us define the 1-form: for the loop $(A_t)_{t\in[0,\varepsilon]} \subset \mathcal{F}_s$ in the space of *pseudodifferential operators,* we set

$$\omega(A_t) = \frac{d}{dt}\{\eta(A_t)\}\Big|_{t=0}.$$

Proposition 5 gives the equality of cohomology classes $-[\omega]$ and $[\mathrm{sf}]$, in other words, one has

$$-\int_{(A_t)_{t\in\mathbb{S}^1}} \omega = \mathrm{sf}(A_t)_{t\in\mathbb{S}^1}.$$

3.2. How to make η homotopy invariant?

The η-invariant for general operators is not homotopy invariant and takes arbitrary real values. However, for special classes of operators it is possible to define homotopy invariants using the η-invariant. To this end, it is necessary to require that both components in (3.3) are equal to zero. The triviality of the spectral flow sf can be achieved in two ways: either we consider only the fractional part of the η-invariant $\{\eta(A)\} \in \mathbb{R}/\mathbb{Z}$ (this is used in [4] when considering invariants of flat bundles, see also Section 3.3) or by requiring that the spectral flow is trivial for the operators considered (such situation appears for the signature operator or for the Dirac operator under the positive scalar curvature assumption, e.g., see [21]). To prove the vanishing of the smooth component of the variation, it is necessary to have a formula for the derivative of the ζ-function. R. Seeley [68] proved (see also [1] and [40]) that the value of the ζ-function at zero can be computed in terms of the principal symbol of the operator. Let us proceed to the formula. Let A be an elliptic self-adjoint nonnegative operator with complete symbol

$$\sigma(A) \sim a_m + a_{m-1} + a_{m-2} + \cdots.$$

Let us introduce the following recurrent family of symbols b_{-m-j}, $j \geq 0$:

$$b_{-m-j}(x,\xi,\lambda)(a_m(x,\xi)-\lambda) + \sum_{\substack{k+l+|\alpha|=j,\\ l>0}} \frac{1}{\alpha!}(-i\partial_\xi)^\alpha b_{-m-k}(x,\xi,\lambda)(-i\partial_x)^\alpha a_{m-l}(x,\xi) = 0. \tag{3.4}$$

The symbols depend on auxiliary parameter λ. Then the ζ-invariant is

$$2\zeta(A) \stackrel{\text{def}}{=} \zeta(A,0) + \dim\ker A$$
$$= \frac{1}{(2\pi)^{\dim M} \operatorname{ord} A} \int_{S^*M} dx d\xi \int_0^\infty b_{-\dim M - \operatorname{ord} A}(x,\xi,-\lambda)\, d\lambda. \quad (3.5)$$

Analyzing the symmetries of this formula, one can find a number of operator classes for which the derivative of the η-invariant is zero. Two such classes are considered in the next sections.

3.3. η-invariants and flat bundles

Recall that a vector bundle $\gamma \in \operatorname{Vect}(M)$ is *flat* if it is defined by locally constant transition functions. Consider an operator

$$A : C^\infty(M, E) \longrightarrow C^\infty(M, F).$$

Then we can define the *operator A with coefficients in the flat bundle*:

$$A \otimes 1_\gamma : C^\infty(M, E \otimes \gamma) \longrightarrow C^\infty(M, F \otimes \gamma).$$

It can be defined by patching together local expressions in coordinate charts for the direct sum of $\dim\gamma$ copies of A using the transition functions. To preserve the self-adjointness, one requires additionally that the transition functions for the flat bundle are unitary. Finally, if A is a pseudodifferential operator, then the operator with coefficients is well defined modulo infinitely smoothing operators.

Example 8. On the circle, the operator $-id/d\varphi + t$ is isomorphic to the operator $-id/d\varphi \otimes 1_\gamma$ with coefficients in γ, where the line bundle γ is defined by the transition function $e^{2\pi i t}$. The isomorphism

$$e^{-ti\varphi}\left(-i\frac{d}{d\varphi}\right)e^{it\varphi} = -i\frac{d}{d\varphi} + t$$

uses the trivialization $e^{it\varphi}$ of γ.

Proposition 6. [4] *The fractional part of the η-invariant is homotopy invariant in the class of direct sums*

$$A \otimes 1_\gamma \oplus (-\dim\gamma A)$$

with a given flat vector bundle γ.

To prove the proposition, one notes that $A \otimes 1_\gamma$ and $\dim\gamma A$ are locally isomorphic. Therefore, we obtain

$$\frac{d}{dt}\{\eta(A_t \otimes 1_\gamma)\} = \frac{d}{dt}\{n\eta(A_t)\}$$

by means of the locality of these derivatives, see (3.5). □

ρ-invariant [3]. Consider an oriented Riemannian manifold M of dimension $4k-1$. There is a self-adjoint Hirzebruch operator

$$A|_{\Lambda^{2p}(M)} = (-1)^{k+p}(d* - *d), \qquad A : \Lambda^{ev}(M) \longrightarrow \Lambda^{ev}(M).$$

In this case, the difference

$$\eta(A \otimes 1_\gamma) - \dim \gamma \eta(A) \in \mathbb{R}$$

defines a homotopy invariant. Indeed, by Hodge theory the kernels of A and $A \otimes 1_\gamma$ coincide with the corresponding cohomology of M (with a local coefficient system γ in the second case); hence their dimensions do not depend on the choice of metric on M. This difference is called the *ρ-invariant of manifold M and flat bundle γ*.

Operators with coefficients in flat bundles have been thoroughly studied already in the classical paper of Atiyah, Patodi, and Singer. Thus, in this paper, we recall only the index formula pertaining to this case.

The index formula in trivialized flat bundles [4]. Suppose that the flat bundle γ is trivial $\gamma \overset{\alpha}{\simeq} \mathbb{C}^n$ and A is an elliptic self-adjoint operator as above.

Then the triple (γ, α, A) defines an elliptic operator in subspaces:

$$\Pi_+(nA)(1 \otimes \alpha^*) : \operatorname{Im}\Pi_+(A \otimes 1_\gamma) \longrightarrow \operatorname{Im}\Pi_+(nA). \tag{3.6}$$

Let us fix the flat bundle with its trivialization and consider the index decomposition problem for operators (3.6) into the sum of contributions of the principal symbol of the operator and the contribution of subspaces. It is not difficult to see that the necessary condition for such decompositions (Theorem 4) is satisfied. Then we can take the difference of the η-invariants

$$\eta(A \otimes 1_\gamma) - n\eta(A)$$

as the contribution of the subspaces. This difference will be referred to as *the relative η-invariant.* The corresponding index theorem in trivialized flat bundles was obtained by Atiyah, Patodi, and Singer.

Theorem 11. *One has*

$$\begin{aligned} \operatorname{ind}(\Pi_+(nA)(1 \otimes \alpha^*) : \operatorname{Im}\Pi_+(A \otimes 1_\gamma) \longrightarrow \operatorname{Im}\Pi_+(nA)) \\ \langle \operatorname{ch}L_+(A)\operatorname{ch}(\gamma,\alpha)Td(T^*M \otimes \mathbb{C}), [S^*M]\rangle + \eta(A \otimes 1_\gamma) - n\eta(A), \end{aligned} \tag{3.7}$$

where $\operatorname{ch}(\gamma, \alpha) \in H^{odd}(M, \mathbb{Q})$ *is the Chern character of the trivialized flat bundle.*

The proof uses the heat equation method.

As a corollary, let us take the fractional part of the index formula. Then we obtain an expression of the fractional part of the relative η-invariant in topological terms. For nontrivial flat bundles, the relative η-invariant was also computed in [4], but the formula in this case is written in K-theoretic terms and is less explicit.

3.4. η-invariant and parity conditions

One more class of examples of η-invariants without continuous component of the variation is related to parity conditions.

Theorem 12. [35] *The fractional part of the η-invariant of an elliptic self-adjoint differential operator A on a manifold M is invariant under homotopies if the following parity condition is satisfied:*

$$\operatorname{ord}A + \dim M \equiv 1(\operatorname{mod}2).$$

Idea of the proof. The homogeneous components of the complete symbol of a differential operator are polynomials. Hence they are even or odd with respect to the involution $\xi \mapsto -\xi$ acting on the covariables. An accurate account of this symmetry in (3.5) shows that the local expression for the derivative of the η-invariant is zero. □

The η-invariant as a dimension functional. It is clear that if the continuous component of the variation of the η-invariant is missing, then the η-invariant can be considered as a dimension functional (compare (1.5) with (3.3)).

Theorem 13. [64] *Let A be an elliptic self-adjoint differential operator of a positive order. Then the η-invariant is equal to the value of the dimension functional of Section* 1.3 *on the spectral subspace $\widehat{L}_+(A)$*

$$\eta(A) = d\left(\widehat{L}_+(A)\right)$$

provided that $\operatorname{ord}A + \dim M \equiv 1(\operatorname{mod}2)$.

To prove the theorem, it suffices to check the normalization condition.

This result shows that *we can substitute the η-invariant for the functional d in the index formulas of Section* 1.3 *provided that the pseudodifferential subspace is defined as the spectral subspace of a differential operator.*

Remark 6. To prove Theorem 13, one has to work with η-invariants in the broader context of *pseudodifferential operators*, for which the statement of Theorem 12 is true. We refer the reader to [64] for the precise statement of the parity condition for this case.

Computation of the fractional part of the η-invariant. If the parity condition is satisfied, then the fractional part $\{\eta(A)\}$ is topologically invariant and can be computed in topological terms. It turns out that this invariant strongly depends on the orientation bundle $\Lambda^n(M)$.

Theorem 14. [63] *The fractional part of twice the η-invariant is equal to the pairing*

$$\{2\eta(A)\} = \left\langle [\sigma(A)], 1 - [\Lambda^n(M)] \right\rangle \in \mathbb{Z}\left[\frac{1}{2}\right]/\mathbb{Z}$$

of the difference element of the operator with the orientation bundle $\Lambda^n(M)$, $n =$ $\dim M$, *where the brackets denote the (nondegenerate) Poincaré duality*

$$\langle,\rangle : \ \mathrm{Tor}K_c^1(T^*M) \times \mathrm{Tor}K^0(M) \quad \longrightarrow \mathbb{Q}/\mathbb{Z}$$

in K*-theory for the torsion subgroups.*

Let us make a couple of remarks concerning this formula.

1. The computation is based on the following property of symbols of subspaces with parity conditions. For N sufficiently large, the sum $2^N L$ can be lifted from the base M. If we choose an isomorphism $\sigma : 2^N L \longrightarrow \pi^* F$, where $F \in \mathrm{Vect}(M)$, then the index formula in subspaces expresses the fractional part of the η-invariant in terms of the index of the corresponding operator

$$\widehat{\sigma} : \widehat{L} \longrightarrow C^\infty(M, F),$$

as a *residue* modulo 2^N. Such an index-residue can be computed in K-theory with coefficients in the group $\mathbb{Z}_{2^N}$ (the corresponding index theory modulo n is discussed in Section 4.2). Finally, the expression in terms of Poincaré duality is none other than a short way of expressing the corresponding K-theoretical formula.

2. The orientation bundle appears naturally in the problem, since the involution $(x, \xi) \leftrightarrow (x, -\xi)$ acts on $K_c^*(T^*M)$ as a product with the virtual bundle $(-1)^{\dim M}[\Lambda^n(M)]$ (see [62]).

Corollary 2. *If the parity condition is satisfied, then the* η*-invariant on an orientable manifold is half-integer. On a nonorientable manifold* M *of dimension* $2k$ *or* $2k+1$, *the following estimate of the denominator of the* η*-invariant holds:*

$$\{2^{k+1}\eta(A)\} = 0. \tag{3.8}$$

Indeed, the orientation bundle $\Lambda^n(M^n)$ has the structure group $\mathbb{Z}_2$. Hence it is induced by the canonical bundle over $\mathbb{RP}^n$. The reduced K-groups of the projective spaces are the torsion groups $\widetilde{K}\left(\mathbb{RP}^{2k}\right) \simeq \widetilde{K}\left(\mathbb{RP}^{2k+1}\right) \simeq \mathbb{Z}_{2^k}$. Hence

$$2^k\left(1 - [\Lambda^n(M^n)]\right) = 0.$$

Substituting this equality into the formula for the η-invariant, we obtain the desired assertion. □

Remark 7. The formula for the fractional part of the η-invariant can be rewritten, by analogy with the Atiyah–Singer formula, in terms of the direct image map

$$\{2\eta(A)\} = f_![\sigma(A)],$$

where $f : M \rightarrow \mathbb{RP}^{2N}$ is the map classifying the orientation bundle. Here we assume the identification $K^1(T^*\mathbb{RP}^{2n}) = \mathbb{Z}_{2^n} \subset \mathbb{Q}/\mathbb{Z}$.

Examples of first-order operators. We have seen that the properties of the η-invariant for operators with Gilkey's parity condition substantially depend on the properties of the manifold. In the orientable case, one can obtain a half-integral η-invariant at most. This possibility is easy to realize, e.g., by the operators $d+\delta$ on all forms:

$$\{\eta(d+\delta)\} = \left\{\frac{\chi(M)}{2}\right\}.$$

The computation is based on the fact that this operator is isomorphic to the matrix $\begin{pmatrix} 0 & D \\ D^* & 0 \end{pmatrix}$ with the Euler operator $D = d+\delta$ acting from even to odd forms. The eigenvalues of this matrix are symmetric with respect to the origin. Therefore, the η-function is zero identically.

It turns out that on *nonorientable manifolds* there exist operators with arbitrary dyadic η-invariants. Examples of such operators were first constructed by P. Gilkey [33].

An operator on $\mathbb{RP}^{2n}$ with a very fractional η-invariant. Let us define a Dirac type operator on an even-dimensional real projective space $\mathbb{RP}^{2n}$. To this end, we consider a set of Hermitian Clifford $2^n \times 2^n$ matrices $e_0, e_1, \ldots, e_{2n}$:

$$e_k e_j + e_j e_k = 2\delta_{kj}.$$

For a vector $v = (v_0, \ldots, v_{2n}) \in \mathbb{R}^{2n+1}$, we define a linear operator

$$e(v) = \sum_{i=0}^{2n} v_i e_i : \mathbb{C}^{2^n} \longrightarrow \mathbb{C}^{2^n}.$$

It is invertible if $v \neq 0$. Consider the Hermitian symbol

$$\sigma(D)(x,\xi) = ie(x)\,e(\xi) : \mathbb{C}^{2^n} \longrightarrow \mathbb{C}^{2^n}$$

on the unit sphere $\mathbb{S}^{2n} \subset \mathbb{R}^{2n+1}$, where ξ is a tangent vector at $x \in \mathbb{S}^{2n}$. The symbol is invariant under the involution $(x,\xi) \to (-x,-\xi)$. Thus it defines a symbol on $\mathbb{RP}^{2n}$.

Theorem 15. [33]

$$\{\eta(D)\} = \frac{1}{2^{n+1}}. \tag{3.9}$$

For simplicity, we will only compute the denominator of the η-invariant.

The reduced K-group of $\mathbb{RP}^{2n}$ is a cyclic group $\widetilde{K}(\mathbb{RP}^{2n}) \simeq \mathbb{Z}_{2^n}$, and the generator is given by the orientation bundle

$$1 - \left[\Lambda^{2n}\left(\mathbb{RP}^{2n}\right)\right] \in \widetilde{K}\left(\mathbb{RP}^{2n}\right).$$

On the other hand, the symbol defines the generator of the isomorphic group

$$[\sigma(D)] \in K_c^1\left(T^*\mathbb{RP}^{2n}\right) = \operatorname{Tor} K_c^1\left(T^*\mathbb{RP}^{2n}\right) \simeq \mathbb{Z}_{2^n}.$$

Hence, by Poincaré duality for torsion groups (see above) the pairing of the generators is nonzero and has the largest possible denominator

$$\left\langle 2^{n-1}\left[\sigma\left(D\right)\right], 1-\left[\Lambda^{2n}\left(\mathbb{RP}^{2n}\right)\right]\right\rangle = \frac{1}{2} \in \mathbb{Q}/\mathbb{Z}.$$

It remains now to express the pairing in terms of the η-invariant. We have

$$\left\{2^n \eta\left(D\right)\right\} = \frac{1}{2}.$$

□

3.5. Examples of second-order operators with nontrivial η-invariants

The problem of nontriviality of the η-invariant for second-order operators was stated by P. Gilkey [35]. For a long time, the main difficulty of the problem was the absence of nontrivial elliptic operators of order two. There was essentially one nontrivial operator $d\delta - \delta d$ acting on differential forms. However, its η-invariant turned out to be integer-valued [64]. From a different point of view, this operator is generated by the operators of de Rham–Hodge theory and is in some sense an analog of the Euler operator. To obtain more interesting operators, one has to define the analog of the Dirac operator.

Such an operator was constructed in [63].

Example 9. We define a second-order differential operator $\mathcal{D}$ on $\mathbb{RP}^{2n} \times \mathbb{S}^1$. To this end, we denote the coordinates by x, φ, the dual coordinates by ξ, τ. On the cylinder $\mathbb{RP}^{2n} \times [0, \pi]$ we define

$$\mathcal{D}' = \begin{pmatrix} 2\sin\varphi\left(-i\frac{\partial}{\partial\varphi}\right)D - i\cos\varphi D & \triangle_x e^{-i\varphi} + \left(-i\frac{\partial}{\partial\varphi}\right)e^{i\varphi}\left(-i\frac{\partial}{\partial\varphi}\right) \\ \triangle_x e^{i\varphi} + \left(-i\frac{\partial}{\partial\varphi}\right)e^{-i\varphi}\left(-i\frac{\partial}{\partial\varphi}\right) & 2\sin\varphi\left(i\frac{\partial}{\partial\varphi}\right)D + i\cos\varphi D \end{pmatrix}, \tag{3.10}$$

where D is the pin^c Dirac operator on the projective space (see previous section), and $\triangle_x = D^2$ is its Laplacian. The operator $\mathcal{D}'$ is symmetric and elliptic. The ellipticity follows from the following formula for the principal symbol

$$\sigma\left(\mathcal{D}'\right)^2(\xi,\tau) = \left(\xi^2+\tau^2\right)^2.$$

(In other words, the operator $\mathcal{D}'$ is the square root of the square of the Laplacian.) Let F be the vector bundle over the product $\mathbb{RP}^{2n} \times \mathbb{S}^1$, obtained by twisting the trivial bundle $\mathbb{C}^{2^n} \oplus \mathbb{C}^{2^n}$ with the matrix-valued function

$$\begin{pmatrix} 0 & 1 \\ -1 & 0 \end{pmatrix}$$

defined on the base $\mathbb{RP}^{2n} \times \{0\}$. Then $\sigma(\mathcal{D})'$ can be considered as acting in F:

$$\sigma(\mathcal{D}') : \pi^* F \longrightarrow \pi^* F, \quad \pi : S^*(\mathbb{RP}^{2n} \times \mathbb{S}^1) \to \mathbb{R}P^{\,2n} \times \mathbb{S}^1.$$

Denote by

$$\mathcal{D} : C^\infty\left(\mathbb{RP}^{2n} \times \mathbb{S}^1, F\right) \longrightarrow C^\infty\left(\mathbb{RP}^{2n} \times \mathbb{S}^1, F\right)$$

the elliptic self-adjoint differential operator obtained by smoothing the coefficients of $\mathcal{D}'$.

The topological formulas for the η-invariant obtained earlier enables us to prove the following result, solving the problem of nontriviality of η-invariants for even-order operators.

Theorem 16. [63] *One has*

$$\{2\eta(\mathcal{D})\} = \frac{1}{2^{n-1}}.$$

The idea of the proof is to interpret the operator $\mathcal{D}$ as an exterior tensor product of an operator on the projective space by an elliptic operator on the circle. Then the η-invariant is also a product of the η-invariant on $\mathbb{RP}^n$ and the index on $\mathbb{S}^1$. Unfortunately, the operator itself does not have this product structure. But K-theoretically such a representation holds:

$$[\sigma(\mathcal{D})] = [\sigma(D)] \cdot [\sigma(D_1)] \in K_c^1\left(T^*\left(\mathbb{RP}^{2n} \times \mathbb{S}^1\right)\right) \tag{3.11}$$

with an elliptic pseudodifferential operator of index two

$$D_1 = \frac{1}{2}\left[e^{-i\varphi}\left(Q + |Q|\right) + e^{i\varphi}\left(|Q| - Q\right)\right], \quad Q = -i\frac{d}{d\varphi}$$

on $\mathbb{S}^1$. To obtain the theorem, it now suffices to substitute (3.11) in the formula for the η-invariant in terms of Poincaré duality and use the multiplicative property of the pairing. □

3.6. Applications to bordisms and embeddings of manifolds

η-invariants on pin^c**-manifolds and bordisms.** The operator on the projective space constructed in Section 3.4 is a specialization of the Dirac operator on a (nonorientable) pin^c-manifold. The definition of this operator can be found in [33]. We note only that the group pin$^c(n)$ is defined in terms of the extension

$$0 \longrightarrow \mathbb{Z}_2 \longrightarrow \mathrm{pin}^c(n) \longrightarrow O(n) \times U(1) \longrightarrow 0.$$

This sequence defines a natural projection pin$^c(n) \to O(n)$. Finally, a manifold M of dimension n is a pin^c-*manifold* if its structure group is reduced to pin$^c(n)$.

On even-dimensional pin^c-manifolds, the Dirac operator, denoted by D, is self-adjoint. Therefore, on such manifolds the fractional topological invariant

$$\{\eta(D)\} \in \mathbb{Z}\left[\frac{1}{2}\right]/\mathbb{Z}$$

is well defined. Moreover, one can also show that this fractional part is invariant under bordisms of pin^c-manifolds.

Theorem 17. [12] pin^c*-manifolds M_1 and M_2 are bordant if and only if they have equal Stiefel–Whitney numbers and the fractional parts of the η-invariants*

$$\{\eta(D_{M_1})\} = \{\eta(D_{M_2})\}.$$

Note that the characteristic property of the theory of pin^c-bordisms is that the bordism group Ω^{pin^c} has nontrivial elements (represented by the projective spaces $\mathbb{RP}^{2k}$) of arbitrarily large order 2^k. To distinguish these elements, the fractional analytic invariant is indispensable.

Application to embeddings. A natural geometric setting when one can consider second-order operators of Section 3.5 was found by P. Gilkey [32]. Let us describe the situation in more detail.

Suppose N is a submanifold with trivial normal bundle in a closed manifold M. Then one can define an elliptic self-adjoint second-order operator that is concentrated in a neighborhood of the submanifold in the following sense. This operator is a sum of Laplacians outside a neighborhood of N. We shall consider for simplicity the codimension one case.

Let us introduce coordinates in a tubular neighborhood U of the boundary, x tangent to the submanifold, and $\varphi \in [0, 2\pi]$ normal to the submanifold. The dual coordinates are ξ, τ.

Consider the quadratic transformation

$$h(\tau, \xi) = \left(\tau^2 - \xi^2, \tau\xi\right) : S^*M|_U \longrightarrow S^*M|_U$$

over U. At a point $(x, \varphi) \in U$, this map is a two-sheeted covering of the sphere. It takes big circles passing through the North pole to big circles passing through the North pole being run through with double speed. Let us define the family of vector bundle homomorphisms

$$\Phi_\varphi : T^*M|_N \longrightarrow \mathbb{R} \oplus T^*M|_N$$

parametrized by $\varphi \in [0, 2\pi]$:

$$\Phi_\varphi(\tau, \xi) = \begin{cases} \left(\cos\varphi\left(\xi^2 + \tau^2\right), \sin\varphi h(\tau, \xi)\right), & \varphi \in [0, \pi], \\ \left(\cos\varphi\left(\xi^2 + \tau^2\right), \sin\varphi\left(\xi^2 + \tau^2\right), 0, \dots\right), & \varphi \in [\pi, 2\pi]. \end{cases}$$

Suppose that $N \times [0, 2\pi]$ is equipped with a pin^c-structure. Consider the corresponding Clifford module

$$c : \mathrm{Cl}(\mathbb{R} \oplus T^*M|_N) \longrightarrow \mathrm{End}(E),$$

where $\mathrm{Cl}(V)$ is the bundle of Clifford algebras of a real vector bundle and E is the spinor bundle of $N \times [0, 2\pi]$. The symbol $\sigma(D)$ of order two is defined in a neighborhood of N as the composition

$$\sigma(D)(\varphi, \tau, \xi) \overset{\text{def}}{=} c(\Phi_\varphi(\tau, \xi)).$$

On the boundary of the neighborhood, the symbol is

$$\sigma(D)(\varphi, \tau, \xi)|_{\varphi=0,2\pi} = c(1, 0, \dots, 0)\left(\xi^2 + \tau^2\right).$$

It coincides with the direct sum of the symbols of Laplacians. Thus, $\sigma(D)$ extends outside U as the direct sum of symbols of Laplacians.

Second-order operators associated with submanifolds with trivial normal bundles enable one to construct some topological invariants.

Proposition 7. *Let M be a closed smooth manifold,* $\dim M = 2k + 1$. *A necessary condition for an embedding*

$$\mathbb{RP}^{2k} \subset M$$

of the projective space $\mathbb{RP}^{2k}$ *with trivial normal bundle is the surjectivity of the direct image map*

$$f_! : \mathrm{Tor} K_c^1(T^*M) \longrightarrow \mathbb{Z}_{2^k} \subset \mathbb{Z}[1/2]/\mathbb{Z},$$

induced by the map $f : M \longrightarrow B\mathbb{Z}_2 = \mathbb{RP}^\infty$ *classifying the orientation bundle* $\Lambda^{2k+1}(M)$. *In particular,* $K_c^1(T^*M)$ *has to have nontrivial torsion elements of order* 2^k.

The proposition can be proved if one notes that on M we have a second-order operator with the η-invariant having denominator 2^{k+1}. On the other hand, the η-invariant is computed by the direct image map corresponding to the classifying space. □

3.7. The Atiyah–Patodi–Singer formula

An expression for the index of spectral boundary value problems was found in [2]. Namely, using the heat equation method [8], the relation

$$\mathrm{ind}(D, \Pi_+(A)) = \int_X a(D) - \eta(A) \tag{3.12}$$

was obtained for the index of a spectral boundary value problem on a manifold X for an elliptic operator of order one that has the decomposition $\partial/\partial t + A$ near the boundary. Here $a(D)$ is by definition the constant term in the local asymptotic expansion

$$\mathrm{tr}(e^{-tD^*D}(x,x)) - \mathrm{tr}(e^{-tDD^*}(x,x))$$

as $t \to 0$. It is defined (as in the case of operators on closed manifolds) as some algebraic expression in the coefficients of the operator and their derivatives. The second term is the *η-invariant* of the tangential operator A.

In the general case, the formula for $a(D)$ is extremely cumbersome. However, for the classical operators (Euler operator, signature operator, etc.) it is described by explicitly computable formulas. For example, if D is the signature operator with coefficients in a bundle E equipped with a connection, we have

$$a(D) = L(X)\mathrm{ch}E,$$

where $L(X) \in \Lambda^{ev}(X)$ stands for the Hirzebruch polynomial [56] in the Pontryagin forms of the Riemannian manifold and $\mathrm{ch}E \in \Lambda^{ev}(X)$ is the Chern character of the bundle computed in terms of the connection via Chern–Weil theory.

A similar expression for the form is valid for the remaining classical operators; one has only to substitute polynomials corresponding to the operators in place of the L-polynomial.

The Atiyah–Patodi–Singer formula has numerous applications ranging from algebraic geometry [5] to quantum field theory [72]. As an explanation of this phenomenon, M. Atiyah points out that for the signature operator the formula (3.12) relates three objects of entirely different nature: a *topological* invariant (the signature) on the left-hand side and a *metric* invariant (the integral of the Pontryagin forms) as well as the *spectral* η-invariant on the right-hand side.

3.8. The index decomposition of the Dirac operator (Kreck-Stolz invariant)

Consider the Dirac operator on a $4k$-dimensional manifold M. In this section, we obtain, following [45], a decomposition of the index of this operator. Strikingly enough, it turns out that the index defect can be defined using the signature operator! The decomposition is made under the assumption that the boundary has trivial Pontryagin classes.

Denote the Dirac operator by $\mathcal{D}$ and its tangential operator by $\mathcal{A}$. By Atiyah–Patodi–Singer theorem, the sum $\operatorname{ind}\mathcal{D} + \eta(\mathcal{A})$ is equal to the integral over the manifold with boundary of the $\widehat{A}$ polynomial in the Pontryagin forms

$$\int_M \widehat{A}(p).$$

Hence to construct an index decomposition we have to decompose this integral into a geometric invariant determined by the boundary and the remainder homotopy invariant term. Such a decomposition is obtained for all decomposable components of the $\widehat{A}$-polynomial (except the top component p_k!) by the following lemma.

Lemma 2. *Let α, β be positive degree forms on M whose restrictions to the boundary are exact. Then*

$$\int_M \alpha \wedge \beta = \int_{\partial M} \widehat{\alpha} \wedge \beta + \left\langle j^{-1}[\alpha] \cup j^{-1}[\beta], [M, \partial M] \right\rangle,$$

where $d\widehat{\alpha} = \alpha|_{\partial M}$, $j^{-1}[\alpha]$ is an arbitrary preimage of the cohomology class $[\alpha] \in H^*(M)$ under the restriction map $j : H^*(M, \partial M) \to H^*(M)$, and $\langle \cdot, [M, \partial M] \rangle$ is the pairing with the fundamental class. Moreover, the terms on the right-hand side of the relation do not depend on the choices.

The proof uses integration by parts. □

Denote the first term in the formula of the lemma by

$$\int_{\partial M} d^{-1}(\alpha \wedge \beta) \stackrel{\text{def}}{=} \int_{\partial M} \widehat{\alpha} \wedge \beta.$$

It only remains now to decompose the integral of the top Pontryagin class. Here we make use of the signature operator: for this operator, the Atiyah–Patodi–Singer formula contains the integral of the L-class. In turn, the L-class also includes the top Pontryagin class. A standard computation shows that the sum $\widehat{A}(p) + a_k L(p)$, where $a_k = (2^{2k+1}(2^{2k-1} - 1))^{-1}$ in degrees $\leq 4k$, contains only products of Pontryagin classes of positive degrees, i.e., does not contain the top class p_k. For example, for an 8-manifold one has

$$\widehat{A}(M) = \frac{1}{5760}(7p_1^2 - 4p_2), \quad L(M) = \frac{1}{45}(7p_2 - (p_1)^2).$$

Further, by rewriting the sum $\operatorname{ind}\mathcal{D} + a_k \operatorname{ind} D$ by Atiyah–Patodi–Singer theorem, we obtain

$$\operatorname{ind}\mathcal{D} + \eta(\mathcal{A}) - a_k\eta(A) - \int_{\partial M} d^{-1}(\widehat{A} + a_k L)(p) = t(M),$$

where $t(M)$ denotes the following topological invariant of manifolds with boundary:

$$t(M) = \langle (\widehat{A} + a_k L)(j^{-1}p(M)), [M, \partial M]\rangle - a_k \operatorname{ind} D.$$

The contribution of the boundary is called the *Kreck–Stolz invariant* $s(\partial M, g)$ of the manifold ∂M with metric g.

Theorem 18. [45] *The index of the Dirac operator on a manifold with boundary having trivial Pontryagin classes has the decomposition*

$$\operatorname{ind} \mathcal{D} = t(M) + s(\partial M, g),$$

where the Kreck–Stolz invariant $s(\partial M, g)$ is a homotopy invariant of the metric in the class of metrics of positive scalar curvature.

4. Elliptic theory "modulo n"

Another field of applications of elliptic theory in subspaces concerns so-called theories with coefficients in finite groups $\mathbb{Z}_n$. The characteristic feature of such theories is that, for some reason, the index in such theories makes sense only as a residue.

In this section, we briefly discuss two versions of this theory: on $\mathbb{Z}_n$-manifolds and on closed manifolds.

4.1. The Freed–Melrose theory on $\mathbb{Z}_k$-manifolds

Definition 11. A *$\mathbb{Z}_k$-manifold* is a compact smooth manifold M with boundary ∂M, which is a disjoint union of k copies of some manifold X

$$\partial M = M_1 \sqcup \cdots \sqcup M_k, \quad M_i \overset{g_i}{\approx} X$$

with fixed diffeomorphisms g_i.

$\mathbb{Z}_k$-manifolds naturally define the singular spaces

$$\overline{M} = M / \{M_i \overset{g_j^{-1} g_i}{\sim} M_j\}, \tag{4.1}$$

identifying points on the components of the boundary (see Fig. 3).

$\mathbb{Z}_k$-manifolds were introduced by Sullivan [71]. One of the motivations indicating the interest in this class of singular spaces is the fact that (in the orientable case) a singular manifold $\overline{M}$ carries a fundamental cycle in homology with coefficients $\mathbb{Z}_k$

$$[\overline{M}] \in H_m\left(\overline{M}, \mathbb{Z}_k\right), \qquad m = \dim M.$$

These singular manifolds were also used as a geometric realization of bordisms with coefficients in $\mathbb{Z}_k$. For further research in this direction, we refer the reader to [20]. On a $\mathbb{Z}_k$-manifold, we fix a collar neighborhood of the boundary

$$U_{\partial M} \approx [0, 1) \times X \times \{1, \ldots, k\}. \tag{4.2}$$

Definition 12. *An operator on a $\mathbb{Z}_k$-manifold* is an operator D on M, which is invariant under the group of permutations of the components of the collar neighborhood (4.2) of the boundary.

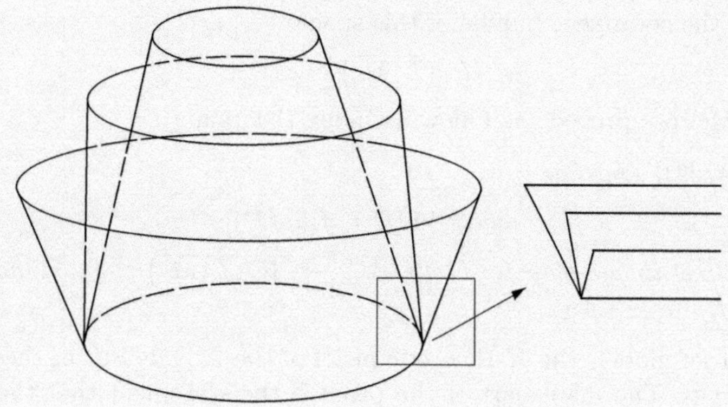

FIGURE 3. Manifold with singularities

We equip elliptic operators D on $\mathbb{Z}_k$-manifolds with spectral boundary conditions.

Proposition 8. *The index residue* $\operatorname{mod} k\text{-}\operatorname{ind}(D, \Pi_+(A))$ *is constant for homotopies of the operator* D.

Indeed, for a continuous homotopy $\{D_t\}_{t\in[0,1]}$ the change of the index is equal to the spectral flow of the family of tangential operators on the boundary

$$\operatorname{ind}(D_1, \Pi_+(A_1)) - \operatorname{ind}(D_0, \Pi_+(A_0)) = -\operatorname{sf}\{A_t\}_{t\in[0,1]}.$$

On the other hand, the family A_t at the boundary is by assumption the direct sum of k copies of some family on X. Therefore, the spectral flow is divisible by k. □

This homotopy invariant index residue was computed in terms of the principal symbol by Freed and Melrose. Let us briefly recall their result.

Theorem of Freed and Melrose. The cotangent bundle T^*M is a noncompact $\mathbb{Z}_k$-manifold, and the principal symbol of operator D defines an element in the K-group

$$[\sigma(D)] \in K_c\left(\overline{T^*M}\right).$$

(Here we use identification (4.1).) It turns out that the direct image mapping in K-theory extends to the category of $\mathbb{Z}_k$-manifolds (the morphisms are by definition those embeddings which map boundary to boundary preserving the $\mathbb{Z}_k$-structure). More precisely, for an embedding $f : M \to N$ one has

$$f_! : K_c\left(\overline{T^*M}\right) \longrightarrow K_c\left(\overline{T^*N}\right).$$

On the other hand, one can construct a universal space for such embeddings (i.e., the space into which any $\mathbb{Z}_k$ manifold can be embedded). The universal space can be obtained from $\mathbb{R}^L$ by deleting k disjoint disks of a sufficiently small radius. We obtain the $\mathbb{Z}_k$-manifold M_k whose boundary is the union of k spheres (diffeomorphisms of spheres are given by parallel translations). It is easy to compute the

K-group of the cotangent bundle of this space

$$K_c\left(\overline{T^*M_k}\right) \simeq \mathbb{Z}_k.$$

Freed and Melrose proved the following index theorem.

Theorem 19. [31] *One has*

$$\operatorname{mod} k\text{-ind}D = f_!\left[\sigma\left(D\right)\right],$$

where the direct image map $f_! : K_c\left(\overline{T^*M}\right) \longrightarrow K_c\left(\overline{T^*M_k}\right) \simeq \mathbb{Z}_k$ *is induced by an embedding* $f : M \longrightarrow M_k$.

The proof models the K-theoretic proof of the Atiyah–Singer theorem based on embeddings. The main part of the proof is the statement that the analytical index is preserved for embeddings, i.e., for an embedding of M in N the following diagram commutes

$$\begin{array}{ccc} K_c(\overline{T^*M}) & \stackrel{f_!}{\longrightarrow} & K_c(\overline{T^*N}) \\ & \searrow \quad \swarrow & \\ & \mathbb{Z}_n. & \end{array}$$

4.2. Index modulo n on closed manifolds

Index-residues also arise on a closed manifold. Consider the following question: what objects of elliptic theory correspond to the elements of K-group $K_c(T^*M, \mathbb{Z}_n)$ with coefficients $\mathbb{Z}_n$?

The answer is given in terms of operators in subspaces

$$D : n\widehat{L} \longrightarrow C^\infty\left(M, F\right). \tag{4.3}$$

Leu us show how symbols of such operators define elements of the K-group with coefficients. To this end, we recall the definition of the latter.

K-theory with coefficients. It is defined as

$$K_c\left(T^*M, \mathbb{Z}_n\right) = K_c\left(T^*M \times \mathbb{M}_n, T^*M \times pt\right); \tag{4.4}$$

where $\mathbb{M}_n$ is the so-called Moore space of the group $\mathbb{Z}_n$. An explicit construction of this space can be found in [3]. We will only use the fact that the reduced K-groups of the Moore space is $\mathbb{Z}_n$ and generated by the difference $1 - [\gamma]$, where γ is a line bundle. We will also fix a trivialization

$$n\gamma \stackrel{\beta}{\simeq} \mathbb{C}^n.$$

Geometric construction of elements of K-groups with coefficients. It follows from definition (4.4) that elements of $K_c\left(T^*M, \mathbb{Z}_n\right)$ can be realized in terms of families of elliptic symbols[3] on M. The family is parametrized by the Moore space. It is

[3]Here we use the difference construction for families (see [11]). It associates element $[\sigma] \in K\left(T^*M \times X\right)$ with a family $\sigma\left(x\right)$, $x \in X$ of elliptic symbols on M parametrized by space X:

$$\sigma\left(x\right) : \pi^*E \longrightarrow \pi^*F, \quad E, F \in \operatorname{Vect}\left(M \times X\right), \ \pi : S^*M \times X \to M \times X.$$

easy to define such a family as a composition:

$$\begin{aligned} C^\infty(M,F) &\overset{D^{-1}}{\longrightarrow} n\widehat{L}, \\ n\widehat{L} &\overset{\beta^{-1}\otimes 1_{\widehat{L}}}{\longrightarrow} \gamma\otimes n\widehat{L}, \\ \gamma\otimes n\widehat{L} &\overset{1_\gamma\otimes D}{\longrightarrow} \gamma\otimes C^\infty(M,F), \end{aligned} \tag{4.5}$$

where D^{-1} is an almost inverse and the last family is obtained by twisting with γ. The family of symbols corresponding to this composition defines the desired element in the K-group with coefficients. Denote it by

$$[\sigma(D)] \in K_c(T^*M, \mathbb{Z}_n).$$

In [60], it is shown that the K-group with coefficients is actually isomorphic to the group of stable homotopy classes of operators (4.3). Let us conclude this section with an index theorem.

Index theorem. Note that the index of operator (4.3) as a residue modulo n

$$\operatorname{mod} n\text{-ind}\, D \in \mathbb{Z}_n$$

is a homotopy invariant of the principal symbol of the operator. The following theorem gives an expression for this index in topological terms.

Theorem 20. *One has*

$$\operatorname{mod} n\text{-ind}\, D = p_!\,[\sigma(D)], \tag{4.6}$$

where the direct image map $p_! : K(T^*M, \mathbb{Z}_n) \longrightarrow \widetilde{K}(pt, \mathbb{Z}_n) = \mathbb{Z}_n$ *in* K*-theory with coefficients is induced by* $p : M \longrightarrow pt$.

Let us apply the Atiyah–Singer index formula for families to compute the index of the composition (4.5). This formula expresses the index as the right-hand side of (4.6). On the other hand, the index of the composition can be computed directly as

$$\operatorname{ind} D([\gamma]-1) \in K(\mathbb{M}_n),$$

i.e., it coincides with the modulo n index of the operator in subspaces in the group $\widetilde{K}(pt, \mathbb{Z}_n) = \mathbb{Z}_n$. □

Acknowledgment

We are grateful to Prof. V.M. Buchstaber and the referee for a number of valuable remarks.

References

[1] M.S. Agranovich. *Elliptic operators on closed manifolds.* In Partial differential equations. VI, volume **63** of *Encycl. Math. Sci.*, pages 1–130, 1994.

[2] M. Atiyah, V. Patodi, and I. Singer. *Spectral asymmetry and Riemannian geometry* I. Math. Proc. Cambridge Philos. Soc., **77** (1975), 43–69.

[3] M. Atiyah, V. Patodi, and I. Singer. *Spectral asymmetry and Riemannian geometry* II. Math. Proc. Cambridge Philos. Soc., **78** (1976), 405–432.

[4] M. Atiyah, V. Patodi, and I. Singer. *Spectral asymmetry and Riemannian geometry* III. Math. Proc. Cambridge Philos. Soc., **79** (1976), 71–99.

[5] M.F. Atiyah. *The logarithm of the Dedekind η-function.* Math. Annalen, **278** (1987), 335–380.

[6] M.F. Atiyah. *K-Theory.* The Advanced Book Program. Addison–Wesley, Inc., second edition, 1989.

[7] M.F. Atiyah and R. Bott. The index problem for manifolds with boundary. In *Bombay Colloquium on Differential Analysis*, pages 175–186, Oxford, 1964.

[8] M.F. Atiyah, R. Bott, and V.K. Patodi. *On the heat equation and the index theorem.* Invent. Math., **19** (1973), 279–330.

[9] M.F. Atiyah and I.M. Singer. *The index of elliptic operators* I. Ann. of Math., **87** (1968), 484–530.

[10] M.F. Atiyah and I.M. Singer. *Index theory for skew-adjoint Fredholm operators.* Publ. Math. IHES, **37** (1969), 5–26.

[11] M.F. Atiyah and I.M. Singer. *The index of elliptic operators* IV. Ann. Math., **93** (1971), 119–138.

[12] A. Bahri and P. Gilkey. *The eta invariant, Pin^c bordism, and equivariant $Spin^c$ bordism for cyclic 2-groups.* Pacific Jour. Math., **128** (1987), no. 1, 1–24.

[13] M.S. Birman and M.Z. Solomyak. *On the subspaces admitting a pseudodifferential projection.* Vestnik LGU, **1** (1982), 18–25.

[14] M.Sh. Birman and M.Z. Solomyak. *The asymptotic behavior of the spectrum of variational problems on solutions of elliptic systems.* Zapiski LOMI, **115** (1982), 23–39.

[15] J.-M. Bismut. Local index theory, eta invariants and holomorphic torsion: A survey. *Surveys in differential geometry*, volume 3, pages 1–76, Cambridge, 1998.

[16] J.-M. Bismut and D.S. Freed. *The analysis of elliptic families.* I. Commun. Math. Phys., **106** (1986), 159–176.

[17] B. Blackadar. *K-Theory for Operator Algebras.* Cambridge University Press, 1998. Second edition.

[18] B. Booß-Bavnbek and K. Wojciechowski. *Elliptic Boundary Problems for Dirac Operators.* Birkhäuser, Boston–Basel–Berlin, 1993.

[19] R. Bott and L. Tu. *Differential Forms in Algebraic Topology*, volume 82 of *Graduate Texts in Mathematics.* Springer–Verlag, Berlin–Heidelberg–New York, 1982.

[20] B. Botvinnik. *Manifolds with singularities accepting a metric of positive scalar curvature.* Geom. Topol., **5** (2001), 683–718.

[21] B. Botvinnik and P. Gilkey. *The Gromov–Lawson–Rosenberg conjecture: The twisted case.* Houston J. Math., **23** (1997), no. 1, 143–160.

[22] B.I. Botvinnik, V.M. Bukhshtaber, S.P. Novikov, and S.A. Yuzvinskiĭ. *Algebraic aspects of multiplication theory in complex cobordisms.* Uspekhi Mat. Nauk, **55** (2000), no. 4, 5–24.

[23] L. Boutet de Monvel and V. Guillemin. *The spectral theory of Toeplitz operators*, volume 99 of *Ann. of Math. Studies.* Princeton University Press, Princeton, 1981.

[24] L. Brown, R. Douglas, and P. Fillmore. *Unitary equivalence modulo the compact operators and extensions of C^*-algebras*, volume 345 of *Lecture Notes in Math.* 1973.

[25] L. Brown, R. Douglas, and P. Fillmore. *Extensions of C^*-algebras and K-homology.* Ann. Math. II, **105** (1977), 265–324.

[26] A.P. Calderón. Boundary value problems for elliptic equations. *Outlines of the Joint Soviet–American Symposium on Partial Differential Equations,* 303–304, 1963.

[27] C. Carvalho. *A K-theory proof of the cobordism invariance of the index.* available at math.KT/0408260, to appear in K-theory.

[28] A.A. Dezin. *Multidimensional Analysis and Discrete Models.* CRC–Press, Boca Raton, Florida, USA, 1995. English transl.: Nauka, Moscow, 1990.

[29] J. Eells Jr. and N.H. Kuiper. *An invariant for certain smooth manifolds.* Ann. Mat. Pura Appl. (4), **60** (1962), 93–110.

[30] D. Freed. *$\mathbb{Z}/k$ manifolds and families of Dirac operators.* Invent. Math., **92** (1988), no. 2, 243–254.

[31] D. Freed and R. Melrose. *A mod k index theorem.* Invent. Math., **107** (1992), no. 2, 283–299.

[32] P.B. Gilkey. *The residue of the global eta function at the origin.* Adv. in Math., **40** (1981), 290–307.

[33] P.B. Gilkey. *The eta invariant for even-dimensional Pin^c manifolds.* Adv. in Math., **58** (1985), 243–284.

[34] P.B. Gilkey. *The eta invariant and non-singular bilinear products on $\mathbf{R}^\mathbf{n}$.* Can. Math. Bull., **30** (1987), 147–154.

[35] P.B. Gilkey. The eta invariant of even order operators. *Lecture Notes in Mathematics,* **1410**:202–211, 1989.

[36] P.B. Gilkey. *Invariance theory, the heat equation, and the Atiyah-Singer index theorem.* CRC Press, Boca Raton, FL, second edition, 1995.

[37] I.Ts. Gohberg and M.G. Krein. *Systems of integral equations on the half-line with kernels depending on the difference of the arguments.* Uspekhi Matem. Nauk, **13** (1958), no. 2, 3–72.

[38] G. Grubb. A resolvent approach to traces and zeta Laurent expansions. *Spectral geometry of manifolds with boundary and decomposition of manifolds, Contemp. Math.,* **366**, pages 67–93. Amer. Math. Soc., Providence, RI, 2005.

[39] G. Grubb and E. Schrohe. *Traces and quasi-traces on the Boutet de Monvel algebra.* Ann. Inst. Fourier (Grenoble), **54** (2004), no. 5, 1641–1696.

[40] G. Grubb and R.T. Seeley. *Zeta and eta functions for Atiyah-Patodi-Singer operators.* J. Geom. Anal., **6** (1996), no. 1, 31–77.

[41] L. Hörmander. *The Analysis of Linear Partial Differential Operators. III.* Springer–Verlag, Berlin Heidelberg New York Tokyo, 1985.

[42] G.G. Kasparov. *The generalized index of elliptic operators.* Funct. Anal. Appl., **7** (1973), 238–240.

[43] J. Kohn and L. Nirenberg. *An algebra of pseudo-differential operators.* Comm. Pure Appl. Math., **18** (1965), 269–305.

[44] M. Kontsevich and S. Vishik. Geometry of determinants of elliptic operators. In *Functional analysis on the eve of the 21'st century,* volume 131 of *Progr. Math.,* pages 173–197. Birkhäuser, Boston, 1995.

[45] M. Kreck and S. Stolz. *Nonconnected moduli spaces of positive sectional curvature metrics.* J. Amer. Math. Soc., **6** (1993), no. 4, 825–850.

[46] R. Lauter and S. Moroianu. *An index formula on manifolds with fibered cusp ends.* J. of Geom. Anal., **15** (2005), no. 2, 261–283.

[47] R. Melrose. *The Atiyah–Patodi–Singer Index Theorem.* Research Notes in Mathematics. A.K. Peters, Boston, 1993.

[48] R. Melrose. *The eta invariant and families of pseudodifferential operators.* Math. Research Letters, **2** (1995), no. 5, 541–561.

[49] O.K. Mironov. *Existence of multiplicative structures in the theory of cobordism with singularities.* Izv. Akad. Nauk SSR Ser. Mat., **39** (1975), no. 5, 1065–1092.

[50] J. Morgan and D. Sullivan. *The transversality characteristic class and linking cycles in surgery theory.* Ann. of Math., II. Ser, **99** (1974), 463–544.

[51] S. Moroianu. *Homology and residues of adiabatic pseudodifferential operators.* Nagoya Math. J., **175** (2004), 171–221.

[52] W. Müller. *Eta-invariant (some recent developments).* Sem. Bourbaki. Astérisque, **227** (1994), 335–364.

[53] V. Nazaikinskii, B.-W. Schulze, B. Sternin, and V. Shatalov. *Spectral boundary value problems and elliptic equations on singular manifolds.* Differents. Uravnenija, **34** (1998), no. 5, 695–708. English trans.: Differential Equations, **34** (1998), no. 5, 696–710.

[54] V. Nistor. *Asymptotics and index for families invariant with respect to a bundle of Lie groups.* Rev. Roum. Math. Pures Appl., **47** (2002), no. 4, 451–483.

[55] S.P. Novikov. Pontrjagin classes, the fundamental group and some problems of stable algebra. In *Essays on Topology and Related Topics (Mémoires dédiés à Georges de Rham)*, pages 147–155. Springer, New York, 1970.

[56] R.S. Palais. *Seminar on the Atiyah–Singer index theorem.* Princeton Univ. Press, Princeton, NJ, 1965.

[57] J. Phillips. *Self-adjoint Fredholm operators and spectral flow.* Canad. Math. Bull., **39** (1996), no. 4, 460–467.

[58] S. Rosenberg. *Nonlocal invariants in index theory.* Bull. AMS, **34** (1997), no. 4, 423–433.

[59] A. Savin, B.-W. Schulze, and B. Sternin. *On invariant index formulas for spectral boundary value problems.* Differentsial'nye uravnenija, **35** (1999), no. 5, 709–718.

[60] A. Savin, B.-W. Schulze, and B. Sternin. *Elliptic Operators in Subspaces and the Eta Invariant.* K-theory, **27** (2002), no. 3, 253–272.

[61] A. Savin and B. Sternin. *To the problem of homotopy classification of the elliptic boundary value problems.* Doklady Mathematics, **63** (2001), no. 2, 174–178.

[62] A. Savin and B. Sternin. *The eta-invariant and Pontryagin duality in K-theory.* Math. Notes, **71** (2002), no. 2, 245–261. arXiv: math/0006046.

[63] A. Savin and B. Sternin. *The eta invariant and parity conditions.* Adv. in Math., **182** (2004), no. 2, 173–203.

[64] A.Yu. Savin and B.Yu. Sternin. *Elliptic operators in even subspaces.* Matem. sbornik, **190** (1999), no. 8, 125–160. English transl.: Sbornik: Mathematics **190** (1999), no. 8, 1195–1228; arXiv: math/9907027.

[65] A.Yu. Savin and B.Yu. Sternin. *Elliptic operators in odd subspaces.* Matem. sbornik, **191** (2000), no. 8, 89–112. English transl.: Sbornik: Mathematics **191** (2000), no. 8, arXiv: math/9907039.

[66] B.-W. Schulze, B. Sternin, and V. Shatalov. *On general boundary value problems for elliptic equations.* Math. Sb., **189** (1998), no. 10, 145–160. English transl.: Sbornik: Mathematics **189** (1998), no. 10, 1573–1586.

[67] R.T. Seeley. *Singular integrals and boundary value problems.* Am. J. Math., **88** (1966), 781–809.

[68] R.T. Seeley. *Complex powers of an elliptic operator.* Proc. Sympos. Pure Math., **10** (1967), 288–307.

[69] R.T. Seeley. Topics in pseudodifferential operators. In *Pseudo-Differential Operators*, 167–305, Roma, 1969.

[70] I.M. Singer. Eigenvalues of the Laplacian and invariants of manifolds. *Proceedings of the International Congress of Mathematicians*, pages 187–199, Vancouver, 1974.

[71] D. Sullivan. *Geometric Topology. Localization, Periodicity and Galois Symmetry.* MIT, Cambridge, Massachusets, 1970.

[72] E. Witten. *Global gravitational anomalies.* Commun. Math. Phys., **100** (1985), 197–229.

[73] K. Wojciechowski. *A note on the space of pseudodifferential projections with the same principal symbol.* J. Operator Theory, **15** (1986), no. 2, 207–216.

Anton Savin
Independent University of Moscow
B. Vlasyevskiy pereulok, d. 11
119002 Moscow, Russia
e-mail: `antonsavin@mail.ru`

Boris Sternin
Independent University of Moscow
B. Vlasyevskiy pereulok, d. 11
119002 Moscow, Russia
e-mail: `sternin@mail.ru`

C^*-algebras and Elliptic Theory
Trends in Mathematics, 291–312

New L^2-invariants of Chain Complexes and Applications

Vladimir V. Sharko

Abstract. We study the homotopy invariants of free cochain complexes and Hilbert complex. This invariants are applied to calculation of exact values of Morse numbers of smooth manifolds.

Mathematics Subject Classification (2000). Primary 05C10, 57R45.

Keywords. Stable rank, chain complex, Hilbert $N(G)$-module, Hilbert complex, manifold, Morse function, Morse numbers.

1. Introduction

Let W^n be a closed smooth manifold, $p : \widehat{W^n} \to W^n$ be the universal covering for W^n. Denote by W_i^n the ith skeleton of some arbitrary triangulation of W^n and let $\widehat{W^n} = p^{-1}(W_i^n)$. Set

$$S^i(W^n) = \mu_s(H^i(\widehat{W_i^n}, \mathbb{Z})) - \mu(H^i(W_i^n, \mathbb{Z})) ,$$

where $H^i(\widehat{W_i^n}, \mathbb{Z})$ is considered as a $\mathbb{Z}[\pi_1(W^n)]$-module. Here the number $\mu_s(H)$ is the stable minimal number of generators of the $\mathbb{Z}[\pi_1(W^n)]$-module H (see definition 2.1 below or [19]) and $\mu(H)$ is the minimal number of generators of the group H. It is known that the numbers $S^i(W^n)$ do not depend on triangulation of W^n and are invariants of homotopy type of W^n [19].

By definition the ith Morse number $\mathcal{M}_i(W^n)$ of a manifold W^n is the minimal number of critical points of index i taken over all Morse functions on W^n. It is known [19] that there is the following estimate for Morse number of index i of W^n:

$$\mathcal{M}_i(W^n) \geq S^i(W^n) + S^{i+1}(W^n) + \mu(H^i(W^n, \mathbb{Z})) + \mu(TorsH^{i+1}(W^n, \mathbb{Z})) .$$

There are examples of manifolds W^n such that an arbitrary Morse function f on them has the number of critical points of index i

$$\mathcal{M}_i(f) > S^i(W^n) + S^{i+1}(W^n) + \mu(H^i(W^n, \mathbb{Z})) + \mu(TorsH^{i+1}(W^n, \mathbb{Z})) .$$

I am grateful to E. Troitsky for his very helpful suggestions.

This causes that in the definition of the numbers $S^i(W^n)$ we can use the stable minimal number of generators of a $\mathbb{Z}[\pi_1(W^n)]$-module.

It is known [2, 10, 19] that for closed smooth manifolds of the dimension greater than 6 the ith Morse numbers are invariant of the homotopy type. There is a very complicated unsolved problem: *to find exact values of Morse numbers for every i* (see [19] for more details).

In this paper we at first study the homotopy invariants of free cochain complexes and Hilbert complex. Next we introduce new numerical invariants of manifolds $\mathbb{D}^i(W^n)$ which allows us to find the exact values of Morse numbers $\mathcal{M}_i(W^n)$ of a smooth closed manifold W^n $(n \geq 8)$ for $4 \leq i \leq n-4$. Denote $\pi = \pi_1(W^n)$. We prove following theorem:

Theorem 7.3. *Let W^n $(n \geq 8)$ be a smooth closed manifold. The following equality holds for the ith Morse number $4 \leq i \leq n-4$:*

$$\mathcal{M}_i(W^n) = \mathbb{D}^i(W^n) + \widehat{S}^i_{(2)}(W^n) + \widehat{S}^{i+1}_{(2)}(W^n) + dim_{N(\pi)}(H^i_{(2)}(W^n, \mathbb{Z}))\ .$$

2. Stable invariants of finite generated modules and Hilbert $N[G]$-modules

We give several definitions and results for finitely generated modules over group rings. Most facts are also valid for modules over a broader class of rings.

Denote the ring of integers by $\mathbb{Z}$ and the field of complex numbers by $\mathbb{C}$. Let G be a discrete group. Denote its integer group ring by $\mathbb{Z}[G]$ and the group ring over the field $\mathbb{C}$ by $\mathbb{C}[G]$. In the group ring there exists an augmentation epimorphism $\varepsilon : \mathbb{Z}[G] \longrightarrow \mathbb{Z}$ $(\varepsilon : \mathbb{C}[G] \longrightarrow \mathbb{C})$ acting by the rule $\varepsilon(\sum_i \alpha_i g_i) = \Sigma_i \alpha_i$. Denote the kernel of the epimorphism ε by $\mathbb{I}[G]$. In the ring $\mathbb{C}[G]$ there exists an involution $* : \mathbb{C}[G] \longrightarrow \mathbb{C}[G]$, $(\sum_i \alpha_i g_i)^* = \sum_i \overline{\alpha}_i g_i^{-1}$, where $\overline{\alpha}$ denotes conjugation in $\mathbb{C}$. This involution satisfies the following conditions for all elements r of the ring $\mathbb{C}[G]$:

a) $(r^*)^* = r$;
b) $(\alpha r_1 + \beta r_2)^* = \overline{\alpha} r_1^* + \overline{\beta} r_2^*, (\alpha, \beta \in C)$;
c) $(r_1 r_2)^* = r_2^* r_1^*$.

We can define the trace $tr : \mathbb{C}[G] \longrightarrow \mathbb{C}$ by the rule $tr(\sum_i^k \alpha_i g_i) = \alpha_1$, where α_1 is the coefficient of $g_1 = e$, which is the identity of the group G. It is obvious that the trace satisfies the following conditions:

a) $tr(e) = 1$;
b) tr is $\mathbb{C}$-linear mapping ;
c) $tr(r_1 r_2) = tr(r_2 r_1)$;
d) $tr(rr^*) \geqq 0$, and if $tr(rr^*) = 0$, then $r = 0$.

In what follows, a module M over a certain associative ring Λ with identity is understood, unless otherwise stated, as a left finitely generated Λ-module. Rings for which the rank of the free module is uniquely defined are called IBN-rings [19]. It is known that the group rings $\mathbb{Z}[G]$ and $\mathbb{C}[G]$ are IBN-rings [19]. In the present paper, we consider only IBN-rings. Note that a submodule of a free module of finite rank over a group ring $\mathbb{Z}[G]$ can be infinitely generated [21] even if the group G is finitely generated. Denoting the minimum number of the generators of the module M by $\mu(M)$, we get $\mu(M \bigoplus F_n) < \mu(M) + n$, where F_n is a free module of rank n. There exist examples (stably-free modules) where the strict inequality holds [19]. Recall that a Λ-module M is called *stably-free* if the direct sum of M and a free Λ-module F_k is free. We assume that if the module M is zero, then $\mu(M) = 0$.

Definition 2.1. For a finite generated module M over IBN-ring Λ let us define the following function (stable minimal generators of the module M)

$$\mu_s(M) = \lim_{n \longrightarrow \infty} (\mu(M \oplus F_n) - n)) .$$

From equality

$$\begin{aligned}\mu_s(M \bigoplus F_k) &= \lim_{n \longrightarrow \infty} (\mu(M \bigoplus F_k \oplus F_n) - n)) \\ &= \lim_{n \longrightarrow \infty} (\mu(M \bigoplus F_{k+n}) - (n+k)) + k = \mu_s(M) + k\end{aligned}$$

it follows that

$$\mu_s(M \bigoplus F_k) = \mu_s(M) + k .$$

If ring Λ is Hopfian then for any Λ-module M

$$\mu_s(M) = 0 ,$$

if and only if $M = 0$. Recall that a ring Λ is called *Hopfian*, if every epimorphism of a free Λ-module F_n on itself is an isomorphism. In fact, suppose that for a certain non-zero module M the equality $\mu_s(M) = 0$ occurs. Then there exists a natural number n, such that for the module $N = M \bigoplus F_n$ we have $\mu(N) = n$. Therefore, there is an epimorphism $f : F_n \to N$ of a free module F_n of rank n onto the module N. In addition, there exists a canonical epimorphism $p : N = M \bigoplus F_n \to F_n$ with the kernel equal M. Consider the kernel K of the epimorphism $p \cdot f : F_n \to F_n$. By the construction $K \neq 0$ and $K \oplus F_n = F_n$. Since Λ is Hopfian, it follows that $K = 0$.

From theorems of Kaplansky and Cockroft [19] it follows that the group rings $\mathbb{Z}[G]$ and $\mathbb{C}[G]$ are Hopfian. It is clear, that for any non-zero module M we have $0 < \mu_s(M) \leqslant \mu(M)$. The difference

$$\mu(M) - \mu_s(M)$$

shows how many times one can add a free module of rank one to the modules $M \bigoplus k\Lambda$ $(k = 0, 1, \dots)$ without increasing by one of the number $\mu(M \bigoplus k\Lambda)$. For every finite generated module M over IBN-ring Λ there is a natural number n such that for the module $N = M \bigoplus n\Lambda$ and all $m \geq 0$ we have $\mu(N \bigoplus m\Lambda) = \mu(N) + m$.

Definition 2.2. For a finitely generated module M over $\mathbb{Z}[G]$ (respectively $\mathbb{C}[G]$) set

$$S_{\mathbb{Z}}(M) = \mu_s(M) - \mu(\mathbb{Z} \bigotimes_{\mathbb{Z}[G]} M) \text{ (respectively } S_{\mathbb{C}}(M) = \mu_s(M) - \mu(\mathbb{C} \bigotimes_{\mathbb{C}[G]} M)).$$

Here $\mathbb{Z}$ (respectively $\mathbb{C}$) are considered as a trivial $\mathbb{Z}[G]$-module (respectively $\mathbb{C}[G]$-module) [19]. It is not difficult to show that $S_{\mathbb{Z}}(\mathbb{I}[G]) \geq 0$. It is known, that for $\mathbb{Z}[G]$-module $\mathbb{I}[G]$, $S_{\mathbb{Z}}(\mathbb{I}[G]) > 0$, if G is perfect group [19]. If $\mathbb{Z}[G]$- or $\mathbb{C}[G]$-modules M and N are such that $M \bigoplus F_n$ and $N \bigoplus F_m$ are isomorphic, then $S_{\mathbb{Z}}(M) = S_{\mathbb{Z}}(N) = S_{\mathbb{Z}}(M \bigoplus F_n) = S_{\mathbb{Z}}(N \bigoplus F_m)$ and $S_{\mathbb{C}}(M) = S_{\mathbb{C}}(N)$ $= S_{\mathbb{C}}(M \bigoplus F_n) = S_{\mathbb{C}}(N \bigoplus F_m)$.

In the ring $\mathbb{C}[G]$ there is an inner product $\langle \sum_i \alpha_i g_i, \sum_i \beta_i g_i \rangle = \sum_i \alpha_i \overline{\beta}_i$. The norm for an element from $\mathbb{C}[G]$ may define by $|r| = \operatorname{tr}(rr^*)^{1/2}$. Consider a completion of the ring $\mathbb{C}[G]$ with respect to this norm and denote it by $L^2(G)$. Then $L^2(G)$ is a Hilbert space (the inner product assigns the same formula as for the group ring $\mathbb{C}[G]$). The Hilbert space $L^2(G)$ has an orthonormal basis consisting of all elements of the group G. Now $\mathbb{C}[G]$ acts faithfully and continuously by left multiplication on $L^2(G)$

$$\mathbb{C}[G] \times L^2(G) \longrightarrow L^2(G),$$

so we may regard $\mathbb{C}[G] \subseteq \mathbf{B}(L^2(G))$, where $\mathbf{B}(L^2(G))$ denotes the set of bounded linear operators on $L^2(G)$. Let $N[G]$ denote the (reduced) group von Neumann algebra of G: thus by definition $N[G]$ is a week closure of $\mathbb{C}[G]$ in $\mathbf{B}(L^2(G))$. Therefore the map $w \to w(e)$ allows us to identify $N[G]$ with a subspace of $L^2(G)$, where $w \in N[G]$ and e is unit element of the group G. Thus algebraically we have $\mathbb{C}[G] \subset N[G] \subset L^2(G)$. The involution and the trace map on $N[G]$ may be defined exactly as for the ring $\mathbb{C}[G]$. For the set $M_n(N[G])$ of $n \times n$ matrices over von Neumann algebra $N[G]$, the trace map can be extended by setting $\operatorname{tr}(W) = \Sigma_{i=1}^n w_{ii}$, where $W = (w_{ij})$ is a matrix with entries in $N[G]$.

Following Cohen [4] we define following Hilbert $N[G]$-module. Let $E = N \bigcup \infty$, where ∞ denotes first infinite cardinal. If $n \in E$ then $L^2(G)^n$ denote the Hilbert direct sum n copies of $L^2(G)$, so $L^2(G)^n$ is a Hilbert space. The von Neumann algebra $N[G]$ acts on $L^2(G)^n$ from the left, so $L^2(G)^n$ is a left $N[G]$-module called a *free Hilbert $N[G]$-module of rank n. The left Hilbert $N[G]$-module* M is a closed left $\mathbb{C}[G]$-submodule of $L^2(G)^n$ for some $n \in E$. If $n \in N$, then Hilbert $N[G]$-module M is called finite generated. Following [4, 12] an *Hilbert $N[G]$-submodule* of M is a closed left $\mathbb{C}[G]$-submodule of M, an $L^2(G)$-ideal is an Hilbert $N[G]$-submodule of $L^2(G)$, and *homomorphism* $f : M \longrightarrow N$ between Hilbert $N[G]$-modules is a continuous left $\mathbb{C}[G]$-map.

Let M be a Hilbert $N[G]$-module and let $p : L^2(G)^n \to L^2(G)^n$ be an orthogonal projection onto $M \subset L^2(G)^n$. *Von Neumann dimension* of Hilbert $N[G]$-module M is called the number $dim_{N[G]}(M) = \operatorname{tr}(p) = \Sigma_{i=1}^n \langle p(e_i), e_i \rangle_{L^2(G)^n}$. Here $e_i = (0, \dots, g, \dots, 0)$ is standard basis in $L^2(G)^n$. It is known that $\dim_{N[G]}(V)$ is non-negative real number [12].

3. Stable invariants of homomorphisms

Consider a Λ-homomorphism $f : F_k \to F_t$, where F_k, F_t are free modules of ranks k and t respectively over ring Λ. We say that homomorphism f is a *splitting along a submodule* $\overline{F}_p \subseteq F_k$, if there is a presentation of f of the form

$$f = f_p \bigoplus f_t : \overline{F}_p \bigoplus \overline{F}_{k-p} \to \widetilde{F}_p \bigoplus \widetilde{F}_{t-p},$$

such that

$$f|_{\overline{F}_p \oplus 0} = f_p : \overline{F}_p \to \widetilde{F}_p, \qquad f|_{0 \oplus \overline{F}_{k-p}} = f_t : \overline{F}_{k-p} \to \widetilde{F}_{t-p},$$

where f_p is an isomorphism. From now in this situation we will suppose that submodules $\overline{F}_p$, $\overline{F}_{k-p}$, $\widetilde{F}_p$, $\widetilde{F}_{t-p}$ are free.

Definition 3.1. The number p above is called the *rank* of a splitting $f = f_p \bigoplus f_t$. The *rank* $R(f)$ of a homomorphism f is the maximal value of possible ranks of splittings of f.

Definition 3.2. *Stabilization* of a homomorphism $f : F_k \to F_t$ by a free module F_p is a homomorphism

$$f_{st}(p) : F_k \bigoplus F_p \to F_t \bigoplus F_p,$$

such that

$$f_{st}(p)|_{F_k \oplus 0} = f, \qquad f_{st}(p)|_{0 \oplus F_p} = Id.$$

A thickening of a homomorphism $f : F_k \to F_t$ by free modules F_m and F_n is the homomorphism

$$f_{th}(m,n) : F_k \bigoplus F_m \to F_t \bigoplus F_n,$$

such that

$$f_{th}(m,n)|_{F_k \oplus 0} = f, \qquad f_{th}(m,n)|_{0 \oplus F_m} = 0.$$

A thickening of a homomorphism $f : F_k \to F_t$ from the *left* (respectively from the *right*) by free module F_m (F_n) is the homomorphism

$$f_{th,l}(m) : F_k \bigoplus F_m \to F_t \qquad \text{(respectively } f_{th,r}(n) : F_k \to F_t \bigoplus F_n),$$

such that

$$f_{th,l}(m)|_{F_k \oplus 0} = f, \qquad f_{th,l}(m)|_{0 \oplus F_m} = 0, \qquad \text{(respectively } f_{th,r}(n) = f).$$

Definition 3.3. The *stable rank* $Sr(f)$ of a homomorphism $f : F_k \to F_t$ is the limit of values of

$$Sr(f) = \lim_{m,n,p \to \infty} (R(f_{th}(m,n)_{st}(p)) - p).$$

Since $Sr(f) \leq \min(k,t)$, this limit always exists. There are examples of stably free modules with $Sr(f) > R(f)$.

Lemma 3.4. *For any homomorphism $f : F_k \to F_t$ the following equality holds:*

$$Sr(f_{st}(v)) = Sr(f) + v\,.$$

Proof. Indeed

$$\begin{aligned} Sr(f_{st}(v)) &= \lim_{m,n,p\to\infty}(R(f_{st}(v)_{th}(m,n)_{st}(p)) - p) \\ &= \lim_{m,n,p\to\infty}(R(f_{st}(v+p)_{th}(m,n)) - (p+v)) + v = Sr(f) + v. \quad \square \end{aligned}$$

Remark 3.5. For an arbitrary homomorphism $f : F_k \to F_t$ there exists a number n_0 such that the stable rank $Sr(f)$ of the homomorphism f can be calculated by the formula

$$Sr(f) = R(f_{th}(m,n)_{st}(p)) - p$$

for any $m \geq n_0, n \geq n_0, p \geq n_0$.

Definition 3.6. The *stable rank from the left* (respectively from the *right*) $Sr_l(f)$ (respectively $Sr_r(f)$) of a homomorphism f is the following limit of values of ranks:

$$Sr_l(f) = \lim_{m,p\to\infty}(R(f_{th,l}(m)_{st}(p)) - p)$$

$$\text{(respectively} \qquad Sr_r(f) = \lim_{n,p\to\infty}(R(f_{th,r}(n)_{st}(p)) - p).$$

Remark 3.7. For the stable rank from the left (from the right) the analogues of Lemma 3.4 and Remark 3.5. hold.

Remark 3.8. For a homomorphism f define the following numbers:

$$\mathbb{D}_r(f) = Sr(f) - Sr_r(f), \qquad \mathbb{D}_l(f) = Sr(f) - Sr_l(f).$$

It is clear that $\mathbb{D}_r(f) = \mathbb{D}_r(f_{st}(p))$ (respectively$\mathbb{D}_l(f) = \mathbb{D}_l(f_{st}(p))$) for any integer p.

Definition 3.9. An epimorphism of Λ-modules $f : F_m \longrightarrow M$ is said to be *minimal* if $m = \mu(M)$.

Lemma 3.10. *Let* $F_n \xrightarrow{g} F_m \xrightarrow{f} M \longrightarrow 0$ *be an exact sequence where* F_m *and* F_n *are free modules. Then* f *is a minimal epimorphism if and only if* $Sr_l(g) = 0$.

Proof. **Necessity.** Let $F_n \xrightarrow{g} F_m \xrightarrow{f} M \longrightarrow 0$ be an exact sequence, where f is a minimal epimorphism. If $Sr_l(g) > 0$, then by Remark 3.7 there exist numbers p and x such that for thickening and the stabilization of the homomorphism g by free modules of ranks p and x respectively we have

$$Sr_l(g) = R(g_{th}(p)_{st}(x)) - x > 0\ .$$

Therefore $Sr_l(g) = R(g_{th}(p)_{st}(x)) = u > x$. Hence the homomorphism $g_{th}(p)_{st}(x)$ is a splitting along a free submodule F_u of rank u. We can delete submodules of rank u which direct summands of the modules $F_n \bigoplus F_p \bigoplus F_x$ and $F_m \bigoplus F_x$ respectively. This allows us to decrease the rank of a free module mapped onto the module M. But this contradicts to the assumption that the map f is minimal.

Sufficiency. First, if for an exact sequence

$$F_n \xrightarrow{g} F_m \xrightarrow{f} M \longrightarrow 0$$

we have $Sr_l(g) = 0$, then $R(g_{th}(p)_{st}(x)) = x$ for any integers p and x.

Let $F_n \xrightarrow{g} F_m \xrightarrow{f} M \longrightarrow 0$ be an exact sequence such that $Sr_l(g) = 0$ and epimorphism f is not minimal. Then $\mu(M) < m$. Consider an arbitrary exact sequence $F_v \xrightarrow{k} F_w \xrightarrow{h} M \longrightarrow 0$ such that h is a minimal epimorphism and $\mu(M) = w$. Then via the stabilization of the homomorphism g via module F_w and the stabilization of the homomorphism k via module F_m the epimorphisms $f_{th,l}(w)$ and $h_{th,l}(m)$ will proved to be equivalent. As noted above, $Sr_l(k) = 0$ (since h is minimal epimorphism), whence $Sr_l(h_{st}h(m)) = R(h) + m = m$. By similar arguments $Sr_l(g_{st}(w)) = Sr_l(g_{st}(w)_{th}(q)) = R(g) + w = w$ for any integer q. Therefore the homomorphism $g_{st}(w)_{th}(q)$ is a splitting along a free module of the rank m and hence $Sr_l(g_{st}(w)_{th}(q)) \geq m > w$. This contradiction proves the lemma. □

4. Stable invariants of a composition of homomorphisms

Consider a composition of homomorphisms of free modules

$$F_m \xrightarrow{\quad f \quad} F_n \xrightarrow{\quad g \quad} F_t,$$

such that

$$g \cdot f = 0. \qquad (\partial)$$

We say, that the homomorphisms f and g are splitting along submodules $\overline{F}_p \subseteq F_m$ and $\overline{F}_q \subseteq F_n$ if there are presentations of f and g of the form

$$\begin{array}{ccccccccc}
0 & \longrightarrow & \overline{F}_p & \xrightarrow{f_1} & \widetilde{F_p} & \longrightarrow & 0 & & \\
 & & \oplus & & \oplus & & & & \\
 & & F_{m-p} & \xrightarrow{f_2} & F_{n-p-q} & \xrightarrow{g_2} & F_{t-q} & & \\
 & & & & \oplus & & \oplus & & \\
 & & 0 & \longrightarrow & \overline{F_q} & \xrightarrow{g_1} & \widetilde{F_q} & \longrightarrow & 0
\end{array}$$

such that

$$f|_{\overline{F}_p \oplus 0} = f_1, \qquad g|_{0 \oplus 0 \oplus \overline{F}_q} = g_1.$$

We admit that the module $\overline{F}_p$ or $\overline{F}_q$ to be zero module. In sequel we will suppose, that submodules $\overline{F}_p$, $\overline{F}_q$, F_{m-p}, F_{t-q}, F_{n-p-q} are free.

Definition 4.1. The number $p + q$ will be called the *common rank* of a splitting of homomorphisms f and g along submodules $\overline{F}_p \subseteq F_m$ and $\overline{F}_q \subseteq F_n$. The *common rank* $Cr(f, g)$ of the homomorphisms f and g is a maximal value of common ranks of a splitting of f and g.

Definition 4.2. The *stabilization* of a composition of homomorphisms of free modules

$$F_m \xrightarrow{f} F_n \xrightarrow{g} F_t,$$

satisfying the condition (∂) by free modules F_p and F_q is the following composition of homomorphisms

$$\begin{array}{ccccccc} 0 & \longrightarrow & F_p & \xrightarrow{id} & F_p & \longrightarrow & 0 & & \\ & & \bigoplus & & \bigoplus & & & & \\ & & F_m & \xrightarrow{f} & F_n & \xrightarrow{g} & F_t & & \\ & & & & \bigoplus & & \bigoplus & & \\ & & 0 & \longrightarrow & F_q & \xrightarrow{id} & F_q & \longrightarrow & 0. \end{array}$$

We will denote it by $(f_{st}(p), g_{st}(q))$.

Definition 4.3. Consider a composition of homomorphisms f and g

$$F_m \xrightarrow{f} F_n \xrightarrow{g} F_t,$$

satisfying the condition (∂). The *thickening* this composition by free modules F_p and F_q is the following composition of homomorphisms

$$F_m \bigoplus F_p \xrightarrow{f_{th}(p)} F_n \xrightarrow{g_{th}(q)} F_t \bigoplus F_q,$$

such that

$$f_{th}(p)|_{F_m \bigoplus 0} = f\,, \quad f_{th}(p)|_{0 \bigoplus F_p} = 0\,, \quad g_{th}(q) = g\,.$$

It will be denoted by $(f_{th}(p), g_{th}(q))$.

The *thickening from the left* (respectively *right*) of this composition of homomorphisms f and g by free modules F_p (respectively F_q) is the following composition of homomorphisms

$$F_m \bigoplus F_p \xrightarrow{f_{th,l}(p)} F_n \xrightarrow{g} F_t$$

$$(\text{respectively } F_m \xrightarrow{f} F_n \xrightarrow{g_{th,r}(q)} F_r \bigoplus F_q),$$

such that

$$f_{th,l}(p)|_{F_m \bigoplus 0} = f, \qquad f_{th,l}(p)|_{0 \bigoplus F_p} = 0, \qquad (\text{respectively } g_{th,r}(q) = g).$$

It will be denoted by $(f_{th,l}(p), g)$ (respectively $f, g_{th,r}(q)$).

Definition 4.4. The *stable common rank* $Scr(f,g)$ *of the composition of homomorphisms* of free modules

$$F_m \xrightarrow{f} F_n \xrightarrow{g} F_t,$$

satisfying the condition (∂) is the limit of values of common ranks

$$Scr(f,g) = \lim_{p,q,v,w \to \infty} (Cr(f_{th}(p)_{st}(v), g_{th}(q)_{st}(w)) - v - w).$$

Since $Scr(f,g) \leq n$, this limit always exists. There are examples of stably free modules showing that $Scr(f,g) \geq Cr(f,g)$.

Lemma 4.5. *For arbitrary composition of homomorphisms f and g*

$$F_m \xrightarrow{\ f\ } F_n \xrightarrow{\ g\ } F_t,$$

satisfying the condition (∂) the following equality holds true:

$$Scr(f_{st}(x), g_{st}(y)) = Scr(f,g) + x + y.$$

Proof. Indeed,

$$\begin{aligned} Scr(f_{st}(x), g_{st}(y)) &= \lim_{p,q,v,w\to\infty} (Cr(f_{st}(x)_{th}(p)_{st}(v), g_{st}(y)_{th}(q)_{st}(w)) - v - w) \\ &= \lim_{p,q,v,w\to\infty} (Cr(f_{st}(x+v)_{th}(p), g_{st}(y+w)_{th}(q)) \\ &\quad - (x+v) - (y+w)) + x + y \\ &= Scr(f,g) + x + y\,. \end{aligned}$$

□

Remark 4.6. For every composition of homomorphisms f and g satisfying the condition (∂) there exists a number n_0 such that the stable common rank $Sr(f)$ can be calculated by the following formula:

$$Scr(f,g) = Cr(f_{th}(p)_{st}(v), g_{th}(q)_{st}(w)) - v - w$$

for any $p \geq n_0, q \geq n_0, v \geq n_0, w \geq n_0$.

Definition 4.7. The *stable common rank from the left* (respectively from the *right*) $Scr_l(f,g)$ (respectively $Scr_r(f,g)$) of the composition of homomorphisms of free modules

$$F_m \xrightarrow{\ f\ } F_n \xrightarrow{\ g\ } F_t,$$

satisfying condition (∂) is the following limit of values of common ranks:

$$Scr_l(f,g) = \lim_{p,v,w\to\infty} (Cr(f_{th,l}(p)_{st}(v), g_{st}(w)) - v - w)$$

$$\text{(respectively } Scr_r(f,g) = \lim_{q,v,w\to\infty} (Cr(f_{st}(v), g_{th,r}(q)_{st}(w)) - v - w)).$$

Remark 4.8. For stable common rank from the left (respectively from the right) $Scr_l(f,g)$ (respectively $Scr_r(f,g)$) of a composition of the homomorphisms satisfying the condition (∂) the analogues of Lemma 4.5 and Remark 4.6 hold.

Definition 4.9. The *defect* $\mathbb{D}(f,g)$ of a composition of homomorphisms of free modules

$$F_m \xrightarrow{\ f\ } F_n \xrightarrow{\ g\ } F_t,$$

satisfying condition (∂) is the following number:

$$\mathbb{D}(f,g) = Sr(f) + Sr(g) - Scr(f,g).$$

Remark 4.10. a) For arbitrary composition of homomorphisms f and g satisfying the condition (∂) there exists a number n_0 such that defect $\mathbb{D}(f,g)$ can be calculated by the formula

$$\mathbb{D}(f,g) = R(f_{th}(p,w)_{st}(v)) + R(g_{th}(v,q)_{st}(w)) + Cr(f_{th}(p)_{st}(v), g_{th}(q)_{st}(w))$$

for any $p \geq n_0, q \geq n_0, v \geq n_0, w \geq n_0$;

b) For any such composition with $F_n/f(F_m)$ being stable free, but non free, one has $\mathbb{D}(f,g) > 0$.

Lemma 4.11. *Consider two compositions of homomorphisms of free modules*

$$F_m \xrightarrow{\ f\ } F_n \xrightarrow{\ g\ } F_t$$

and

$$\begin{array}{ccccccc} 0 \longrightarrow & F_v & \xrightarrow{id} & F_v & \longrightarrow & 0 & \\ & \bigoplus & & \bigoplus & & \bigoplus & \\ & F_m \bigoplus F_p & \xrightarrow{f_{th,l}(p)} & F_n & \xrightarrow{g_{th,r}(q)} & F_t \bigoplus F_q & \\ & \bigoplus & & \bigoplus & & \bigoplus & \\ & 0 & \longrightarrow & F_w & \xrightarrow{id} & F_w & \longrightarrow 0 \end{array}$$

satisfying the condition (∂). Then the following equality holds:

$$\mathbb{D}(f,g) = \mathbb{D}(f_{th,l}(p)_{st}(v), g_{th,r}(q)_{st}(w)).$$

Proof. This lemma can be proved using Lemma 3.4 and Lemma 4.5. □

Definition 4.12. The *defect from the left* (respectively from the *right*) $\mathbb{D}_l(f,g)$ (respectively $\mathbb{D}_r(f,g)$) of a composition of homomorphisms of free modules

$$F_m \xrightarrow{\ f\ } F_n \xrightarrow{\ g\ } F_t,$$

satisfying condition (∂) is the following number

$$\mathbb{D}_l(f,g) = Sr_l(f) + Sr(g) - Scr_l(f,g)$$

$$(\text{respectively} \quad \mathbb{D}_r(f,g) = Sr(f) + Sr_r(g) - Scr_r(f,g)).$$

Remark 4.13. For the defect from the left (respectively from the right) $\mathbb{D}_l(f,g)$ (respectively $\mathbb{D}_r(f,g)$) of a composition of homomorphisms f and g satisfying condition (∂) the analogues of Lemma 4.11 and Remark 4.10 hold true.

5. Homotopy invariants of cochain complexes

The following statement can be found in [3]:

Proposition 5.1 (Cockroft-Swan). *Let $f = f_n : (C,d) \longrightarrow (\overline{C},\overline{d})$, $n \geq 0$ be a cochain mapping between the free cochain complexes (C,d) and $(\overline{C},\overline{d})$ that induces an isomorphism in cohomology. Then there exist contractible free cochain* *complexes* (D, ∂) *and* $(\overline{D}, \overline{\partial})$ *such that the cochain complexes* $(C \bigoplus D, d \bigoplus \partial)$ *and* $(\overline{C} \bigoplus \overline{D}, \overline{d} \bigoplus \overline{\partial})$*, are cochain-isomorphic.*

If $(C, d) : C^0 \xrightarrow{d^0} C^1 \xrightarrow{d^1} \cdots \xrightarrow{d^{n-1}} C^n$ is a free cochain complex over a ring Λ, then the numbers $\mathbb{D}_r(d^0)$, $\mathbb{D}_l(d^{n-1})$, $\mathbb{D}_r(d^0, d^1)$, $\mathbb{D}_l(d^{n-2}, d^{n-1})$, $\mathbb{D}(d^i, d^{i+1})$ are defined for $1 \le i \le n-3$. The next lemma shows that they are invariants of the homotopy type of a cochain complex (C, d).

Lemma 5.2. *Let* $(C, d)_\Lambda$ *be the class of free cochain complexes over ring* Λ *homotopy equivalent to cochain complex* $(C, d) : C^0 \xrightarrow{d^0} C^1 \xrightarrow{d^1} \cdots \xrightarrow{d^{n-1}} C^n$*. Then for any cochain complex* $(D, \partial) : D^0 \xrightarrow{\partial^0} D^1 \xrightarrow{\partial^1} \cdots \xrightarrow{\partial^{n-1}} D^n$ *belonging to the class* $(C, d)_\Lambda$ $(n \ge 4)$ *the following equalities hold:*

$$\begin{aligned} \mathbb{D}_r(d^0) &= \mathbb{D}_r(\partial^0), \\ \mathbb{D}_r(d^{n-1}) &= \mathbb{D}_r(\partial^{n-1}), \\ \mathbb{D}_r(d^0, d^1) &= \mathbb{D}_r(\partial^0, \partial^1), \\ \mathbb{D}_l(d^{n-2}, d^{n-1}) &= \mathbb{D}_l(\partial^{n-2}, \partial^{n-1}), \\ \mathbb{D}(d^i, d^{i+1}) &= \mathbb{D}(\partial^i, \partial^{i+1}) \end{aligned}$$

for $1 \le i \le n-3$.

Proof. This is a consequence of Proposition 5.1, Lemma 4.11 and Remark 3.7. □

Definition 5.3. A free cochain complex $(C, d) : C^0 \xrightarrow{d^0} C^1 \xrightarrow{d^1} \cdots \xrightarrow{d^{n-1}} C^n$ is called *minimal in dimension* i if for arbitrary free cochain complex $(D, \partial) : D^0 \xrightarrow{\partial^0} D^1 \xrightarrow{\partial^1} \cdots \xrightarrow{\partial^{n-1}} D^n$ that is homotopically equivalent to (C, d), one has $\mu(C^i) \le \mu(D^i)$, where $\mu(C^i)$ is the rank of the free module C^i. A free cochain complex (C, d) is called *minimal* if it is minimal in all dimensions.

It is obvious that, for every i in homotopy class of arbitrary free cochain complex (C, d) always exists a minimal free cochain complex in dimension i. The following lemma gives a necessary and sufficient condition for a free cochain complex over the ring Λ to be minimal in dimension i.

Lemma 5.4. *Let* $(C, d) : C^0 \xrightarrow{d^0} C^1 \xrightarrow{d^1} \cdots \xrightarrow{d^{n-1}} C^n$ *be a free cochain complex over a ring* Λ $(n \ge 4)$*. For the cochain complex* (C, d) *to be minimal in dimension* i *it is necessary and sufficient that*

$$\begin{aligned} Sr_r(d^0) &= 0, && \textit{for} \quad i = 0, \\ Scr_r(d^0, d^1) &= 0, && \textit{for} \quad i = 1, \\ Scr(d^{i-1}, d^i) &= 0, && \textit{for} \quad 2 \le i \le n-2, \\ Scr_l(d^{n-2}, d^{n-1}) &= 0, && \textit{for} \quad i = n-1, \\ Sr_l(d^{n-1}) &= 0, && \textit{for} \quad i = n. \end{aligned}$$

Proof. We will consider only the case, when $1 \leq i \leq n-3$. Other cases can be proved by similar arguments.

Necessity. Let $(D,\partial) : D^0 \xrightarrow{\partial^0} D^1 \xrightarrow{\partial^1} \cdots \xrightarrow{\partial^{n-1}} D^n$ be a minimal cochain complex in dimension i. If $Scr(\partial^i, \partial^{i+1}) > 0$, then by Remark 4.6 there exist numbers p, q, v, w such that the stabilization of the homomorphisms ∂^{i-1}, ∂^i, ∂^{i+1}, ∂^{i+2} by free modules of ranks p, q, v, w respectively the number $Scr(\partial^i, \partial^{i+1})$ can be calculated by the following formula

$$Scr(\partial^i, \partial^{i+1}) = Cr(\partial^i_{th}(p)_{st}(v), \partial^{i+1}_{th}(q)_{st}(w)) - v - w\,.$$

The obtained cochain complex will be denoted by $(\overline{D}, \overline{\partial})$. Then

$$Scr(\partial^i, \partial^{i+1}) = Cr(\partial^i_{th}(p)_{st}(v), \partial^{i+1}_{th}(q)_{st}(w)) > v + w\,.$$

Using operations of reduction of cochain complex $(\overline{D}, \overline{\partial})$ we can decrease the rank of the cochain module in dimension i by the number equal to $Scr(\partial^i, \partial^{i+1}) = Cr(\partial^i_{th}(p)_{st}(v), \partial^{i+1}_{th}(q)_{st}(w)) > v+w$. But this contradicts to the assumption that the cochain complex (D,∂) is minimal in dimension i.

Sufficiency. First, notice that if for the composition of homomorphisms of free modules

$$F_m \xrightarrow{f} F_n \xrightarrow{g} F_t$$

the equalities $g \cdot f = 0$ and $Scr(f,g) = 0$ hold, then for any numbers p, q, v, w we have

$$Cr(f_{th}(p)_{st}(v), g_{th}(q)_{st}(w)) = v + w.$$

Let $(C,d) : C^0 \xrightarrow{d^0} C^1 \xrightarrow{d^1} \cdots \xrightarrow{d^{n-1}} C^n$ be a free cochain complex such that $Scr(d^i, d^{i+1}) = 0$ and $(D,\partial) : D^0 \xrightarrow{\partial^0} D^1 \xrightarrow{\partial^1} \cdots \xrightarrow{\partial^{n-1}} D^n$ be a minimal cochain complex in dimension i (by previous $Scr(\partial^i, \partial^{i+1}) = 0$). Suppose that $\mu(C^i) > \mu(D^i)$. By Proposition 5.1 using stabilization of these chain complexes we can make them isomorphic. In other words, there are numbers $k, l, r, s, \overline{k}, \overline{l}, \overline{r}, \overline{s}$ such that the sequences

$$\begin{array}{ccccccccc}
0 & \longrightarrow & F_r & \xrightarrow{id} & F_r & \longrightarrow & 0 & & \\
 & & \bigoplus & & \bigoplus & & \bigoplus & & \\
 & & C^{i-1} \bigoplus F_k & \xrightarrow{d^{i-1}_{th}(k)} & C^i & \xrightarrow{d^i_{th}(l)} & C^{i+1} \bigoplus F_l & & \\
 & & \bigoplus & & \bigoplus & & \bigoplus & & \\
 & & 0 & \longrightarrow & F_s & \xrightarrow{id} & F_s & \longrightarrow & 0
\end{array}$$

and

$$\begin{array}{ccccccc}
0 & \longrightarrow & F_{\overline{r}} & \xrightarrow{id} & F_{\overline{r}} & \longrightarrow & 0 \\
 & & \bigoplus & & \bigoplus & & \bigoplus \\
 & & D^{i-1} \bigoplus F_{\overline{k}} & \xrightarrow{\partial_{th}^{i-1}(\overline{k})} & D^i & \xrightarrow{\partial_{th}^{i}(\overline{l})} & D^{i+1} \bigoplus F_{\overline{l}} \\
 & & \bigoplus & & \bigoplus & & \bigoplus \\
 & & 0 & \longrightarrow & F_{\overline{s}} & \xrightarrow{id} & F_{\overline{s}} & \longrightarrow 0
\end{array}$$

are isomorphic. Therefore these sequences have the same common ranks, i.e., $r + s = \overline{r} + \overline{s}$, whence $\mu(C^i) = \mu(D^i)$. □

However in the homotopy class of an arbitrary free cochain complex (C, d) may be no minimal free cochain complex, because of the existence of stably free modules.

Definition 5.5. Let (C, d): $C^0 \xrightarrow{d^0} C^1 \to \cdots \xrightarrow{d^{n-1}} C^n$ be a free cochain complex. Then cochain complex $(C(i), d(i)) : C^0 \xrightarrow{d^0} C^1 \to \cdots \xrightarrow{d^{i-1}} C^i$ is called ith skeleton of cochain complex (C, d).

Let $(C, d) : C^0 \xrightarrow{d^0} C^1 \to \cdots \xrightarrow{d^{n-1}} C^n$ be a free cochain complex over a group ring $\mathbb{Z}[G]$ ($\mathbb{C}[G]$). Consider a cochain complex of finite generated free abelian groups (vector space over field $\mathbb{C}$)

$$(\mathbb{Z} \bigotimes_{\mathbb{Z}[G]} C, id \bigotimes d) : \mathbb{Z} \bigotimes_{\mathbb{Z}[G]} C^0 \xrightarrow{id \otimes d^0} \mathbb{Z} \bigotimes_{\mathbb{Z}[G]} C^1 \to \cdots \xrightarrow{id \otimes d^{n-1}} \mathbb{Z} \bigotimes_{\mathbb{Z}[G]} C^n$$

$$(\mathbb{C} \bigotimes_{\mathbb{C}[G]} C, id \bigotimes d) : \mathbb{C} \bigotimes_{\mathbb{C}[G]} C^0 \xrightarrow{id \otimes d^0} \mathbb{C} \bigotimes_{\mathbb{C}[G]} C^1 \to \cdots \xrightarrow{id \otimes d^{n-1}} \mathbb{C} \bigotimes_{\mathbb{C}[G]} C^n).$$

Here $\mathbb{Z}$ (respectively $\mathbb{C}$) is considered as a trivial $\mathbb{Z}[G]$-module (respectively $\mathbb{C}[G]$-module). Let us consider the ith skeletons of these cochain complexes
$(C(i), d(i)) : C^0 \xrightarrow{d^0} C^1 \to \cdots \xrightarrow{d^{i-1}} C^i$ and

$$(\mathbb{Z} \bigotimes_{\mathbb{Z}[G]} C(i), id \bigotimes d(i)) : \mathbb{Z} \bigotimes_{\mathbb{Z}[G]} C^0 \xrightarrow{id \otimes d^0} \mathbb{Z} \bigotimes_{\mathbb{Z}[G]} C^1 \to \cdots \xrightarrow{id \otimes d^{i-1}} \mathbb{Z} \bigotimes_{\mathbb{Z}[G]} C^i,$$

$$(\mathbb{C} \bigotimes_{\mathbb{C}[G]} C(i), id \bigotimes d(i)) : \mathbb{C} \bigotimes_{\mathbb{C}[G]} C^0 \xrightarrow{id \otimes d^0} \mathbb{Z} \bigotimes_{\mathbb{C}[G]} C^1 \to \cdots \xrightarrow{id \otimes d^{i-1}} \mathbb{C} \bigotimes_{\mathbb{C}[G]} C^i).$$

Denote $\Gamma^i = C^i/d^{i-1}(C^{i-1})$. Clearly, the module Γ^i coincides with ith cohomology module of i-skeleton $(C(i), d(i))$. Moreover, the abelian group (respectively vector space) $C \bigotimes_{Z[G]} \Gamma^i$ (respectively $C \bigotimes_{\mathbb{C}[G]} \Gamma^i$) may be interpreted as the ith cohomology of the ith skeleton

$$(\mathbb{Z} \bigotimes_{\mathbb{Z}[G]} C(i), id \bigotimes d(i)) \qquad (\text{resp.} (\mathbb{C} \bigotimes_{\mathbb{C}[G]} C(i), id \bigotimes d(i))).$$

Definition 5.6. For a free cochain complex (C, d): $C^0 \xrightarrow{d^0} C^1 \to \cdots \xrightarrow{d^{n-1}} C^n$ over group ring $\mathbb{Z}[G]$ (respectively $\mathbb{C}[G]$), we set

$$S^i_{\mathbb{Z}}(C, d) = S_{\mathbb{Z}}(\Gamma^i) \text{ (respectively } S^i_{\mathbb{C}}(C, d) = S_{\mathbb{C}}(\Gamma^i)),$$

where the numbers $S_{\mathbb{Z}}(\Gamma)$ (respectively $S_{\mathbb{C}}(\Gamma)$) for finite generated $\mathbb{Z}[G]$ (respectively $\mathbb{C}[G]$)-module Γ is defined in Section 2).

If (C, d) and (D, ∂) are two homotopy equivalent free cochain complexes over group ring $\mathbb{Z}[G]$ (respectively $\mathbb{C}[G]$) then $S^i_{\mathbb{Z}}(C, d) = S^i_{\mathbb{Z}}(D, \partial)$ (respectively $S^i_{\mathbb{C}}(C, d) = S^i_{\mathbb{C}}(D, \partial)$). This is a consequence of Proposition 5.1.

Remark 5.7. If (C, d) is a free cochain complex then the number $S^i_{\mathbb{Z}}(C, d)$ (respectively $S^i_{\mathbb{C}}(C, d)$) estimates from below the rank of the homomorphism $id \bigotimes d^{i-1}$ of the cochain complex $(\mathbb{Z} \bigotimes_{\mathbb{Z}[G]} C, id \bigotimes d)$ (respectively $(\mathbb{C} \bigotimes_{\mathbb{C}[G]} C, id \bigotimes d)$).

Let $(C^*, d^*))$: $C_0 \xrightarrow{d_1} C_1 \to \cdots \xrightarrow{d_n} C_n$, be a sequence of free Hilbert $N[G]$-modules and bounded $\mathbb{C}[G]$-map such that $d_{i+1} \circ d_i = 0$. It is called a Hilbert complex. *The reduced cohomology of Hilbert complex* (C^*, d^*), it is a collection of are $L^2(G)$-modules $\overline{H^i}_{(2)}(C^*, d^*) = \operatorname{Ker} d^i / \overline{\Im d^{i-1}}$.

Definition 5.8. Consider a free cochain complex over $\mathbb{Z}[G]$ (respectively $\mathbb{C}[G]$)

$$(C^*, d^*) : C^0 \xrightarrow{d^0} C^1 \to \cdots \xrightarrow{d^{n-1}} C^n.$$

Hilbert complex

$$(L^2(G) \bigotimes_{\mathbb{Z}[G]} C^*, Id \bigotimes_{\mathbb{Z}[G]} d^*) :$$

$$L^2(G) \bigotimes_{\mathbb{Z}[G]} C^0 \xrightarrow{Id \bigotimes_{\mathbb{Z}[G]} d^0} L^2(G) \bigotimes_{\mathbb{Z}[G]} C^1 \to \cdots \xrightarrow{Id \bigotimes_{\mathbb{Z}[G]} d^{n-1}} L^2(G) \bigotimes_{\mathbb{Z}[G]} C^n$$

(respectively $(L^2(G) \bigotimes_{\mathbb{C}[G]} C^*, Id \bigotimes_{\mathbb{C}[G]} d^*)$:

$$L^2(G) \bigotimes_{\mathbb{C}[G]} C^0 \xrightarrow{Id \bigotimes_{\mathbb{C}[G]} d^0} L^2(G) \bigotimes_{\mathbb{C}[G]} C^1 \to \cdots \xrightarrow{Id \bigotimes_{\mathbb{C}[G]} d^{n-1}} L^2(G) \bigotimes_{\mathbb{C}[G]} C^n)$$

of free Hilbert $N[G]$-modules is the *Hilbert complex generated by* $\mathbb{Z}[G]$*- (respectively* $\mathbb{C}[G]$*)-cochain comlex* (C^*, d^*).

Consider the ith skeletons of these complexes

$$(C^*(i), d^*(i)) : C^0 \xrightarrow{d^0} C^1 \to \cdots \xrightarrow{d^{i-1}} C^i,$$

$$L^2(G) \bigotimes_{\mathbb{Z}[G]} C^0 \xrightarrow{Id \bigotimes_{\mathbb{Z}[G]} d^0} L^2(G) \bigotimes_{\mathbb{Z}[G]} C^1 \to \cdots \xrightarrow{Id \bigotimes_{\mathbb{Z}[G]} d^{i-1}} L^2(G) \bigotimes_{\mathbb{Z}[G]} C^i,$$

(respectively

$$L^2(G) \bigotimes_{\mathbb{C}[G]} C^0 \xrightarrow{Id \bigotimes_{\mathbb{C}[G]} d^0} L^2(G) \bigotimes_{\mathbb{C}[G]} C^1 \to \cdots \xrightarrow{Id \bigotimes_{\mathbb{C}[G]} d^{i-1}} L^2(G) \bigotimes_{\mathbb{C}[G]} C^i).$$

Set $\Gamma^i = C^i/d^{i-1}(C^{i-1})$. It is clear that

$$\widehat{\Gamma^i} = L^2(G) \bigotimes_{\mathbb{Z}[G]} C^i / \overline{Id \bigotimes_{\mathbb{Z}[G]} d^{i-1}(L^2(G) \bigotimes_{\mathbb{Z}[G]} C^{i-1})}$$

$$\text{(respectively} \qquad \widetilde{\Gamma^i} = L^2(G) \bigotimes_{\mathbb{C}[G]} C^i / \overline{Id \bigotimes_{\mathbb{C}[G]} d^{i-1}(L^2(G) \bigotimes_{\mathbb{C}[G]} C^{i-1})}).$$

is the ith Hilbert $N[G]$-module of reduced cohomology of the ith skeleton of the Hilbert complex

$$(L^2(G) \bigotimes_{\mathbb{Z}[G]} C^*(i), Id \bigotimes_{\mathbb{Z}[G]} d^*(i) \bigotimes id)$$

$$\text{(respectively} \quad (L^2(G) \bigotimes_{\mathbb{C}[G]} C^*(i), Id \bigotimes_{\mathbb{C}[G]} d^*(i) \bigotimes id)).$$

Definition 5.9. For the cochain complex (C^*, d^*) over $\mathbb{Z}[G]$ (respectively $\mathbb{C}[G]$) set

$$\widehat{S}^i_{(2)}(C^*, d^*) = \mu_s(\Gamma^i) - \dim_{N[G]} \widehat{\Gamma^i}$$

$$\text{(respectively} \qquad \widetilde{S}^i_{(2)}(C^*, d^*) = \mu_s(\Gamma^i) - \dim_{N[G]} \widetilde{\Gamma^i}).$$

If (C^*, d^*) and (D^*, ∂^*) are two homotopy equivalent free cochain complexes over the group ring $\mathbb{Z}[G]$ (respectively $\mathbb{C}[G]$) then

$$\widehat{S}^i_{(2)}(C^*, d^*) = \widehat{S}^i_{(2)}(D^*, \partial^*) \qquad \text{(respectively} \qquad \widetilde{S}^i_{(2)}(C^*, d^*) = \widetilde{S}^i_{(2)}(D^*, \partial^*)).$$

This is a consequence of Proposition 5.1 and the additivity of $\mu_s(\Gamma)$ and von Neumann dimension.

Lemma 5.10. *The numbers $\widehat{S}^i_{(2)}(C^*, d^*)$ and $\widetilde{S}^i_{(2)}(C^*, d^*)$ are non-negative for every i.*

Proof. We give the proof only for the case $\widehat{S}^i_{(2)}(C^*, d^*)$. The case $\widetilde{S}^i_{(2)}(C^*, d^*)$ is similar. We can assume without loss of generality that cochain complex (C^*, d^*) is such that $Sr(d^{i-1}) = 0$. Therefore according to Lemma 3.10 the epimorphism $p : C^i \longrightarrow \Gamma^i$ is minimal. We can also assume without loss of generality that cochain complex (C^*, d^*) is such that $\mu(\overline{\Gamma}^{(i)}) = \mu_s(\overline{\Gamma}^{(i)})$ and therefore $\mu(C^i) = \mu(\Gamma^i)$. Thus for the calculation of the number $\widehat{S}^i_{(2)}(C^*, d^*)$ we can use the following formula:

$$\widehat{S}^i_{(2)}(C^*, d^*) = \mu(\Gamma^i) - \dim_{N[G]} \widehat{\Gamma^i}.$$

By the construction $\mu(C^i) \geq \dim_{N[G]} \widehat{\Gamma^i}$. Therefore $\widehat{S}^i_{(2)}(C^*, d^*)$ is non-negative. □

6. Morse numbers

Definition 6.1. The *ith homotopy Morse number* of a cochain complex (C,d) over a ring Λ is the number $\mathcal{M}_i(C,d) = \mu(D_i)$, where $(D,\partial): D^0 \xrightarrow{\partial^0} D^1 \xrightarrow{\partial^1} \cdots \xrightarrow{\partial^{n-1}} D^n$ is the minimal cochain complex in dimension i, which is homotopy equivalent to (C,d).

Theorem 6.2. *Let (C,d): $C^0 \xrightarrow{d^0} C^1 \to \cdots \xrightarrow{d^{n-1}} C^n$ be a free cochain complex over a group ring $\mathbb{C}[G]$ $(n \geq 4)$. Its ith homotopy Morse numbers satisfy the following equalities:*

$$\mathcal{M}_0(C,d) = \mathbb{D}_r(d^0) + S^1_{\mathbb{C}}(C,d) + \dim_C(H^0(\mathbb{C} \bigotimes_{\mathbb{C}[G]} C, id \bigotimes d)),$$

$$\mathcal{M}_1(C,d) = \mathbb{D}_l(d^0,d^1) + S^1_{\mathbb{C}}(C,d) + S^2_{\mathbb{C}}(C,d) + \dim_C(H^1(\mathbb{C} \bigotimes_{\mathbb{C}[G]} C, id \bigotimes d)),$$

$$\mathcal{M}_i(C,d) = \mathbb{D}(d^{i-1},d^i) + S^i_{\mathbb{C}}(C,d) + S^{i+1}_{\mathbb{C}}(C,d) + \dim_C(H^i(\mathbb{C} \bigotimes_{\mathbb{C}[G]} C, id \bigotimes d))$$

for $2 \leq i \leq n-2$,

$$\mathcal{M}_{n-1}(C,d) = \mathbb{D}_l(d^{n-2},d^{n-1}) + S^{n-1}_{\mathbb{C}}(C,d) + \mu(H^n(C,d))$$

$$+ \dim_C(H^{n-1}(C \bigotimes_{\mathbb{C}[G]} C, id \bigotimes d)) - \dim_C(H^n(\mathbb{C} \bigotimes_{C[G]} C, id \bigotimes d)),$$

$$\mathcal{M}_n(C,d) = \mu(H^n(C,D)),$$

where $H^i(\mathbb{C} \otimes_{\mathbb{C}[G]} C, id \otimes d)$ is the cohomology of the cochain complex $(\mathbb{C} \otimes_{\mathbb{C}[G]} C, id \otimes d)$.

Remark 6.3. a) The number $\mathbb{D}(d^{i-1},d^i)$ arises in this theorem because in definition of the number $S^i(C,d)$ we take the number $\mu_s(\Gamma^i)$ but not the number $\mu(\Gamma^i)$. For example, in view of Remark 4.10 if the module $C^i/d^{i-1}(C^{i-1})$ is stable free but non free, then $\mathbb{D}(d^{i-1},d^i) > 0$.

b) The similar formulas hold for cochain complex (C,d) over the ring $\mathbb{Z}[G]$ but we need to use the cohomology groups of the cochain complex $(\mathbb{Z} \otimes_{\mathbb{Z}[G]} C, id \otimes d)$

Proof of Theorem 6.2. We consider only the case when $2 \leq i \leq n-2$. For the other cases the arguments are similar. Let $(D,\partial): D^0 \xrightarrow{\partial^0} D^1 \xrightarrow{\partial^1} \cdots \xrightarrow{\partial^{n-1}} D^n$ be a minimal cochain complex in dimension i, which is homotopy equivalent to (C,d). To prove the theorem it suffices to calculate the rank of the group $\mathbb{C} \otimes_{\mathbb{C}[G]} D^i$. By Lemma 5.4 we have $Scr(\partial^{i-1},\partial^i) = 0$. Without loss of generality, we can assume that $Cr(\partial^{i-1},\partial^i) = 0$ and $Sr(\partial^i) = R(\partial^i) = 0$. It is clear, that the rank of the homomorphism $id \otimes \partial^{i-1}$ is $\mathbb{D}(d^{i-1},d^i) + S^i_{\mathbb{C}}(C,d)$. Moreover, the rank of the homomorphism $id \bigotimes \partial^i$ is equal to $S^{i+1}_{\mathbb{C}}(C,d)$. Hence, for the calculation of $S^{i+1}(C,d)$ we can use the formula

$$S^{i+1}_{\mathbb{C}}(C,d) = \mu(\Gamma^{i+1}) - \mu(\mathbb{C} \bigotimes_{\mathbb{C}[G]} \Gamma^i).$$

By Lemma 3.10 the epimorphism $p : D^{i+1} \longrightarrow \Gamma^{i+1}$ is minimal. Then using stabilization of the homomorphism ∂^{i+1} we can made the module Γ^{i+1} arbitrarily large. We can now easily show that the dimension of the vector space $\mathbb{C} \bigotimes_{\mathbb{C}[G]} D^i$ is equal to

$$\mathcal{M}_i(C,d) = \mathbb{D}(d^{i-1}, d^i) + S^i_{\mathbb{C}}(C,d) + S^{i+1}_{\mathbb{C}}(C,d) + \dim_C(H^i(\mathbb{C} \bigotimes_{\mathbb{C}[G]} C, id \bigotimes d).$$

□

Remark 6.4. Let (C,d) be the free cochain complex over $\mathbb{C}[G]$. The ith homotopy Morse numbers of the Hilbert complex $(L^2(G) \bigotimes_{\mathbb{C}[G]} C, Id \bigotimes_{\mathbb{C}[G]} d)$ satisfy the following equalities:

$$\mathcal{M}_0(C,d) = \mathbb{D}_r(d^0) + \widetilde{S}^1_{(2)}(C,d) + \dim_{N[G]}(H^0_{(2)}(L^2(G) \bigotimes_{\mathbb{C}[G]} C, Id \bigotimes_{\mathbb{C}[G]} d))$$

$$\mathcal{M}_1(C,d) = \mathbb{D}_l(d^0, d^1) + \widetilde{S}^1_{(2)}(C,d) + \widetilde{S}^2_{(2)}(C,d)$$
$$+ \dim_{N[G]}(H^1_{(2)}(L^2(G) \bigotimes_{\mathbb{C}[G]} C, Id \bigotimes_{\mathbb{C}[G]} d)$$

$$\mathcal{M}_i(C,d) = \mathbb{D}(d^{i-1}, d^i) + \widetilde{S}^i_{(2)}(C,d) + \widetilde{S}^{i+1}_{(2)}(C,d)$$
$$+ \dim_{N[G]}(H^i_{(2)}(L^2(G) \bigotimes_{\mathbb{C}[G]} C, Id \bigotimes_{\mathbb{C}[G]} d))$$

for $2 \le i \le n-2$,

$$\mathcal{M}_{n-1}(C,d) = \mathbb{D}_l(d^{n-2}, d^{n-1}) + \widetilde{S}^{n-1}_{(2)}(C,d) + \mu(H^n(C,d))$$
$$+ \dim_{N[G]}(H^{n-1}_{(2)}(L^2(G) \bigotimes_{\mathbb{C}[G]} C, Id \bigotimes_{\mathbb{C}[G]} d)) - \dim_{N[G]}(H^n_{(2)}(L^2(G) \bigotimes_{\mathbb{C}[G]} C, Id \bigotimes_{\mathbb{C}[G]} d))$$

$$\mathcal{M}_n(C,d) = \mu(H^n(C,D)),$$

where $H^i_{(2)}(L^2(G) \bigotimes_{\mathbb{C}[G]} C, Id \bigotimes_{\mathbb{C}[G]} d)$ is Hilbert $N[G]$-module of the reduced cohomology of the Hilbert complex $(L^2(G) \bigotimes_{\mathbb{C}[G]} C, Id \bigotimes_{\mathbb{C}[G]} d)$.

The similar formulas hold for cochain complex (C,d) over the ring $\mathbb{Z}[G]$, but we need to use the numbers $\widehat{S}^i_{(2)}(C,d)$.

A free cochain complex (C,d) over a group ring $\mathbb{C}[G]$ generates two sequences of numbers: $\widetilde{S}^i(C,d)$, $\widetilde{S}^i_{(2)}(C,d)$ which are invariants of homotopy type of cochain complex (C,d). Moreover, they related to each other by the following lemma:

Lemma 6.5. *For a cochain complex* (C,d) *the following equalities hold:*

$$\widetilde{S}^i_{(2)}(C,d)+\sum_{j=0}^{i-1}(-1)^{i-j-1}\dim_{N[G]}(\overline{H}^j_{(2)}(L^2(G)\bigotimes_{\mathbb{C}[G]}C,Id\bigotimes_{\mathbb{C}[G]}d)$$

$$=S^i_C(C,d)+\sum_{j=0}^{i-1}(-1)^{i-j-1}\dim_C H^j(\mathbb{C}\bigotimes_{\mathbb{C}[G]}C,Id\bigotimes_{\mathbb{C}[G]}d).$$

Proof. Without loss of generality suppose that (C,d) is a minimal cochain complex in dimension i. We have the equality (i):

$$\begin{aligned}&\mathcal{M}_i(C,d)\\&=\mathbb{D}(d^{i-1},d^i)+\widetilde{S}^i_{(2)}(C,d)+\widetilde{S}^{i+1}_{(2)}(C,d)+\dim_{N[G]}(H^i_{(2)}(L^2(G)\bigotimes_{\mathbb{C}[G]}C,id\bigotimes_{\mathbb{C}[G]}d))\\&=\mathbb{D}(d^{i-1},d^i)+S^i_{\mathbb{C}}(C,d)+S^{i+1}_{\mathbb{C}}(C,d)+\dim_C(H^i(\mathbb{C}\bigotimes_{\mathbb{C}[G]}C,id\bigotimes d)).\end{aligned}$$

Hence by successive finding of the value of $S^{i+1}_{\mathbb{C}}(C,d)$ from equality (i) and substituting of it in the equality $(i+1)$ for $i=0,1,\dots,n-1$ we shall obtain necessary equalities. □

The similar formulas hold for a cochain complex (C,d) over ring $\mathbb{Z}[G]$, but we need to use the numbers $\widehat{S}^i_{(2)}(C,d)$ and ranges of cohomology groups of cochain complex $(\mathbb{Z}\bigotimes_{\mathbb{Z}[G]}C,id\bigotimes d)$.

7. Applications

Let K be a topological space with a structure of finite CW-complex and with nonzero fundamental group $\pi=\pi_1(K)$. Consider the universal covering $p:\widetilde{K}\to K$ of K. Using the map π, lift the structure of CW-complex from K to $\widetilde{K}$. On the universal covering space $\widetilde{K}$ there is a free action of the fundamental group $\pi=\pi_1(K)$ preserving the cell structure. This action equips each chain group $C_i(\widetilde{K},\mathbb{Z})$ with the structure of a left module over the group ring $\mathbb{Z}[\pi]$. It is evident that the resulting chain module $C_i(\widetilde{K},\mathbb{Z})$ is free and finite by generated with the i-cells K as a generators. As a result we obtain a free chain complex over the ring $\mathbb{Z}[\pi]$

$$C_*(\widetilde{K}):C_0(\widetilde{K},\mathbb{Z})\xleftarrow{d_1}C_1(\widetilde{K},\mathbb{Z})\leftarrow\cdots\xleftarrow{d_n}C_n(\widetilde{K},\mathbb{Z}).$$

Denote by $w:\pi_1(K)\to\mathbb{Z}_2$ the homomorphism of orientation (the first Stifel-Whitney class). Define an involution on the group ring $\mathbb{Z}[\pi]$ by the formula $g\to w(g)g^{-1}$. This involution allows us to regard every right $\mathbb{Z}[\pi]$-module as a left $\mathbb{Z}[\pi]$-module. In particular,

$$C^i(\widetilde{K},\mathbb{Z})=\mathrm{Hom}_{\mathbb{Z}[\pi]}(C_i(\widetilde{K},\mathbb{Z}),\mathbb{Z}[\pi])$$

is a left module as well. Consider the free cochain complex

$$C^*(\widetilde{K}) : C^0(\widetilde{K}, Z) \xrightarrow{d^0} C^1(\widetilde{K}, Z) \to \cdots \xrightarrow{d^{n-1}} C^n(\widetilde{K}, Z).$$

Cohomology modules of this cochain complex coincide with the cohomology with compact support of CW-complex $\widetilde{K}$. Taking the tensor product of $C^*(\widetilde{K})$ and $L^2(\pi)$ as $\mathbb{Z}[\pi]$-module we obtain the Hilbert complex

$$C^*_{(2)}(\widetilde{K}) : L^2(\pi) \bigotimes_{\mathbb{Z}[\pi]} C^0(\widetilde{K}, \mathbb{Z}) \xrightarrow{id \otimes d^0} L^2(\pi) \bigotimes_{\mathbb{Z}[\pi]} C^1(\widetilde{K}, \mathbb{Z}) \to \cdots$$
$$\cdots \xrightarrow{id \otimes d^{n-1}} L^2(G) \bigotimes_{Z[\pi]} C^n(\widetilde{K}, Z).$$

The Hilbert $N[\pi]$-module of cohomology of this Hilbert complex are Hilbert $N[\pi]$-module of cohomology of the space K. Since the cochain complex $C^*(\widetilde{K})$ is constructed from cellular structure of the space $\widetilde{K}$, we see that the segments of cochain complexes

$$C^*(\widetilde{K})(i) : C^0(\widetilde{K}, \mathbb{Z}) \xrightarrow{d^0} C^1(\widetilde{K}, \mathbb{Z}) \to \cdots \xrightarrow{d^{i-1}} C^i(\widetilde{K}, \mathbb{Z});$$
$$C^*_{(2)}(\widetilde{K})(i) : L^2(\pi) \bigotimes_{\mathbb{Z}[\pi]} C^0(\widetilde{K}, \mathbb{Z}) \xrightarrow{id \otimes d^0} L^2(\pi) \bigotimes_{\mathbb{Z}[\pi]} C^1(\widetilde{K}, \mathbb{Z}) \to \cdots$$
$$\cdots \xrightarrow{id \otimes d^{i-1}} L^2(\pi) \bigotimes_{\mathbb{Z}[\pi]} C^i(\widetilde{K}, \mathbb{Z})$$

are evidently the cochain complexes of the ith skeleton of the cellular decomposition of $\widetilde{K}$ and K respectively. Therefore a $Z[\pi]$-module

$$\widehat{\Gamma}^i(\widetilde{K}) = C^i(\widetilde{K}, \mathbb{Z})/d^{i-1}(C^{i-1}(\widetilde{K}, \mathbb{Z})),$$

(similarly Hilbert $N[\pi]$-module)

$$\Gamma^i(K) = L^2(\pi) \bigotimes_{\mathbb{Z}[\pi]} C^i(\widetilde{K}, \mathbb{Z})/id \bigotimes d^{i-1}(L^2(\pi) \bigotimes_{Z[\pi]} C^{i-1}(\widetilde{K}, \mathbb{Z}))$$

can be interpreted as the ith cohomology module with compact support (the ith $L^2(\pi)$-module of cohomology) of the ith skeleton of $\widetilde{K}$ (the ith skeleton of K).

Definition 7.1. For a cell complex K, set

$$\widehat{S}^i_{(2)}(K) = \mu_s(\widehat{\Gamma}^i(\widetilde{K})) - \dim_{N[\pi]}(\Gamma^i(K)),$$
$$\mathbb{D}^i(K) = \mathbb{D}(d^{i-1}, d^i).$$

It is well known that all chain complexes constructed from cellular decompositions of a topological space K have the same homotopy type. Therefore it follows directly from the previous consideration or from [11,19] that the numbers $\widehat{S}^i_{(2)}(W)$ and $\mathbb{D}^i(K)$ are invariants of the homotopy type of the topological space K. For a smooth manifold W there is an approach to construction of cochain complex via Morse functions. The details can be found in [17]. Let $(W^n, V_0^{n-1}, V_1^{n-1})$ be a compact smooth manifold with the boundary $\partial W^n = V_0^{n-1} \cup V_1^{n-1}$ (one of V_i^{n-1} or both may be empty). Let $\pi = \pi_1(W^n)$ be the fundamental group of the manifold W^n. Denote by

$$p : (\widetilde{W}^n, \widetilde{V}_0^{n-1}, \widetilde{V}_1^{n-1}) \to (W^n, V_0^{n-1}, V_1^{n-1})$$

the universal covering. Here $\widetilde{V}_i^{n-1} = p^{-1}(V_i^{n-1})$. Let us choose on W^n an ordered Morse function

$$f : W^n \to [0,1], \qquad f^{-1}(0) = V_0^{n-1}, \qquad f^{-1}(1) = V_1^{n-1}$$

and a gradient-like vector field ξ [19]. Using the mapping p, lift f and ξ to $\widetilde{W}^n$, and denote the lifted function and the vector field by $\widetilde{f}$and $\widetilde{\xi}$ respectively. Using f, ξ and $\widetilde{f}, \widetilde{\xi}$ construct chain complexes of abelian groups:

$$C_*(W^n, f, \xi) : C_0 \xleftarrow{d_1} C_1 \leftarrow \cdots \xleftarrow{d_n} C_n;$$

$$C_*(\widetilde{W}^n, \widetilde{f}, \widetilde{\xi}) : \widetilde{C}_0 \xleftarrow{\widetilde{d}_1} \widetilde{C}_1 \leftarrow \cdots \xleftarrow{\widetilde{d}_n} \widetilde{C}_n,$$

where

$$C_i = H_i(W_i, W_{i-1}, C), \qquad \widetilde{C}_i = H_i(\widetilde{W}_i, \widetilde{W}_{i-1}, C);$$

and

$$\widetilde{W}_i = \widetilde{f}^{-1}[0, a_i] \qquad W_i = f^{-1}[0, a_i]$$

are submanifolds containing all critical points of indices less or equal i. For the generators of the chain groups C_i (respectively $\widehat{C_i}$) constructed with the help of the vector field ξ (respectively($\widehat{\xi}$) one can take middle disks of critical points of index i. The fundamental group $\pi = \pi_1(W^n)$ acts on manifolds $\widetilde{W^n}$. This action equips the chain groups $\widetilde{C}_i$ with the structure of finitely generated modules over the ring $\mathbb{Z}[\pi]$. Making use of the involution, we turn the right $\mathbb{Z}[\pi]$-module

$$C^{(i)} = \mathrm{Hom}_{\mathbb{Z}[\pi]}(C_i, \mathbb{Z}[\pi])$$

into a left one and construct the following free cochain complex

$$C^*(\widetilde{W}^n, \widetilde{f}, \widetilde{\xi}) : \widetilde{C}^{(0)} \xrightarrow{\widetilde{d}^{(0)}} \widetilde{C}^{(1)} \to \cdots \xrightarrow{\widetilde{d}^{(n-1)}} \widetilde{C}^{(n)}.$$

Taking the tensor product of $C^*(\widetilde{W}^n, \widetilde{f}, \widetilde{\xi})$ and $L^2(\pi)$ as $\mathbb{Z}[\pi]$-module, we obtain the cochain complex of abelian groups which can be used for the definition the numbers $\widehat{S}^i_{(2)}(W^n)$ and $\mathbb{D}^i(W^n)$. It is proved in [11] that the chain complexes constructed from Morse functions on the manifold W^n via different cellular decomposition of W^n have the same homotopy type. This means that the values of the numbers $\widehat{S}^i_{(2)}(W^n)$ and $\mathbb{D}^i(W^n)$ do not depend on the method of constructing of a chain complex.

Definition 7.2. The ith Morse number $\mathcal{M}_i(W^n)$ of a manifold W^n is the minimal number of critical points of index i taken over all Morse functions on W^n.

It is known [2, 10, 19] that for closed smooth manifolds of dimension greater than 6 the ith Morse numbers are invariants of the homotopy type.

Theorem 7.3. *Let W^n $(n \geq 8)$ be a smooth closed manifold. The following equality holds for the ith Morse number $4 \leq i \leq n-4$:*

$$\mathcal{M}_i(W^n) = \mathbb{D}^i(W^n) + \widehat{S}^i_{(2)}(W^n) + \widehat{S}^{i+1}_{(2)}(W^n) + \dim_{N(\pi)}(H^i_{(2)}(W^n, \mathbb{Z})).$$

Proof. Let f be an arbitrary ordered Morse function, ξ a gradient-like vector field on W^n, and

$$C_*(\widetilde{W^n}, \widetilde{f}, \widetilde{\xi}) : \widetilde{C}_0 \xleftarrow{\widetilde{d}_1} \widetilde{C}_1 \leftarrow \cdots \xleftarrow{\widetilde{d}_n} \widetilde{C}_n,$$

the chain complex associated with them. Denote by

$$C^*(\widetilde{W^n}, \widetilde{f}, \widetilde{\xi}) : \widetilde{C}^{(0)} \xrightarrow{\widetilde{d}^{(0)}} \widetilde{C}^{(1)} \rightarrow \cdots \xrightarrow{\widetilde{d}^{(n-1)}} \widetilde{C}^{(n)}$$

the cochain complex constructed starting from a chain complex $C_*(\widetilde{W^n}, \widetilde{f}, \widetilde{\xi})$. It is clear that if the chain complex $C_*(\widetilde{W^n}, \widetilde{f}, \widetilde{\xi})$ is minimal in dimension i then cochain complex $C^*(\widetilde{W^n}, \widetilde{f}, \widetilde{\xi})$ is minimal in dimension i as well. It is known that the operation of stabilization of the homomorphisms d_i can be realized by changing Morse function and gradient-like vector field on W^n. But the inverse operation, the elimination of contractible free chain complex of the form $0 \longrightarrow \overline{C}_i \longrightarrow \overline{C}_{i+1} \longrightarrow 0$ from the chain complex $C^*(\widetilde{W^n}, \widetilde{f}, \widetilde{\xi})$ can not always be realized by a change of Morse function and gradient-like vector field on W^n. It is possible if $4 \leq i \leq n-4$ and $n \geq 8$ [19]. Let $(\overline{C}, \overline{d})$ be a minimal chain complex in dimension i homotopy equivalent to the chain complex $C_*(\widetilde{W^n}, \widetilde{f}, \widetilde{\xi})$. By Proposition 5.1 there exist contractible free chain complexes (D, ∂) and $(\overline{D}, \overline{\partial})$ such that the chain complexes $(C^*(\widetilde{W^n}, \widetilde{f}, \widetilde{\xi} \bigoplus D, d \bigoplus \partial))$ and $(\overline{C} \bigoplus \overline{D}, \overline{d} \bigoplus \overline{\partial})$, are chain-isomorphic. The previous notice ensures the existence of a Morse function g and gradient-like vector field that realize the complex $(C^*(\widetilde{W^n}, \widetilde{f}, \widetilde{\xi} \bigoplus D, d \bigoplus \partial))$. Using elimination of contractible free chain complexes of the form $0 \longrightarrow \overline{C}_i \longrightarrow \overline{C}_{i+1} \longrightarrow 0$ and $0 \longrightarrow \overline{C}_{i-1} \longrightarrow \overline{C}_i \longrightarrow 0$ from the chain complex $(C^*(\widetilde{W^n}, \widetilde{f}, \widetilde{\xi} \bigoplus D, d \bigoplus \partial))$ we can obtain a chain complex $(\widehat{C}, \widehat{d})$ that is minimal in dimension i. The conditions that $4 \leq i \leq n-4$ and $n \geq 8$ ensure the existence of a Morse function g and gradient-like vector field η that realize the complex $(\widehat{C}, \widehat{d})$. The number of critical points of Morse function g can be computed using previous formulas. □

The estimate for Morse numbers was studied in papers [1, 5–9, 12–20, 22], where some other approaches were used as well. In next papers we shall calculate the values of Morse numbers for some other values of i.

References

[1] D. Andrica, *Functions with a Minimal Number of Critical Points.* Preprint 1995, 95-8, "Babes-Bolyai" University.

[2] O.I. Bogoyavlenskiy, *A "nice" function on manifold.* Mat. Zametki **8** (1970), 77–83.

[3] W. Cockroft, R. Swan, *On homotopy type of certain two-dimensional complexes.* Proc. London Math. Sos., **11** (1961), 193–202.

[4] J. Cohen, *Von Neumann dimension and homology of covering spaces.* Quart. J. Math. Oxford, **30** (1979), 133–142. r.A - 41 - p. 485–521.

[5] A. Damian, *On stable Morse number of a closed manifold.* Bulletin of the London Math. Society, **34** (2002), 420–430.

[6] R. Forman, *Witten-Morse theory for cell complexes.* Topology, **37** (1998), 945–979.

[7] Y. Eliashberg, M. Gromov, *Lagrangian Intersections and Stable Morse Theory.* Bollettino U.M.I, **(7)11-B** (1997), 289–326.

[8] M. Farber, *Homological algebra of Novikov-Shubin invariants and Morse inequalities.* Geometric and Functional Analysis, **6** (1996), 628–665.

[9] M. Gromov, M. Shubin, *Near-cohomology of Hilbert complexes and topology of non-simply connected manifolds.* Astérisque, **210** (1992), 283–294.

[10] B. Hajduk, *Comparing handle decomposition of homotopy equivalent manifolds.* Fundamenta Mathematica, **95** (1997), 35–47.

[11] R.C. Kirby, L.C. Siebenmann, *Foundational essays on topological manifolds, smoothings and triangulatiions.* Ann. of Math. Stud., vol. 88, Princeton Univ. Press, Princeton, NJ, 1977.

[12] W. Lück, *L^2-Invariants: Theory and Applications to Geometry and K-Theory.* Ergebnisse der Mathematik und ihrer Grenzgebiete, 44, Springer, 2002, 620 p.

[13] V. Mathai, M. Shubin, *Twisted L^2-invariants of non-simply connected manifolds and asymptotic L^2 Morse inequalities.* Russian J. Math. Phys., **4** (1996), 499–526.

[14] S.P. Novikov, M.A. Shubin, *Morse inequalities and von Neumamm algebras.* Uspekhi Mat. Nauk, **41** (1986), 163–164.

[15] S.P. Novikov, M.A. Shubin, *Morse inequalities and von Neumamm II_1-factors.* Soviet Math. Dokl., **34** (1987), 79–82.

[16] S.P. Novikov, M.A. Shubin, *Morse theory and von Neumann invariants of non-simply connected manifolds.* Uspekhi Mat. Nauk, **41** (1986), 222–223.

[17] A. Pajitniov, *On the asymptotics Morse Numbers.* Topology, **38** (1999), 529–541.

[18] V. Sharko, *L^2-invariants of manifolds and cell complexes.* Proc. Ukrainian Math. Congress, 2001, Topology and Geometry, Inst. of Math. Nat. Acad. of Sci. of Ukraine, Kyiv, 2003, 103–135.

[19] V.V. Sharko, *Functions on Manifolds: Algebraic and topological aspects.* Translations of Mathematical Monographs, 131, American Mathematical Society, 1993.

[20] V.V. Sharko, *The invariants of chain complexes and Morse numbers of manifolds* (to appear).

[21] J. Stallings, *A finitely presented group whose 3-dimensional integral homology is not finitely generated.* Amer. Journal Math., **85** (1963), 541–543.

[22] E. Witten, *Supersymmetry and Morse theory.* Journal of Differential Geometry, **17** (1982), 661–692.

Vladimir V. Sharko
Institute of Mathematics, National Academy of Sciences of Ukraine,
Tereshchenkivska 3,
01601 Kyiv-4, Ukraine
e-mail: `sharko@imath.kiev.ua`, `sharko@ukrpack.net`

C^*-algebras and Elliptic Theory
Trends in Mathematics, 313–327

Bundles of C^*-algebras and the $KK(X;-,-)$-bifunctor

Ezio Vasselli

Abstract. An overview about C^*-algebra bundles with a $\mathbb{Z}$-grading is presented, with particular emphasis on classification questions. In particular, we discuss the role of the representable $KK(X;-,-)$-bifunctor introduced by Kasparov. As an application, we consider Cuntz-Pimsner algebras associated with vector bundles, and give a classification in terms of K-theoretical invariants in the case in which the base space is an n-sphere.

Mathematics Subject Classification (2000). Primary 46L05; Secondary 46L80.

Keywords. KK-theory; Continuous bundle; Vector Bundle; Cuntz-Pimsner-algebra.

1. Introduction

The classification of C^*-algebras by K-theoretical invariants is a rich and interesting topic; the relative program particularly succeeded in the case of simple, nuclear, purely infinite C^*-algebras ([12, 19]).

In order to extend such results to the case of non-simple C^*-algebras, it is natural to consider a particular class, namely C^*-algebra bundles over a locally compact Hausdorff space X. In order to find good invariants, in this case the better-behaved tool is the representable $KK(X;-,-)$-theory introduced by Kasparov in [11], which takes into account the bundle structure of the underlying C^*-algebra. $KK(X;-,-)$-theory has been recently extended to the case in which X is a T_0 space, in order to consider primitive ideal spectra of C^*-algebras ([13]).

Aim of the present paper is to present an overview about C^*-algebra bundles with a $\mathbb{Z}$-grading, and their associated $KK(X;-,-)$-theoretical invariants. Our main motivation arises from the case of the universal C^*-algebra of a vector bundle $\mathcal{E} \to X$, which is constructed as the Cuntz-Pimsner algebra associated with the bimodule of continuous sections of $\mathcal{E}$. Such a C^*-algebra has a natural structure of a $\mathbb{Z}$-graded bundle over X, with fibre the well-known Cuntz algebra. We are interested to classify such C^*-algebras in terms of properties of the underlying vector bundles.

All the material presented in the present paper appeared elsewhere (in some different form), with the exception of our main result Thm. 5.10, where we classify C^*-algebras of vector bundles in the case in which the base space is an n-sphere.

It is aim of the present work to be self-contained: the reader is assumed to be familiar at an elementary level with C^*-algebra theory ([18]), and K-theory ([2, 4]). In the case of results proved elsewhere, the proofs will be sketched or omitted. Some of the results exposed in the present paper appear in [24].

2. Bundles and $C_0(X)$-algebras

Let X be a locally compact Hausdorff space, $C_0(X)$ the C^*-algebra of complex-valued, continuous, vanishing at infinity functions on X. A *continuous bundle* of C^*-algebras over X is a C^*-algebra $\mathcal{F}$, equipped with a faithful family of epimorphisms $\{\pi_x : \mathcal{F} \to \mathcal{F}_x\}_{x \in X}$ such that, for every $a \in \mathcal{F}$, the *norm function* $\{x \mapsto \|\pi_x(a)\|\}$ belongs to $C_0(X)$; furthermore, $\mathcal{F}$ is required to be a nondegenerate $C_0(X)$-module w.r.t. pointwise multiplication $f, a \mapsto \{f(x) \cdot \pi_x(a)\}$, $f \in C_0(X)$. If X is compact, we consider the analogous notion by using the C^*-algebra $C(X)$ of continuous functions on X.

Example 1. Let $\mathcal{A}$ be a C^*-algebra. Then, the C^*-algebra tensor product $C_0(X) \otimes \mathcal{A}$ is a continuous bundle, called the **trivial bundle**. To be more concise, we define

$$X\mathcal{A} := C_0(X) \otimes \mathcal{A} . \tag{2.1}$$

The above notion of continuous bundle has been given in [14]: it is a simplified version of the classical notion of *continuous field* (see [7, §10]). We refer to the last-cited reference for the notions of *restriction* ([7, 10.1.7] and *local triviality* ([7, 10.1.8]), which are the analogues to well-known notions in the setting of topological bundles.

Let $\mathcal{A}$ be a C^*-algebra. To be more concise, we will call $\mathcal{A}$-*bundle* a locally trivial continuous bundle $\mathcal{F}$ with fibre $\mathcal{F}_x \equiv \mathcal{A}$, $x \in X$.

A $C_0(X)$-*algebra* is a C^*-algebra $\mathcal{A}$, equipped with a nondegenerate morphism from $C_0(X)$ into the centre of the multiplier algebra $M(\mathcal{A})$ ([11, §2]); in the sequel, we will identify $C_0(X)$ with the image in $M(\mathcal{A})$. C^*-algebra morphisms commuting with the $C_0(X)$-actions are called $C_0(X)$-*morphisms*. We denote by $\mathbf{aut}_X\mathcal{A}$ the group of $C_0(X)$-automorphisms of $\mathcal{A}$. It is proved in [16] that the category of $C_0(X)$-algebras is equivalent to the one of 'upper semicontinuous bundles'; the fibre of $\mathcal{A}$ over $x \in X$ is defined as follows: we consider the closed ideal $I_x := C_0(X - \{x\}) \cdot \mathcal{A} \subset \mathcal{A}$, and define $\mathcal{A}_x := \mathcal{A}/I_x$. In particular, every continuous bundle is a $C_0(X)$-algebra. We will denote by $\otimes_X$ the minimal tensor product with coefficients in $C_0(X)$ ([5]).

3. Hilbert bimodules and Cuntz-Pimsner algebras

For basic notions and terminology about *Hilbert bimodules*, we refer to [4, §13].

Let $\mathcal{B}$ be a C^*-algebra, $\mathcal{M}$ a right Hilbert $\mathcal{B}$-module. We denote by $L(\mathcal{M})$ the C^*-algebra of (bounded) adjointable right $\mathcal{B}$-module operators on $\mathcal{M}$, and by $K(\mathcal{M})$ the ideal of compact right $\mathcal{B}$-module operators of the type

$$\theta_{\psi,\psi'}\varphi := \psi \cdot \langle \psi', \varphi \rangle \quad , \tag{3.1}$$

where $\psi, \psi', \varphi \in \mathcal{M}$ and $\langle \cdot, \cdot \rangle$ denotes the $\mathcal{B}$-valued scalar product.

Let $\mathcal{M}$ be a Hilbert $\mathcal{A}$-$\mathcal{B}$-bimodule. In the sequel, if not specified, we will identify elements of $\mathcal{A}$ with the corresponding operators in $L(\mathcal{M})$, by assuming that the morphism $\mathcal{A} \to L(\mathcal{M})$ is injective.

Definition 3.1. Let $\mathcal{A}$, $\mathcal{B}$ be a C^*-algebras, $\mathcal{M}$ a Hilbert $\mathcal{A}$-bimodule, $\mathcal{N}$ a Hilbert $\mathcal{B}$-bimodule. A **covariant morphism** from $\mathcal{M}$ into $\mathcal{N}$ is a pair (β, η), where $\beta : \mathcal{M} \to \mathcal{N}$ is a Banach space map, $\eta : \mathcal{A} \to \mathcal{B}$ is a C^*-algebra morphism, and the following properties are satisfied for $a \in \mathcal{A}$, $\psi, \psi' \in \mathcal{M}$:

$$\beta(a\psi) = \eta(a)\beta(\psi) \quad , \quad \beta(\psi a) = \beta(\psi)\eta(a) \quad , \quad \langle \beta(\psi), \beta(\psi') \rangle = \eta \langle \psi, \psi' \rangle \ ,$$

where $\langle \cdot, \cdot \rangle$ denotes the $\mathcal{A}$-valued (resp. $\mathcal{B}$-valued) scalar product of $\mathcal{M}$ (resp. $\mathcal{N}$).

Example 2. Let $\alpha : \mathcal{A} \to \mathcal{B}$ be a C^*-algebra isomorphism, $\mathcal{M}$ a Hilbert $\mathcal{A}$-bimodule. We introduce a Hilbert $\mathcal{B}$-bimodule $\mathcal{M}_\alpha$, defined as the set $\mathcal{M}_\alpha := \{\underline{\psi}, \psi \in \mathcal{M}\}$ endowed with the vector space structure induced by $\mathcal{M}$. The Hilbert $\mathcal{B}$-bimodule structure is defined as follows:

$$b\underline{\psi} := \underline{\alpha^{-1}(b)\psi} \quad , \quad \underline{\psi}b := \underline{\psi\alpha^{-1}(b)} \quad , \quad \langle \underline{\psi}, \underline{\psi}' \rangle := \alpha \langle \psi, \psi' \rangle \ .$$

We call $\mathcal{M}_\alpha$ the *pullback bimodule* of $\mathcal{M}$. Let now $\beta(\psi) := \underline{\psi}$, $\psi \in \mathcal{M}$; it is clear that the pair (β, α) is a covariant isomorphism from $\mathcal{M}$ onto $\mathcal{M}_\alpha$. Viceversa, if (β, α) is a covariant isomorphism from a Hilbert $\mathcal{A}$-bimodule $\mathcal{M}$ into a Hilbert $\mathcal{B}$-bimodule $\mathcal{N}$, then $\mathcal{M}_\alpha$ is isomorphic to $\mathcal{N}$ as a Hilbert $\mathcal{B}$-bimodule.

Let $\mathcal{A}$ be a C^*-algebra, $\mathcal{M}$ a Hilbert $\mathcal{A}$-bimodule. The *Cuntz-Pimsner C^*-algebra* (*CP-algebra*, in the sequel) associated with $\mathcal{M}$ has been introduced in [20]; it is obtained by a universal construction, and supplies a generalization of crossed products by $\mathbb{Z}$ (see Ex. 6 below) and the well-known Cuntz algebras $\mathcal{O}_d$, $d \in \mathbb{N}$ ([6]). We will denote by $\mathcal{O}_\mathcal{M}$ the CP-algebra associated with $\mathcal{M}$.

In order to simplify the exposition, we give a description of $\mathcal{O}_\mathcal{M}$ in the case in which $\mathcal{A}$ has identity 1 and $\mathcal{M}$ is finitely generated as a right Hilbert $\mathcal{A}$-module. Let $\{\psi_l\}_{l=1}^n \subset \mathcal{M}$ be a finite set of generators, $\langle \cdot, \cdot \rangle$ the $\mathcal{A}$-valued scalar product; then, for every index l, $a \in \mathcal{A}$, we find $a\psi_l = \sum_m \psi_m a_{ml}$, $a_{ml} := \langle \psi_m, a\psi_l \rangle \in \mathcal{A}$. We consider the universal $*$-algebra ${}^0\mathcal{O}_\mathcal{M}$ generated by $\{\psi_l\}$, $\mathcal{A}$, with relations

$$\psi_l^* \psi_m = \langle \psi_l, \psi_m \rangle \quad , \quad a\psi_l = \sum_m \psi_m a_{ml} \quad , \quad \sum_l \psi_l \psi_l^* = 1 \ . \tag{3.2}$$

Note that every $\psi \in \mathcal{M}$ appears as an element of ${}^0\mathcal{O}_\mathcal{M}$, in fact $\psi = \sum_l \psi_l(\psi_l^*\psi)$. It can be proved that there exists a *unique* (faithful) C^*-norm on ${}^0\mathcal{O}_\mathcal{M}$ such that the *circle action*

$$\alpha_z(\psi) := z\psi \quad , \quad z \in \mathbb{T}, \psi \in \mathcal{M} \tag{3.3}$$

extends to an (isometric) automorphic action. The resulting C^*-algebra is the CP-algebra $\mathcal{O}_\mathcal{M}$, naturally endowed with the action $\alpha : \mathbb{T} \to \mathbf{aut}\mathcal{O}_\mathcal{M}$. We introduce the notation

$$\mathcal{O}_\mathcal{M}^k := \{t \in \mathcal{O}_\mathcal{M} : \alpha_z(t) = z^k t\} \quad , \quad k \in \mathbb{Z} . \tag{3.4}$$

Example 3. The Cuntz algebra $\mathcal{O}_d$, $d \in \mathbb{N}$, is obtained in the case $\mathcal{A} = \mathbb{C}$, $\mathcal{M} := \mathbb{C}^d$. Note that (3.2) take the form $\psi_h^*\psi_k = \delta_{hk}1$, $\sum_h \psi_h\psi_h^* = 1$, where δ_{hk} is the Kronecker symbol.

Definition 3.2. ([8, §1]) Let $\mathcal{A} \subset \mathcal{B}$ be a C^*-algebra inclusion. A closed vector space $\mathcal{M} \subset \mathcal{B}$ is called **Hilbert $\mathcal{A}$-bimodule in $\mathcal{B}$** if

1. $\mathcal{M}$ is stable w.r.t. left and right multiplication by elements of $\mathcal{A}$;
2. $t^*t' \in \mathcal{A}$, for every $t, t' \in \mathcal{M}$.

We say that $\mathcal{M}$ has **support** 1 if $\mathcal{M}\mathcal{M}^* :=$ closed span $\{t't^* \ : \ t, t' \in \mathcal{M}\}$ contains an approximate unit for $\mathcal{B}$.

Note that if $\mathcal{M}$ is a Hilbert $\mathcal{A}$-bimodule in $\mathcal{B}$, then the map $t, t' \mapsto t^*t'$ can be regarded as an $\mathcal{A}$-valued scalar product; moreover, there is a natural identification $\mathcal{M}\mathcal{M}^* \simeq K(\mathcal{M})$, $t't^* \mapsto \theta_{t,t'}$. The following proposition is a consequence of the universality of the CP-algebra (see [20, Thm. 3.12]).

Proposition 3.3. *Covariant morphisms between Hilbert bimodules give rise to C^*-algebra morphisms between the associated CP-algebras. In particular, $\mathcal{O}_\mathcal{M}$ is isomorphic to $\mathcal{O}_{\mathcal{M}_\alpha}$ for every pullback bimodule $\mathcal{M}_\alpha$ (Ex. 2). If $\mathcal{B}$ is a unital C^*-algebra, and $\mathcal{M}$ is a Hilbert $\mathcal{A}$-bimodule in $\mathcal{B}$ with support 1, then there is a canonical morphism $\mathcal{O}_\mathcal{M} \to \mathcal{B}$.*

Let X be a locally compact Hausdorff space, $\mathcal{E} \to X$ a rank d vector bundle, $d \in \mathbb{N}$. Moreover, let $\widehat{\mathcal{E}}$ be the Hilbert $C_0(X)$-bimodule of continuous, vanishing at infinity sections of $\mathcal{E}$, endowed with coinciding left and right $C_0(X)$-module actions. We denote by $\mathcal{O}_\mathcal{E}$ the CP-algebra associated with $\widehat{\mathcal{E}}$. For X compact (so that $C(X)$ is unital and $\widehat{\mathcal{E}}$ is finitely generated), (3.2) take the form

$$\psi_l^*\psi_m = \langle \psi_l, \psi_m \rangle \quad , \quad f\psi_l = \psi_l f \quad , \quad \sum_l \psi_l\psi_l^* = 1 \quad ,$$

$f \in C(X)$. It is proved in [24, Prop. 4.2] that $\mathcal{O}_\mathcal{E}$ is a locally trivial continuous bundle over X, with fibre the Cuntz algebra $\mathcal{O}_d$. Moreover, it is clear that the circle action (3.3) is by $C_0(X)$-automorphisms: $\alpha : \mathbb{T} \to \mathbf{aut}_X\mathcal{O}_\mathcal{E}$. In particular, if $\mathcal{L} \to X$ is a line bundle, then the fibre of $\mathcal{O}_\mathcal{L}$ is the C^*-algebra $C(S^1)$ of continuous functions over the circle; CP-algebras associated with line bundles have been classified in [23, Prop.4.3]. In the rest of the present paper, we will consider only vector bundles with rank > 1.

The main motivation of the present paper is the classification of the C^*-algebras $\mathcal{O}_\mathcal{E}$ in terms of topological properties of the underlying vector bundles.

4. Representable KK-theory

Let $\mathbf{C^*alg}$ denote the category of C^*-algebras, $\mathbf{Ab}$ the category of abelian groups. Kasparov constructed a bifunctor $KK_0 : \mathbf{C^*alg} \times \mathbf{C^*alg} \to \mathbf{Ab}$, assigning to the pair $(\mathcal{A}, \mathcal{B})$ an abelian group $KK_0(\mathcal{A}, \mathcal{B})$. KK_0 depends contravariantly on the first variable, and covariantly on the second one. Let $K_0(\mathcal{A})$ denote the K-theory of $\mathcal{A}$, $K^0(\mathcal{A})$ the K-homology; it turns out that there are isomorphisms $KK(\mathbb{C}, \mathcal{A}) \simeq K_0(\mathcal{A})$, $KK(\mathcal{A}, \mathbb{C}) \simeq K^0(\mathcal{A})$.

Let X be a locally compact Hausdorff space, $\mathcal{A}$, $\mathcal{B}$ $C_0(X)$-algebras. A *$C_0(X)$-Hilbert $\mathcal{A}$-$\mathcal{B}$-bimodule* is a Hilbert $\mathcal{A}$-$\mathcal{B}$-bimodule $\mathcal{M}$ such that $(af)\psi b = a\psi(fb)$ for every $f \in C_0(X)$, $\psi \in \mathcal{M}$, $a \in \mathcal{A}$, $b \in \mathcal{B}$.

Roughly speaking, a $C_0(X)$-Hilbert $\mathcal{A}$-$\mathcal{B}$-bimodule can be regarded as the space of sections of a bundle, having as fibres Hilbert $\mathcal{A}_x$-$\mathcal{B}_x$-bimodules, $x \in X$.

Example 4. Let $\mathcal{A}$ be a C^*-algebra, X a locally compact Hausdorff space, $\mathcal{E} \to X$ a vector $\mathcal{A}$-bundle in the sense of Mishchenko ([15]). Then, the module of continuous, vanishing at infinity sections of $\mathcal{E}$ has an obvious structure of $C_0(X)$-Hilbert $C_0(X)$-$(X\mathcal{A})$-bimodule.

Remark 4.1. In the sequel, we will make use of the following two notions of tensor product of $C_0(X)$-Hilbert bimodules.

1. Let $\mathcal{M}$ be a $C_0(X)$-Hilbert $\mathcal{A}$-$\mathcal{B}$-bimodule, $\mathcal{N}$ a $C_0(X)$-Hilbert $\mathcal{B}$-$\mathcal{C}$-bimodule. We consider the algebraic tensor product $\mathcal{M} \odot_\mathcal{B} \mathcal{N}$ with coefficients in $\mathcal{B}$, and denote by $\mathcal{M} \otimes_\mathcal{B} \mathcal{N}$ the completion w.r.t. the $\mathcal{C}$-valued scalar product $\langle \psi \otimes \varphi, \psi' \otimes \varphi' \rangle := \langle \varphi, \langle \psi, \psi' \rangle_\mathcal{M} \varphi' \rangle_\mathcal{N}$, $\psi, \psi' \in \mathcal{M}$, φ, $\varphi' \in \mathcal{N}$; here $\langle \cdot, \cdot \rangle_\mathcal{N}$ (resp. $\langle \cdot, \cdot \rangle_\mathcal{M}$) denotes the scalar product on $\mathcal{N}$ (resp, $\mathcal{M}$); note that $\langle \psi, \psi' \rangle_\mathcal{M} \in \mathcal{B}$, so that it makes sense to consider $\langle \psi, \psi' \rangle_\mathcal{M} \varphi'$. $\mathcal{M} \otimes_\mathcal{B} \mathcal{N}$ is a $C_0(X)$-Hilbert $\mathcal{A}$-$\mathcal{C}$-bimodule in a natural way, and is called the **internal tensor product** of $\mathcal{M}$ and $\mathcal{N}$.
2. Let $\mathcal{M}'$ be a $C_0(X)$-Hilbert $\mathcal{A}'$-$\mathcal{B}'$-bimodule. The algebraic tensor product $\mathcal{M} \odot_{C_0(X)} \mathcal{M}'$ with coefficients in $C_0(X)$ is endowed with a natural left $(\mathcal{A} \otimes_X \mathcal{A}')$-module action, and with natural $(\mathcal{B} \otimes_X \mathcal{B}')$-valued scalar product and right action. The corresponding completion $\mathcal{M} \otimes_X \mathcal{M}'$ is a $C_0(X)$-Hilbert $(\mathcal{A} \otimes_X \mathcal{A}')$-$(\mathcal{B} \otimes_X \mathcal{B}')$-bimodule, and is called the **external tensor product** of $\mathcal{M}$ and $\mathcal{M}'$.

Let X be a σ-compact metrisable space. Motivated by the Novikov conjecture, Kasparov generalized the construction of $KK_0(-,-)$ to the category of $C_0(X)$-algebras ([11, 2.19]); the corresponding bifunctor is called *representable* KK-theory. We will denote it by the notation $KK(X;-,-)$ (note that in [11] the notation $\mathcal{R}KK(X;-,-)$ is used). The rest of the present section is devoted to a brief exposition of the construction of $KK(X;-,-)$.

Definition 4.2. Let $\mathcal{A}$, $\mathcal{B}$ be separable $C_0(X)$-algebras. A **Kasparov $\mathcal{A}$-$\mathcal{B}$-module** is a pair $(\mathcal{M}, F)$, where $\mathcal{M}$ is a countably generated $C_0(X)$-Hilbert $\mathcal{A}$-$\mathcal{B}$-bimodule, and $F = F^* \in L(\mathcal{M})$ is an operator such that $[F, a], a(F^2 - 1) \in K(\mathcal{M})$ for every $a \in \mathcal{A}$. We denote by $\mathbb{E}(X; \mathcal{A}, \mathcal{B})$ the set of Kasparov $\mathcal{A}$-$\mathcal{B}$-modules.

It is customary to consider a $\mathbb{Z}_2$-grading on Kasparov modules ([4, §14]). Since we do not need such a structure, we will assume that every C^*-algebra (Hilbert bimodule) is endowed with the trivial $\mathbb{Z}_2$-grading.

Example 5. Let $\mathcal{M}$ be a countably generated $C_0(X)$-Hilbert $\mathcal{A}$-$\mathcal{B}$-bimodule. Let us suppose that $a \in K(\mathcal{M})$ for every $a \in \mathcal{A}$; then $(\mathcal{M}, 0) \in \mathbb{E}(X; \mathcal{A}, \mathcal{B})$, where 0 is the zero operator. In particular, if every element of $K(\mathcal{M})$ is the image of some element of $\mathcal{A}$ w.r.t. the left $\mathcal{A}$-module action (so that, there is an isomorphism $\mathcal{A} \simeq K(\mathcal{M})$), then $\mathcal{M}$ is called **imprimitivity $\mathcal{A}$-$\mathcal{B}$-bimodule** (see [3]).

Example 6. Let $\phi : \mathcal{A} \to \mathcal{B}$ be a nondegenerate $C_0(X)$-algebra morphism. We endow $\mathcal{B}$ with the $C_0(X)$-Hilbert $\mathcal{A}$-$\mathcal{B}$-bimodule structure

$$a, \psi \mapsto \phi(a)\psi \quad , \quad \psi, b \mapsto \psi b \quad , \quad \langle \psi, \psi' \rangle := \psi^* \psi' \ ,$$

$a \in \mathcal{A}$, $b, \psi, \psi' \in \mathcal{B}$, and denote by $\mathcal{B}_\phi$ the associated $C_0(X)$-Hilbert $\mathcal{A}$-$\mathcal{B}$-bimodule. Now, it is clear that $K(\mathcal{B}_\phi) \simeq \mathcal{B}$; thus, $\phi(a) \in K(\mathcal{B}_\phi)$ for every $a \in \mathcal{A}$. If $\mathcal{B}$ is σ-unital ([4, 12.3]), then $(\mathcal{B}_\phi, 0) \in \mathbb{E}(\mathcal{A}, \mathcal{B})$ (in fact, $\mathcal{B}_\phi$ is countably generated by an approximate unit $\{u_n\}_{n \in \mathbb{N}} \subset \mathcal{B}$). If $\mathcal{A} = \mathcal{B}$ and $\phi \in \mathbf{aut}_X \mathcal{A}$, then the CP-algebra $\mathcal{O}_{\mathcal{A}_\phi}$ is isomorphic to the crossed product $\mathcal{A} \rtimes_\phi \mathbb{Z}$ ([20, §2]). If $\iota : \mathcal{A} \to \mathcal{A}$ is the identity automorphism, we define $[1]_{\mathcal{A}} := (\mathcal{A}_\iota, 0) \in \mathbb{E}(X; \mathcal{A}, \mathcal{A})$.

There are natural notions of *homotopy* and *direct sum* over $\mathbb{E}(X; \mathcal{A}, \mathcal{B})$. The *representable KK-theory group* $KK(X; \mathcal{A}, \mathcal{B})$ is constructed in the same way as the usual KK_0-group, by endowing the set of homotopy classes of Kasparov $\mathcal{A}$-$\mathcal{B}$-modules with the operation of direct sum. The bifunctor $KK(X; -, -)$ is *stable*, i.e., $KK(X; \mathcal{A}, \mathcal{B})$ is invariant by tensoring $\mathcal{A}$ or $\mathcal{B}$ by the C^*-algebra $\mathcal{K}$ of compact operators over a separable Hilbert space.

With an abuse of the notation, we will identify the elements of $\mathbb{E}(X; \mathcal{A}, \mathcal{B})$ with the corresponding classes in $KK(X; \mathcal{A}, \mathcal{B})$. We use the notation $+$ to denote the operation of direct sum in $KK(X; \mathcal{A}, \mathcal{B})$. When $X = \bullet$ reduces to a single point, then $KK(\bullet, \mathcal{A}, \mathcal{B})$ is the usual KK-group $KK_0(\mathcal{A}, \mathcal{B})$.

Let $\mathcal{A}$, $\mathcal{B}$, $\mathcal{A}'$, $\mathcal{B}'$, $\mathcal{C}$ be separable $C_0(X)$-algebras. We recall that the *Kasparov product* ([11, §2.21]) induces bilinear maps

$$\begin{cases} \times_{\mathcal{B}} : KK(X; \mathcal{A}, \mathcal{B}) \otimes KK(X; \mathcal{B}, \mathcal{C}) \to KK(X; \mathcal{A}, \mathcal{C}) \ , \\ \times : KK(X; \mathcal{A}, \mathcal{B}) \otimes KK(X; \mathcal{A}', \mathcal{B}') \to KK(X; \mathcal{A} \otimes_X \mathcal{A}', \mathcal{B} \otimes_X \mathcal{B}') \ . \end{cases}$$

Remark 4.3. In some particular cases, the Kasparov product takes a simple form; in fact, with the notation of Ex. 5, Ex. 6, we find that

1. $(\mathcal{M}, 0) \times_{\mathcal{B}} (\mathcal{N}, 0) = (\mathcal{M} \otimes_{\mathcal{B}} \mathcal{N}, 0)$, where $(\mathcal{M}, 0) \in KK(X; \mathcal{A}, \mathcal{B})$, $(\mathcal{N}, 0) \in KK(X; \mathcal{B}, \mathcal{C})$, and $\otimes_{\mathcal{B}}$ denotes the internal tensor product (Rem. 4.1);

2. $(\mathcal{M},0) \times (\mathcal{M}',0) = (\mathcal{M} \otimes_X \mathcal{M}',0)$, where $(\mathcal{M},0) \in KK(X;\mathcal{A},\mathcal{B})$, $(\mathcal{M}',0) \in KK(X;\mathcal{A}',\mathcal{B}')$, and $\otimes_X$ denotes the external tensor product (Rem. 4.1);
3. Let $\phi : \mathcal{A} \to \mathcal{B}$, $\eta : \mathcal{B} \to \mathcal{C}$ be $C_0(X)$-algebra morphisms. Then $(\mathcal{B}_\phi,0) \times_{\mathcal{B}} (\mathcal{C}_\eta,0) = (\mathcal{C}_{\eta\circ\phi},0)$.

Thus, $(KK(X;\mathcal{A},\mathcal{A})\ ,\ +\ ,\ \times_{\mathcal{A}})$ is a ring with identity the class $[1]_{\mathcal{A}}$ defined in Ex. 6. Let us now consider the ring

$$RK^0(X) := KK(X;C_0(X),C_0(X)) \ ;$$

it is proven in [11, 2.19] that $(RK^0(X),+)$ is isomorphic to the representable K-theory group introduced by Segal in [21]. If X is compact, it is verified that $(RK^0(X),+)$ coincides with the topological K-theory $K^0(X)$. In order for a more concise notation, we denote by $[1]_X \in RK^0(X)$ the class $([C_0(X)]_\iota,0)$ defined in Ex. 6.

Lemma 4.4. *Let X be a σ-compact Hausdorff space, $d \in \mathbb{N}$, $\mathcal{E} \to X$ a rank d vector bundle. Then, $C_0(X)$ acts on the left on $\widehat{\mathcal{E}}$ by elements of $K(\widehat{\mathcal{E}})$, and the pair $(\widehat{\mathcal{E}},0)$ is a Kasparov module with class $[\mathcal{E}] := (\widehat{\mathcal{E}},0) \in RK^0(X)$.*

Proof. Let 1 be the identity over $\widehat{\mathcal{E}}$, $\theta_{\psi,\psi'} \in K(\widehat{\mathcal{E}})$, $\psi,\psi' \in \widehat{\mathcal{E}}$, the operator defined in (3.1). X being σ-compact, there is a sequence $\{K_n\}_n$ of compact subsets covering X. Let $\{\lambda_n\}$ be a partition of unity with $\overline{\mathrm{supp}\lambda_n} = K_n$, $n \in \mathbb{N}$. By the Serre-Swan theorem, the bimodule of continuous sections of the restriction $\mathcal{E}|_{K_n}$ is finitely generated by a set $\{\ \varphi_{n,k}\ \}_k$; we define $\psi_{n,k} := \lambda_n \varphi_{n,k} \in \widehat{\mathcal{E}}$. Let now $u_n := \sum_k \theta_{\psi_{n,k},\psi_{n,k}} \in K(\widehat{\mathcal{E}})$. Note that $u_n = \lambda_n^2 \sum_k \theta_{\varphi_{n,k},\varphi_{n,k}} = \lambda_n^2$. Thus, the sequence $U_m := \sum_n^m u_n = \sum_n^m \lambda_n^2$ converges to 1 in the strict topology. We conclude that $\widehat{\mathcal{E}}$ is countably generated as a right Hilbert $C_0(X)$-module by the set $\{\psi_{n,k}\}$. Let now $f \in C_0(X)$. We regard at f as an element of $L(\widehat{\mathcal{E}})$. Now, $\|f - f\sum_n^m u_n\| = \|f - \sum_n^m \lambda_n^2 f\| \stackrel{m}{\to} 0$; thus f is norm limit of elements of $K(\widehat{\mathcal{E}})$, and $C_0(X)$ acts on the left over $\widehat{\mathcal{E}}$ by elements of $K(\widehat{\mathcal{E}})$. We conclude that the pair $(\widehat{\mathcal{E}},0)$ defines a class in $RK^0(X)$. □

Let $\mathcal{A}$ be a $C_0(X)$-algebra, $\mathcal{E} \to X$ a vector bundle. We define $\widehat{\mathcal{E}} \otimes_X \mathcal{A}$ as the external tensor product $\widehat{\mathcal{E}} \otimes_X \mathcal{A}_\iota$ (where $\mathcal{A}_\iota$ is defined in Ex. 6). The Kasparov product induces a natural structure of $RK^0(X)$-bimodule on $KK(X;\mathcal{A},\mathcal{B})$ ([11, 2.19]). In particular, there is a morphism of unital rings

$$i_{\mathcal{A}} : RK^0(X) \to KK(X;\mathcal{A},\mathcal{A}) \ \ , \ \ i_{\mathcal{A}}(\mathcal{M},F) := (\mathcal{M},F) \times [1]_{\mathcal{A}} \ . \qquad (4.1)$$

If $\mathcal{E} \to X$ is a vector bundle, then it turns out that $i_{\mathcal{A}}[\mathcal{E}] = (\widehat{\mathcal{E}} \otimes_X \mathcal{A},0)$.

Let $\mathcal{A}$ be a $C_0(X)$-algebra, $\mathrm{Pic}(X;\mathcal{A})$ the set of isomorphism classes of imprimitivity $C_0(X)$-Hilbert $\mathcal{A}$-bimodules (Ex. 5). We endow $\mathrm{Pic}(X;\mathcal{A})$ with the operation of internal tensor product $\otimes_{\mathcal{A}}$; note that the bimodule $\mathcal{A}_\iota$ defined in Ex. 6 is a unit for $\mathrm{Pic}(X;\mathcal{A})$. By applying the argument of [3, §3], it is verified that if $\mathcal{M}$ is an imprimitivity $C_0(X)$-Hilbert $\mathcal{A}$-bimodule, and $\overline{\mathcal{M}}$ is the conjugate bimodule, then $\overline{\mathcal{M}}$ is an imprimitivity bimodule, and the map

$$\mathcal{M} \otimes_{\mathcal{A}} \overline{\mathcal{M}} \to K(\mathcal{M}) \simeq \mathcal{A}_\iota \quad , \quad \psi' \otimes \overline{\psi} \mapsto \theta_{\psi',\psi}$$

defines an isomorphism of $C_0(X)$-Hilbert $\mathcal{A}$-bimodules ($\theta_{\psi',\psi} \in K(\mathcal{M})$ is defined by (3.1)). Thus, $\mathrm{Pic}(X;\mathcal{A})$ is a group, called the *Picard group* of $\mathcal{A}$.

Let $\mathbf{out}_X\mathcal{A}$ denote the group of $C_0(X)$-automorphisms of $\mathcal{A}$ modulo inner automorphisms induced by unitaries in $M(\mathcal{A})$. If $\mathcal{A}$ is σ-unital, then by [3, Cor. 3.5] we obtain a group anti-isomorphism

$$\theta : \mathrm{Pic}(X;\mathcal{A}\otimes\mathcal{K}) \xrightarrow{\simeq} \mathbf{out}_X(\mathcal{A}\otimes\mathcal{K}) \quad , \quad \mathcal{M} \mapsto \theta_{\mathcal{M}} \ . \tag{4.2}$$

The previous isomorphism has to be intended in the sense that every imprimitivity $(\mathcal{A}\otimes\mathcal{K})$-bimodule is isomorphic to a bimodule of the type described in Ex. 6.

Example 7. Let X be a paracompact Hausdorff space. Then $\mathrm{Pic}(X;C_0(X))$ is isomorphic to the Cech cohomology group $H^2(X,\mathbb{Z})$. In fact, imprimitivity $C_0(X)$-Hilbert $C_0(X)$-bimodules correspond to modules of continuous sections of line bundles over X; on the other hand, it is well known that the set of line bundles, endowed with the operation of tensor product, is isomorphic as a group to $H^2(X,\mathbb{Z})$ (see [3, §3]).

We denote by $KK(X;\mathcal{A},\mathcal{A})^{-1}$ the multiplicative group of invertible elements of $KK(X;\mathcal{A},\mathcal{A})$. The argument of Rem. 4.3 implies that there is a group morphism

$$\pi_{\mathcal{A}} : \mathrm{Pic}(X;\mathcal{A}) \to KK(X;\mathcal{A},\mathcal{A})^{-1} \quad , \quad \mathcal{M} \mapsto (\mathcal{M},0) \ . \tag{4.3}$$

Example 8. We refer to Ex. 7. Let X be a locally compact, paracompact Hausdorff space. Then, we have a group morphism

$$\pi_X : H^2(X,\mathbb{Z}) \to RK^0(X)^{-1} \ ,$$

assigning to the isomorphism class of a line bundle the corresponding class in K-theory.

5. Graded $\mathcal{A}$-bundles

Aim of the present section is to assign KK-theoretical invariants to $\mathcal{A}$-bundles carrying a suitable circle action.

5.1. Circle actions

Let $\mathcal{A}$ be a C^*-algebra carrying an automorphic action $\alpha : \mathbb{T} \to \mathbf{aut}\mathcal{A}$. The C^*-dynamical system $(\mathcal{A},\mathbb{T})$ is said *full* if $\mathcal{A}$ is generated as a C^*-algebra by the disjoint union of the *spectral subspaces*

$$\mathcal{A}^k := \left\{ a \in \mathcal{A} : \alpha_z(a) = z^k a \ , \ z \in \mathbb{T} \right\} \quad , \quad k \in \mathbb{Z} \ .$$

Note that $\mathcal{A}^h \cdot \mathcal{A}^k \subseteq \mathcal{A}^{h+k}$, $(\mathcal{A}^k)^* = \mathcal{A}^{-k}$, $h,k \in \mathbb{Z}$. In particular, every $\mathcal{A}^k$ is a *Hilbert $\mathcal{A}^0$-bimodule in $\mathcal{A}$* (Def. 3.2). Note that $\mathcal{A}^k$ is full as a right Hilbert $\mathcal{A}^0$-module if and only if $\mathcal{A}^{-k}\cdot\mathcal{A}^k = \mathcal{A}^0$. Moreover, there is a natural map $\mathcal{A}^k\cdot\mathcal{A}^{-k} \to K(\mathcal{A}^k)$, $t't^* \mapsto \theta_{t',t}$.

A C^*-dynamical system $(\mathcal{A}, \mathbb{T})$ is said *semi-saturated* if $\mathcal{A}$ is generated as a C^*-algebra by $\mathcal{A}^0$, $\mathcal{A}^1$ (see [9, 1]). It is clear that if $\mathcal{A}$ is semi-saturated, then $\mathcal{A}$ is full. From the above considerations, we obtain the following lemma.

Lemma 5.1. *Suppose $\mathcal{A}^0 = \mathcal{A}^1 \cdot \mathcal{A}^{-1} = \mathcal{A}^{-1} \cdot \mathcal{A}^1$. Then, $\mathcal{A}^1$ is an imprimitivity $\mathcal{A}^0$-bimodule; if $\mathcal{A}^1$ is countably generated, the class $\delta_1(\mathcal{A}) := (\mathcal{A}^1, 0) \in KK_0(\mathcal{A}^0, \mathcal{A}^0)$ is defined.*

Example 9. Let $\mathcal{M}$ be a full Hilbert $\mathcal{A}$-bimodule, $\mathcal{O}_\mathcal{M}$ the associated CP-algebra. Then, $\mathcal{O}_\mathcal{M}$ is semi-saturated w.r.t. the circle action (3.3), so that every $\mathcal{O}_\mathcal{M}^k$, $k \in \mathbb{Z}$, is an imprimitivity bimodule over the zero-grade algebra $\mathcal{O}_\mathcal{M}^0$.

The previous example is universal, as we will show in the next lemma.

Let us introduce the following terminology: if $(\mathcal{A}, \mathbb{T}, \alpha)$, $(\mathcal{B}, \mathbb{T}, \beta)$ are C^*-dynamical systems, a *graded morphism* is a C^*-algebra morphism $\phi : \mathcal{A} \to \mathcal{B}$ such that $\phi(\mathcal{A}^k) \subseteq \mathcal{B}^k$, $k \in \mathbb{Z}$. Graded morphisms will be denoted by the notation $\phi : (\mathcal{A}, \mathbb{Z}) \to (\mathcal{B}, \mathbb{Z})$.

Lemma 5.2. ([1, Thm. 3.1]) *Let $(\mathcal{A}, \mathbb{T})$ be semi-saturated, and $\mathcal{A}^1$ full as a Hilbert $\mathcal{A}^0$-bimodule. Then, there is a graded isomorphism $(\mathcal{A}, \mathbb{Z}) \simeq (\mathcal{O}_{\mathcal{A}^1}, \mathbb{Z})$, where $\mathcal{O}_{\mathcal{A}^1}$ is the CP-algebra associated with $\mathcal{A}^1$.*

Proof. It is a direct consequence of Prop. 3.3: in fact, $\mathcal{A}^1$ is a Hilbert $\mathcal{A}^0$-bimodule in $\mathcal{A}$ with support 1, and generates $\mathcal{A}$ as a C^*-algebra. □

5.2. Graded Bundles

As usual, we denote by X a locally compact Hausdorff space.

Definition 5.3. Let $(\mathcal{A}, G, \alpha)$ be a C^*-dynamical system. An $\mathcal{A}$-bundle $(\mathcal{F}, (\pi_x : \mathcal{F} \to \mathcal{A})_{x \in X})$ has a **global G-action** if there exists an action $\alpha^X : G \to \mathbf{aut}_X \mathcal{F}$, such that $\pi_x \circ \alpha^X = \alpha \circ \pi_x$ for every $x \in X$.

Let $\mathcal{F}$ be a $\mathcal{A}$-bundle carrying a global $\mathbb{T}$-action. Then, every spectral subspace $\mathcal{F}^k$, $k \in \mathbb{Z}$ has an additional structure of $C_0(X)$-Hilbert $\mathcal{F}^0$-bimodule, in fact $ft = tf$, $t \in \mathcal{F}^k$, $f \in C_0(X)$. We say in such a case that $\mathcal{F}$ is a *graded* $\mathcal{A}$-bundle.

Proposition 5.4. *Let $(\mathcal{A}, \mathbb{T}, \alpha)$ be a semi-saturated C^*-dynamical system, with $\mathcal{A}^1$ full as a Hilbert $\mathcal{A}^0$-bimodule. Moreover, let X be paracompact. Then, for every graded $\mathcal{A}$-bundle $\mathcal{F}$ over X there is an isomorphism $(\mathcal{F}, \mathbb{Z}) \simeq (\mathcal{O}_{\mathcal{F}^1}, \mathbb{Z})$. If $\mathcal{B}$ is a graded $\mathcal{A}$-bundle, there is an isomorphism $(\mathcal{F}, \mathbb{Z}) \simeq (\mathcal{B}, \mathbb{Z})$ if and only if the Hilbert bimodules $\mathcal{F}^1$, $\mathcal{B}^1$ are covariantly isomorphic.*

Proof. We prove that $(\mathcal{F}, \mathbb{T}, \alpha^X)$ is semi-saturated, and that $\mathcal{F}^1$ is full as a Hilbert $\mathcal{F}^0$-bimodule. Let $\mathcal{U} := \{U \subseteq X\}$ be an open (locally finite) trivializing cover (i.e., every restriction $\mathcal{F}_U := C_0(U)\mathcal{F}$ is isomorphic to $C_0(U) \otimes \mathcal{A}$, $U \in \mathcal{U}$). Since $\alpha^X(t) \in \mathcal{F}_U$, $t \in \mathcal{F}_U$, for every $U \in \mathcal{U}$ we obtain a global action $\alpha_U : \mathbb{T} \to \mathbf{aut}_U \mathcal{F}_U$. Since $\mathcal{F}_U$ is a trivial bundle, it is clear that $(\mathcal{F}_U, \mathbb{T}, \alpha^U)$ is semi-saturated, and that $\mathcal{F}_U^1$ is full as a Hilbert $\mathcal{F}_U^0$-bimodule. We now consider a partition of unit $\{\lambda_U \in C_0(X)\}$ subordinate to $\mathcal{U}$; since every $t \in \mathcal{F}$ admits a decomposition $t = \sum_U \lambda_U t$, with

$\lambda_U t \in \mathcal{F}_U$, we conclude that $(\mathcal{F}, \mathbb{T}, \alpha^X)$ is semi-saturated, and that $\mathcal{F}^1$ is full. By applying Lemma 5.2, we obtain the isomorphism $(\mathcal{F}, \mathbb{Z}) \simeq (\mathcal{O}_{\mathcal{F}^1}, \mathbb{Z})$. The second assertion is an immediate consequence of the $\mathbb{Z}$-grading defined on $\mathcal{F}$, $\mathcal{B}$. □

Corollary 5.5. *With the above notation, every $\mathcal{F}^r$ is an imprimitivity $\mathcal{F}^0$-bimodule, $r \in \mathbb{N}$.*

Proof. It suffices to consider the identifications $K(\mathcal{F}^r) \simeq \mathcal{F}^r \cdot (\mathcal{F}^r)^* \simeq \mathcal{F}^0$. □

From Prop. 5.4, we have an interpretation of the set of isomorphism classes of graded $\mathcal{A}$-bundles in terms of covariant isomorphism classes of $C_0(X)$-Hilbert bimodules. Thus, a description in terms of $KK(X; -, -)$-groups becomes natural. As a first step, we consider the zero grade algebra; for every C^*-algebra $\mathcal{A}$, we denote by $H^1(X, \mathbf{aut}\mathcal{A})$ the set of isomorphism classes of $\mathcal{A}$-bundles over X (see for example [23, Thm. 2.1] for a justification of such a notation). $H^1(X, \mathbf{aut}\mathcal{A})$ has a distinguished element, called 0, corresponding to the trivial $\mathcal{A}$-bundle.

Let $\mathcal{F}$ be a graded $\mathcal{A}$-bundle over X. We define

$$\delta_0(\mathcal{F}) := [\mathcal{F}^0 \otimes \mathcal{K}] \in H^1(X, \mathbf{aut}(\mathcal{A}^0 \otimes \mathcal{K})) \ . \tag{5.1}$$

Thus, the equality $\delta_0(\mathcal{F}) = \delta_0(\mathcal{B})$ is intended in the sense that $\mathcal{F}^0$, $\mathcal{B}^0$ are stably isomorphic as $\mathcal{A}^0$-bundles.

Remark 5.6. Let X be a pointed, compact, connected CW-complex such that the pair (X, x_0), $x_0 \in X$, is a homotopy-cogroup. We denote by SX the (reduced) suspension. In order for more compact notations, we define $X^\bullet := X - \{x_0\}$. It follows from a result by Nistor ([17, §5]) that

$$H^1(SX, \mathbf{aut}\mathcal{O}_d^0) \simeq [X, \mathbf{aut}\mathcal{O}_d^0]_{x_0} \simeq KK_1(C\mathcal{O}_d^0 \ , \ X^\bullet \mathcal{O}_d^0) \ ,$$

where $C\mathcal{O}_d^0 := \{(z, a) \in \mathbb{C} \oplus C_0([0,1), \mathcal{O}_d^0) \ : \ a(0) = z1\}$ is the mapping cone. We also find

$$H^1(SX, \mathbf{aut}(\mathcal{O}_d^0 \otimes \mathcal{K})) = [X, \mathbf{aut}(\mathcal{O}_d^0 \otimes \mathcal{K})]_{x_0} \simeq KK_0(\mathcal{O}_d^0 \ , \ X^\bullet \mathcal{O}_d^0) \ .$$

In particular, when X is the $(n-1)$-sphere, we obtain $H^1(S^n, \mathbf{aut}\mathcal{O}_d^0) = \{0\}$, as proved also in [22, Thm. 1.15].

Let X be a σ-compact metrisable space, $\mathcal{A}$ separable and σ-unital; then, every graded $\mathcal{A}$-bundle $\mathcal{F}$ over X is separable and σ-unital, and $\mathcal{F}^1$ is countably generated as a Hilbert $\mathcal{F}^0$-bimodule (in fact, $\mathcal{F}$ is countably generated over compact subsets). Moreover, Cor. 5.5 implies that $\mathcal{F}^0$ acts on the left on $\mathcal{F}^1$ by elements of $K(\mathcal{F}^1)$. Thus, we define

$$\delta_1(\mathcal{F}) := (\mathcal{F}^1, 0) \in KK(X; \mathcal{F}^0, \mathcal{F}^0) \ . \tag{5.2}$$

With an abuse of notation, in the sequel we will denote by $\delta_1(\mathcal{F})$ also the class of $(\mathcal{F}^1, 0)$ in $KK_0(\mathcal{F}^0, \mathcal{F}^0)$ obtained by *forgetting* the $C_0(X)$-structure. Note that since $\mathcal{F}^1$ is an imprimitivity bimodule, we find that $\delta_1(\mathcal{F})$ is invertible; thus, the Kasparov product by $\delta_1(\mathcal{F})$ defines an automorphism on $KK_0(\mathcal{B}, \mathcal{F}^0)$ for every C^*-algebra $\mathcal{B}$. In particular, $\delta_1(\mathcal{F}) \in \mathbf{aut}K_0(\mathcal{F}^0)$.

From [20, Thm. 4.9] and Prop. 5.4, we get an exact sequence for the KK-theory of $\mathcal{F}$. It is clear that in the case in which $\mathcal{F}$ is the CP-algebra of a vector bundle, we may directly apply [20, Thm. 4.9] by replacing $\mathcal{F}^0$ with $C_0(X)$.

Corollary 5.7. *For every separable C^*-algebra $\mathcal{B}$, and graded $\mathcal{A}$-bundle $\mathcal{F}$, the following exact sequence holds:*

$$\begin{array}{ccccc} KK_0(\mathcal{B},\mathcal{F}^0) & \xrightarrow{1-\delta_1(\mathcal{F})} & KK_0(\mathcal{B},\mathcal{F}^0) & \xrightarrow{i_0} & KK_0(\mathcal{B},\mathcal{F}) \\ {\scriptstyle \delta_1}\uparrow & & & & \downarrow{\scriptstyle \delta_0} \\ KK_1(\mathcal{B},\mathcal{F}) & \xleftarrow{i_1} & KK_1(\mathcal{B},\mathcal{F}^0) & \xleftarrow{1-\delta_1(\mathcal{F})} & KK_1(\mathcal{B},\mathcal{F}^0) \end{array} \tag{5.3}$$

where i_* are the morphisms induced by the inclusion $\mathcal{F}^0 \hookrightarrow \mathcal{F}$, and δ_* are the connecting maps induced by the KK-equivalence between $\mathcal{F}^0$, $\mathcal{T}_{\mathcal{F}^1}$.

We introduce a notation. Let $\mathcal{F}$, $\mathcal{B}$ be graded $\mathcal{A}$-bundles with $\delta_0(\mathcal{F}) = \delta_0(\mathcal{B})$; then, there is a $C_0(X)$-algebra isomorphism $\alpha : \mathcal{F}^0 \otimes \mathcal{K} \to \mathcal{B}^0 \otimes \mathcal{K}$, and a ring isomorphism $\alpha_* : KK(X;\mathcal{F}^0,\mathcal{F}^0) \to KK(X;\mathcal{B}^0,\mathcal{B}^0)$ is defined. We write

$$\delta(\mathcal{F}) = \delta(\mathcal{B}) \quad \Leftrightarrow \quad \delta_0(\mathcal{F}) = \delta_0(\mathcal{B}) \;\text{ and }\; \alpha_*\delta_1(\mathcal{F}) = \delta_1(\mathcal{B}) \; ; \tag{5.4}$$

note that we used the stability of $KK(X;-,-)$, so that we identified $(\mathcal{F}^1,0) \in KK(X;\mathcal{F}^0,\mathcal{F}^0)$ with $(\mathcal{F}^1 \otimes \mathcal{K},0) \in KK(X;\mathcal{F}^0 \otimes \mathcal{K},\mathcal{F}^0 \otimes \mathcal{K})$. The tensor product of $\mathcal{F}^1$ by $\mathcal{K}$ is understood as the external tensor product of Hilbert bimodules. Note that $\alpha_*\delta_1(\mathcal{F}) = ((\mathcal{F}^1 \otimes \mathcal{K})_\alpha, 0)$, where $(\mathcal{F}^1 \otimes \mathcal{K})_\alpha$ is the pullback bimodule defined as in Ex. 2. Let us now consider the natural $\mathbb{Z}$-gradings on $\mathcal{F} \otimes \mathcal{K}$, $\mathcal{B} \otimes \mathcal{K}$ induced by $\mathcal{F}$, $\mathcal{B}$; if there is an isomorphism $\alpha : (\mathcal{F} \otimes \mathcal{K}, \mathbb{Z}) \to (\mathcal{B} \otimes \mathcal{K}, \mathbb{Z})$, then $\delta(\mathcal{F}) = \delta(\mathcal{B})$.

Example 10. Let $\mathcal{F} := X\mathcal{A}$; then, $\delta_1(\mathcal{F}) = \delta_1(\mathcal{A}) \times [1]_X$ (we used the notation (2.1)).

Let $\mathcal{F}$ be a graded $\mathcal{A}$-bundle. In general, it is clear that elements of $KK(X;\mathcal{F}^0,\mathcal{F}^0)$ do not arise from grade-one components of graded $\mathcal{A}$-bundles. Anyway, they can be recognized by considering any open trivializing cover $\mathcal{U} := \{U \subseteq X\}$ for $\mathcal{F}$, and by noting that the conditions $\delta_0(\mathcal{F}^0_U) = 0$, $\delta_1(\mathcal{F}^1_U) = \delta_1(\mathcal{A}) \times [1]_U$ hold (see previous example).

Remark 5.8. Let $\mathcal{F}$, $\mathcal{B}$ be graded $\mathcal{A}$-bundles over a σ-compact Hausdorff space X, with a fixed $C_0(X)$-isomorphism $\alpha : (\mathcal{F}^0 \otimes \mathcal{K}, \mathbb{Z}) \to (\mathcal{B}^0 \otimes \mathcal{K}, \mathbb{Z})$. The pullback bimodule $(\mathcal{F}^1 \otimes \mathcal{K})_\alpha$ has the same class as $\mathcal{B}^1 \otimes \mathcal{K}$ in $\mathrm{Pic}(X;\mathcal{B}^0 \otimes \mathcal{K})$ if and only if there is a $C_0(X)$-isomorphism $(\mathcal{F} \otimes \mathcal{K}, \mathbb{Z}) \to (\mathcal{B} \otimes \mathcal{K}, \mathbb{Z})$ (see Prop. 5.4). Let $\theta(\mathcal{F}) := \theta_{\mathcal{F}^1 \otimes \mathcal{K}} \in \mathbf{aut}_X(\mathcal{F}^0 \otimes \mathcal{K})$ be defined according to (4.2), up to inner automorphisms. The CP-algebra associated with $(\mathcal{F}^0 \otimes \mathcal{K})_{\theta(\mathcal{F})}$ is graded isomorphic to $(\mathcal{F}^0 \otimes \mathcal{K}) \rtimes_{\theta(\mathcal{F})} \mathbb{Z}$ (see Ex. 6); by universality, we obtain the isomorphism

$$(\mathcal{F} \otimes \mathcal{K}, \mathbb{Z}) \simeq ((\mathcal{F}^0 \otimes \mathcal{K}) \rtimes_{\theta(\mathcal{F})} \mathbb{Z}, \mathbb{Z}) \; .$$

Thus, $\mathrm{Pic}(X;\mathcal{F}^0\otimes\mathcal{K})$ describes the set of graded isomorphism classes of stabilized $\mathcal{A}$-bundles. The map

$$\mathrm{Pic}(X;\mathcal{F}^0)\to KK(X;\mathcal{F}^0,\mathcal{F}^0)\ ,\ \ \theta(\mathcal{F})\mapsto\delta_1(\mathcal{F})$$

gives a measure of the accuracy of the class δ_1 in describing the set of graded isomorphism classes of stabilized $\mathcal{A}$-bundles.

Example 11. Let $\mathcal{E}\to X$ be a rank d vector bundle. We denote by $\alpha^X:\mathbb{T}\to\mathbf{aut}_X\mathcal{O}_\mathcal{E}$ the circle action (3.3). Let now $\pi_x:\mathcal{O}_\mathcal{E}\to\mathcal{O}_d$, $x\in X$, be the fibre epimorphisms of $\mathcal{O}_\mathcal{E}$ as an $\mathcal{O}_d$-bundle; we denote by $\alpha_x:\mathbb{T}\to\mathbf{aut}\mathcal{O}_d$ the circle action (3.3) on the Cuntz algebra. By definition, it turns out that $\pi_x\circ\alpha^X=\alpha_x\circ\pi_x$, thus α^X is a global $\mathbb{T}$-action. Moreover, α^X is full and semi-saturated, according to Ex. 9. Thus, the previous considerations apply with $\mathcal{F}=\mathcal{O}_\mathcal{E}$, $\mathcal{A}=\mathcal{O}_d$; in particular, $\mathcal{O}_\mathcal{E}^1$ is an imprimitivity $C_0(X)$-Hilbert $\mathcal{O}_\mathcal{E}^0$-bimodule.

5.3. The KK-class for the CP-algebra of a vector bundle

Let $\mathcal{E}\to X$ be a rank d vector bundle over a σ-compact Hausdorff space X. We denote by $i:RK^0(X)\to KK(X;\mathcal{O}_\mathcal{E}^0,\mathcal{O}_\mathcal{E}^0)$ the structure morphism (4.1). In order to simplify the notation, we write $\mathbf{1}_d:=[1]_{\mathcal{O}_d^0}\in KK_0(\mathcal{O}_d^0,\mathcal{O}_d^0)$. The following result has been proved in [24, Thm. 5.6].

Theorem 5.9. *With the above notation,*

$$\delta_1(\mathcal{O}_\mathcal{E})=i[\mathcal{E}]=[\mathcal{E}]\times[1]_{\mathcal{O}_\mathcal{E}^0}\ . \tag{5.5}$$

In particular, $\delta_1(X\mathcal{O}_d)=d[1]_X\times\mathbf{1}_d$.

We now discuss the properties of the class $\delta_1(\mathcal{O}_\mathcal{E})$ in the case in which the base space is an even sphere S^{2n}; for this purpose, recall that $K^0(S^{2n})=\mathbb{Z}^2$, $K^1(S^{2n})=0$. We will also make use of the ring structure of $K^0(S^{2n})$, induced by the operation of tensor product: it turns out that there is a ring isomorphism $K^0(S^{2n})\simeq\mathbb{Z}^2[\lambda]/(\lambda^2)$, i.e., the elements of $K^0(S^{2n})$ are polynomials of the type $z+\lambda z'$, $z,z'\in\mathbb{Z}$, with the relation $\lambda^2=0$ ([10, Chp. 11]). Tensoring by a vector bundle $\mathcal{E}\to S^{2n}$ with class $d+\lambda c\in K^0(S^{2n})$ defines an endomorphism $\lambda_\mathcal{E}\in\mathbf{end}K^0(S^{2n})$, $\lambda_\mathcal{E}(z+\lambda z'):=dz+\lambda(dz'+cz)$. We also denote by ι: $\iota(z+\lambda z'):=z+\lambda z'$ the identity automorphism on $K^0(S^{2n})$. Note that $\iota-\lambda_\mathcal{E}$ is injective (for $d>1$).

We can now compute the K-theory of $\mathcal{O}_\mathcal{E}$ over even spheres: the exact sequence [20, Thm. 4.9], and the above considerations, imply

$$\begin{array}{ccccc}
\mathbb{Z}^2 & \xrightarrow{\iota-\lambda_\mathcal{E}} & \mathbb{Z}^2 & \xrightarrow{i_0} & K_0(\mathcal{O}_\mathcal{E}) \\
{\scriptstyle 0}\uparrow & & & & \downarrow{\scriptstyle 0} \\
K_1(\mathcal{O}_\mathcal{E}) & \xleftarrow{i_1} & 0 & \xleftarrow{0} & 0
\end{array}$$

so that, we have the K-groups

$$K_0(\mathcal{O}_{\mathcal{E}}) = \frac{\mathbb{Z}^2}{(\iota - \lambda_{\mathcal{E}})\mathbb{Z}^2} \quad , \quad K_1(\mathcal{O}_{\mathcal{E}}) = 0 \quad .$$

In the case in which $d-1$ and c are relatively prime, some elementary computations show that $K_0(\mathcal{O}_{\mathcal{E}}) = \mathbb{Z}_{(d-1)^2}$. This shows that $\mathcal{O}_{\mathcal{E}}$ is non-trivial as an $\mathcal{O}_d$-bundle: in fact, the Kunneth theorem ([4, §23]) implies $K_0(S^{2n}\mathcal{O}_d) = \mathbb{Z}_{d-1} \oplus \mathbb{Z}_{d-1}$.

We now pass to describe the class $\delta_1(\mathcal{O}_{\mathcal{E}})$. For every rank d vector bundle $\mathcal{E} \to S^{2n}$ we find $\delta_0(\mathcal{O}_{\mathcal{E}}) = 0$, in fact $\mathcal{O}_{\mathcal{E}}^0 \simeq S^{2n}\mathcal{O}_d^0$ (see Rem. 5.6); we denote by

$$\alpha_{\mathcal{E}} : KK(S^{2n}; \mathcal{O}_{\mathcal{E}}^0, \mathcal{O}_{\mathcal{E}}^0) \to KK(S^{2n} \ ; \ S^{2n}\mathcal{O}_d^0, S^{2n}\mathcal{O}_d^0)$$

the induced ring isomorphism. The Kunneth theorem implies

$$\begin{cases} K_0(S^{2n}\mathcal{O}_d^0) = K^0(S^{2n}) \otimes K_0(\mathcal{O}_d^0) = \mathbb{Z}^2 \otimes \mathbb{Z}\left[\frac{1}{d}\right] \\ KK_0(S^{2n}\mathcal{O}_d^0, S^{2n}\mathcal{O}_d^0) = \mathbf{end}\left(\mathbb{Z}^2 \otimes \mathbb{Z}\left[\frac{1}{d}\right]\right) \end{cases}$$

where $\mathbb{Z}\left[\frac{1}{d}\right]$ is the group of d-adic integers ([4, 10.11.8]). We denote by

$$(z + \lambda z') \otimes q \quad , \quad z, z' \in \mathbb{Z} \ , \ q \in \mathbb{Z}\left[\frac{1}{d}\right] \ ,$$

the generic "elementary tensor" in $K_0(S^{2n}\mathcal{O}_d^0)$. Let now

$$\delta_1'(\mathcal{O}_{\mathcal{E}}) := \alpha_{\mathcal{E}} \circ \delta_1(\mathcal{O}_{\mathcal{E}}) = \alpha_{\mathcal{E}} \circ i[\mathcal{E}] \in KK(S^{2n} \ ; \ S^{2n}\mathcal{O}_d^0 \ , \ S^{2n}\mathcal{O}_d^0) \ ;$$

it is clear that $\delta_1'(\mathcal{O}_{\mathcal{E}}) = [\mathcal{E}] \times [1]_{S^{2n}\mathcal{O}_d^0}$ is the KK-class associated with the bimodule $\widehat{\mathcal{E}} \otimes_X (S^{2n}\mathcal{O}_d^0)$. We denote by $\beta_{\mathcal{E}} \in \mathbf{aut}K_0(S^{2n}\mathcal{O}_d^0)$ the automorphism induced by $\delta_1'(\mathcal{O}_{\mathcal{E}})$; it follows from the above considerations that $\beta_{\mathcal{E}} = \lambda_{\mathcal{E}} \times \mathbf{1}_d \in \mathbf{end}\left(\mathbb{Z}^2 \otimes \mathbb{Z}\left[\frac{1}{d}\right]\right)$. Thus, we find

$$\beta_{\mathcal{E}}((z + \lambda z') \otimes q) = \lambda_{\mathcal{E}}(z + \lambda z') \otimes q = [dz + \lambda(dz' + cz)] \otimes q \ . \tag{5.6}$$

Theorem 5.10. *Let $\mathcal{E}, \mathcal{E}' \to S^{2n}$ be rank d vector bundles. Then, the following are equivalent:*

1. *$\delta_1'(\mathcal{O}_{\mathcal{E}}) = \delta_1'(\mathcal{O}_{\mathcal{E}'}) \in KK(S^{2n} \ ; \ S^{2n}\mathcal{O}_d^0 \ , \ S^{2n}\mathcal{O}_d^0)$;*
2. *there is a $C(S^{2n})$-isomorphism $(\mathcal{O}_{\mathcal{E}} \otimes \mathcal{K}, \mathbb{Z}) \to (\mathcal{O}_{\mathcal{E}'} \otimes \mathcal{K}, \mathbb{Z})$;*
3. *$[\mathcal{E}] = [\mathcal{E}'] \in K^0(S^{2n})$.*

Proof. 1) $\Rightarrow$ 3): we have $\delta_1'(\mathcal{O}_{\mathcal{E}}) = \delta_1'(\mathcal{O}_{\mathcal{E}'})$ if and only if $\beta_{\mathcal{E}} = \beta_{\mathcal{E}'}$; thus, (5.6) implies $\beta_{\mathcal{E}}(1 \otimes 1) = (d + \lambda c) \otimes 1 = (d + \lambda c') \otimes 1 = \beta_{\mathcal{E}'}(1 \otimes 1)$, where $[\mathcal{E}] := d + \lambda c$, $[\mathcal{E}'] := d + \lambda c'$. From the equality $(d + \lambda c) \otimes 1 = (d + \lambda c') \otimes 1$, we conclude $[\mathcal{E}] = [\mathcal{E}']$ (note that we know *a priori* that $\mathcal{E}$, $\mathcal{E}'$ have the same rank d).

3) $\Rightarrow$ 2) follows from [24, Prop.5.10].

2) $\Rightarrow$ 1) is trivial by definition of δ_1. □

Let $\mathcal{E}, \mathcal{E}' \to S^{2n}$ be rank d vector bundles with classes $[\mathcal{E}] = d + \lambda c$, $[\mathcal{E}'] = d + \lambda c'$, such that $d-1$, c and $d-1$,c' are relatively prime; then, $K_0(\mathcal{O}_{\mathcal{E}}) = K_0(\mathcal{O}_{\mathcal{E}'}) = \mathbb{Z}_{(d-1)^2}$. If $c \neq c'$, then the previous theorem implies that $\mathcal{O}_{\mathcal{E}}$ is not graded stably isomorphic to $\mathcal{O}_{\mathcal{E}'}$, and $\delta_1(\mathcal{O}_{\mathcal{E}}) \neq \delta_1(\mathcal{O}_{\mathcal{E}'})$. This shows that δ_1 is a more detailed invariant w.r.t. the K-theory of $\mathcal{O}_{\mathcal{E}}$.

We conclude with a remark about odd spheres: in this case $K^0(S^{2n+1}) = \mathbb{Z}$, so that for every rank d vector bundle $\mathcal{E} \to S^{2n+1}$ we find $[\mathcal{E}] = d$. Thus, [24, Prop. 5.10] implies $(\mathcal{O}_{\mathcal{E}} \otimes \mathcal{K}, \mathbb{Z}) \simeq (S^{2n+1}\mathcal{O}_d \otimes \mathcal{K}, \mathbb{Z})$.

References

[1] B. Abadie, S. Eilers, R. Exel, *Morita Equivalence for Generalized Crossed Products.* Trans. Amer. Math. Soc. **350** (1998), 3043–3054.

[2] M.F. Atiyah, *K-Theory.* Benjamin, 1967.

[3] L.G. Brown, P. Green, M.A. Rieffel, *Stable isomorphisms and strong Morita equivalence of C^*-algebras.* Pac. J. Math. **71** (1977), 349–363.

[4] B. Blackadar, *K-Theory of Operator Algebras.* MSRI Publications, 1995.

[5] E. Blanchard, *Tensor Products of $C(X)$-algebras over $C(X)$* Astérisqque **232** (1995), 81–92.

[6] J. Cuntz, *Simple C^*-algebras Generated by Isometries.* Comm. Math. Phys. **57** (1977), 173–185.

[7] J. Dixmier, *C^*-algebras.* North-Holland Publishing Company, 1977.

[8] S. Doplicher, C. Pinzari, R. Zuccante, *The C^*-alĝebra of a Hilbert bimodule.* Bollettino UMI, Serie VIII **1B** (1998), 263–282.

[9] R. Exel, *Circle Actions on C^*-Algebras, Partial Automorphisms and a Generalized Pimsner-Voiculescu Exact Sequence.* J. Funct. Anal. **122** (1994), 361–401.

[10] D. Husemoller, *Fibre Bundles.* Mc Graw-Hill Series in Mathematics, 1966.

[11] G. Kasparov, *Equivariant KK-Theory and the Novikov Conjecture.* Invent. Math. **91** (1988), 147–201.

[12] E. Kirchberg, *The classification of purely infinite C^*-algebras using Kasparov's theory.* in: *C^*-algebras.* Springer, 2000.

[13] E. Kirchberg, *Das nicht-kommutative Michael-Auswahlprinzip und die Klassifikation nicht-einfacher Algebren.* Geometrische Strukturen in der Mathematik, Preprint-Reihe SFB **478** (2000).

[14] E. Kirchberg, S. Wassermann, *Operations on Continuous Bundles of C^*-algebras.* Mathematische Annalen **303** (1995), 677–697.

[15] A.S. Mishchenko, A.T. Fomenko, *The Index of Elliptic Operators over C^*-algebras.* Math. USSR Izvestija **15**(1) (1980), 87–112.

[16] M. Nilsen, *C^*-Bundles and $C_0(X)$-algebras.* Indiana Univ. Math. J. **45** (1996), 463–477.

[17] V. Nistor, *Fields of AF-algebras.* J. Oper. Theory **28** (1992), 3–25.

[18] G.K. Pedersen: *C^*-algebras and their automorphism groups.* Academic Press, 1979.

[19] N.C. Phillips, *A classification theorem for nuclear purely infinite simple C^*-algebras.* Preprint ArXiv **funct-an/9506010**.

[20] M. Pimsner, *A Class of C^*-algebras generalizing both Cuntz-Krieger algebras and Cross Products by $\mathbb{Z}$.* in: *Free Probability Theory.* AMS, 1993.

[21] G. Segal, *Fredholm Complexes.* Quart. J. Math. Oxford **21** (1970), 385–402.

[22] K. Thomsen, *The Homotopy Type of the Group of Automorphisms of a UHF-algebra.* J. Funct. Anal. **72** (1987), 182–207.

[23] E. Vasselli, *Continuous Fields of C^*-algebras Arising from Extensions of Tensor C^*-Categories.* J. Funct. Anal. **199** (2003), 122–152.

[24] E. Vasselli: *The C^*-algebra of a vector bundle and fields of Cuntz algebras.* ArXiv math.OA/0404166, J. Funct. Anal. **222** (2005), 491–502.

Ezio Vasselli
Dipartimento di Matematica
Università La Sapienza di Roma
P.le Aldo Moro, 2
I-00185 Roma, Italy
(c/o Sergio Doplicher)
e-mail: `vasselli@mat.uniroma2.it`

GPSR Compliance

The European Union's (EU) General Product Safety Regulation (GPSR) is a set of rules that requires consumer products to be safe and our obligations to ensure this.

If you have any concerns about our products, you can contact us on ProductSafety@springernature.com

In case Publisher is established outside the EU, the EU authorized representative is:

Springer Nature Customer Service Center GmbH
Europaplatz 3
69115 Heidelberg, Germany

Batch number: 10732470

Printed by Printforce, the Netherlands